LONDON MATHEMATICAL SOCIETY LECTURE NOTE SERIES

Managing Editor: Professor N.J. Hitchin, Mathematical Institute,
University of Oxford, 24–29 St Giles, Oxford OX1 3LB, United Kingdom

The titles below are available from booksellers, or, in case of difficulty, from Cambridge University Press.

London Mathematical Society Lecture Note Series. 283

Nonlinear Elasticity: Theory and Applications

Edited by

Y. B. Fu
University of Keele

R. W. Ogden
University of Glasgow

CAMBRIDGE
UNIVERSITY PRESS

PUBLISHED BY THE PRESS SYNDICATE OF THE UNIVERSITY OF CAMBRIDGE
The Pitt Building, Trumpington Street, Cambridge, United Kingdom

CAMBRIDGE UNIVERSITY PRESS
The Edinburgh Building, Cambridge, CB2 2RU, UK
40 West 20th Street, New York, NY 10011–4211, USA
10 Stamford Road, Oakleigh, VIC 3166, Australia
Ruiz de Alarcón 13, 28014 Madrid, Spain
Dock House, The Waterfront, Cape Town 8001, South Africa

http://www.cambridge.org

First published 2001

A catalogue record for this book is available from the British Library

ISBN   0 521 79695 4 paperback

Transferred to digital printing 2002

Contents

Preface

The subject of Finite Elasticity (or Nonlinear Elasticity), although many of its ingredients were available much earlier, really came into its own as a discipline distinct from the classical theory of linear elasticity as a result of the important developments in the theory from the late 1940s associated with Rivlin and the collateral developments in general Continuum Mechanics associated with the Truesdell school during the 1950s and 1960s. Much of the impetus for the theoretical developments in Finite Elasticity came from the rubber industry because of the importance of (natural) rubber in many engineering components, not least car tyres and bridge and engine mountings. This impetus is maintained today with an ever increasing use of rubber (natural and synthetic) and other polymeric materials in a broader and broader range of engineering products. The importance of gaining a sound theoretically-based understanding of the thermomechanical behaviour of rubber was only too graphically illustrated by the role of the rubber O-ring seals in the Challenger shuttle disaster. This extreme example serves to underline the need for detailed characterization of the mechanical properties of different rubberlike materials, and this requires not just appropriate experimental data but also the rigorous theoretical framework for analyzing those data. This involves both elasticity theory *per se* and extensions of the theory to account for inelastic effects.

Over the last few years the applications of the theory have extended beyond the traditional regime of rubber mechanics and they now embrace other materials capable of large elastic strains. These include, in particular, biological tissue such as skin, arterial walls and the heart. This is an important new development and it is increasingly recognized by medical researchers and clinicians that understanding of the mechanics of such tissue is of fundamental importance in developing improved intervention treatments (such as balloon angioplasty) and artificial replacement parts.

While understanding of Finite Elasticity is in itself important the theory also provides a gateway towards the understanding of more complex (non-elastic) material behaviour in the large deformation regime, such as finite deformation plasticity and nonlinear viscoelasticity, and it has an underpinning role in such theories.

Additionally, because of its intrinsic nonlinearity, the equations of Finite Elasticity provide a rich basis for purely mathematical studies in, for example, nonlinear analysis. Indeed, developments, in particular in the theory of partial differential equations, have been stimulated by work in Finite Elasticity, and these *mathematical* developments in turn have had an influence on research in the *mechanical* aspects of Finite Elasticity. Thus, the subject is wide ranging in both its theoretical and application perspectives.

The turn of the century is an appropriate time to assess the state of development of the subject of Finite Elasticity and its potential for further applications and further theoretical development. With this in mind this volume aims to provide an overview of the theory from the perspectives of twelve researchers who have contributed to the subject over the last few years. It is hoped that it will provide a foundation and springboard for possible future developments. The material covered in this volume is necessarily selective and not by any means exhaustive. Thus, some important topics (such as fracture mechanics and the mechanics of composite materials) are not addressed. Nevertheless, there is nothing presently available in the literature that covers such a broad range of topics within the general framework of Finite/Nonlinear Elasticity as that presented here. The various chapters combine concise theory with a number of important applications, and the emphasis is directed more towards understanding of mechanical phenomena and problem solving rather than development of the theory for its own sake.

Different chapters deal with, on the one hand, a number of classical research directions concerned with, for example, exact solutions, universal relations and the effect of internal constraints, and, on the other hand, with recent developments associated with phase transitions and pseudo-elasticity. New ideas from nonlinear analysis, such as nonlinear bifurcation analysis and dynamical systems theory are also featured.

Chapter 1 provides the basic theory required for use in the other chapters. Chapters 2–6 deal with different aspects of the solution of boundary-value problems for unconstrained and internally constrained materials, while Chapters 7 and 8 are concerned with the related topics of membrane theory and the theory of elastic surfaces. Chapter 9 deals with the important topic of non-uniqueness of solution using the tools of singularity theory and bifurcation theory and Chapter 10 examines some related aspects concerned with nonlinear stability analysis based on methods of perturbation theory. Nonlinear dynamics is discussed in Chapter 11, which is concerned with nonlinear wave propagation in an elastic rod. Chapters 12 and 13 are based on different notions of pseudo-elasticity theory: Chapter 12 develops a theory of phase transitions using non-convex strain-energy functions, while Chapter 13 is concerned with the effect of changing the (elastic) constitutive law during the deformation process.

Most of the contributors to this volume participated in an International

Workshop on Nonlinear Elasticity held at City University, Hong Kong, in April 2000. Whilst the chapters in this volume do not form the proceedings of that Workshop it is important to emphasize that this volume and the Workshop were planned in parallel and that the Workshop served to focus ideas for the volume. We are pleased to acknowledge the generous support for the Workshop from the Liu Bie Ju Centre for Mathematical Sciences, City University of Hong Kong, and the personal support and encouragement of Professor Roderick S. C. Wong, Director of the Centre. We are very grateful to all the contributors for their enthusiastic response to this project and for the timely production of their chapters.

Y.B. Fu
R.W. Ogden
December 2000

1

Elements of the theory of finite elasticity

R.W. Ogden

Department of Mathematics
University of Glasgow, Glasgow G12 8QW, U.K.
Email: rwo@maths.gla.ac.uk

In this chapter we provide a brief overview of the main ingredients of the nonlinear theory of elasticity in order to establish the basic background material as a reference source for the other, more specialized, chapters in this volume.

1.1 Introduction

In this introductory chapter we summarize the basic equations of nonlinear elasticity theory as a point of departure and as a reference source for the other articles in this volume which are concerned with more specific topics.

There are several texts and monographs which deal with the subject of nonlinear elasticity in some detail and from different standpoints. The most important of these are, in chronological order of the publication of the first edition, Green and Zerna (1954, 1968, 1992), Green and Adkins (1960, 1970), Truesdell and Noll (1965), Wang and Truesdell (1973), Chadwick (1976, 1999), Marsden and Hughes (1983, 1994), Ogden (1984a, 1997), Ciarlet (1988) and Antman (1995). See also the textbook by Holzapfel (2000), which deals with viscoelasticity and other aspects of nonlinear solid mechanics as well as containing an extensive treatment of nonlinear elasticity. These books may be referred to for more detailed study. Subsequently in this chapter we shall refer to the most recent editions of these works. The review articles by Spencer (1970) and Beatty (1987) are also valuable sources of reference.

Section 1.2 of this chapter is concerned with laying down the basic equations of elastostatics and it includes a summary of the relevant geometry of deformation and strain, an account of stress and stress tensors, the equilibrium equations and boundary conditions and an introduction to the formulation of constitutive laws for elastic materials, with discussion of the important notions of objectivity and material symmetry. Some emphasis is placed on the special case of isotropic elastic materials, and the constitutive laws for anisotropic

1

material consisting of one or two families of fibres are also discussed. The modifications to the constitutive laws when internal constraints such as incompressibility and inextensibility are present are provided. The general boundary-value problem of nonlinear elasticity is then formulated and the circumstances when this can be cast in a variational structure are discussed briefly.

In Section 1.3 some basic examples of boundary-value problems are given. Specifically, the equations governing some homogeneous deformations are highlighted, with the emphasis on incompressible materials. Other chapters in this volume will discuss a range of different boundary-value problems involving nonhomogeneous deformations so here we focus attention on just one problem as an exemplar. This is the problem of extension and inflation of a thick-walled circular cylindrical tube. The analysis is given for an incompressible isotropic elastic solid and also for a material with two mechanically equivalent symmetrically disposed families of fibres in order to illustrate some differences between isotropic and anisotropic response.

The (linearized) equations of incremental elasticity associated with small deformations superimposed on a finite deformation are summarized in Section 1.4. The incremental constitutive law for an elastic material is used to identify the (fourth-order) tensor of elastic moduli associated with the stress and deformation variables used in the formulation of the governing equations, and explicit expressions for the components of this tensor are given in the case of an isotropic material. For the two-dimensional specialization, necessary and sufficient conditions on these components for the strong ellipticity inequalities to hold are given for both unconstrained and incompressible materials. A brief discussion of incremental uniqueness and stability is then given in the context of the dead-load boundary-value problem and the associated local inequalities are given explicit form for an isotropic material, again for both unconstrained and incompressible materials. A short discussion of global aspects of nonuniqueness for an isotropic material sets the incremental results in a broader context.

In Section 1.5 the equations of incremental deformations and equilibrium given in Section 1.4 are specialized to the plane strain context in order to provide a formulation for the analysis of incremental plane strain boundary-value problems. Specifically, we provide an example of a typical incremental boundary-value problem by considering bifurcation of a uniformly deformed half-space from a homogeneously deformed configuration into a non-homogeneous local mode of deformation. An explicit bifurcation condition is given for this problem and the results are illustrated for two forms of strain-energy function.

Finally, in Section 1.6 we summarize the equations associated with the (nonlinear) dynamics of an elastic body at finite strain. The (linearized) equations for small motions superimposed on a static finite deformation are then given and these are applied to the analysis of plane waves propagating in a homogeneously deformed material.

References are given throughout the text but these are not intended to provide an exhaustive list of original sources. Where appropriate we mention papers and books where more detailed citations can be found. Also, where a topic is to be dealt with in detail in one of the other chapters of this volume the appropriate citations are included there.

1.2 Elastostatics

In this section we summarize the basic equations of the static theory of nonlinear elasticity, including the kinematics of deformation, the analysis of stress and the governing equations of equilibrium, and we introduce the various forms of constitutive law for an elastic material, including a discussion of isotropy and anisotropy. We then formulate the basic boundary-value problem of nonlinear elasticity. The development here is a synthesis of the essential material taken from the book by Ogden (1997) with some minor differences and additions.

1.2.1 Deformation and strain

We consider a continuous body which occupies a connected open subset of a three-dimensional Euclidean point space, and we refer to such a subset as a *configuration* of the body. We identify an arbitrary configuration as a *reference configuration* and denote this by $\mathcal{B}_r$. Let points in $\mathcal{B}_r$ be labelled by their position vectors $\mathbf{X}$ relative to an arbitrarily chosen origin and let $\partial\mathcal{B}_r$ denote the boundary of $\mathcal{B}_r$. Now suppose that the body is deformed quasi-statically from $\mathcal{B}_r$ so that it occupies a new configuration, $\mathcal{B}$ say, with boundary $\partial\mathcal{B}$. We refer to $\mathcal{B}$ as the *current* or *deformed configuration* of the body. The deformation is represented by the mapping $\chi : \mathcal{B}_r \to \mathcal{B}$ which takes points $\mathbf{X}$ in $\mathcal{B}_r$ to points $\mathbf{x}$ in $\mathcal{B}$. Thus,

$$\mathbf{x} = \chi(\mathbf{X}), \qquad \mathbf{X} \in \mathcal{B}_r, \tag{2.1}$$

where $\mathbf{x}$ is the position vector of the point $\mathbf{X}$ in $\mathcal{B}$. The mapping χ is called the *deformation* from $\mathcal{B}_r$ to $\mathcal{B}$. We require χ to be one-to-one and we write its inverse as χ^{-1}, so that

$$\mathbf{X} = \chi^{-1}(\mathbf{x}), \qquad \mathbf{x} \in \mathcal{B}. \tag{2.2}$$

Both χ and its inverse are assumed to satisfy appropriate regularity conditions. Here, it suffices to take χ to be twice continuously differentiable, but different requirements may be specified in other chapters of this volume.

For simplicity we consider only Cartesian coordinate systems and let $\mathbf{X}$ and $\mathbf{x}$ respectively have coordinates X_α and x_i, where $\alpha, i \in \{1,2,3\}$, so that $x_i = \chi_i(X_\alpha)$. Greek and Roman indices refer, respectively, to $\mathcal{B}_r$ and $\mathcal{B}$ and the usual summation convention for repeated indices is used.

The *deformation gradient tensor*, denoted $\mathbf{F}$, is given by

$$\mathbf{F} = \mathrm{Grad}\,\mathbf{x} \qquad (2.3)$$

and has Cartesian components $F_{i\alpha} = \partial x_i / \partial X_\alpha$, Grad being the gradient operator in $\mathcal{B}_r$. Local invertibility of χ requires that $\mathbf{F}$ be non-singular, and we adopt the usual convention that $\det \mathbf{F} > 0$. Similarly, for the inverse deformation gradient

$$\mathbf{F}^{-1} = \mathrm{grad}\,\mathbf{X}, \quad (\mathbf{F}^{-1})_{\alpha i} = \frac{\partial X_\alpha}{\partial x_i}, \qquad (2.4)$$

where grad is the gradient operator in $\mathcal{B}$. With use of the notation defined by

$$J = \det \mathbf{F} \qquad (2.5)$$

we then have

$$0 < J < \infty. \qquad (2.6)$$

The equation

$$\mathrm{d}\mathbf{x} = \mathbf{F}\,\mathrm{d}\mathbf{X} \qquad (2.7)$$

(in components $\mathrm{d}x_i = F_{i\alpha}\mathrm{d}X_\alpha$) describes how an infinitesimal *line element* $\mathrm{d}\mathbf{X}$ of material at the point $\mathbf{X}$ transforms *linearly* under the deformation into the line element $\mathrm{d}\mathbf{x}$ at $\mathbf{x}$.

We now set down how elements of surface area and volume transform. Let $\mathrm{d}\mathbf{A} \equiv \mathbf{N}\mathrm{d}A$ denote a vector surface area element on $\partial\mathcal{B}_r$, where $\mathbf{N}$ is the unit outward normal to the surface, and $\mathrm{d}\mathbf{a} \equiv \mathbf{n}\mathrm{d}a$ the corresponding area element on $\partial\mathcal{B}$. Then, the area elements are connected according to *Nanson's formula*

$$\mathbf{n}\mathrm{d}a = J\mathbf{F}^{-T}\mathbf{N}\mathrm{d}A, \qquad (2.8)$$

where $\mathbf{F}^{-T} = (\mathbf{F}^{-1})^T$ and T denotes the transpose. Note that, unlike a line element, the normal vector is not embedded in the material, i.e. $\mathbf{n}$ is not in general aligned with the same line element of material as $\mathbf{N}$.

If $\mathrm{d}V$ and $\mathrm{d}v$ denote volume elements in $\mathcal{B}_r$ and $\mathcal{B}$ respectively then we also have

$$\mathrm{d}v = J\mathrm{d}V. \qquad (2.9)$$

For a volume preserving (*isochoric*) deformation we have

$$J = \det \mathbf{F} = 1. \tag{2.10}$$

A material for which (2.10) is constrained to be satisfied for all deformation gradients $\mathbf{F}$ is said to be *incompressible*.

The identities

$$\text{Div}\,(J\mathbf{F}^{-1}) = \mathbf{0}, \quad \text{div}\,(J^{-1}\mathbf{F}) = \mathbf{0} \tag{2.11}$$

are important tools in transformations between equations associated with the reference and current configurations, where Div and div are the divergence operators in $\mathcal{B}_r$ and $\mathcal{B}$ respectively. The first identity in (2.11) can readily be established by integrating (2.8) over an arbitrary closed surface in $\mathcal{B}$ and applying the divergence theorem and the second similarly by integrating $\mathbf{N}dA$ over an arbitrary closed surface in $\mathcal{B}_r$.

From (2.7) we have

$$|d\mathbf{x}|^2 = (\mathbf{FM}) \cdot (\mathbf{FM})\,|d\mathbf{X}|^2 = (\mathbf{F}^T\mathbf{F}\mathbf{M}) \cdot \mathbf{M}\,|d\mathbf{X}|^2, \tag{2.12}$$

where we have introduced the unit vector $\mathbf{M}$ in the direction of $d\mathbf{X}$ and $\cdot$ signifies the scalar product of two vectors. Then, the ratio $|d\mathbf{x}|/|d\mathbf{X}|$ of the lengths of a line element in the deformed and reference configurations is given by

$$\frac{|d\mathbf{x}|}{|d\mathbf{X}|} = |\mathbf{FM}| = [\mathbf{M} \cdot (\mathbf{F}^T\mathbf{F}\mathbf{M})]^{1/2} \equiv \lambda(\mathbf{M}). \tag{2.13}$$

Equation (2.13) defines the *stretch* $\lambda(\mathbf{M})$ *in the direction* $\mathbf{M}$ at $\mathbf{X}$, and we note that it is restricted according to the inequalities

$$0 < \lambda(\mathbf{M}) < \infty. \tag{2.14}$$

If there is no stretch in the direction $\mathbf{M}$ then $\lambda(\mathbf{M}) = 1$ and hence

$$(\mathbf{F}^T\mathbf{F}\mathbf{M}) \cdot \mathbf{M} = 1. \tag{2.15}$$

If there is no stretch in any direction, i.e. (2.15) holds for all $\mathbf{M}$, then the material is said to be *unstrained* at $\mathbf{X}$, and it follows that $\mathbf{F}^T\mathbf{F} = \mathbf{I}$, where $\mathbf{I}$ is the identity tensor. A suitable tensor measure of strain is therefore $\mathbf{F}^T\mathbf{F} - \mathbf{I}$ since this tensor vanishes when the material is unstrained. This leads to the definition of the *Green strain tensor*

$$\mathbf{E} = \frac{1}{2}(\mathbf{F}^T\mathbf{F} - \mathbf{I}), \tag{2.16}$$

where the $1/2$ is a normalization factor. If, for a given $\mathbf{M}$, equation (2.15) holds for all possible deformation gradients $\mathbf{F}$ then the considered material is said to be *inextensible* in the direction $\mathbf{M}$.

The deformation gradient can be decomposed according to the *polar decompositions*

$$\mathbf{F} = \mathbf{RU} = \mathbf{VR}, \tag{2.17}$$

where $\mathbf{R}$ is a proper orthogonal tensor and $\mathbf{U}$, $\mathbf{V}$ are positive definite and symmetric tensors. Each of the decompositions in (2.17) is unique. Respectively, $\mathbf{U}$ and $\mathbf{V}$ are called the *right* and *left stretch tensors*.

These stretch tensors can also be put in spectral form. For $\mathbf{U}$ we have the *spectral decomposition*

$$\mathbf{U} = \sum_{i=1}^{3} \lambda_i \mathbf{u}^{(i)} \otimes \mathbf{u}^{(i)}, \tag{2.18}$$

where $\lambda_i > 0$, $i \in \{1, 2, 3\}$, are the *principal stretches*, $\mathbf{u}^{(i)}$, the (unit) eigenvectors of $\mathbf{U}$, are called the *Lagrangian principal axes* and $\otimes$ denotes the tensor product. Note that $\lambda(\mathbf{u}^{(i)}) = \lambda_i$ in accordance with the definition (2.13). Similarly, $\mathbf{V}$ has the spectral decomposition

$$\mathbf{V} = \sum_{i=1}^{3} \lambda_i \mathbf{v}^{(i)} \otimes \mathbf{v}^{(i)}, \tag{2.19}$$

where

$$\mathbf{v}^{(i)} = \mathbf{R}\mathbf{u}^{(i)}, \quad i \in \{1, 2, 3\}. \tag{2.20}$$

It follows from (2.5), (2.17) and (2.18) that

$$J = \lambda_1 \lambda_2 \lambda_3. \tag{2.21}$$

Using the polar decompositions (2.17) for the deformation gradient $\mathbf{F}$, we may also form the following tensor measures of deformation:

$$\mathbf{C} = \mathbf{F}^T \mathbf{F} = \mathbf{U}^2, \qquad \mathbf{B} = \mathbf{F}\mathbf{F}^T = \mathbf{V}^2. \tag{2.22}$$

These define $\mathbf{C}$ and $\mathbf{B}$, which are called, respectively, the *right* and *left Cauchy-Green deformation tensors*.

More general classes of strain tensors, i.e. tensors which vanish when there is no strain, can be constructed on the basis that $\mathbf{U} = \mathbf{I}$ when the material is unstrained. Thus, for example, we define Lagrangian strain tensors

$$\mathbf{E}^{(m)} = \frac{1}{m}(\mathbf{U}^m - \mathbf{I}), \quad m \neq 0, \tag{2.23}$$

$$\mathbf{E}^{(0)} = \ln \mathbf{U}, \quad m = 0, \tag{2.24}$$

where m is a real number (not necessarily an integer). Eulerian strain tensors based on the use of $\mathbf{V}$ may be constructed similarly. See, for example, Doyle and Ericksen (1956), Seth (1964) and Hill (1968, 1970, 1978). Note that for $m = 2$ equation (2.23) reduces to the Green strain tensor (2.16). For discussion of the logarithmic strain tensor (2.24) we refer to, for example, Hoger (1987).

Let ρ_r and ρ be the *mass densities* in $\mathcal{B}_r$ and $\mathcal{B}$ respectively. Then, since the material in the volume element dV is the same as that in dv the mass is conserved, i.e. $\rho dv = \rho_r dV$, and hence, from (2.9), we may express the *mass conservation equation* in the form

$$\rho_r = \rho J. \tag{2.25}$$

1.2.2 Stress tensors and equilibrium equations

The surface force per unit area (or *stress vector*) on the vector area element $d\mathbf{a}$ is denoted by $\mathbf{t}$. It depends on $\mathbf{n}$ according to the formula

$$\mathbf{t} = \sigma^T \mathbf{n}, \tag{2.26}$$

where σ, a second-order tensor independent of $\mathbf{n}$, is called the *Cauchy stress tensor*.

By means of (2.8) the force on $d\mathbf{a}$ may be written as

$$\mathbf{t} da = \mathbf{S}^T \mathbf{N} dA, \tag{2.27}$$

where the *nominal stress tensor* $\mathbf{S}$ is related to σ by

$$\mathbf{S} = J\mathbf{F}^{-1}\sigma. \tag{2.28}$$

The *first Piola-Kirchhoff stress tensor*, denoted here by π, is given by $\pi = \mathbf{S}^T$ and this will be used in preference to $\mathbf{S}$ in some parts of this volume.

Let $\mathbf{b}$ denote the body force per unit mass. Then, in integral form, the *equilibrium equation* for the body may be written with reference either to $\mathcal{B}$ or $\mathcal{B}_r$. Thus,

$$\int_{\mathcal{B}} \rho \mathbf{b} dv + \int_{\partial \mathcal{B}} \sigma^T \mathbf{n} da = \int_{\mathcal{B}_r} \rho_r \mathbf{b} dV + \int_{\partial \mathcal{B}_r} \mathbf{S}^T \mathbf{N} dA = \mathbf{0}. \tag{2.29}$$

On use of the divergence theorem equations (2.29) yield the equivalent equilibrium equations

$$\operatorname{div} \sigma + \rho \mathbf{b} = \mathbf{0}, \tag{2.30}$$

$$\operatorname{Div} \mathbf{S} + \rho_r \mathbf{b} = \mathbf{0}, \tag{2.31}$$

where again div and Div denote the divergence operators in $\mathcal{B}$ and $\mathcal{B}_r$ respectively. The derivation of the pointwise equations (2.30) and (2.31) requires that the left-hand sides of these equations are continuous (in $\mathcal{B}$ and $\mathcal{B}_r$ respectively). Note that on use of (2.11) and (2.25) equation (2.31) may be converted immediately to (2.30). In components, (2.31) has the form

$$\frac{\partial S_{\alpha i}}{\partial X_\alpha} + \rho_r b_i = 0, \tag{2.32}$$

and similarly for (2.30), where $S_{\alpha i}$ are the components of $\mathbf{S}$ and b_i those of $\mathbf{b}$.

Balance of the moments of the forces acting on the body yields simply $\sigma^T = \sigma$, which may also be expressed as

$$\mathbf{S}^T \mathbf{F}^T = \mathbf{F S}. \tag{2.33}$$

The Lagrangian formulation based on the use of $\mathbf{S}$ and equation (2.31), with $\mathbf{X}$ as the independent variable, is normally preferred in nonlinear elasticity to the Eulerian formulation based on use of σ and equation (2.30) with $\mathbf{x}$ as the independent variable since the initial geometry is known, whereas $\mathbf{x}$ depends on the deformation to be determined.

We now consider the work done by the surface and body forces in a virtual displacement $\delta\mathbf{x}$ from the current configuration $\mathcal{B}$. By using the divergence theorem and equation (2.31) we obtain the *virtual work* equation

$$\int_{\mathcal{B}_r} \rho_r \mathbf{b} \cdot \delta\mathbf{x}\, dV + \int_{\partial\mathcal{B}_r} (\mathbf{S}^T \mathbf{N}) \cdot \delta\mathbf{x}\, dA = \int_{\mathcal{B}_r} \mathrm{tr}\,(\mathbf{S}\delta\mathbf{F})\, dV, \tag{2.34}$$

where the left-hand side of (2.34) represents the virtual work of the body and surface forces and in the integrand on the right-hand side tr denotes the trace of a second-order tensor and $\delta\mathbf{F} = \mathrm{Grad}\,\delta\mathbf{x}$. The term on the right-hand side is the virtual work of the stresses in the bulk of the material. For a conservative system this latter work is recoverable and is stored as elastic strain energy (this will be discussed in Section 1.2.5.1) but in general it includes a dissipative part. In either case the integrand, which represents the virtual work increment per unit volume in $\mathcal{B}_r$, may be expressed in many alternative forms using different deformation and strain measures.

For example, using (2.16), (2.17) and the symmetry (2.33), we obtain

$$\mathrm{tr}\,(\mathbf{S}\delta\mathbf{F}) = \mathrm{tr}\,(\mathbf{T}^{(1)}\delta\mathbf{U}) = \mathrm{tr}\,(\mathbf{T}^{(2)}\delta\mathbf{E}), \tag{2.35}$$

in which we have defined the *Biot stress tensor* $\mathbf{T}^{(1)}$ (Biot, 1965) and the *second Piola-Kirchhoff stress tensor* $\mathbf{T}^{(2)}$ (both symmetric) by

$$\mathbf{T}^{(1)} = \frac{1}{2}(\mathbf{SR} + \mathbf{R}^T \mathbf{S}^T), \quad \mathbf{T}^{(2)} = \mathbf{SF}^{-T} = J\mathbf{F}^{-1}\sigma\mathbf{F}^{-T}. \tag{2.36}$$

We note the connection

$$\mathbf{T}^{(1)} = \frac{1}{2}(\mathbf{T}^{(2)}\mathbf{U} + \mathbf{U}\mathbf{T}^{(2)}). \tag{2.37}$$

More generally, the expression in (2.35) may be written in terms of the strain tensors $\mathbf{E}^{(m)}$ given by (2.23) and (2.24) and their (symmetric) *conjugate stress tensors* $\mathbf{T}^{(m)}$ as

$$\text{tr}\left(\mathbf{T}^{(m)}\delta\mathbf{E}^{(m)}\right). \tag{2.38}$$

Note that the examples $m = 1$ and $m = 2$ from (2.35) are included in (2.38) as special cases. The notion of conjugate stress and strain tensors was introduced by Hill (1968) and applies more generally than to the special class of strain tensors (2.23). A more detailed discussion can be found in Ogden (1997). We observe that the definition of conjugate stress and strain tensors is independent of any choice of material constitutive law.

1.2.3 Elasticity

The constitutive equation of an elastic material is given in the form

$$\boldsymbol{\sigma} = \mathbf{G}(\mathbf{F}), \tag{2.39}$$

where $\mathbf{G}$ is a *symmetric tensor-valued function*, defined on the space of deformation gradients $\mathbf{F}$. In general the form of $\mathbf{G}$ depends on the choice of reference configuration and $\mathbf{G}$ is referred to as the *response function* of the material *relative* to $\mathcal{B}_r$. For a given $\mathcal{B}_r$, therefore, the stress in $\mathcal{B}$ at a (material) point $\mathbf{X}$ depends only on the deformation gradient at $\mathbf{X}$ and not on the history of deformation. A material whose constitutive law has the form (2.39) is generally referred to as a *Cauchy elastic material*. Its specialization to the situation when there exists a strain-energy function will be considered in Section 1.2.4.

If the stress vanishes in $\mathcal{B}_r$ then

$$\mathbf{G}(\mathbf{I}) = \mathbf{O}, \tag{2.40}$$

and $\mathcal{B}_r$ is called a *natural configuration*. If the stress does not vanish in $\mathcal{B}_r$ then there is said to be *residual stress* in this configuration. In a residually-stressed configuration the traction must vanish at all points of the boundary, so that *a fortiori* residual stress is inhomogeneous in character. For detailed discussion of residual stress we refer to the work of Hoger and co-workers (see, for example, Hoger, 1985, 1986, 1993a, b and Johnson and Hoger 1993, 1995, 1998).

1.2.3.1 Objectivity

Suppose that a rigid-body deformation

$$\mathbf{x}^* = \mathbf{Q}\mathbf{x} + \mathbf{c} \tag{2.41}$$

is superimposed on the deformation $\mathbf{x} = \chi(\mathbf{X})$, where $\mathbf{Q}$ and $\mathbf{c}$ are constants, $\mathbf{Q}$ being a rotation tensor and $\mathbf{c}$ a translation vector. Then, the resulting deformation gradient, $\mathbf{F}^*$ say, is given by

$$\mathbf{F}^* = \mathbf{Q}\mathbf{F}. \tag{2.42}$$

For an elastic material with response function $\mathbf{G}$ relative to $\mathcal{B}_r$, the Cauchy stress tensor, σ^* say, associated with the deformation gradient $\mathbf{F}^*$ is $\sigma^* = \mathbf{G}(\mathbf{F}^*)$.

Under the transformation (2.41) σ transforms according to the formula

$$\sigma^* = \mathbf{Q}\sigma\mathbf{Q}^T. \tag{2.43}$$

The response function $\mathbf{G}$ must therefore satisfy the *invariance requirement*

$$\mathbf{G}(\mathbf{F}^*) \equiv \mathbf{G}(\mathbf{Q}\mathbf{F}) = \mathbf{Q}\mathbf{G}(\mathbf{F})\mathbf{Q}^T \tag{2.44}$$

for each deformation gradient $\mathbf{F}$ and *all* rotations $\mathbf{Q}$. This expresses the fact that the constitutive law (2.39) is *objective*. The terminology *material frame-indifference* is also used for this concept of objectivity (see, for example, Truesdell and Noll, 1965). In essence, this means that material properties are independent of superimposed rigid-body deformations.

A second-order Eulerian tensor, such as σ, which satisfies the transformation rule (2.43) is said to be an (Eulerian) *objective second-order tensor*. We now expand on this notion slightly. Let $\phi, \mathbf{u}, \mathbf{T}$ be (Eulerian) scalar, vector and (second-order) tensor functions defined on $\mathcal{B}$. Let $\phi^*, \mathbf{u}^*, \mathbf{T}^*$ be the corresponding functions defined on $\mathcal{B}^*$, where $\mathcal{B}^*$ is obtained from $\mathcal{B}$ by the rigid deformation (2.41). The functions are said to be (Eulerian) *objective scalar, vector and tensor functions* (or fields) if, for all such deformations,

$$\phi^* = \phi, \quad \mathbf{u}^* = \mathbf{Q}\mathbf{u}, \quad \mathbf{T}^* = \mathbf{Q}\mathbf{T}\mathbf{Q}^T. \tag{2.45}$$

We observe that the density ρ is an example of an objective scalar function and that the normal vector $\mathbf{n}$, which appears in (2.8), and the traction vector $\mathbf{t}$, given by (2.26), are examples of objective vector functions, while the left Cauchy-green deformation tensor $\mathbf{B}$ is an objective tensor function.

It is important to distinguish between the behaviour of Lagrangian and Eulerian vector and tensor functions as far the definition of objectivity is concerned. The vector function $\mathbf{N}$, which is related to $\mathbf{n}$ by (2.8), and the right Cauchy-Green deformation tensor $\mathbf{C}$, given by (2.22), for example, are unchanged under the transformation (2.41). They are Lagrangian functions defined on $\mathcal{B}_r$. Thus, objectivity may equally well be defined in terms of Lagrangian functions. An objective Lagrangian (scalar, vector or tensor) function is one which is *unchanged* by the transformation (2.41). Other examples of objective Lagrangian tensors are the Biot and second Piola-Kirchhoff stress tensors defined in (2.36). Objective mixed tensors, such as $\mathbf{F}$, which are partly Lagrangian and partly Eulerian, change either as in (2.42) or its transpose. Thus, the nominal stress tensor $\mathbf{S}$, given by (2.28), transforms like $\mathbf{S}^* = \mathbf{SQ}^T$ (for more detailed discussion, see Ogden, 1984b).

We mention here that Lagrangian vectors and tensors can be transformed into Eulerian vectors and tensors by appropriate 'push-forward' operations and this process is reversed by 'pull-back' transformations in the sense described in Marsden and Hughes (1994); see also Holzapfel (2000). The form of the push-forward and pull-back transformations depends on whether the vectors and tensors in question have covariant or contravariant character. For example, the push forward of the (covariant) Green strain tensor $\mathbf{E}$ is $\mathbf{F}^{-T}\mathbf{E}\mathbf{F}^{-1}$, which is an Eulerian strain tensor, while the push forward of the (contravariant) second Piola-Kirchhoff stress tensor $\mathbf{T}^{(2)}$ is $\mathbf{F}\mathbf{T}^{(2)}\mathbf{F}^T$, which is just J times the (Eulerian) Cauchy stress tensor. Partial push forward or pull back can be applied to either type of tensor to obtain mixed tensors or to mixed tensors to obtain Lagrangian or Eulerian tensors.

1.2.3.2 Material symmetry

Let σ be the stress in configuration $\mathcal{B}$, and let $\mathbf{F}$ and $\mathbf{F}'$ be the deformation gradients in $\mathcal{B}$ relative to two different reference configurations, $\mathcal{B}_r$ and $\mathcal{B}'_r$ respectively. We denote by $\mathbf{G}$ and $\mathbf{G}'$ the response functions relative to $\mathcal{B}_r$ and $\mathcal{B}'_r$, so that

$$\sigma = \mathbf{G}(\mathbf{F}) = \mathbf{G}'(\mathbf{F}'). \qquad (2.46)$$

Let $\mathbf{P} = \operatorname{Grad} \mathbf{X}'$ be the deformation gradient of $\mathcal{B}'_r$ relative to $\mathcal{B}_r$, where $\mathbf{X}'$ is the position vector of a point in $\mathcal{B}'_r$. Then

$$\mathbf{F} = \mathbf{F}'\mathbf{P}. \qquad (2.47)$$

Substitution of (2.47) into (2.46) then gives $\mathbf{G}(\mathbf{F}'\mathbf{P}) = \mathbf{G}'(\mathbf{F}')$.

In general, the response of the material relative to $\mathcal{B}'_r$ differs from that relative to $\mathcal{B}_r$, i.e $\mathbf{G}' \neq \mathbf{G}$. However, for specific $\mathbf{P}$ we may have $\mathbf{G}' = \mathbf{G}$, in which case

$$\mathbf{G}(\mathbf{F}'\mathbf{P}) = \mathbf{G}(\mathbf{F}') \qquad (2.48)$$

for all deformation gradients $\mathbf{F}'$ and for all such $\mathbf{P}$. Equation (2.46) then gives $\sigma = \mathbf{G}(\mathbf{F}) = \mathbf{G}(\mathbf{F}')$, and, in order to calculate σ, it is not necessary to distinguish between $\mathcal{B}_r$ and $\mathcal{B}'_r$.

The set of tensors $\mathbf{P}$ for which (2.48) holds forms a multiplicative group, called the *symmetry group of the material relative to* $\mathcal{B}_r$. This group characterizes the physical symmetry properties of the material.

Let $\mathbf{P}$ be the deformation gradient $\mathcal{B}_r \to \mathcal{B}'_r$, and now we do *not* assume that $\mathbf{P}$ is a member of the symmetry group. Then, if $\mathcal{G}$ is the symmetry group of the material relative to $\mathcal{B}_r$ and $\mathcal{G}'$ that relative to $\mathcal{B}'_r$, these groups are related according to *Noll's rule*

$$\mathcal{G}' = \mathbf{P}\mathcal{G}\mathbf{P}^{-1}. \tag{2.49}$$

Clearly, for the special case in which $\mathbf{P} \in \mathcal{G}$, we have $\mathcal{G}' = \mathcal{G}$.

1.2.3.3 Isotropic elasticity

If $\mathcal{G}$ is the proper orthogonal group then the material is said to be *isotropic relative to* $\mathcal{B}_r$, and then

$$\sigma = \mathbf{G}(\mathbf{FQ}) = \mathbf{G}(\mathbf{F}) \tag{2.50}$$

for all proper orthogonal $\mathbf{Q}$ and for every deformation gradient $\mathbf{F}$. Physically, this means that the response of a small specimen of material is independent of its orientation in $\mathcal{B}_r$.

Before proceeding further we require some definitions and results relating to isotropic functions of a second-order tensor. Firstly, the scalar function $\phi(\mathbf{T})$ of a *symmetric* second-order tensor $\mathbf{T}$ is said to be an *isotropic function* of $\mathbf{T}$ if

$$\phi(\mathbf{QTQ}^T) = \phi(\mathbf{T}) \tag{2.51}$$

for all orthogonal tensors $\mathbf{Q}$. An isotropic scalar-valued function of $\mathbf{T}$ is also called a *scalar invariant* of $\mathbf{T}$. It may easily be checked that the *principal invariants* of $\mathbf{T}$, defined by

$$I_1(\mathbf{T}) = \mathrm{tr}\,(\mathbf{T}), \quad I_2(\mathbf{T}) = \frac{1}{2}[I_1(\mathbf{T})^2 - \mathrm{tr}\,(\mathbf{T}^2)], \quad I_3(\mathbf{T}) = \det \mathbf{T}, \tag{2.52}$$

are scalar invariants in accordance with the definition (2.51). It may be shown that $\phi(\mathbf{T})$ is a scalar invariant of $\mathbf{T}$ if and only if it is expressible as a function of $I_1(\mathbf{T}), I_2(\mathbf{T}), I_3(\mathbf{T})$.

Secondly, suppose that $\mathbf{G}(\mathbf{T})$ is a symmetric second-order tensor function of $\mathbf{T}$. Then, $\mathbf{G}(\mathbf{T})$ is said to be an *isotropic tensor function* of $\mathbf{T}$ if

$$\mathbf{G}(\mathbf{QTQ}^T) = \mathbf{QG}(\mathbf{T})\mathbf{Q}^T \tag{2.53}$$

for all orthogonal $\mathbf{Q}$. Consequences of this are (i) if $\mathbf{G}(\mathbf{T})$ is isotropic then its eigenvalues are scalar invariants of $\mathbf{T}$, (ii) $\mathbf{G}(\mathbf{T})$ is coaxial with $\mathbf{T}$, i.e.

$$\mathbf{G}(\mathbf{T})\mathbf{T} = \mathbf{T}\mathbf{G}(\mathbf{T}), \tag{2.54}$$

and (iii) $\mathbf{G}(\mathbf{T})$ is isotropic if and only if it has the representation

$$\mathbf{G}(\mathbf{T}) = \phi_0\mathbf{I} + \phi_1\mathbf{T} + \phi_2\mathbf{T}^2, \tag{2.55}$$

where ϕ_0, ϕ_1, ϕ_2 are scalar invariants of $\mathbf{T}$ and hence functions of $I_1(\mathbf{T})$, $I_2(\mathbf{T})$, $I_3(\mathbf{T})$.

The choice $\mathbf{Q} = \mathbf{R}^T$ and use of the polar decomposition $\mathbf{F} = \mathbf{VR}$ in (2.50) gives

$$\boldsymbol{\sigma} = \mathbf{G}(\mathbf{V}). \tag{2.56}$$

We then obtain

$$\mathbf{QG}(\mathbf{V})\mathbf{Q}^T = \mathbf{G}(\mathbf{QVQ}^T) \tag{2.57}$$

for all proper orthogonal $\mathbf{Q}$. In fact, since $\mathbf{Q}$ occurs twice on each side of (2.57), allowing $\mathbf{Q}$ to be improper orthogonal does not affect (2.57), which then states that $\mathbf{G}(\mathbf{V})$ is an isotropic function of $\mathbf{V}$ in accordance with the definition (2.53).

In particular, for an *isotropic elastic material*, $\boldsymbol{\sigma} = \mathbf{G}(\mathbf{V})$ *is coaxial with* $\mathbf{V}$, i.e. with the Eulerian principal axes, and we therefore have

$$\boldsymbol{\sigma} = \mathbf{G}(\mathbf{V}) = \phi_0\mathbf{I} + \phi_1\mathbf{V} + \phi_2\mathbf{V}^2, \tag{2.58}$$

where ϕ_0, ϕ_1, ϕ_2 are scalar invariants of $\mathbf{V}$, i.e. functions of

$$i_1 = I_1(\mathbf{V}) = \mathrm{tr}\,(\mathbf{V}) = \lambda_1 + \lambda_2 + \lambda_3, \tag{2.59}$$

$$i_2 = I_2(\mathbf{V}) = \frac{1}{2}[i_1^2 - \mathrm{tr}\,(\mathbf{V}^2)] = \lambda_2\lambda_3 + \lambda_3\lambda_1 + \lambda_1\lambda_2, \tag{2.60}$$

$$i_3 = I_3(\mathbf{V}) = \det \mathbf{V} = \lambda_1\lambda_2\lambda_3, \tag{2.61}$$

where the expressions have also been given in terms of the principal stretches and the notation i_1, i_2, i_3 has been introduced specifically for the principal invariants of $\mathbf{V}$ (and hence of $\mathbf{U}$). Alternatively, we may write

$$\boldsymbol{\sigma} = \sum_{i=1}^{3} \sigma_i \mathbf{v}^{(i)} \otimes \mathbf{v}^{(i)}, \tag{2.62}$$

where

$$\sigma_i = \phi_0 + \phi_1\lambda_i + \phi_2\lambda_i^2 \qquad i \in \{1,2,3\}, \tag{2.63}$$

and this allows us to introduce the *scalar response function* g, such that

$$\sigma_i = g(\lambda_i, \lambda_j, \lambda_k) = g(\lambda_i, \lambda_k, \lambda_j) \equiv \phi_0 + \phi_1\lambda_i + \phi_2\lambda_i^2, \tag{2.64}$$

where (i, j, k) is permutation of $(1, 2, 3)$.

The expansion (2.58) may be written, equivalently, in terms of $\mathbf{B} = \mathbf{V}^2$. For example,

$$\boldsymbol{\sigma} = \alpha_0\mathbf{I} + \alpha_1\mathbf{B} + \alpha_2\mathbf{B}^2, \tag{2.65}$$

or

$$\boldsymbol{\sigma} = \beta_0\mathbf{I} + \beta_1\mathbf{B} + \beta_{-1}\mathbf{B}^{-1}, \tag{2.66}$$

where $\alpha_0, \alpha_1, \alpha_2, \beta_0, \beta_1, \beta_{-1}$ are scalar invariants of $\mathbf{B}$ (and hence of $\mathbf{V}$); see, for example, Beatty (1987). Connections between these different coefficients are determined by using the Cayley-Hamilton theorem in the form

$$\mathbf{V}^3 - i_1\mathbf{V}^2 + i_2\mathbf{V} - i_3\mathbf{I} = \mathbf{O} \tag{2.67}$$

or its counterpart for $\mathbf{B}$. It is convenient in what follows to use the standard notation I_1, I_2, I_3 for the principal invariants of $\mathbf{B}$ (also of $\mathbf{C}$). Thus, specifically, we write

$$I_1 = I_1(\mathbf{B}) = \text{tr}\,(\mathbf{B}) = \lambda_1^2 + \lambda_2^2 + \lambda_3^2, \tag{2.68}$$

$$I_2 = I_2(\mathbf{B}) = \frac{1}{2}[I_1^2 - \text{tr}\,(\mathbf{B}^2)] = \lambda_2^2\lambda_3^2 + \lambda_3^2\lambda_1^2 + \lambda_1^2\lambda_2^2, \tag{2.69}$$

$$I_3 = I_3(\mathbf{B}) = \det\mathbf{B} = \lambda_1^2\lambda_2^2\lambda_3^2. \tag{2.70}$$

In view of the connection (2.28) between $\mathbf{S}$ and σ we may also define the response function, $\mathbf{H}$ say, associated with $\mathbf{S}$ (relative to $\mathcal{B}_r$) by

$$\mathbf{S} = \mathbf{H}(\mathbf{F}) \equiv J\mathbf{F}^{-1}\mathbf{G}(\mathbf{F}). \tag{2.71}$$

The objectivity requirement (2.44) then becomes

$$\mathbf{H}(\mathbf{QF}) = \mathbf{H}(\mathbf{F})\mathbf{Q}^T. \tag{2.72}$$

A corresponding change for the material symmetry transformation (2.48) can be written down, and, in particular, for an isotropic elastic solid, we have

$$\mathbf{H}(\mathbf{FQ}) = \mathbf{Q}^T\mathbf{H}(\mathbf{F}). \tag{2.73}$$

Moreover, it follows from (2.73) that

$$\mathbf{H}(\mathbf{F}) = \mathbf{H}(\mathbf{U})\mathbf{R}^T = \mathbf{R}^T\mathbf{H}(\mathbf{V}), \tag{2.74}$$

with $\mathbf{H}(\mathbf{U})$ being symmetric and coaxial with $\mathbf{U}$.

1.2.3.4 Internal constraints

In Section 1.2.1 the (internal) constraints of incompressibility and inextensibility were mentioned. More generally, a single constraint may be written in the form

$$C(\mathbf{F}) = 0, \tag{2.75}$$

where C is a scalar function. Equation (2.75) holds for all possible deformation gradients $\mathbf{F}$. For the incompressibility and inextensibility constraints we have, respectively,

$$C(\mathbf{F}) = \det\mathbf{F} - 1, \qquad C(\mathbf{F}) = \mathbf{M} \cdot (\mathbf{F}^T\mathbf{F}\mathbf{M}) - 1. \tag{2.76}$$

Since any constraint is unaffected by a superimposed rigid deformation, C must be an objective scalar function, so that

$$C(\mathbf{QF}) = C(\mathbf{F}) \tag{2.77}$$

for all rotations $\mathbf{Q}$. In particular, the choice $\mathbf{Q} = \mathbf{R}^T$ yields

$$C(\mathbf{F}) = C(\mathbf{U}). \tag{2.78}$$

For incompressibility the $C(\mathbf{U})$ given by $(2.76)_1$ is a scalar invariant of $\mathbf{U}$, but this is not the case for a general constraint function $C(\mathbf{U})$.

The constraint (2.75) defines a hypersurface in the (nine-dimensional) space of deformation gradients. Any stress in the normal direction to the surface (i.e. the direction $\partial C/\partial \mathbf{F}$) does no work in any (virtual) incremental deformation $\delta \mathbf{x}$ compatible with the constraint since $\mathrm{tr}\,[(\partial C/\partial \mathbf{F})\delta \mathbf{F}] = 0$. The stress is therefore determined by the constitutive law, in the form (2.71) for example, only to within an additive contribution parallel to the normal. Thus, for a constrained material the stress-deformation relation (2.71) is replaced by

$$\mathbf{S} = \mathbf{H}(\mathbf{F}) + q\frac{\partial C}{\partial \mathbf{F}}, \tag{2.79}$$

or, in terms of Cauchy stress,

$$\boldsymbol{\sigma} = \mathbf{G}(\mathbf{F}) + qJ^{-1}\mathbf{F}\frac{\partial C}{\partial \mathbf{F}}, \tag{2.80}$$

where q is an arbitrary (Lagrange) multiplier. The term in q is referred to as the *constraint stress* since it arises from the constraint and is not otherwise derivable from the material properties.

For incompressibility we have $\partial C/\partial \mathbf{F} = \mathbf{F}^{-1}$ since $J = 1$ and hence

$$\mathbf{S} = \mathbf{H}(\mathbf{F}) + q\mathbf{F}^{-1}, \qquad \boldsymbol{\sigma} = \mathbf{G}(\mathbf{F}) + q\mathbf{I}, \tag{2.81}$$

$\mathbf{I}$ again being the identity tensor, while for inextensibility $\partial C/\partial \mathbf{F} = 2\mathbf{M} \otimes \mathbf{FM}$ and

$$\mathbf{S} = \mathbf{H}(\mathbf{F}) + 2q\mathbf{M} \otimes \mathbf{FM}, \qquad \boldsymbol{\sigma} = \mathbf{G}(\mathbf{F}) + 2qJ^{-1}\mathbf{FM} \otimes \mathbf{FM}. \tag{2.82}$$

In the case of (2.81), if the material is isotropic then $\mathbf{G}(\mathbf{F})$ is given by (2.58) but with the term in ϕ_0 omitted since it may be absorbed into q and ϕ_1 and ϕ_2 being functions of the two remaining independent invariants.

Another constraint, called the *Bell constraint*, is the focus of Chapter 2 and will not therefore be discussed here. If there is more than one constraint then an additive constraint stress has to be included in the expression for the stress in respect of each constraint. However, the constraints must be mutually compatible since, as illustrated in Chapter 2, the incompressibility and Bell constraints are not compatible.

1.2.4 Hyperelasticity

As mentioned at the beginning of Section 1.2.3, the notion of elasticity introduced there is referred to as *Cauchy elasticity*. From the point of view of both theory and applications a more useful concept of elasticity, which is a special case of Cauchy elasticity, is *hyperelasticity* (or *Green elasticity*). In this theory there exists a *strain-energy function* (or *stored-energy function*), denoted $W = W(\mathbf{F})$, defined on the space of deformation gradients such that (for an unconstrained material)

$$\mathbf{S} = \mathbf{H}(\mathbf{F}) = \frac{\partial W}{\partial \mathbf{F}}, \quad \boldsymbol{\sigma} = \mathbf{G}(\mathbf{F}) = J^{-1}\mathbf{F}\frac{\partial W}{\partial \mathbf{F}}. \tag{2.83}$$

The work increment in (2.35) is then converted into stored energy and is simply $\mathrm{tr}\,(\mathbf{S}\delta\mathbf{F}) = \delta W$. Equation (2.83) is the *stress-deformation relation* or *constitutive relation* for an elastic material which possesses a strain-energy function, W being defined per unit volume in $\mathcal{B}_r$ and representing the work done per unit volume at $\mathbf{X}$ in changing the deformation gradient from $\mathbf{I}$ to $\mathbf{F}$. In components, the first equation in (2.83) is written $S_{\alpha i} = \partial W/\partial F_{i\alpha}$, which provides the convention for ordering of the indices in the partial derivative with respect to $\mathbf{F}$.

Henceforth in this chapter we restrict attention to hyperelasticity and regard an elastic material as characterized by the existence of a strain-energy function such that (2.83) hold. We take W and the stress to vanish in $\mathcal{B}_r$, so that

$$W(\mathbf{I}) = 0, \quad \frac{\partial W}{\partial \mathbf{F}}(\mathbf{I}) = \mathbf{O}, \tag{2.84}$$

the latter being consistent with (2.40).

We note here the modification of (2.83) appropriate for incompressibility. From (2.81) we obtain

$$\mathbf{S} = \frac{\partial W}{\partial \mathbf{F}} - p\mathbf{F}^{-1}, \quad \boldsymbol{\sigma} = \mathbf{F}\frac{\partial W}{\partial \mathbf{F}} - p\mathbf{I}, \quad \det \mathbf{F} = 1, \tag{2.85}$$

where q has been replaced by $-p$, with p, in standard notation, then referred to as an *arbitrary hydrostatic pressure*. Equations (2.85) are *stress-deformation relations for an incompressible elastic material*. The corresponding expressions for the constraint of inextensibility may be read off from (2.82).

1.2.4.1 Objectivity and Material Symmetry

The elastic stored energy is required to be independent of superimposed rigid deformations of the form (2.41) and it therefore follows that

$$W(\mathbf{QF}) = W(\mathbf{F}) \tag{2.86}$$

for *all* rotations $\mathbf{Q}$. A strain-energy function satisfying this requirement is said to be *objective*.

Use of the polar decomposition (2.17) and the choice $\mathbf{Q} = \mathbf{R}^T$ in (2.86) shows that

$$W(\mathbf{F}) = W(\mathbf{U}). \tag{2.87}$$

Thus, W depends on $\mathbf{F}$ only through the stretch tensor $\mathbf{U}$ and may therefore be defined on the class of positive definite symmetric tensors. Since $\mathbf{E}^{(1)} = \mathbf{U} - \mathbf{I}$, as defined in (2.23), is conjugate to the Biot stress tensor $\mathbf{T}^{(1)}$, which we write henceforth as $\mathbf{T}$, we have

$$\mathbf{T} = \frac{\partial W}{\partial \mathbf{U}} \tag{2.88}$$

for an unconstrained material and

$$\mathbf{T} = \frac{\partial W}{\partial \mathbf{U}} - p\mathbf{U}^{-1}, \qquad \det \mathbf{U} = 1 \tag{2.89}$$

for an incompressible material. Note that when expressed as a function of $\mathbf{U}$ the strain energy automatically satisfies the objectivity requirement.

Mathematically, there is no restriction so far other than (2.84) and (2.86) on the form that the function W may take, but the predicted stress-strain behaviour based on the form of W must on the one hand be acceptable for the description of the elastic behaviour of real materials and on the other hand make mathematical sense.

Further restrictions on the form of W arise if the material possesses symmetries in the configuration $\mathcal{B}_r$. For a hyperelastic material the symmetry requirement (2.48) is replaced by

$$W(\mathbf{F}'\mathbf{P}) = W(\mathbf{F}') \tag{2.90}$$

for *all* deformation gradients $\mathbf{F}'$. This states that the strain-energy function is unaffected by a change of reference configuration with deformation gradient $\mathbf{P}$ which is a member of the symmetry group of the material relative to $\mathcal{B}_r$.

1.2.4.2 Isotropic hyperelasticity

To be specific we now consider *isotropic elastic materials*, for which the symmetry group is the *proper orthogonal group*. Then, we have

$$W(\mathbf{FQ}) = W(\mathbf{F}) \qquad (2.91)$$

for *all* rotations $\mathbf{Q}$. Bearing in mind that the $\mathbf{Q}$'s appearing in (2.86) and (2.91) are independent the combination of these two equations yields

$$W(\mathbf{QUQ}^T) = W(\mathbf{U}) \qquad (2.92)$$

for all rotations $\mathbf{Q}$, or, equivalently, $W(\mathbf{QVQ}^T) = W(\mathbf{V})$. Equation (2.92) states that W is an *isotropic function* of $\mathbf{U}$. It follows from the spectral decomposition (2.18) that W depends on $\mathbf{U}$ only through the principal stretches $\lambda_1, \lambda_2, \lambda_3$. To avoid introducing additional notation we express this dependence as $W(\lambda_1, \lambda_2, \lambda_3)$; by selecting appropriate values for $\mathbf{Q}$ in (2.92) we may deduce that W depends symmetrically on $\lambda_1, \lambda_2, \lambda_3$, i.e.

$$W(\lambda_1, \lambda_2, \lambda_3) = W(\lambda_1, \lambda_3, \lambda_2) = W(\lambda_2, \lambda_1, \lambda_3). \qquad (2.93)$$

A consequence of isotropy is that $\mathbf{T}$ is *coaxial* with $\mathbf{U}$ and hence, in parallel with (2.18), we have

$$\mathbf{T} = \sum_{i=1}^{3} t_i \mathbf{u}^{(i)} \otimes \mathbf{u}^{(i)}, \qquad (2.94)$$

where t_i, $i \in \{1,2,3\}$ are the *principal Biot stresses*. For an unconstrained material,

$$t_i = \frac{\partial W}{\partial \lambda_i}, \qquad (2.95)$$

and for an incompressible material this is replaced by

$$t_i = \frac{\partial W}{\partial \lambda_i} - p\lambda_i^{-1}, \qquad \lambda_1\lambda_2\lambda_3 = 1. \qquad (2.96)$$

For later reference, we note here that the principal Cauchy stresses σ_i, $i \in \{1,2,3\}$, are given by

$$J\sigma_i = \lambda_i \frac{\partial W}{\partial \lambda_i} \qquad (2.97)$$

and

$$\sigma_i = \lambda_i \frac{\partial W}{\partial \lambda_i} - p, \qquad \lambda_1\lambda_2\lambda_3 = 1 \qquad (2.98)$$

for unconstrained and incompressible materials respectively. Note that in (2.97) and (2.98) there is no summation over the repeated index i.

With reference to (2.63) it can be deduced from (2.97), by regarding W as a function of i_1, i_2, i_3, that the coefficients ϕ_0, ϕ_1, ϕ_2 are given by

$$\phi_0 = \frac{\partial W}{\partial i_3}, \quad \phi_1 = i_3^{-1}\left(\frac{\partial W}{\partial i_1} + i_1\frac{\partial W}{\partial i_2}\right), \quad \phi_2 = -i_3^{-1}\frac{\partial W}{\partial i_2}. \tag{2.99}$$

Similarly, the coefficients $\alpha_0, \alpha_1, \alpha_2$ in (2.65) are given by

$$\alpha_0 = 2I_3^{1/2}\frac{\partial W}{\partial I_3}, \quad \alpha_1 = 2I_3^{-1/2}\left(\frac{\partial W}{\partial I_1} + I_1\frac{\partial W}{\partial I_2}\right), \quad \alpha_2 = -2I_3^{-1/2}\frac{\partial W}{\partial I_2}, \tag{2.100}$$

where W is now regarded as a function of I_1, I_2, I_3. For an incompressible material the term in ϕ_0, or α_0 as appropriate, is absorbed into p and $i_3 = I_3 = 1$ in the remaining terms in (2.99) and (2.100).

We emphasize that, as follows from (2.74), *for an isotropic elastic material* **SR** *is symmetric* and we have

$$\mathbf{S} = \mathbf{TR}^T = \sum_{i=1}^{3} t_i \mathbf{u}^{(i)} \otimes \mathbf{v}^{(i)}. \tag{2.101}$$

The first equation in (2.101) is a polar decomposition of **S** analogous to (2.17)$_1$ except that it is not unique. This equation has important consequences for considerations of uniqueness and stability and will be discussed briefly in Section 1.4.4.

1.2.4.3 Examples of strain-energy functions

There are numerous specific forms of strain-energy function in the literature both for compressible and incompressible materials, mainly isotropic, and we make no attempt to catalogue them here. Many will be used in the various chapters of this volume. Here we just mention, for purposes of illustration, a few examples of strain-energy functions for *incompressible isotropic* materials. Many, but not all, of the compressible strain-energy functions used in the literature are obtained from their incompressible counterparts by addition of a function of the volume measure J and multiplication of the other terms by powers of J. Examples of compressible strain-energy functions are contained in Chapter 4, for example.

As already noted the strain energy of an isotropic elastic solid can be regarded either as a symmetric function of the principal stretches or as a function of three independent invariants, such as i_1, i_2, i_3 or I_1, I_2, I_3. For an incompressible material (2.10) holds, $i_3 = I_3 = 1$ and hence the strain energy depends on only two independent invariants. An important example is the *Mooney-Rivlin* form of strain energy, defined by

$$W = C_1(I_1 - 3) + C_2(I_2 - 3) \equiv C_1(\lambda_1^2 + \lambda_2^2 + \lambda_3^2 - 3) + C_2(\lambda_1^{-2} + \lambda_2^{-2} + \lambda_3^{-2} - 3),$$
$$(2.102)$$

where C_1, C_2 are constants and $\lambda_1\lambda_2\lambda_3 = 1$. When $C_2 = 0$ this reduces to the so-called *neo-Hookean* strain energy

$$W = \frac{1}{2}\mu(I_1 - 3) \equiv \frac{1}{2}\mu(\lambda_1^2 + \lambda_2^2 + \lambda_3^2 - 3), \qquad (2.103)$$

where C_1 has been replaced by $\mu/2$, $\mu\,(> 0)$ being the shear modulus of the material in the undeformed configuration. These two forms of energy function played key roles in the development of the subject of finite elasticity, particularly in respect of its connection with rubber elasticity. For reviews of this aspect we refer to Ogden (1982, 1986) in which more details are given of different forms of strain-energy functions appropriate for rubberlike solids. Equation (2.102) constitutes the linear terms in a polynomial expansion of W in terms of $I_1 - 3$ and $I_2 - 3$, special cases of which are used extensively in the literature.

Another special form of strain energy worthy of mention is the Varga form, defined by

$$W = 2\mu(i_1 - 3) \equiv 2\mu(\lambda_1 + \lambda_2 + \lambda_3 - 3), \qquad (2.104)$$

which is used extensively in basic stress-strain analysis (Varga, 1966). See also Chapter 5 in this volume.

Each of the strain-energy functions (2.102)–(2.104) is of the *separable form*

$$W = w(\lambda_1) + w(\lambda_2) + w(\lambda_3), \qquad (2.105)$$

which was introduced by Valanis and Landel (1967). Equivalent to (2.105) is the expansion

$$W = \sum_{m=1}^{\infty} \mu_m(\lambda_1^{\alpha_m} + \lambda_2^{\alpha_m} + \lambda_3^{\alpha_m} - 3)/\alpha_m \qquad (2.106)$$

in terms of powers of the stretches, where each μ_m and α_m is a material constant, the latter not necessarily being integers (Ogden, 1972, 1982). For practical purposes the sum in (2.106) is restricted to a finite number of terms, while, for consistency with the classical theory, the constants must satisfy the requirement

$$\sum_{m=1}^{N} \mu_m\alpha_m = 2\mu, \qquad (2.107)$$

where N is a positive integer and μ is again the shear modulus of the material in the natural configuration. The counterpart of (2.107) for the Valanis-Landel material is

$$w''(1) + w'(1) = 2\mu. \tag{2.108}$$

In respect of (2.105) the principal Cauchy stresses are obtained from (2.98) in the form

$$\sigma_i = \lambda_i w'(\lambda_i) - p \tag{2.109}$$

and the specializations appropriate for (2.102)–(2.104) are then easily read off. For the neo-Hookean solid, for example, we have

$$\sigma = \mu \mathbf{B} - p\mathbf{I}, \tag{2.110}$$

where $\mathbf{B}$ is the left Cauchy-Green deformation tensor defined in (2.22).

1.2.4.4 Anisotropy: fibre-reinforced materials

For a general discussion of anisotropic elasticity, including the crystal classes, we refer to Green and Adkins (1970) or Truesdell and Noll (1965), for example. Here, we illustrate the structure of the strain-energy function of an anisotropic elastic solid for the example of *transverse isotropy*, in which there is a single preferred direction, and the extension of this to the case of two preferred directions. These are important examples in practical applications to fibre-reinforced materials, such as high-pressure hoses and soft biological tissues.

Firstly, we consider transverse isotropy. Let the unit vector $\mathbf{M}$ be a preferred direction in the reference configuration of the material. The material response is then indifferent to arbitrary rotations about the direction $\mathbf{M}$ and by replacement of $\mathbf{M}$ by $-\mathbf{M}$. Such a material can be characterized with a strain energy which depends on $\mathbf{F}$ and the tensor $\mathbf{M} \otimes \mathbf{M}$, as described by Spencer (1972, 1984); see, also, Rogers (1984a) and Holzapfel (2000). Thus, we write $W(\mathbf{F}, \mathbf{M} \otimes \mathbf{M})$ and the required symmetry reduces W to dependence on the five invariants

$$I_1, \ I_2, \ I_3, \ I_4 = \mathbf{M} \cdot (\mathbf{CM}), \ I_5 = \mathbf{M} \cdot (\mathbf{C}^2 \mathbf{M}), \tag{2.111}$$

where I_1, I_2, I_3 are defined in (2.68)–(2.70). The resulting nominal stress tensor is given by

$$\mathbf{S} = 2W_1 \mathbf{F}^T + 2W_2(I_1\mathbf{I} - \mathbf{C})\mathbf{F}^T + 2I_3 W_3 \mathbf{F}^{-1} + 2W_4 \mathbf{M} \otimes \mathbf{FM}$$
$$+ 2W_5(\mathbf{M} \otimes \mathbf{FCM} + \mathbf{CM} \otimes \mathbf{FM}), \tag{2.112}$$

where $W_i = \partial W/\partial I_i, i = 1, \ldots, 5$. For an isotropic material the terms in W_4 and W_5 are omitted. Equation (2.112) describes the response of a fibre-reinforced material with the fibre direction corresponding to $\mathbf{M}$ locally in the reference configuration. For an incompressible material the dependence on I_3 is omitted and the Cauchy stress tensor is given by

$$\boldsymbol{\sigma} = -p\mathbf{I} + 2W_1\mathbf{B} + 2W_2(I_1\mathbf{B} - \mathbf{B}^2) + 2W_4\mathbf{FM} \otimes \mathbf{FM}$$
$$+ 2W_5(\mathbf{FM} \otimes \mathbf{BFM} + \mathbf{BFM} \otimes \mathbf{FM}), \qquad (2.113)$$

from which the symmetry of $\boldsymbol{\sigma}$ can be seen immediately. Note that in (2.113) the left Cauchy-Green tensor $\mathbf{B}$ has been used.

When there are two families of fibres corresponding to two preferred directions in the reference configuration, $\mathbf{M}$ and $\mathbf{M}'$ say, then, in addition to (2.111), the strain energy depends on the invariants

$$I_6 = \mathbf{M}' \cdot (\mathbf{CM}'), \quad I_7 = \mathbf{M}' \cdot (\mathbf{C}^2\mathbf{M}'), \quad I_8 = \mathbf{M} \cdot (\mathbf{CM}'), \qquad (2.114)$$

and also on $\mathbf{M} \cdot \mathbf{M}'$ (which does not depend on the deformation); see Spencer (1972, 1984) for details. Note that I_8 involves interaction between the two preferred directions, but the term $\mathbf{M} \cdot (\mathbf{C}^2\mathbf{M}')$, which might be expected to appear in the list (2.114), is omitted since it depends on the other invariants and on $\mathbf{M} \cdot \mathbf{M}'$. It suffices here to give the Cauchy stress tensor for an incompressible material. This is

$$\boldsymbol{\sigma} = -p\mathbf{I} + 2W_1\mathbf{B} + 2W_2(I_1\mathbf{B} - \mathbf{B}^2) + 2W_4\mathbf{FM} \otimes \mathbf{FM}$$
$$+ 2W_5(\mathbf{FM} \otimes \mathbf{BFM} + \mathbf{BFM} \otimes \mathbf{FM}) + 2W_6\mathbf{FM}' \otimes \mathbf{FM}'$$
$$+ 2W_7(\mathbf{FM}' \otimes \mathbf{BFM}' + \mathbf{BFM}' \otimes \mathbf{FM}')$$
$$+ W_8(\mathbf{FM} \otimes \mathbf{FM}' + \mathbf{FM}' \otimes \mathbf{FM}), \qquad (2.115)$$

where the notation $W_i = \partial W / \partial I_i$ now applies for $i = 1, \ldots, 8$. In the context of finite deformation theory very few boundary-value problems have been solved for either transversely isotropic materials or for materials with two preferred directions. An account of such results is given in Green and Adkins (1970), but relatively little progress has been made since the publication of this book as far as obtaining closed-form solutions is concerned. For materials with one or two families of *inextensible* fibres some basic results are contained in Spencer (1972, 1984) and Rogers (1984b). However, there is very little in the literature concerned with *specific* forms of strain-energy function based on the invariants (2.111) or (2.111) combined with (2.114), although, in the context of membrane biomechanics, special forms of (2.113) have been used; see, for example, Humphrey (1995) for references.

1.2.5 Boundary-value problems

We now consider the equilibrium equation (2.31) together with the stress-deformation relation (2.83)$_1$ for an unconstrained material, and the deformation gradient (2.3) coupled with (2.1). Thus,

$$\text{Div}\left(\frac{\partial W}{\partial \mathbf{F}}\right) + \rho_r \mathbf{b} = 0, \quad \mathbf{F} = \text{Grad}\,\mathbf{x}, \quad \mathbf{x} = \chi(\mathbf{X}), \quad \mathbf{X} \in \mathcal{B}_r. \quad (2.116)$$

A boundary-value problem is obtained by supplementing (2.116) with appropriate boundary conditions. Typical boundary conditions arising in problems of nonlinear elasticity are those in which $\mathbf{x}$ is specified on part of the boundary, $\partial \mathcal{B}_r^x \subset \partial \mathcal{B}_r$ say, and the stress vector on the remainder, $\partial \mathcal{B}_r^\tau$, so that $\partial \mathcal{B}_r^x \cup \partial \mathcal{B}_r^\tau = \partial \mathcal{B}_r$ and $\partial \mathcal{B}_r^x \cap \partial \mathcal{B}_r^\tau = \emptyset$. We write

$$\mathbf{x} = \boldsymbol{\xi}(\mathbf{X}) \quad \text{on } \partial \mathcal{B}_r^x, \quad (2.117)$$
$$\mathbf{S}^T \mathbf{N} = \boldsymbol{\tau}(\mathbf{F}, \mathbf{X}) \quad \text{on } \partial \mathcal{B}_r^\tau, \quad (2.118)$$

where $\boldsymbol{\xi}$ and $\boldsymbol{\tau}$ are specified functions. In general, $\boldsymbol{\tau}$ may depend on the deformation and we have indicated this in (2.118) by showing the dependence of $\boldsymbol{\tau}$ on the deformation gradient $\mathbf{F}$. If the surface traction defined by (2.118) is independent of $\mathbf{F}$ it is referred to as a *dead-load traction*. If the boundary traction in (2.118) is associated with a hydrostatic pressure, P say, so that $\boldsymbol{\sigma}\mathbf{n} = -P\mathbf{n}$, then $\boldsymbol{\tau}$ depends on the deformation in the form

$$\boldsymbol{\tau} = -J\mathbf{P}\mathbf{F}^{-T}\mathbf{N} \quad \text{on } \partial \mathcal{B}_r^\tau. \quad (2.119)$$

In general, when the dependence on $\mathbf{F}$ is retained, (2.118) is referred to as a *configuration dependent loading* (Sewell, 1967). The basic boundary-value problem of nonlinear elasticity is characterized by (2.116)–(2.118).

In components, the equilibrium equation in (2.116) can be written

$$\mathcal{A}^1_{\alpha i \beta j} \frac{\partial^2 x_j}{\partial X_\alpha \partial X_\beta} + \rho_r b_i = 0, \quad (2.120)$$

for $i \in \{1, 2, 3\}$, where the coefficients $\mathcal{A}^1_{\alpha i \beta j}$ are defined by

$$\mathcal{A}^1_{\alpha i \beta j} = \mathcal{A}^1_{\beta j \alpha i} = \frac{\partial^2 W}{\partial F_{i\alpha} \partial F_{j\beta}}, \quad (2.121)$$

the pairwise symmetry of the indices thereby being emphasized.

When coupled with suitable boundary conditions, equation (2.120) forms a coupled system of three second-order quasi-linear partial differential equations for $x_i = \chi_i(X_\alpha)$. The coefficients $\mathcal{A}^1_{\alpha i \beta j}$ are, in general, nonlinear functions of the components of the deformation gradient. We emphasize that here we are using Cartesian coordinates; expressions for the equilibrium equations in other coordinate systems are not given here but can be found in Ogden (1997), for example.

For *unconstrained materials* very few explicit solutions have been obtained for boundary-value problems, and these arise for very special choices of the form of W and for relatively simple geometries. References are given in Ogden (1997), for example, and an up-to-date account is contained in Chapter 4. For *incompressible materials* the corresponding equations, obtained by substituting $(2.85)_1$ into (2.31) to give

$$\mathcal{A}^1_{\alpha i \beta j} \frac{\partial^2 x_j}{\partial X_\alpha \partial X_\beta} - \frac{\partial p}{\partial x_i} + \rho_r b_i = 0 \qquad (2.122)$$

subject to (2.10), where the coefficients are again given by (2.121). These have yielded more success, and we refer to Green and Zerna (1968), Green and Adkins (1970) and Ogden (1997) for details of the solutions, many of which are based on the pioneering work of Rivlin (see, for example, Rivlin, 1948a, b, 1949a, b; references to further work by Rivlin and co-workers can be found in, for example, Truesdell and Noll, 1965 and Green and Zerna, 1968; the edited papers of Rivlin are provided in the volume by Barenblatt and Joseph, 1996). See also Chapters 3, 4, and 5 in this volume for further discussion of boundary-value problems. Note that, exceptionally, for the neo-Hookean form of strain energy (2.103), the coefficients $\mathcal{A}^1_{\alpha i \beta j}$ are constant and given by

$$\mathcal{A}^1_{\alpha i \beta j} = \mu \delta_{ij} \delta_{\alpha \beta}. \qquad (2.123)$$

The equations (2.122), although appearing linear in this case, are in general nonlinear because of the term in p.

In order to analyze such boundary-value problems additional information about the nature of the function W is required. This information may come from the construction of special forms of strain-energy function based on comparison of theory with experiment for particular materials, may arise naturally in the course of solution of particular problems, or may be derived from mathematical considerations relating to the properties that W should possess in order for existence of solutions to be guaranteed, for example. This important aspect of the theory is not examined to any great extent in this chapter, but reference can be made to, for example, Ciarlet (1988) or Ogden (1997) for discussion of this matter. In this connection, however, we now examine certain aspects of the structure of the equations.

Equations (2.120) are said to be *strongly elliptic* if the inequality

$$\mathcal{A}^1_{\alpha i \beta j} m_i m_j N_\alpha N_\beta > 0 \qquad (2.124)$$

holds for all non-zero vectors $\mathbf{m}$ and $\mathbf{N}$. Note that this inequality is independent of any boundary conditions. Strong ellipticity ensures, in particular, that in an infinite medium infinitesimal motions superimposed on a finite deformation do not grow exponentially (see Section 1.6.3.1).

For an incompressible material the strong ellipticity condition associated with (2.122) again has the form (2.124) but the incompressibility constraint now imposes the restriction

$$\mathbf{m} \cdot \mathbf{n} = 0 \tag{2.125}$$

on the (non-zero) vectors $\mathbf{m}$ and $\mathbf{n}$, where $\mathbf{n}$, the push forward of $\mathbf{N}$, is defined by

$$\mathbf{F}^T \mathbf{n} = \mathbf{N}. \tag{2.126}$$

Note that $\mathbf{N}$ and $\mathbf{n}$ here are not related to the surface normal vectors defined in Section 1.2.1.

In terms of $\mathbf{n}$ the strong ellipticity condition (2.124) may be written

$$\mathcal{A}^1_{0piqj} m_i m_j n_p n_q > 0, \tag{2.127}$$

where $\mathcal{A}^1_{0piqj}$ are related to $\mathcal{A}^1_{\alpha i \beta j}$ by

$$\mathcal{A}^1_{0piqj} = J^{-1} F_{p\alpha} F_{q\beta} \mathcal{A}^1_{\alpha i \beta j}. \tag{2.128}$$

For an isotropic material necessary and sufficient conditions for strong ellipticity to hold *in two dimensions* for unconstrained and incompressible materials will be given in Section 1.4.2.2. The corresponding conditions for three dimensions are quite complicated and are not therefore given here. We refer to Zee and Sternberg (1983) for incompressible materials and Simpson and Spector (1983), Rosakis (1990) and Wang and Aron (1996) for unconstrained materials. See also Hill (1979).

Failure of ellipticity in the sense that equality holds in (2.127) for some specific $\mathbf{m}$ and $\mathbf{n}$ plays a very important role in connection with the emergence of solutions with discontinuous deformation gradients. For discussion of this see, for example, Knowles and Sternberg (1975, 1977) and Chapter 12 in this volume, which contains more detailed references to work on this topic.

There are some important situations where simplifications of the governing equations (2.120) or (2.122) arise. For example, for *plane strain deformations*, equation (2.120) reduces to a pair of equations for $x_\alpha = \chi_\alpha(X_\beta)$, where the Greek indices have the range $\{1, 2\}$. A further simplification arises for *anti-plane shear deformations*, for which (2.120) reduces to a single quasi-linear equation for a scalar function $x = \chi(X_\alpha)$, $\alpha \in \{1, 2\}$. For a review of anti-plane shear see, for example, Horgan (1995).

1.2.5.1 Variational structure

We now take the body force in (2.34) to be conservative so that we may write

$$\mathbf{b} = -\operatorname{grad} \phi, \tag{2.129}$$

where ϕ is a scalar field defined on points in B and grad denotes the gradient operator in B. Since we are considering an elastic material with strain-energy function W, it follows that the virtual work equation (2.34) can be expressed in the form

$$\delta \int_{B_r} (W + \rho_r \phi) \, dV - \int_{\partial B_r} (\mathbf{S}^T \mathbf{N}) \cdot \delta \mathbf{x} \, dA = 0. \tag{2.130}$$

In view of the boundary condition (2.117) we have $\delta \mathbf{x} = \mathbf{0}$ on ∂B_r^x, and (2.130) becomes

$$\delta \int_{B_r} (W + \rho_r \phi) \, dV - \int_{\partial B_r^\tau} \boldsymbol{\tau} \cdot \delta \mathbf{x} \, dA = 0. \tag{2.131}$$

For illustrative purposes and for simplicity we now take $\boldsymbol{\tau}$ to be independent of the deformation and we write $\boldsymbol{\tau} = \operatorname{grad}(\boldsymbol{\tau} \cdot \mathbf{x})$, so that (2.131) becomes

$$\delta \left\{ \int_{B_r} (W + \rho_r \phi) \, dV - \int_{\partial B_r^\tau} \boldsymbol{\tau} \cdot \mathbf{x} \, dA \right\} = 0. \tag{2.132}$$

If we regard $\delta \chi$ as a variation of the function χ, then (2.132) provides a variational formulation of the boundary-value problem (2.116)–(2.118) and can be written $\delta E = 0$, where E is the *functional* defined by

$$E\{\chi\} = \int_{B_r} \{W(\operatorname{Grad} \chi) + \rho_r \phi(\chi)\} \, dV - \int_{\partial B_r^\tau} \boldsymbol{\tau} \cdot \chi \, dA. \tag{2.133}$$

In (2.133) χ is taken in some appropriate class of mappings (for example, the twice continuously differentiable mappings we have agreed to consider in this chapter) and to satisfy the boundary condition (2.117). Similarly for the admissible variations $\delta \chi$ subject to $\delta \chi = \mathbf{0}$ on ∂B_r^x.

Then, starting from (2.133) with χ in this class, we obtain, after use of the divergence theorem and the boundary conditions on ∂B_r^x,

$$\delta E = - \int_{B_r} (\operatorname{Div} \mathbf{S} + \rho_r \mathbf{b}) \cdot \delta \chi \, dV + \int_{\partial B_r^\tau} (\mathbf{S}^T \mathbf{N} - \boldsymbol{\tau}) \cdot \delta \chi \, dA.$$

The variational statement then takes the form $\delta E = 0$ for all admissible $\delta \chi$ if and only if χ is such that, with $\mathbf{F} = \operatorname{Grad} \chi$ and $\mathbf{S}$ given by (2.83)$_1$ (for the unconstrained case), the equilibrium equations (2.31) and the boundary conditions (2.118) are satisfied. In other words, E is stationary if and only if χ is an actual solution (not necessarily unique) of the boundary-value problem.

As will be discussed in Section 1.4.3, the energy functional plays an important role in the analysis of stability for the dead-load problem. Detailed discussion of variational principles, including stationary energy, complementary energy and mixed principles in the context of nonlinear elasticity is contained in Ogden (1997); technical mathematical aspects are discussed in Ciarlet (1988), Marsden and Hughes (1983) and the paper by Ball (1977), for example.

1.3 Examples of boundary-value problems

1.3.1 Homogeneous deformations

1.3.1.1 Isotropic materials

We consider first some elementary problems in which the deformation is *homogeneous*, that is for which the deformation gradient $\mathbf{F}$ is constant.

A *pure homogeneous strain* is a deformation of the form

$$x_1 = \lambda_1 X_1, \quad x_2 = \lambda_2 X_2, \quad x_3 = \lambda_3 X_3, \tag{3.1}$$

where $\lambda_1, \lambda_2, \lambda_3$ are the principal stretches, and, since the deformation is homogeneous, they are constants. For this deformation $\mathbf{F} = \mathbf{U} = \mathbf{V}, \mathbf{R} = \mathbf{I}$ and the principal axes of the deformation coincide with the Cartesian coordinate directions and are fixed as the values of the stretches change. For an unconstrained isotropic elastic material the associated principal Biot stresses are given by (2.95). These equations serve as a basis for determining the form of W from triaxial experimental tests in which $\lambda_1, \lambda_2, \lambda_3$ and t_1, t_2, t_3 are measured. If biaxial tests are conducted on a thin sheet of material which lies in the (X_1, X_2)-plane with no force applied to the faces of the sheet then equations (2.95) reduce to

$$t_1 = \frac{\partial W}{\partial \lambda_1}(\lambda_1, \lambda_2, \lambda_3), \quad t_2 = \frac{\partial W}{\partial \lambda_2}(\lambda_1, \lambda_2, \lambda_3), \quad t_3 = \frac{\partial W}{\partial \lambda_3}(\lambda_1, \lambda_2, \lambda_3) = 0,$$
$$\tag{3.2}$$

and the third equation gives λ_3 implicitly in terms of λ_1 and λ_2 when W is known.

The biaxial test is more important in the context of the incompressibility constraint

$$\lambda_1 \lambda_2 \lambda_3 = 1 \tag{3.3}$$

since then only two stretches can be varied independently and biaxial tests are sufficient to obtain a characterization of W. The counterpart of (2.95) for the incompressible case is given by (2.96), or, in terms of the principal Cauchy stresses, (2.98). It is convenient to make use of (3.3) to express the strain energy as a function of two independent stretches and for this purpose we define

$$\hat{W}(\lambda_1, \lambda_2) = W(\lambda_1, \lambda_2, \lambda_1^{-1}\lambda_2^{-1}). \tag{3.4}$$

This enables p to be eliminated from equation (2.98) and leads to

$$\sigma_1 - \sigma_3 = \lambda_1 \frac{\partial \hat{W}}{\partial \lambda_1}, \quad \sigma_2 - \sigma_3 = \lambda_2 \frac{\partial \hat{W}}{\partial \lambda_2}. \tag{3.5}$$

It is important to note that, because of the incompressibility constraint, equation (3.5) is unaffected by the superposition of an arbitrary hydrostatic stress. Thus, without loss of generality, we may set $\sigma_3 = 0$ in (3.5). In terms of the principal Biot stresses we then have simply

$$t_1 = \frac{\partial \hat{W}}{\partial \lambda_1}, \quad t_2 = \frac{\partial \hat{W}}{\partial \lambda_2}, \tag{3.6}$$

which provides two equations relating λ_1, λ_2 and t_1, t_2 and therefore a basis for characterizing $\hat{W}$ from measured biaxial data.

There are several special cases of the biaxial test which are of interest, but we just give the details for *simple tension*, for which we set $t_2 = 0$. By symmetry, the incompressibility constraint then yields $\lambda_2 = \lambda_3 = \lambda_1^{-1/2}$. The strain energy may now be treated as a function of just λ_1, and we write

$$\tilde{W}(\lambda_1) = \hat{W}(\lambda_1, \lambda_1^{-1/2}), \tag{3.7}$$

and (3.6) reduces to

$$t_1 = \tilde{W}'(\lambda_1), \tag{3.8}$$

where the prime indicates differentiation with respect to λ_1.

Next we consider the *simple shear* deformation defined by

$$x_1 = X_1 + \gamma X_2, \quad x_2 = X_2, \quad x_3 = X_3, \tag{3.9}$$

where γ is a constant, called the *amount of shear*. This is a plane strain deformation and it provides an illustration of a deformation in which the orientation of the principal axes of strain varies with the magnitude of the deformation (in this case with γ).

For an incompressible elastic solid the Cauchy stress tensor is given by the second equation in (2.85). If this is expressed in terms of the invariants I_1, I_2 defined by (2.68)–(2.69) then, for an isotropic material, the components of $\boldsymbol{\sigma}$

are

$$\sigma_{11} = -p + 2(1 + \gamma^2)W_1 + 2(2 + \gamma^2)W_2,$$
$$\sigma_{22} = -p + 2W_1 + 4W_2, \quad \sigma_{12} = 2\gamma(W_1 + W_2), \tag{3.10}$$
$$\sigma_{33} = -p + 2W_1 + 2(2 + \gamma^2)W_2, \quad \sigma_{13} = \sigma_{23} = 0,$$

evaluated for $I_1 = I_2 = 3 + \gamma^2$.

In the (X_1, X_2)-plane the Eulerian principal axes $\mathbf{v}^{(1)}, \mathbf{v}^{(2)}$ are given by

$$\mathbf{v}^{(1)} = \cos\phi \, \mathbf{e}_1 + \sin\phi \, \mathbf{e}_2, \quad \mathbf{v}^{(2)} = -\sin\phi \, \mathbf{e}_1 + \cos\phi \, \mathbf{e}_2, \tag{3.11}$$

where $\mathbf{e}_1, \mathbf{e}_2$ are the Cartesian axes and the angle ϕ is given by

$$\tan 2\phi = 2/\gamma. \tag{3.12}$$

Since, for an isotropic material, σ is coaxial with the Eulerian principal axes its components may, alternatively, be given in terms of its principal values by

$$\sigma_{11} = \sigma_1 \cos^2\phi + \sigma_2 \sin^2\phi, \quad \sigma_{22} = \sigma_1 \sin^2\phi + \sigma_2 \cos^2\phi, \tag{3.13}$$
$$\sigma_{12} = (\sigma_1 - \sigma_2)\sin\phi\cos\phi, \quad \sigma_{33} = \sigma_3. \tag{3.14}$$

The principal Cauchy stresses are given in terms of the principal stretches, which, for the considered deformation, are $\lambda_1, \lambda_2 = \lambda_1^{-1}, \lambda_3 = 1$, where λ_1 is related to the amount of shear γ by

$$\lambda_1 - \lambda_1^{-1} = \gamma \tag{3.15}$$

and we have taken $\lambda_1 \geq 1$ to correspond to $\gamma \geq 0$.

Instead of regarding W as a function of I_1 and I_2 or of the stretches we may take it to be a function of γ and define

$$\bar{W}(\gamma) = \hat{W}(\lambda_1, \lambda_1^{-1}), \tag{3.16}$$

subject to (3.15). Then, we have simply

$$\sigma_{12} = \bar{W}'(\gamma), \quad \sigma_{11} - \sigma_{22} = \gamma\sigma_{12}. \tag{3.17}$$

The second equation in (3.17) is an example of a *universal relation*, i.e. a connection between the components of stress which holds irrespective of the specific form of strain-energy function considered (within, in this case, the class of incompressible isotropic elastic solids). A detailed treatment of universal relations, including discussion of Ericksen's celebrated results (Ericksen, 1954, 1955), is contained in Chapter 3 in this volume.

Simple shear is an important example of a finite deformation. A general discussion of *shear* in the finite deformation context is provided in Chapter 6.

1.3.1.2 Fibre-reinforced materials

Again we consider the pure homogeneous strain defined by (3.1) and now we include two fibre directions, symmetrically disposed in the (X_1, X_2)-plane and given by

$$\mathbf{M} = \cos \alpha \, \mathbf{e}_1 + \sin \alpha \, \mathbf{e}_2, \quad \mathbf{M}' = \cos \alpha \, \mathbf{e}_1 - \sin \alpha \, \mathbf{e}_2, \qquad (3.18)$$

where the angle α is constant and $\mathbf{e}_1, \mathbf{e}_2$ denote the Cartesian coordinate directions. The invariants I_1, I_2 are given by (2.68) and (2.69) subject to the incompressibility constraint (3.3), while with the definitions in (2.111) and (2.114) we use (3.18) to obtain

$$I_4 = I_6 = \lambda_1^2 \cos^2 \alpha + \lambda_2^2 \sin^2 \alpha, \quad I_5 = I_7 = \lambda_1^4 \cos^2 \alpha + \lambda_2^4 \sin^2 \alpha,$$

$$I_8 = \lambda_1^2 \cos^2 \alpha - \lambda_2^2 \sin^2 \alpha. \qquad (3.19)$$

From (2.115) we then calculate the components of $\boldsymbol{\sigma}$ as

$$\begin{aligned}
\sigma_{11} = {} & -p + 2W_1 \lambda_1^2 + 2W_2(I_1 \lambda_1^2 - \lambda_1^4) + 2(W_4 + W_6 + W_8)\lambda_1^2 \cos^2 \alpha \\
& + 4(W_5 + W_7)\lambda_1^4 \cos^2 \alpha, \qquad (3.20)
\end{aligned}$$

$$\begin{aligned}
\sigma_{22} = {} & -p + 2W_1 \lambda_2^2 + 2W_2(I_1 \lambda_2^2 - \lambda_2^4) + 2(W_4 + W_6 - W_8)\lambda_2^2 \sin^2 \alpha \\
& + 4(W_5 + W_7)\lambda_2^4 \sin^2 \alpha, \qquad (3.21)
\end{aligned}$$

$$\sigma_{12} = 2[W_4 - W_6 + (W_5 - W_7)(\lambda_1^2 + \lambda_2^2)]\lambda_1 \lambda_2 \sin \alpha \cos \alpha, \qquad (3.22)$$

$$\sigma_{33} = -p + 2W_1 \lambda_3^2 + 2W_2(I_1 \lambda_3^2 - \lambda_3^4), \quad \sigma_{13} = \sigma_{23} = 0. \qquad (3.23)$$

In general, since $\sigma_{12} \neq 0$, shear stresses are required to maintain the pure homogeneous deformation and the principal axes of stress do not coincide with the Cartesian axes. However, in the special case in which the two families of fibres are mechanically equivalent the strain energy must be symmetric with respect to interchange of I_4 and I_6 and of I_5 and I_7. Since, for the considered deformation, we have $I_4 = I_6, I_5 = I_7$ it follows that $W_4 = W_6, W_5 = W_7$ and hence that $\sigma_{12} = 0$. The principal axes of stress then coincide with the Cartesian axes and $\sigma_{11}, \sigma_{22}, \sigma_{33}$ are the principal Cauchy stresses.

Since I_1, I_2 depend symmetrically on $\lambda_1, \lambda_2, \lambda_3$ and I_4, I_5, I_8 depend on λ_1, λ_2 and α, we may regard the strain energy as a function of $\lambda_1, \lambda_2, \lambda_3$, subject to (3.3), and α. We write $W(\lambda_1, \lambda_2, \lambda_3, \alpha)$, but it should be emphasized that, unlike in the isotropic case, W is *not* symmetric with respect to interchange of any pair of the stretches. It is straightforward to check that

$$\sigma_{11} \equiv \sigma_1 = \lambda_1 \frac{\partial W}{\partial \lambda_1} - p, \quad \sigma_{22} \equiv \sigma_2 = \lambda_2 \frac{\partial W}{\partial \lambda_2} - p, \quad \sigma_{33} \equiv \sigma_3 = \lambda_3 \frac{\partial W}{\partial \lambda_3} - p.$$
$$(3.24)$$

As in the isotropic case we make use of (3.3) to recast the strain energy as a function of λ_1 and λ_2 and define

$$\hat{W}(\lambda_1, \lambda_2, \alpha) = W(\lambda_1, \lambda_2, \lambda_1^{-1}\lambda_2^{-1}, \alpha), \tag{3.25}$$

which is not symmetric in λ_1, λ_2 in general. Then, we obtain from (3.24)

$$\sigma_1 - \sigma_3 = \lambda_1 \frac{\partial \hat{W}}{\partial \lambda_1}, \quad \sigma_2 - \sigma_3 = \lambda_2 \frac{\partial \hat{W}}{\partial \lambda_2}, \tag{3.26}$$

which are identical *in form* (but not in content) to equations (3.5).

For the simple shear deformation (3.9) the situation is more complicated. The invariants are given by

$$\begin{aligned}
&I_1 = I_2 = 3 + \gamma^2, \quad I_3 = 1, \\
&I_4 = 1 + \gamma \sin 2\alpha + \gamma^2 \sin^2 \alpha, \quad I_6 = 1 - \gamma \sin 2\alpha + \gamma^2 \sin^2 \alpha, \\
&I_5 = (1+\gamma^2)\cos^2 \alpha + 2\gamma(2+\gamma^2)\sin\alpha\cos\alpha + (\gamma^4 + 3\gamma^2 + 1)\sin^2 \alpha, \quad (3.27)\\
&I_7 = (1+\gamma^2)\cos^2 \alpha - 2\gamma(2+\gamma^2)\sin\alpha\cos\alpha + (\gamma^4 + 3\gamma^2 + 1)\sin^2 \alpha, \\
&I_8 = \cos^2 \alpha - (1+\gamma^2)\sin^2 \alpha.
\end{aligned}$$

The components of the Cauchy stress tensor are calculated as

$$\begin{aligned}
\sigma_{11} = \ & -p + 2W_1(1+\gamma^2) + 2W_2(2+\gamma^2) \\
& + 2[W_4 + W_6 + W_8 + 2(W_5 + W_7)(1+\gamma^2)]\cos^2 \alpha \\
& + 4[W_4 - W_6 + (W_5 - W_7)(3+\gamma^2)]\gamma \sin\alpha\cos\alpha \\
& + 2[W_4 + W_6 - W_8 + 2(W_5 + W_7)(2+\gamma^2)]\gamma^2 \sin^2 \alpha, \\
\sigma_{22} = \ & -p + 4W_1 + 4W_2 + 2(W_4 + W_6 - W_8)\sin^2 \alpha \qquad\qquad (3.28)\\
& + 4(W_5 - W_7)\gamma \sin\alpha\cos\alpha + 4(W_5 + W_7)(1+\gamma^2)\sin^2 \alpha, \\
\sigma_{12} = \ & 2(W_1 + W_2)\gamma + 2(W_4 - W_6)\sin\alpha\cos\alpha + 2(W_4 + W_6 \\
& - W_8)\gamma \sin^2 \alpha + 2(W_5 + W_7)\gamma[\cos^2 \alpha + (3+\gamma^2)\sin^2 \alpha], \\
\sigma_{33} = \ & -p + 2W_1\lambda_3^2 + 2W_2(I_1\lambda_3^2 - \lambda_3^4), \quad \sigma_{13} = \sigma_{23} = 0.
\end{aligned}$$

By defining $\bar{W}(\gamma, \alpha)$ analogously to (3.18) through the invariants (3.27) with (3.17), it is straightforward to show that

$$\sigma_{12} = \frac{\partial \bar{W}}{\partial \gamma}, \tag{3.29}$$

exactly as in the isotropic case. Here, however, the universal relation $(3.17)_2$ does not hold. This is a reflection of the fact that the principal axes of Cauchy stress do not coincide with the Eulerian axes. Indeed, if ϕ^* denotes the analogue of ϕ, which is given by (3.12), for the principal stress axes then, in general, ϕ^* is given by

$$\tan 2\phi^* = \frac{2\sigma_{12}}{\sigma_{11} - \sigma_{22}}, \tag{3.30}$$

and, since the right-hand side of (3.30) is not equal to $2/\gamma$, we see that $\phi^* \neq \phi$.

1.3.2 Extension and inflation of a thick-walled tube

In this section we examine an example of a *non-homogeneous* deformation. Other examples can be found in the texts cited in Section 1.1. We consider a thick-walled circular cylindrical tube whose initial geometry is defined by

$$A \leq R \leq B, \quad 0 \leq \Theta \leq 2\pi, \quad 0 \leq Z \leq L, \qquad (3.31)$$

where A, B, L are positive constants and R, Θ, Z are cylindrical polar coordinates. The tube is deformed so that the circular cylindrical shape is maintained, and the material of the tube is taken to be incompressible. The resulting deformation is then described by the equations

$$r^2 - a^2 = \lambda_z^{-1}(R^2 - A^2), \quad \theta = \Theta, \quad z = \lambda_z Z, \qquad (3.32)$$

where r, θ, z are cylindrical polar coordinates in the deformed configuration, λ_z is the axial stretch and a is the internal radius of the deformed tube.

The principal stretches $\lambda_1, \lambda_2, \lambda_3$ are associated respectively with the radial, azimuthal and axial directions and are written

$$\lambda_1 = \lambda^{-1}\lambda_z^{-1}, \quad \lambda_2 = \frac{r}{R} = \lambda, \quad \lambda_3 = \lambda_z, \qquad (3.33)$$

wherein the notation λ is introduced. It follows from (3.32) and (3.33) that

$$\lambda_a^2 \lambda_z - 1 = \frac{R^2}{A^2}(\lambda^2 \lambda_z - 1) = \frac{B^2}{A^2}(\lambda_b^2 \lambda_z - 1), \qquad (3.34)$$

where

$$\lambda_a = a/A, \quad \lambda_b = b/B. \qquad (3.35)$$

For a fixed value of λ_z the inequalities

$$\lambda_a^2 \lambda_z \geq 1, \quad \lambda_a \geq \lambda \geq \lambda_b \qquad (3.36)$$

hold during inflation of the tube, with equality holding if and only if $\lambda = \lambda_z^{-1/2}$ for $A \leq R \leq B$.

We use the notation (3.4) for the strain energy but with $\lambda_2 = \lambda$ and $\lambda_3 = \lambda_z$ as the independent stretches, so that

$$\hat{W}(\lambda, \lambda_z) = W(\lambda^{-1}\lambda_z^{-1}, \lambda, \lambda_z). \qquad (3.37)$$

Hence

$$\sigma_2 - \sigma_1 = \lambda \hat{W}_\lambda, \quad \sigma_3 - \sigma_1 = \lambda_z \hat{W}_{\lambda_z}, \qquad (3.38)$$

where the subscripts indicate partial derivatives.

The equilibrium equation (2.30) in the absence of body forces reduces to the single scalar equation

$$\frac{d\sigma_1}{dr} + \frac{1}{r}(\sigma_1 - \sigma_2) = 0 \tag{3.39}$$

in terms of the principal Cauchy stresses, and to this are adjoined the boundary conditions

$$\sigma_1 = \begin{cases} -P & \text{on } R = A \\ 0 & \text{on } R = B \end{cases} \tag{3.40}$$

corresponding to pressure $P (\geq 0)$ on the inside of the tube and zero traction on the outside.

By making use of (3.32)–(3.35) the independent variable may be changed from r to λ and integration of (3.39) and application of the boundary conditions (3.40) yields

$$P = \int_{\lambda_b}^{\lambda_a} (\lambda^2 \lambda_z - 1)^{-1} \frac{\partial \hat{W}}{\partial \lambda} d\lambda. \tag{3.41}$$

Since, from (3.34), λ_b depends on λ_a, equation (3.41) provides an expression for P as a function of λ_a when λ_z is fixed. In order to hold λ_z fixed an axial load, N say, must be applied to the ends of the tube. This is given by

$$N/\pi A^2 = (\lambda_a^2 \lambda_z - 1) \int_{\lambda_b}^{\lambda_a} (\lambda^2 \lambda_z - 1)^{-2} \left(2\lambda_z \frac{\partial \hat{W}}{\partial \lambda_z} - \lambda \frac{\partial \hat{W}}{\partial \lambda} \right) \lambda d\lambda + P\lambda_a^2. \tag{3.42}$$

For a thin-walled (membrane) tube the above results may be simplified since integration through the wall thickness is no longer needed. A general account of nonlinear elastic *membrane theory* is contained in Chapter 7, while Chapter 8 is concerned with *elastic surfaces*.

By analogy with the analysis in Section 1.3.1.2, for fibre reinforced materials with the fibre directions $\mathbf{M}$ and $\mathbf{M}'$ locally in the (Θ, Z)-plane symmetrically disposed with respect to the axial direction the strain energy may be written in the form

$$\hat{W}(\lambda, \lambda_z, \alpha), \tag{3.43}$$

where, we recall, $\hat{W}$ is not symmetric in λ and λ_z. Furthermore the formulas (3.41) and (3.42) again apply and are valid if the fibre directions depend on the radius, i.e. if α depends on R.

1.4 Incremental equations

1.4.1 Incremental deformation

Suppose that a solution χ to the boundary-value problem (2.116)–(2.118) is known and consider the problem of finding solutions near to χ when the boundary conditions are perturbed. Let χ' be a solution for the perturbed problem and write $\mathbf{x}' = \chi'(\mathbf{X})$. Also, we write

$$\dot{\mathbf{x}} = \mathbf{x}' - \mathbf{x} = \chi'(\mathbf{X}) - \chi(\mathbf{X}) \equiv \dot{\chi}(\mathbf{X}) \tag{4.1}$$

for the difference of the solutions, so that

$$\operatorname{Grad}\dot{\chi} = \operatorname{Grad}\chi' - \operatorname{Grad}\chi \equiv \dot{\mathbf{F}}. \tag{4.2}$$

Note that $\dot{\mathbf{F}}$ is linear in $\dot{\chi}$ and that in this expression no approximation has been made. In what follows, however, we shall consider approximations in which all terms are linearized in the incremental deformation $\dot{\mathbf{x}}$ and its gradient (4.2). Quantities with a superposed dot indicate the appropriate linearization.

From (2.17) we calculate the increment $\dot{\mathbf{F}}$ in terms of the increments of $\mathbf{R}$ and $\mathbf{U}$ in the form

$$\mathbf{R}^T\dot{\mathbf{F}} = \mathbf{R}^T\dot{\mathbf{R}}\mathbf{U} + \dot{\mathbf{U}}. \tag{4.3}$$

Next, we introduce the notations

$$\mathbf{\Omega}^R = \mathbf{R}^T\dot{\mathbf{R}}, \quad \mathbf{\Omega}^L = \sum_{i=1}^{3} \dot{\mathbf{u}}^{(i)} \otimes \mathbf{u}^{(i)} \tag{4.4}$$

for the incremental rotations associated with $\mathbf{R}$ and with the Lagrangian principal axes respectively. Then, from (2.18), we calculate

$$\dot{\mathbf{U}} = \sum_{i=1}^{3} \dot{\lambda}_i\mathbf{u}^{(i)} \otimes \mathbf{u}^{(i)} + \mathbf{\Omega}^L\mathbf{U} - \mathbf{U}\mathbf{\Omega}^L, \tag{4.5}$$

and hence the components of $\dot{\mathbf{U}}$ and $\mathbf{\Omega}^R$ on the Lagrangian principal axes are obtained in the form

$$(\dot{\mathbf{U}})_{ii} = \dot{\lambda}_i, \quad (\dot{\mathbf{U}})_{ij} = \Omega^L_{ij}(\lambda_j - \lambda_i) \quad i \neq j, \tag{4.6}$$

$$(\mathbf{\Omega}^R\mathbf{U})_{ij} = \Omega^R_{ij}\lambda_j, \quad (\mathbf{\Omega}^R\mathbf{U})_{ji} = -\Omega^R_{ij}\lambda_i \quad i \neq j. \tag{4.7}$$

1.4.2 Incremental stress and equilibrium

The nominal stress difference is

$$\dot{\mathbf{S}} = \mathbf{S}' - \mathbf{S} = \frac{\partial W}{\partial \mathbf{F}}(\mathbf{F}') - \frac{\partial W}{\partial \mathbf{F}}(\mathbf{F}), \tag{4.8}$$

which has the linear approximation

$$\dot{\mathbf{S}} = \mathcal{A}^1 \dot{\mathbf{F}}, \tag{4.9}$$

where

$$\mathcal{A}^1 = \frac{\partial^2 W}{\partial \mathbf{F}^2} \tag{4.10}$$

is the (fourth-order) tensor of *elastic moduli* associated with the pair $(\mathbf{S}, \mathbf{F})$ of stress and deformation tensors. In components, (4.10) is given by (2.121). The component form of (4.9) is

$$\dot{S}_{\alpha i} = A^1_{\alpha i \beta j} \dot{F}_{j\beta}, \tag{4.11}$$

and this serves to define the product appearing in (4.9).

For an incompressible material it follows by taking the increment of the first equation in (2.85) that the counterpart of (4.9) is

$$\dot{\mathbf{S}} = \mathcal{A}^1 \dot{\mathbf{F}} - \dot{p}\mathbf{F}^{-1} + p\mathbf{F}^{-1}\dot{\mathbf{F}}\mathbf{F}^{-1}, \tag{4.12}$$

and this is coupled with the (linearized) incremental form

$$\text{tr}\,(\dot{\mathbf{F}}\mathbf{F}^{-1}) = 0 \tag{4.13}$$

of the incompressibility condition $\det \mathbf{F} = 1$, where $\dot{p}$ is the (linearized) incremental form of p. For an incompressible material the definition (4.10) remains valid subject to the constraint $\det \mathbf{F} = 1$.

From the equilibrium equation (2.31) and its counterpart for χ', we obtain, by subtraction,

$$\text{Div}\,\dot{\mathbf{S}} + \rho_r \dot{\mathbf{b}} = \mathbf{0}. \tag{4.14}$$

This is *exact*, but in the linear approximation $\dot{\mathbf{S}}$ is replaced by either (4.9), or (4.12) with (4.13), as appropriate and with $\dot{\mathbf{b}}$ linearized in $\dot{\chi}$.

Let $\dot{\xi}$ and $\dot{\tau}$ be the prescribed data for the incremental deformation $\dot{\chi}$. Then, the incremental versions of the boundary conditions (2.117) and (2.118) are written

$$\dot{\mathbf{x}} = \dot{\xi} \quad \text{on } \partial \mathcal{B}_r^x, \qquad \dot{\mathbf{S}}^T \mathbf{N} = \dot{\tau} \quad \text{on } \partial \mathcal{B}_r^\tau. \tag{4.15}$$

Together, the equations (4.14), with (4.9) for an unconstrained material, or (4.12) with (4.13) for an incompressible material, (4.2) and the boundary conditions (4.15) constitute the basic boundary-value problem of incremental elasticity given that the underlying deformation χ is known. The equations are also referred to as the equations of small deformations superimposed on a finite (or large) deformation. The linearized equations constitute the first-order terms associated with a formal perturbation expansion in the incremental deformation. The higher-order (nonlinear) terms are required for weakly non-linear analysis of the stability of finitely deformed configurations and this topic is addressed in Chapter 10 in this volume; related work is discussed in the papers by Fu and Rogerson (1994), Fu (1995, 1998), Ogden and Fu (1996) and Fu and Ogden (1999), for example. For discussion of the mathematical structure of the incremental equations see, for example, Hayes and Horgan (1974).

In Section 1.5 the incremental equations (4.14) and boundary conditions (4.15) will be specialized for plane strain deformations and for an incompressible material in order to discuss an illustrative prototype example of a problem involving incremental deformations and bifurcation from a finitely deformed configuration.

1.4.2.1 Elastic moduli for an isotropic material

For the important special case of an isotropic material it is useful to give explicit expressions for the components of $\mathcal{A}^1$. These are obtained by referring equation (4.9) to principal axes and making use of equations (4.3)–(4.7). Thus, for an isotropic elastic material, the (non-zero) components of $\mathcal{A}^1$ referred to the principal axes $\mathbf{u}^{(i)}$ and $\mathbf{v}^{(i)}$ are given by

$$\mathcal{A}^1_{iijj} = W_{ij}, \tag{4.16}$$

$$\mathcal{A}^1_{ijij} - \mathcal{A}^1_{ijji} = \frac{W_i + W_j}{\lambda_i + \lambda_j} \quad i \neq j, \tag{4.17}$$

$$\mathcal{A}^1_{ijij} + \mathcal{A}^1_{ijji} = \frac{W_i - W_j}{\lambda_i - \lambda_j} \quad i \neq j, \lambda_i \neq \lambda_j, \tag{4.18}$$

$$\mathcal{A}^1_{ijij} + \mathcal{A}^1_{ijji} = W_{ii} - W_{ij} \quad i \neq j, \lambda_i = \lambda_j, \tag{4.19}$$

where $W_i = \partial W/\partial \lambda_i$, $W_{ij} = \partial^2 W/\partial \lambda_i \partial \lambda_j$, $i, j \in \{1, 2, 3\}$, and no summation is implied by the repetition of indices. In (4.16)–(4.19) the convention of using Greek letters for indices relating to Lagrangian components has been dropped. For details of the derivation of these components we refer to Ogden (1997). Equations (4.16)–(4.19) apply for both compressible and incompressible materials subject, in the latter case, to the constraint (3.3). Expressions analogous to (4.16)–(4.19) for the components of the tensors of moduli associated with conjugate variables based on the class of strain tensors (2.23)–(2.24) are obtainable in a similar way, but details are not given here. The appropriate calculations are illustrated in, for example, Chadwick and Ogden (1971) and Ogden (1974a, b, 1997) for first- and second-order moduli, while in Fu and Ogden (1999) the corresponding calculations for the first-, second- and third-order moduli are provided relative to the current configuration. For fibre-reinforced materials an expression for the elastic modulus tensor associated with the Green strain tensor $\mathbf{E}$ defined in (2.16) and its conjugate (second Piola-Kirchhoff) stress $\mathbf{T}^{(2)}$ defined in (2.36) is given by Holzapfel and Gasser (2000), in which the theory is extended to consideration of viscoelasticity.

In the classical theory of elasticity, corresponding to the situation in which there is no underlying deformation or stress, the components of $\mathcal{A}^1$ can be written compactly in the form

$$\mathcal{A}^1_{ijkl} = \lambda \delta_{ij} \delta_{kl} + \mu(\delta_{ik}\delta_{jl} + \delta_{il}\delta_{jk}), \qquad (4.20)$$

where λ and μ are the classical Lamé moduli of elasticity and δ_{ij} is the Kronecker delta. Note that λ is not used subsequently in this chapter so no confusion with the notation used for the stretches should arise. The values of W_{ij} when $\lambda_i = 1$ for $i \in \{1, 2, 3\}$ are simply $W_{ii} = \lambda + 2\mu$, $W_{ij} = \lambda$, $i \neq j$. Also, we take $W_i = 0$ when $\lambda_j = 1$ for $i, j \in \{1, 2, 3\}$ so that the configuration $\mathcal{B}_r$ is stress free (a *natural configuration*).

The counterpart of (4.20) for an incompressible material is

$$\mathcal{A}^1_{iiii} = \mathcal{A}^1_{ijij} = \mu, \quad \mathcal{A}^1_{iijj} = \mathcal{A}^1_{ijji} = 0 \quad i \neq j, \qquad (4.21)$$

and $W_{ii} = W_i = \mu$, $W_{ij} = 0$, where μ is the *shear modulus* in $\mathcal{B}_r$. These expressions are not unique because they depend on the point at which $\lambda_1 \lambda_2 \lambda_3$ is set to unity in the differentiation. The differences between (4.21) and any alternative expressions are accounted for by the incremental Lagrange multiplier $\dot{p}$ in (4.12). In terms of the strain-energy function $\hat{W}$ defined in (3.4) the restrictions required in $\mathcal{B}_r$ may be written

$$\hat{W}(1,1) = 0, \quad \hat{W}_\alpha(1,1) = 0, \quad \hat{W}_{12}(1,1) = 2\mu, \quad \hat{W}_{\alpha\alpha}(1,1) = 4\mu, \qquad (4.22)$$

where the index α is 1 or 2.

1.4.2.2 Strong ellipticity for an isotropic material

We now give an explicit form for the strong ellipticity inequality in two dimensions for both compressible and incompressible materials. For this purpose we refer $\mathcal{A}^1$ to principal axes and make use of the expressions (4.16)–(4.19) and the formulas (4.3)–(4.7). By restricting attention to the $(1, 2)$ principal plane it can be seen that necessary and sufficient conditions for the resulting quadratic form in the incremental quantities to be strictly positive are jointly

$$W_{11} > 0, \quad W_{22} > 0, \quad \frac{\lambda_1 W_1 - \lambda_2 W_2}{\lambda_1 - \lambda_2} > 0, \qquad (4.23)$$

$$(W_{11}W_{22})^{1/2} - W_{12} + \frac{W_1 + W_2}{\lambda_1 + \lambda_2} > 0, \qquad (4.24)$$

$$(W_{11}W_{22})^{1/2} + W_{12} + \frac{W_1 - W_2}{\lambda_1 - \lambda_2} > 0. \qquad (4.25)$$

See, for example, Knowles and Sternberg (1977), Hill (1979), Dowaikh and Ogden (1991), Davies (1991) and Wang and Aron (1996).

For an incompressible material we then use (2.125) in two dimensions to write $m_1 = n_2, m_2 = -n_1$, so that (2.127) reduces to

$$\mathcal{A}^1_{01212}n_1^4 + (\mathcal{A}^1_{01111} + \mathcal{A}^1_{02222} - 2\mathcal{A}^1_{01122} - 2\mathcal{A}^1_{02112})n_1^2 n_2^2 + \mathcal{A}^1_{02121}n_2^4 > 0, \quad (4.26)$$

and on use of (4.16)–(4.18) and the specialization of (2.128) to principal axes with $J = 1$, we obtain

$$\mathcal{A}^1_{01111} = \lambda_1^2 W_{11}, \quad \mathcal{A}^1_{01122} = \lambda_1\lambda_2 W_{12}, \quad \mathcal{A}^1_{02222} = \lambda_2^2 W_{22}, \quad (4.27)$$

$$\lambda_1^{-2}\mathcal{A}^1_{01212} = \lambda_2^{-2}\mathcal{A}^1_{02121} = \frac{\lambda_1 W_1 - \lambda_2 W_2}{\lambda_1^2 - \lambda_2^2}, \quad \mathcal{A}^1_{02112} = \frac{\lambda_2 W_1 - \lambda_1 W_2}{\lambda_1^2 - \lambda_2^2}\lambda_1\lambda_2.$$
$$(4.28)$$

Necessary and sufficient conditions for (4.26) to hold are then easily seen to be

$$\mathcal{A}^1_{01212} > 0, \quad \mathcal{A}^1_{01111} + \mathcal{A}^1_{02222} - 2\mathcal{A}^1_{01122} - 2\mathcal{A}^1_{02112} > -2\sqrt{\mathcal{A}^1_{01212}\mathcal{A}^1_{02121}}, \quad (4.29)$$

and we note that these are independent of p.

In terms of the strain-energy function the inequalities (4.29) may be written

$$\frac{\lambda_1 W_1 - \lambda_2 W_2}{\lambda_1 - \lambda_2} > 0, \quad \lambda_1^2 W_{11} - 2\lambda_1\lambda_2 W_{12} + \lambda_2^2 W_{22} + 2\lambda_1\lambda_2\frac{W_1 + W_2}{\lambda_1 + \lambda_2} > 0.$$
$$(4.30)$$

1.4.3 Incremental uniqueness and stability

We next examine the question of uniqueness of solution of the incremental problem and the associated question of stability of the deformation χ. We focus first on the theoretical development for unconstrained materials, and for simplicity we restrict attention to the dead-load boundary-value problem with no body forces. We also take the boundary conditions to be homogeneous, so that $\dot{\xi} = 0, \dot{\tau} = 0$ in (4.15). The appropriate specialization of (4.14) and (4.15) is then

$$\text{Div}\,\dot{\mathbf{S}} = \mathbf{0} \quad \text{in } \mathcal{B}_r, \quad \dot{\chi} = \mathbf{0} \quad \text{on } \partial\mathcal{B}_r^x, \quad \dot{\mathbf{S}}^T\mathbf{N} = \mathbf{0} \quad \text{on } \partial\mathcal{B}_r^\tau. \quad (4.31)$$

One solution of this is $\dot{\chi} = \mathbf{0}$. We therefore wish to determine whether this is the only solution. With this aim in mind we consider next the change in the energy functional $E\{\chi\}$, defined by (2.133), due to the change in the deformation from χ to χ', the body-force term being omitted. On use of the boundary condition

$(4.31)_2$ and the divergence theorem this is seen to be

$$E\{\chi'\} - E\{\chi\} = \int_{B_r} \{W(\mathbf{F}') - W(\mathbf{F}) - \operatorname{tr}(\mathbf{S}\dot{\mathbf{F}})\} \, dV. \tag{4.32}$$

By application of the Taylor expansion to $W(\mathbf{F}')$ this then becomes

$$E\{\chi'\} - E\{\chi\} = \frac{1}{2} \int_{B_r} \operatorname{tr}(\dot{\mathbf{S}}\dot{\mathbf{F}}) \, dV \equiv \frac{1}{2} \int_{B_r} \operatorname{tr}\{(\mathcal{A}^1\dot{\mathbf{F}})\dot{\mathbf{F}}\} \, dV, \tag{4.33}$$

correct to the second order in $\dot{\mathbf{F}}$.

If the inequality

$$\int_{B_r} \operatorname{tr}(\dot{\mathbf{S}}\dot{\mathbf{F}}) \, dV \equiv \int_{B_r} \operatorname{tr}\{(\mathcal{A}^1\dot{\mathbf{F}})\dot{\mathbf{F}}\} \, dV > 0 \tag{4.34}$$

holds for all $\dot{\chi} \not\equiv \mathbf{0}$ in B_r satisfying $\dot{\chi} = \mathbf{0}$ on ∂B_r^x then, to the second order in $\dot{\mathbf{F}}$, (4.33) implies that $E\{\chi'\} > E\{\chi\}$ for all admissible $\dot{\chi} \not\equiv \mathbf{0}$ satisfying $(4.31)_2$. This inequality states that χ is *locally stable* with respect to perturbations $\dot{\chi}$ from χ, and that χ is a *local minimizer* of the energy functional. Furthermore, if (4.34) holds in the configuration B then the only solution of the homogeneous incremental problem is the trivial solution $\dot{\chi} \equiv \mathbf{0}$, i.e. non-trivial solutions are excluded.

This can be seen by noting that if $\dot{\chi} \not\equiv \mathbf{0}$ is a non-trivial solution then, by the divergence theorem and use of equations (4.31), it follows that

$$\int_{B_r} \operatorname{tr}(\dot{\mathbf{S}}\dot{\mathbf{F}}) \, dV = 0 \tag{4.35}$$

necessarily holds. The inequality (4.34) is referred to as the *exclusion condition* (for the dead-load traction boundary condition). The trivial solution is then the unique solution. Thus, on a path of deformation corresponding to mixed dead-load and placement boundary conditions bifurcation of solutions is excluded provided (4.34) holds. This exclusion condition requires modification if τ is allowed to depend on the deformation. Generally, the exclusion condition will involve both a surface integral and a volume integral. For a discussion of connections between incremental (infinitesimal) stability and uniqueness we refer to Hill (1957), Truesdell and Noll (1965) and Beatty (1987).

For the all-round dead-load problem, $\partial B_r^x = \emptyset, \partial B_r = \partial B_r^\tau$. If the underlying deformation is homogeneous then $\mathbf{F}$ is independent of $\mathbf{X}$ and so, therefore, is $\mathcal{A}^1$. It then follows that (4.34) is equivalent to the local condition

$$\operatorname{tr}(\dot{\mathbf{S}}\dot{\mathbf{F}}) \equiv \operatorname{tr}\{(\mathcal{A}^1\dot{\mathbf{F}})\dot{\mathbf{F}}\} > 0 \tag{4.36}$$

for all $\dot{\mathbf{F}} \neq \mathbf{0}$, i.e. $\mathcal{A}^1$ is positive definite at each point $\mathbf{X} \in B_r$. It is well known that the inequality (4.36) *cannot* hold in all configurations, and hence $\mathcal{A}^1$ is singular in certain configurations when regarded as a linear mapping on the (nine-dimensional) space of increments $\dot{\mathbf{F}}$. In other words, in the incremental stress-deformation relation $\dot{\mathbf{S}} = \mathcal{A}^1\dot{\mathbf{F}}$ the incremental stress $\dot{\mathbf{S}}$ can vanish for non-zero $\dot{\mathbf{F}}$ and bifurcation of the deformation path can occur. The reader is referred to, for example, Ogden (1991, 1997, 2000) for detailed analysis of the singularities of $\mathcal{A}^1$ and their implications for bifurcation in the dead-load problem.

1.4.3.1 Specialization to isotropy

In the case of an isotropic material the local stability inequality (4.36) can be given an explicit form in terms of the derivatives of the strain-energy function with respect to the stretches. Use of the expressions (4.3)–(4.7) and (4.16)–(4.18) in (4.36) referred to principal axes leads to

$$\operatorname{tr}\{(\mathcal{A}^1\dot{\mathbf{F}})\dot{\mathbf{F}}\} = \sum_{i,j=1}^{3} W_{ij}\dot{\lambda}_i\dot{\lambda}_j + \sum_{i\neq j}(W_i - W_j)(\lambda_i - \lambda_j)(\Omega_{ij}^L + \tfrac{1}{2}\Omega_{ij}^R)^2$$

$$+ \frac{1}{4}\sum_{i\neq j}(W_i + W_j)(\lambda_i + \lambda_j)(\Omega_{ij}^R)^2. \tag{4.37}$$

Since $\dot{\lambda}_i,\ \Omega_{ij}^L,\ \Omega_{ij}^R$ are independent, necessary and sufficient conditions for (4.36) are therefore

$$\text{matrix } (W_{ij}) \text{ is positive definite,} \tag{4.38}$$

$$W_i + W_j > 0 \qquad i \neq j, \tag{4.39}$$

$$\frac{W_i - W_j}{\lambda_i - \lambda_j} > 0 \qquad i \neq j \tag{4.40}$$

jointly for $i,j \in \{1,2,3\}$. Note that when $\lambda_i = \lambda_j$, $i \neq j$ (4.40) reduces to $W_{ii} - W_{ij} > 0$ and that (4.38) and (4.40) hold in the natural configuration provided the usual inequalities $\mu > 0$, $3\lambda + 2\mu > 0$ satisfied by the Lamé moduli hold.

On use of (4.12) and (4.13) we may deduce that the analogue of the stability inequality (4.36) for an incompressible material is

$$\operatorname{tr}\{(\mathcal{A}^1\dot{\mathbf{F}})\dot{\mathbf{F}}\} + p\operatorname{tr}(\mathbf{F}^{-1}\dot{\mathbf{F}}\mathbf{F}^{-1}\dot{\mathbf{F}}) > 0, \tag{4.41}$$

subject to (4.13). Note that this does not depend on $\dot{p}$.

In terms of the modified strain-energy function $\hat{W}(\lambda_1, \lambda_2)$ defined by (3.4), with p eliminated in favour of σ_3, the inequality (4.41) becomes explicitly

$$(\lambda_1^2\hat{W}_{11} - 2\sigma_3)\left(\frac{\dot{\lambda}_1}{\lambda_1}\right)^2 + (\lambda_2^2\hat{W}_{22} - 2\sigma_3)\left(\frac{\dot{\lambda}_2}{\lambda_2}\right)^2 + 2(\lambda_1\lambda_2\hat{W}_{12} - \sigma_3)\frac{\dot{\lambda}_1\dot{\lambda}_2}{\lambda_1\lambda_2}$$

$$+ \sum_{i \neq j}(t_i - t_j)(\lambda_i - \lambda_j)(\Omega_{ij}^L + \frac{1}{2}\Omega_{ij}^R)^2 + \frac{1}{4}\sum_{i \neq j}(t_i + t_j)(\lambda_i + \lambda_j)(\Omega_{ij}^R)^2 > 0, \quad (4.42)$$

where the subscripts on $\hat{W}$ denote partial derivatives with respect to λ_1 and λ_2 and implicitly the connections (3.5) have been used.

The counterparts of (4.38)–(4.40) in this case are then

$$\text{matrix} \quad \begin{bmatrix} \lambda_1^2 \hat{W}_{11} - 2\sigma_3 & \lambda_1\lambda_2 \hat{W}_{12} - \sigma_3 \\ \lambda_1\lambda_2 \hat{W}_{12} - \sigma_3 & \lambda_2^2 \hat{W}_{22} - 2\sigma_3 \end{bmatrix} \quad \text{is positive definite,} \quad (4.43)$$

$$t_i + t_j > 0 \quad i \neq j, \quad (4.44)$$

$$\frac{t_i - t_j}{\lambda_i - \lambda_j} > 0 \quad i \neq j. \quad (4.45)$$

Note that p occurs in (4.43)–(4.45) implicitly through σ_3 and t_i, $i \in \{1,2,3\}$.

In general the stability inequality (4.36) is stronger than the strong ellipticity inequality (2.124). This can be seen by making the specialization $\dot{\mathbf{F}} = \mathbf{m} \otimes \mathbf{N}$ in (4.36), which then (in component form) reduces to (2.124). This is also the case for incompressible materials, for which use of (2.125) and (2.126) enables the term in p to be removed from (4.41).

1.4.4 Global non-uniqueness

The singularities of $\mathcal{A}^1$ mentioned above are local manifestations of the global non-uniqueness in the relationship between $\mathbf{S}$ and $\mathbf{F}$ expressed through the constitutive equation

$$\mathbf{S} = \frac{\partial W}{\partial \mathbf{F}}(\mathbf{F}) \quad (4.46)$$

for an unconstrained material, or $(2.85)_1$ in the case of an incompressible material. For an isotropic material we recall from (2.101) the polar decomposition

$$\mathbf{S} = \mathbf{T}\mathbf{R}^T. \quad (4.47)$$

As mentioned in Section 1.2.4.2 this decomposition is not unique since $\mathbf{T}$ is not sign definite. We summarize briefly the extent of non-uniqueness and refer the reader to Ogden (1977, 1997, 2000) for detailed discussion. For a given $\mathbf{S}$ there are *four* distinct polar decompositions of the form (4.47) if $t_i^2 \neq t_j^2$, $i, j \in \{1,2,3\}$, and infinitely many when $t_i^2 = t_j^2$, $i \neq j$, where t_i are the principal Biot stresses. However, at most one of these satisfies the stability inequalities (4.39) or (4.44). For each such polar decomposition each $\mathbf{R}$ and $\mathbf{T}$ pair is determined uniquely if $t_i^2 \neq t_j^2$, $i,j \in \{1,2,3\}$ and to within an arbitrary rotation about the $\mathbf{u}^{(k)}$ principal axis if $t_i^2 = t_j^2$, $i \neq j$, where (i,j,k) is a permutation of $(1,2,3)$.

Then, for each $\mathbf{T}$, $\mathbf{U}$ can in principle be found by inverting (2.88), or (2.89) for an incompressible material. Since, from (2.94), we have

$$\mathbf{T} = \sum_{i=1}^{3} t_i \mathbf{u}^{(i)} \otimes \mathbf{u}^{(i)}, \qquad (4.48)$$

and $\mathbf{T}$ is coaxial with $\mathbf{U}$, this inversion is equivalent to inverting the scalar equations (2.95) or (2.96) for λ_i, $i \in \{1,2,3\}$, when t_i, $i \in \{1,2,3\}$, are given. The resulting deformation gradients are then calculated from $\mathbf{F} = \mathbf{RU}$ since $\mathbf{u}^{(i)}$, $i \in \{1,2,3\}$, are determined from $\mathbf{T}$. In general, however, these inversions are not unique and the extent of their non-uniqueness is a separate question from that of non-uniqueness of the polar decomposition (4.47).

For an incompressible material, for example, elimination of p from (2.96) yields the equations

$$\lambda_1 t_1 - \lambda_1 \frac{\partial W}{\partial \lambda_1} = \lambda_2 t_2 - \lambda_2 \frac{\partial W}{\partial \lambda_2} = \lambda_3 t_3 - \lambda_3 \frac{\partial W}{\partial \lambda_3}, \qquad \lambda_1 \lambda_2 \lambda_3 = 1, \quad (4.49)$$

from which $\lambda_1, \lambda_2, \lambda_3$ can, in principle, be determined when t_1, t_2, t_3 are prescribed. Examples illustrating non-uniqueness of the inversion of (4.49) are given in Ogden (1991, 1997, 2000).

The associated physical problem is that of a (dead-load) pure strain $\mathbf{F} = \mathbf{U}$ ($\mathbf{R} = \mathbf{I}$) in which the principal directions $\mathbf{u}^{(i)}$ are fixed as the load increases. The prototype of this problem is the Rivlin cube problem (Rivlin, 1974), for which $t_1 = t_2 = t_3$ and a cube of elastic material is subjected to equal normal forces per unit reference area on its three pairs of faces. Several variants of this problem and related problems are analyzed in Sawyers (1976), Ball and Schaeffer (1983), Ogden (1984c, 1985, 1987, 1997), MacSithigh (1986), Kearsley (1986) Chen (1987, 1988, 1995, 1996) and MacSithigh and Chen (1992a, b) amongst others. For a general discussion of bifurcation we refer to Chapter 9 in this volume.

1.5 Incremental boundary-value problems

Recalling the notation $\dot{\mathbf{x}} = \dot{\chi}(\mathbf{X})$ defined in (4.1) for the increment in $\mathbf{x}$ we now change the independent variable from $\mathbf{X}$ to $\mathbf{x}$ and introduce the *incremental displacement vector* $\mathbf{u}$ defined, through (2.2), as a function of $\mathbf{x}$ by

$$\mathbf{u}(\mathbf{x}) = \dot{\chi}(\chi^{-1}(\mathbf{x})). \qquad (5.1)$$

The *displacement gradient* grad $\mathbf{u}$, which we denote by $\boldsymbol{\Gamma}$ is then given by

$$\boldsymbol{\Gamma} = \dot{\mathbf{F}} \mathbf{F}^{-1}, \qquad (5.2)$$

so that it is the push forward (from $\mathcal{B}_r$ to $\mathcal{B}$) of the increment in $\mathbf{F}$. From (4.13) it follows that the incremental incompressibility condition may be written

$$\text{tr}\,(\dot{\mathbf{F}}\mathbf{F}^{-1}) \equiv \text{tr}\,(\mathbf{\Gamma}) \equiv \text{div}\,\mathbf{u} = 0. \tag{5.3}$$

The corresponding update (or push forward) from $\mathcal{B}_r$ to $\mathcal{B}$ of the incremental nominal stress, denoted $\mathbf{\Sigma}$, is defined by $\dot{\mathbf{S}}^T \mathbf{N}dA = \mathbf{\Sigma}^T \mathbf{n}da$, where $\mathbf{N}dA$ and $\mathbf{n}da$ are surface elements in $\mathcal{B}_r$ and $\mathcal{B}$, respectively. With the aid of (4.12) and Nanson's formula (2.8), we obtain

$$\mathbf{\Sigma} = J^{-1}\mathbf{F}\dot{\mathbf{S}} = \mathcal{A}_0^1 \mathbf{\Gamma} + p\,\mathbf{\Gamma} - \dot{p}\,\mathbf{I}, \tag{5.4}$$

where $\mathcal{A}_0^1$ is the fourth-order (Eulerian) tensor whose components are given in terms of those of $\mathcal{A}^1$ by (2.128). With this updating the incremental equilibrium equation (4.14) becomes

$$\text{div}\,\mathbf{\Sigma} = \mathbf{0}, \tag{5.5}$$

when body forces are omitted, where use has been made of $(2.11)_2$.

In component form equations (5.4) and (5.3) are combined to give

$$\Sigma_{ji} = \mathcal{A}_{0jilk}^1 u_{k,l} + p\,u_{j,i} - \dot{p}\,\delta_{ij}, \quad u_{i,i} = 0. \tag{5.6}$$

The incremental traction $\mathbf{\Sigma}^T \mathbf{n}$ per unit area on a surface in $\mathcal{B}$ with unit normal $\mathbf{n}$ has components

$$\Sigma_{ji}n_j = (\mathcal{A}_{0jilk}^1 + p\,\delta_{jk}\delta_{il})u_{k,l}n_j - \dot{p}\,n_i. \tag{5.7}$$

1.5.1 Plane incremental deformations

We now restrict attention to plane incremental deformations so that $u_3 = 0$ and u_1 and u_2 depend only on x_1 and x_2. Furthermore, we take the Cartesian axes to coincide with the Eulerian principal axes of the finite deformation associated with $\mathbf{F}$. With this restriction we deduce form $(5.6)_2$ the existence of a scalar function, ψ say, of x_1 and x_2 such that

$$u_1 = \psi_{,2}, \quad u_2 = -\psi_{,1}. \tag{5.8}$$

Substitution of (5.8) into the equilibrium equation (5.5) with $(5.6)_1$ appropriately specialized then leads, after elimination of the terms in $\dot{p}$, to an equation for ψ. If the finite deformation is homogeneous and the material is isotropic, so that the components $\mathcal{A}_{0jilk}^1$ are constants and given by (4.27) and (4.28) then this equation has the compact form

$$\alpha\psi_{,1111} + 2\beta\psi_{,1122} + \gamma\psi_{,2222} = 0, \tag{5.9}$$

where the coefficients are defined by

$$\alpha = \mathcal{A}^1_{01212}, \quad 2\beta = \mathcal{A}^1_{01111} + \mathcal{A}^1_{02222} - 2\mathcal{A}^1_{01122} - 2\mathcal{A}^1_{01221}, \quad \gamma = \mathcal{A}^1_{02121}. \quad (5.10)$$

For details of the derivation we refer to Dowaikh and Ogden (1990).

This is the incremental equilibrium equation for plane incremental deformations of an incompressible isotropic elastic solid in the $(1,2)$ principal plane for an arbitrary homogeneous deformation. For a specific incremental boundary-value problem appropriate boundary conditions need to be given. In order to illustrate these we now concentrate attention on the problem of incremental deformations of a homogeneously pure strained half-space.

1.5.1.1 Surface deformations of a half-space subject to pure homogeneous strain

We now consider the pure homogeneous strain

$$x_1 = \lambda_1 X_1, \quad x_2 = \lambda_2 X_2, \quad x_3 = \lambda_3 X_3 \quad (5.11)$$

as in (3.1) and we take the deformed half-space $\mathcal{B}$ to be defined by $x_2 < 0$ with boundary $x_2 = 0$. On this boundary we set the boundary conditions to correspond to vanishing incremental traction. Specialization of (5.7) with $n_1 = 0, n_2 = 1, n_3 = 0$ then yields two equations. When expressed in terms of ψ after elimination of $\dot{p}$ by differentiation along the boundary and use of the equilibrium equation we obtain

$$\gamma(\psi_{,22} - \psi_{,11}) + \sigma_2\psi_{,11} = 0 \quad \text{on } x_2 = 0, \quad (5.12)$$

$$(2\beta + \gamma - \sigma_2)\psi_{,112} + \gamma\psi_{,222} = 0 \quad \text{on } x_2 = 0, \quad (5.13)$$

where σ_2 is the normal stress on the boundary (and uniform through $x_2 < 0$) associated with the underlying homogeneous deformation. Again we refer to Dowaikh and Ogden (1990) for details. Note that the coefficients (material constants) α, β, γ appearing in (5.9) also feature in the boundary conditions (5.12) and (5.13) but additionally σ_2 arises in the boundary conditions as a separate independent parameter.

In the notation (5.10) the strong ellipticity conditions (4.29) then take on the simple forms

$$\alpha > 0, \quad \beta > -\sqrt{\alpha\gamma}, \quad (5.14)$$

and these inequalities arise naturally in consideration of stability of the homogeneous deformation of the half-space, as we see below.

An incremental *surface deformation* must decay as $x_2 \to -\infty$. We consider ψ to have the form

$$\psi = A\exp(skx_2 - ikx_1), \quad (5.15)$$

where A, s and k are constants. This is periodic in the x_1 direction. In general s is complex and is determined by substitution of (5.15) into (5.9), which yields the quadratic equation

$$\gamma s^4 - 2\beta s^2 + \alpha = 0 \tag{5.16}$$

for s^2. This equation has two solutions for s with positive real part, and we denote these by s_1, s_2. The general solution for ψ with the required decay properties is then

$$\psi = (Ae^{s_1 k x_2} + Be^{s_2 k x_2})e^{-i k x_1}, \tag{5.17}$$

where A and B are constants.

After substitution of (5.17) into the boundary conditions (5.12) and (5.13) we deduce that there is a non-trivial solution for A and B (i.e. an incremental deformation is possible) if and only if the material constants α, β, γ and the normal stress σ_2 satisfy the equation

$$\alpha\gamma + 2\sqrt{\alpha\gamma}(\beta + \gamma - \sigma_2) - (\gamma - \sigma_2)^2 = 0. \tag{5.18}$$

Subject to the incompressibility condition (3.3) this identifies values of σ_2 and the stretches $\lambda_1, \lambda_2, \lambda_3$ for which bifurcation from the homogeneous deformation into an inhomogeneous mode of incremental deformation can occur.

If $\sigma_2 = 0$ in (5.18) it reduces to

$$\gamma[\alpha - \gamma + 2\sqrt{\alpha\gamma}(\beta + \gamma)] = 0. \tag{5.19}$$

Since, by (5.10) and (4.21), $\alpha = \gamma = \beta = \mu$ in the undeformed configuration both factors in (5.19) are positive there. Thus, incremental surfaces deformations are excluded in this configuration, and, by continuity, on any path of pure homogeneous strain from this configuration such that the inequalities

$$\gamma > 0, \quad \alpha - \gamma + 2\sqrt{\alpha\gamma}(\beta + \gamma) > 0 \tag{5.20}$$

hold. It is easy to show that the strong ellipticity inequalities (5.14) follow from (5.20). Note that (5.20) are weaker than the exclusion condition appropriate for all-round dead loading obtained by specializing (4.43)–(4.45) to the present two-dimensional situation since the boundary conditions considered here are different.

For $\sigma_2 \neq 0$ the exclusion condition is then seen to be

$$\alpha\gamma + 2\sqrt{\alpha\gamma}(\beta + \gamma - \sigma_2) - (\gamma - \sigma_2)^2 > 0, \tag{5.21}$$

which restricts σ_2 to a range of values dependent on the other parameters, which are subject to (5.20).

To illustrate the results graphically we restrict attention to plane strain with

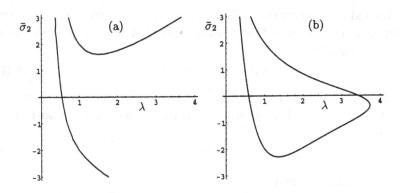

Fig. 1. Plot of the stable region in $(\lambda, \bar{\sigma}_2)$-space for (a) the neo-Hookean strain energy and (b) the strain energy (5.24).

$\lambda_3 = 1$ and set $\lambda_1 = \lambda, \lambda_2 = \lambda^{-1}$ and define the strain energy in terms of λ through

$$\breve{W}(\lambda) = W(\lambda, \lambda^{-1}, 1). \qquad (5.22)$$

Then, (5.21) reduces to

$$\lambda^4 \breve{W}''(\lambda) + \lambda \breve{W}'(\lambda) - 2(\lambda^2 - 1)\sigma_2 - (\lambda^4 - 1)\sigma_2^2/\lambda \breve{W}'(\lambda) > 0, \qquad (5.23)$$

which puts restrictions on the allowable values of λ and σ_2. The curves in Figure 1 show the boundaries of the region defined by (5.23) in respect of the neo-Hookean strain-energy function (2.103) and a single-term strain-energy function in the class (2.106) given, in the present plane strain specialization, by

$$\breve{W}(\lambda) = 8\mu(\lambda^{1/2} + \lambda^{-1/2} - 2). \qquad (5.24)$$

In (a) the stable region is to the right of the left-hand curve and below the upper curve. In (b) the stable region is the area within the loop formed by the curve. In each case the natural configuration $\lambda = 1, \bar{\sigma}_2 = 0$ is within the stable region. If $\bar{\sigma}_2 = 0$ then the half-space is stable for $\lambda \geq 1$ but can become unstable in compression at a critical value of $\lambda < 1$ for the neo-Hookean material, while for the strain energy (5.24) stability may be lost in either compression or tension.

The problem discussed here and some specializations have been examined by a number of authors, and we refer, in particular, to Nowinski (1969a, b), Usmani and Beatty (1974), Reddy (1982, 1983) and Dowaikh and Ogden (1990). For a compressible material the corresponding analysis is given by Dowaikh and Ogden (1991). See also Biot (1965). The analysis of stability of a thick plate subject to a pure homogeneous finite strain has been discussed by several authors and we mention, in particular, Sawyers and Rivlin (1974, 1982) and Sawyers (1977). Stability results are obtained in the static specialization of the problem of vibration of a finitely deformed plate by Ogden and Roxburgh (1993) and Roxburgh and Ogden (1994) for incompressible and unconstrained materials respectively, while the influence of a finite simple shear on stability has been examined on the same basis by Ogden and Connor (1995) and Connor and Ogden (1996).

References to the analysis of stability for problems involving the inflation of a thick-walled sphere or the extension and inflation of a thick-walled circular cylindrical tube can be found in Ogden (1997), for example. Recently, the stability of a rectangular block deformed into a sector of a circular cylindrical tube has been analyzed by Haughton (1999) and Dryburgh and Ogden (1999) from different points of view.

A comprehensive list of references to contributions on linear stability analysis is contained in an appendix at the end of Chapter 10 in this volume.

1.6 Elastodynamics

1.6.1 Kinematics

We now extend the analysis of the previous sections by allowing the deformation to depend on time. As before we take $\mathcal{B}_r$ to denote a fixed (time independent) reference configuration of the body (which may, but need not, be a configuration occupied by the body at some specific time). Let $t \in I \subset \mathbb{R}$ denote time, where I is an interval in $\mathbb{R}$. With each $t \in I$ we associate a unique configuration $\mathcal{B}_t$ of the body. The (one-parameter) family of configurations $\{\mathcal{B}_t : t \in I\}$ is then called a *motion* of the body. We assume that as the body moves continuously then $\mathcal{B}_t$ changes continuously. As in Section 1.2, a point of $\mathcal{B}_r$ is labelled by its position vector $\mathbf{X}$. Let $\mathbf{x}$ be its position vector in the configuration $\mathcal{B}_t$ at time t, which is referred to as the *current configuration*.

Since $\mathcal{B}_t$ depends on t we write

$$\mathbf{x} = \chi_t(\mathbf{X}), \quad \mathbf{X} = \chi_t^{-1}(\mathbf{x}) \tag{6.1}$$

instead of (2.1) and (2.2), or

$$\mathbf{x} = \chi(\mathbf{X}, t) \quad \text{for all } \mathbf{X} \in \mathcal{B}_r, \, t \in I \tag{6.2}$$

in order to make the dependence on t explicit. It is usual to assume that $\chi(\mathbf{X}, t)$ is suitably regular, and for many purposes it may be taken to be twice continuously differentiable with respect to position and time.

The *velocity*, denoted $\mathbf{v}$, and *acceleration*, denoted $\mathbf{a}$, of a material point $\mathbf{X}$ are defined by

$$\mathbf{v} \equiv \mathbf{x}_{,t} = \frac{\partial}{\partial t}\chi(\mathbf{X}, t), \quad \mathbf{a} \equiv \mathbf{v}_{,t} \equiv \mathbf{x}_{,tt} = \frac{\partial^2}{\partial t^2}\chi(\mathbf{X}, t), \tag{6.3}$$

respectively. We emphasize that $\partial/\partial t$ is the *material time derivative*, i.e. the time derivative at fixed $\mathbf{X}$, and it is denoted by $_{,t}$ when the independent variables are understood to be $\mathbf{X}$ and t.

Any scalar, vector or tensor field may be expressed in either the Eulerian description (as a function of $\mathbf{x}$ and t) or, equivalently, in the Lagrangian description (as a function of $\mathbf{X}$ and t) through the motion (6.2) or its inverse.

Thus, if the velocity $\mathbf{v}$ is regarded as a function of $\mathbf{x}$ and t, the *velocity gradient tensor*, denoted $\mathbf{L}$, is an Eulerian tensor defined as

$$\mathbf{L} = \operatorname{grad} \mathbf{v}, \quad L_{ij} = \frac{\partial v_i}{\partial x_j}. \tag{6.4}$$

It follows that

$$\operatorname{Grad} \mathbf{x}_{,t} = \mathbf{F}_{,t} = (\operatorname{grad} \mathbf{v})\mathbf{F} = \mathbf{LF}, \tag{6.5}$$

where the deformation gradient $\mathbf{F}$ is defined as in (2.3) but now depends on t. Equation (6.5) is analogous to the formula (5.2) in the context of incremental deformations.

Using the standard result

$$(\det \mathbf{F})_{,t} = (\det \mathbf{F})\operatorname{tr}(\mathbf{F}^{-1}\mathbf{F}_{,t}), \tag{6.6}$$

we deduce, using (6.5), that

$$J_{,t} \equiv (\det \mathbf{F})_{,t} = (\det \mathbf{F})\operatorname{tr}(\mathbf{L}) = J\operatorname{div} \mathbf{v}, \tag{6.7}$$

where $J = \det \mathbf{F}$, as in (2.5). Thus, $\operatorname{div} \mathbf{v}$ is a measure of the rate at which volume changes during the motion. For an *isochoric* motion $J \equiv 1$ and (6.7) reduces to

$$\operatorname{div} \mathbf{v} = 0. \tag{6.8}$$

Equation (6.8) is the analogue of the incompressibility condition (5.3) arising in the *linearized* incremental theory, but we note while (5.3) is a linear approximation equation (6.8) is exact in the dynamic context.

A rate counterpart of the mass conservation equation (2.25), which also applies when ρ and J depend on t, is obtained by differentiating (2.25) with respect to t and making use of (6.7) to give

$$\rho_{,t} + \rho\operatorname{div} \mathbf{v} = 0. \tag{6.9}$$

1.6.2 Equations of motion

The equation of motion in the form analogous to the equilibrium equation (2.31) is

$$\text{Div}\, \mathbf{S} + \rho_r \mathbf{b} = \rho_r \mathbf{a} \equiv \rho_r \mathbf{x}_{,tt}, \tag{6.10}$$

where $\mathbf{S}$ is given by $(2.83)_1$ for an unconstrained material and $(2.85)_1$ for an incompressible material, with $\mathbf{F}$ and p now depending on t. Equation (6.10) is an equation for the motion (6.2) subject to appropriate boundary and initial conditions, which are not listed here. In components equation (6.10) has the forms

$$\mathcal{A}^1_{\alpha i \beta j} x_{j,\alpha\beta} + \rho_r b_i = \rho_r x_{i,tt} \tag{6.11}$$

and

$$\mathcal{A}^1_{\alpha i \beta j} x_{j,\alpha\beta} - p_{,i} + \rho_r b_i = \rho_r x_{i,tt}, \quad \det(x_{i,\alpha}) = 1 \tag{6.12}$$

for unconstrained and incompressible materials respectively.

There are very few exact finite amplitude solutions of the dynamic equations (6.11) or (6.12) available in the literature. Some of these are outlined in the text by Eringen and Suhubi (1974) and we refer to this work for references up to the date of its publication. We mention the analysis of radial oscillations of a cylindrical tube by Knowles (1960, 1962), radial oscillations of a spherical shell by Guo and Solecki (1963a, b) and of a spherical cavity in an infinite medium by Knowles and Jakub (1965). More recent work includes the derivation of exact solutions by Rajagopal *et al.* (1989), Boulanger and Hayes (1989), Hayes and Rajagopal (1992) and Andreadou *et al.* (1993). For discussion of the propagation of finite amplitude waves we refer to, for example, Carroll (1978), Boulanger and Hayes (1992) and Boulanger *et al.* (1994). Further references are contained in these papers. Some aspects of nonlinear elastic wave propagation are discussed in Chapter 11 in this volume.

1.6.3 Incremental motions

We now consider incremental motions superimposed on a finite motion. Let

$$\dot{\mathbf{x}} = \dot{\chi}(\mathbf{X}, t) \tag{6.13}$$

be the dynamic counterpart of the increment defined in (4.1). Then, the dynamic counterpart of the incremental equilibrium equation (4.14) is

$$\text{Div}\, \dot{\mathbf{S}} + \rho_r \dot{\mathbf{b}} = \rho_r \dot{\mathbf{x}}_{,tt}, \tag{6.14}$$

where a superposed dot again signifies an incremental quantity. In (6.14) no approximation has been made. However, when the equation is linearized in the incremental quantities it becomes, for an unconstrained material,

$$\mathrm{Div}\,(\mathcal{A}^1\dot{\mathbf{F}}) + \rho_r\dot{\mathbf{b}} = \rho_r\dot{\mathbf{x}}_{,tt}, \tag{6.15}$$

where $\dot{\mathbf{b}}$ has been linearized in $\dot{\mathbf{x}}$.

In components equation (6.15) may be written

$$\mathcal{A}^1_{\alpha i \beta j}\dot{x}_{j,\alpha\beta} + \mathcal{A}^2_{\alpha i \beta j \gamma k}\dot{x}_{j,\beta}x_{k,\alpha\gamma} + \rho_r\dot{b}_i = \rho_r\dot{x}_{i,tt}, \tag{6.16}$$

where $\mathcal{A}^2_{\alpha i \beta j \gamma k}$ are the components of the (sixth-order) tensor $\mathcal{A}^2$ of second-order moduli defined by

$$\mathcal{A}^2 = \frac{\partial^3 W}{\partial \mathbf{F}^3}, \quad \mathcal{A}^2_{\alpha i \beta j \gamma k} = \frac{\partial^3 W}{\partial F_{i\alpha}\partial F_{j\beta}\partial F_{k\gamma}}. \tag{6.17}$$

For the special case in which the incremental motion is superimposed on a *homogeneous static* finite deformation and body forces are omitted, equation (6.16) reduces to

$$\mathcal{A}^1_{\alpha i \beta j}\dot{x}_{j,\alpha\beta} = \rho_r\dot{x}_{i,tt}, \tag{6.18}$$

where the coefficients $\mathcal{A}^1_{\alpha i \beta j}$ are now constants. This equation may be updated to the static finitely-deformed configuration to give

$$\mathcal{A}^1_{0piqj}u_{j,pq} = \rho u_{i,tt}, \tag{6.19}$$

where

$$\mathbf{u}(\mathbf{x},t) = \mathbf{u}(\chi(\mathbf{X},t),t) \equiv \dot{\chi}(\mathbf{X},t). \tag{6.20}$$

The counterpart of (6.19) for an incompressible material is

$$\mathcal{A}^1_{0piqj}u_{j,pq} - \dot{p}_{,i} = \rho u_{i,tt}, \quad u_{i,i} = 0. \tag{6.21}$$

Many specific problems have been examined on the basis of the incremental equations and we mention here just a selection of these. The problem of surface (Rayleigh) waves propagating on a homogeneously pre-strained half-space, the dynamic counterpart of the problem discussed in Section 1.5.1.1, was examined by Hayes and Rivlin (1961a), Flavin (1963) and Chadwick and Jarvis (1979). It was reconsidered from the point of view of its connection with stability of the underlying finite deformation by Dowaikh and Ogden (1990, 1991) for incompressible and unconstrained materials respectively. The surface wave problem for an underlying deformation corresponding to simple shear was examined by Connor and Ogden (1995). Wave propagation in a pre-strained plate was discussed by Ogden and Roxburgh (1993), Roxburgh and Ogden (1994) and Connor and Ogden (1996), again with reference to the underlying stability problem. References to related work can be found in these papers. References to papers dealing with wave propagation in pre-stressed cylinders are given by Eringen and Suhubi (1974) and Haughton (1982), but there is very little in the literature for problems with other underlying geometries.

1.6.3.1 Incremental plane waves

In this final section we consider the propagation of plane waves given in the form

$$u = mf(n \cdot x - ct), \qquad (6.22)$$

where the unit vector m is the *polarization vector*, c is the *wave speed* and f is a twice continuously differentiable function. The unit vector n, when real, defines the *direction of propagation* of the wave, which is then a *homogeneous plane wave*; in general n may be complex, in which case the wave is referred to as an *inhomogeneous plane wave*. In many applications f is taken to be an exponential function, but this is not necessary in general.

Substitution of (6.22) into the equation of motion (6.19) and (6.21) yields, after some manipulations,

$$Q_0(n)m = \rho c^2 m, \qquad (6.23)$$

and

$$[Q_0(n) - n \otimes Q_0(n)n]m = \rho c^2 m, \quad m \cdot n = 0, \qquad (6.24)$$

respectively, where the tensor $Q_0(n)$, which depends on n, is defined (in component form) by

$$[Q_0(n)]_{ij} = \mathcal{A}^1_{0piqj} n_p n_q. \qquad (6.25)$$

In view of its connection with wave propagation $Q_0(n)$ is called the *acoustic tensor*.

Equations (6.23) and (6.24), for unconstrained and incompressible materials respectively, are referred to as *propagation conditions*. They determine possible wave speeds and polarization vectors for which plane waves with a given direction n can propagate. The wave speeds are obtained as solutions of the *characteristic equation*

$$\det[Q_0(n) - \rho c^2 I] = 0 \qquad (6.26)$$

(for an unconstrained material), or

$$\det[Q_0(n) - n \otimes Q_0(n)n - \rho c^2 I] = 0 \qquad (6.27)$$

(for an incompressible material), I being the identity tensor.

From either (6.23) or (6.24) we obtain

$$\rho c^2 = [\mathbf{Q}_0(\mathbf{n})\mathbf{m}] \cdot \mathbf{m}. \qquad (6.28)$$

By expressing the right-hand side of (6.28) in component form it is then apparent from (2.127) that $\rho c^2 > 0$ follows from the strong ellipticity condition, and this provides an interpretation of the latter condition in terms of wave propagation. For further discussion of this connection and references see Ogden (1997).

The propagation of infinitesimal plane waves in a homogeneously deformed elastic material was first considered by Hayes and Rivlin (1961b). Recently, an extensive analysis of the reflection of plane waves at the boundary of a pre-stressed half-space subject to pure homogeneous strain has been carried out by Ogden and Sotiropoulos (1997, 1998) for incompressible and unconstrained materials respectively, and a corresponding analysis for a half-space subject to a simple shear deformation by Hussain and Ogden (1999). References to related works are contained in these papers.

References

Andreadou, A., Parker, D.F. and Spencer, A.J.M. 1993 Some exact dynamic solutions in nonlinear elasticity. *Int. J. Engng Sci.* **31**, 695–718.

Antman, S.S. 1995 *Problems of Nonlinear Elasticity*. New York: Springer.

Ball, J.M. 1977 Convexity conditions and existence theorems in non-linear elasticity. *Arch. Ration. Mech. Anal.* **63**, 337–403.

Ball, J.M. and Schaeffer, D.G. 1983 Bifurcation and stability of homogeneous equilibrium configurations of an elastic body under dead-load tractions. *Math. Proc. Camb. Phil. Soc.* **94**, 315–339.

Barenblatt, G.I. and Joseph, D.D. 1996 Editors, *Collected Papers of R.S. Rivlin*. New York: Springer.

Beatty, M.F. 1987 Topics in finite elasticity: hyperelasticity of rubber, elastomers and biological tissues—with examples. *Appl. Mech. Rev.* **40**, 1699–1734.

Biot, M.A. 1965 *Mechanics of Incremental Deformations*. New York: Wiley.

Boulanger, P. and Hayes, M.A. 1989 Finite amplitude motions in some non-linear elastic media. *Proc. R. Ir. Acad.* **89**A, 135–146.

Boulanger, P. and Hayes, M.A. 1992 Finite-amplitude waves in deformed Mooney-Rivlin materials. *Quart. J. Mech. Appl. Math.* **45**, 59–77.

Boulanger, P., Hayes, M.A. and Trimarco, C. 1994 Finite-amplitude plane waves in deformed Hadamard materials. *Geophys. J. Int.* **118**, 447–458.

Carroll, M.M. 1978 Finite amplitude standing waves in compressible elastic solids. *J. Elasticity* **8**, 323–328.

Chadwick, P. 1976 *Continuum Mechanics*. London: George Allen & Unwin.

Chadwick, P. 1999 *Continuum Mechanics*. New York: Dover Publications.

Chadwick, P. and Jarvis, D.A. 1979 Surface waves in a pre-stressed elastic body. *Proc. R. Soc. Lond.* A**366**, 517–536.

Chadwick, P. and Ogden, R.W. 1971 On the definition of elastic moduli. *Arch. Ration. Mech. Anal.* **44**, 54–68.

Chen, Y.-C. 1987 Stability of homogeneous deformations of an incompressible elastic body under dead-load surface tractions. *J. Elasticity* **17**, 223–248.

Chen, Y.-C. 1988 Stability of pure homogeneous deformations of an elastic plate with fixed edges. *Quart. J. Mech. Appl. Math* **41**, 249–264.

Chen, Y.-C. 1995 Stability of homogeneous deformations in nonlinear elasticity. *J. Elasticity* **40**, 75–94.

Chen, Y.-C. 1996 Stability and bifurcation of homogeneous deformations of a compressible elastic body under pressure load. *Math. Mech. Solids* **1**, 57–72.

Ciarlet, P.G. 1988 *Mathematical Elasticity Vol. 1: Three-dimensional Elasticity.* Amsterdam: North Holland.

Connor, P. and Ogden, R.W. 1995 The effect of shear on the propagation of elastic surface waves. *Int. J. Engng Sci.* **33**, 973–982.

Connor, P. and Ogden, R.W. 1996 The influence of shear strain and hydrostatic stress on stability and waves in a layer. *Int. J. Engng Sci.* **34**, 375–397.

Davies, P.J. 1991 A simple derivation of necessary and sufficient conditions for the strong ellipticity of isotropic hyperelastic materials in plane strain. *J. Elasticity* **26**, 291–296.

Dowaikh, M.A. and Ogden, R.W. 1990 On surface waves and deformations in a pre-stressed incompressible elastic solid. *IMA J. Appl. Math.* **44**, 261–284.

Dowaikh, M.A. and Ogden, R.W. 1991 On surface waves and deformations in a compressible elastic half-space. *Stability Appl. Anal. Continuous Media* **1**, 27–45.

Doyle, T.C. and Ericksen, J.L. 1956 Non-linear elasticity. In *Advances in Applied Mechanics* **4**, 53–115.

Dryburgh, G. and Ogden, R.W. 1999 Bifurcation of an elastic surface-coated incompressible isotropic elastic block subject to bending. *ZAMP* **50**, 822–838.

Ericksen, J.L. 1954 Deformations possible in every isotropic, incompressible, perfectly elastic body. *ZAMP* **5**, 466–488.

Ericksen, J.L. 1955 Deformations possible in every isotropic, compressible, perfectly elastic body. *J. Math. Phys.* **34**, 126–128.

Eringen, A.C. and Suhubi, E.S. 1974 *Elastodynamics Vol. 1. Finite Motions.* New York: Academic Press.

Flavin, J.N. 1963 Surface waves in a pre-stressed Mooney material. *Quart. J. Mech. Appl. Mech.* **16**, 441–449.

Fu, Y.B. 1995 Resonant-triad instability of a pre-stressed elastic plate. *J. Elasticity* **41**, 13–37.

Fu, Y.B. 1998 A nonlinear stability analysis of an incompressible elastic plate subjected to an all-round tension. *J. Mech. Phys. Solids* **46**, 2261–2282.

Fu, Y.B. and Ogden, R.W. 1999 Nonlinear stability analysis of pre-stressed elastic bodies. *Cont. Mech. Thermodyn.* **11**, 141–172.

Fu, Y.B. and Rogerson, G.A. 1994 A nonlinear analysis of instability of a pre-stressed incompressible elastic plate. *Proc. R. Soc. Lond.* **A446**, 233–254.

Green, A.E. and Adkins, J.E. 1960 *Large Elastic Deformations.* Oxford: University Press.

Green, A.E. and Adkins, J.E. 1970 *Large Elastic Deformations*, 2nd edition. Oxford: University Press.

Green, A.E. and Zerna, W. 1954 *Theoretical Elasticity.* Oxford: University Press.

Green, A.E. and Zerna, W. 1968 *Theoretical Elasticity*, 2nd edition. Oxford: University Press.

Green, A.E. and Zerna, W. 1992 *Theoretical Elasticity.* New York: Dover Publications.

54 *R.W. Ogden*

Guo, Z.-H. and Solecki, R. 1963a Free and forced finite amplitude oscillations of an elastic thick-walled hollow sphere made of incompressible material. *Arch. Mech.* **15**, 427–433.

Guo, Z.-H. and Solecki, R. 1963b Free and forced finite amplitude oscillations of a thick-walled sphere of incompressible material. *Bull. Acad. Polon. Sci. Ser. Sci. Tech.* **11**, 47–52.

Haughton, D.M. 1982 Wave speeds in rotating elastic cylinders at finite deformation. *Quart. J. Mech. Appl. Math.* **35**, 125–139.

Haughton, D.M. 1999 Flexure and compression of incompressible elastic plates. *Int. J. Engng Sci.* **37**, 1693–1708.

Hayes, M.A. and Horgan, C.O. 1974 On the Dirichlet problem for incompressible elastic materials. *J. Elasticity* **4**, 17–25.

Hayes, M.A. and Rajagopal, K.R. 1992 Inhomogeneous finite amplitude motions in a neo-Hookean solid. *Proc. R. Ir. Acad.* **92**A, 137–147.

Hayes, M.A. and Rivlin, R.S. 1961a Surface waves in deformed elastic materials. *Arch. Ration. Mech. Anal.* **8**, 358–380.

Hayes, M.A. and Rivlin, R.S. 1961b Propagation of a plane wave in an isotropic elastic material subjected to pure homogeneous deformation. *Arch. Ration. Mech. Anal.* **8**, 15–22.

Hill, R. 1957 On uniqueness and stability in the theory of finite elastic strain. *J. Mech. Phys. Solids* **5**, 229–241.

Hill, R. 1968 On constitutive inequalities for simple materials. *J. Mech. Phys. Solids* **16**, 229–242.

Hill, R. 1970 Constitutive inequalities for isotropic elastic solids under finite strain. *Proc. R. Soc. Lond.* A**314**, 457–472.

Hill, R. 1978 Aspects of invariance in solid mechanics. *Adv. Appl. Mech.* **18**, 1–75.

Hill, R. 1979 On the theory of plane strain in finitely deformed compressible materials. *Math. Proc. Camb. Phil. Soc.* **86**, 161–178.

Hoger, A. 1985 On the residual stress possible in an elastic body with material symmetry. *Arch. Ration. Mech. Anal.* **88**, 271–290.

Hoger, A. 1986 On the determination of residual stress in an elastic body. *J. Elasticity* **16**, 303–324.

Hoger, A. 1987 The stress conjugate to logarithmic strain. *Int. J. Solids Structures* **23**, 1645–1656.

Hoger, A. 1993a The elasticity tensors of a residually stressed material. *J. Elasticity* **31**, 219–237.

Hoger, A. 1993b The constitutive equation for finite deformations of a transversely isotropic hyperelastic material with initial stress. *J. Elasticity* **33**, 107–118.

Holzapfel, G.A. 2000 *Nonlinear Solid Mechanics*. Chichester: Wiley.

Holzapfel, G.A. and Gasser, C.T. 2000 An anisotropic viscoelastic model for materials at finite strains: continuum basis and computational aspects. *Comp. Methods Appl. Mech. Engng*, in press.

Horgan, C.O. 1995 Anti-plane shear deformations in linear and nonlinear solid mechanics. *SIAM Review* **37**, 53–81.

Humphrey, J.D. 1995 Mechanics of the arterial wall: review and directions. *Critical Reviews in Biomedical Engineering* **23**, 1–162.

Hussain, W. and Ogden, R.W. 1999 On the reflection of plane waves at the boundary of an elastic half-space subject to simple shear. *Int. J. Engng Sci.* **37**, 1549–1576.

Johnson, B.E. and Hoger, A. 1993 The dependence of the elasticity tensor on residual stress. *J. Elasticity* **33**, 145–165.

Johnson, B.E. and Hoger, A. 1995 The use of a virtual configuration in formulating constitutive equations for residually stressed elastic materials. *J. Elasticity* **41**, 177–215.

Johnson, B.E. and Hoger, A. 1998 The use of strain energy to quantify the effect of residual stress on mechanical behaviour. *Math. Mech. Solids* **3**, 447–470.

Kearsley, E.A. 1986 Asymmetric stretching of a symmetrically loaded elastic sheet. *Int. J. Solids Structures* **22**, 111–119.

Knowles, J.K. 1960 Large amplitude oscillations of a tube of incompressible elastic material. *Quart. Appl. Math.* **18**, 71–77.

Knowles, J.K. 1962 On a class of oscillations in the finite deformation theory of elasticity. *J. Appl. Mech.* **29**, 283–286.

Knowles, J.K. and Jakub, M.T. 1965 Finite dynamic deformations of an incompressible elastic medium containing a spherical cavity. *Arch. Ration. Mech. Anal.* **18**, 367–378.

Knowles, J.K. and Sternberg, E. 1975 On the ellipticity of the equations of non-linear elastostatics for a special material. *J. Elasticity* **5**, 341–361.

Knowles, J.K. and Sternberg, E. 1977 On the failure of ellipticity of the equations for finite elastostatic plane strain. *Arch. Ration. Mech. Anal.* **63**, 321–336.

MacSithigh, G.P. 1986 Energy-minimal finite deformations of a symmetrically loaded elastic sheet. *Quart. J. Mech. Appl. Math.* **39**, 111–123.

MacSithigh G.P. and Chen, Y.C. 1992a Bifurcation and stability of an incompressible elastic body under homogeneous dead loads with symmetry. 1. General isotropic materials. *Quart. J. Mech. Appl. Math.* **45**, 277–291.

MacSithigh G.P. and Chen, Y.C. 1992b Bifurcation and stability of an incompressible elastic body under homogeneous dead loads with symmetry. 2. Mooney-Rivlin materials. *Quart. J. Mech. Appl. Math.* **45**, 293–313.

Marsden, J.E. and Hughes, T.J.R. 1983 *Mathematical Foundations of Elasticity.* Englewood Cliffs, NJ: Prentice Hall.

Marsden, J.E. and Hughes, T.J.R. 1994 *Mathematical Foundations of Elasticity.* New York: Dover Publications.

Nowinski, J.L. 1969a On the surface instability of an isotropic highly elastic half-space. *Indian J. Math. Mech.* **18**, 1–10.

Nowinski, J.L. 1969b Surface instability of a half-space under high two-dimensional compression. *J. Franklin Inst.* **288**, 367–376.

Ogden, R.W. 1972 Large deformation isotropic elasticity: on the correlation of theory and experiment for incompressible rubberlike solids. *Proc. R. Soc. Lond.* **A326**, 565–584.

Ogden, R.W. 1974a On stress rates in solid mechanics with applications to elasticity theory. *Proc. Camb. Phil. Soc.* **75**, 303–319.

Ogden, R.W. 1974b On isotropic tensors and elastic moduli. *Proc. Camb. Phil. Soc.* **75**, 427–436.

Ogden, R.W. 1977 Inequalities associated with the inversion of elastic stress-deformation relations and their implications. *Math. Proc. Camb. Phil. Soc.* **81**, 313–324.

Ogden, R.W. 1982 Elastic deformations of rubberlike solids. In *Mechanics of Solids*, the Rodney Hill 60th Anniversary Volume, Hopkins, H.G. and Sewell, M.J., eds. Oxford: Pergamon Press. 499–537.

Ogden, R.W. 1984a *Non-linear Elastic Deformations.* Chichester: Ellis Horwood.

Ogden, R.W. 1984b On Eulerian and Lagrangian objectivity in continuum mechanics. *Arch. Mech.* **36**, 207–218.

Ogden, R.W. 1984c On non-uniqueness in the traction boundary-value problem for a compressible elastic solid. *Quart. Appl. Math.* **42**, 337–344.

Ogden, R.W. 1985 Local and global bifurcation phenomena in plane strain finite elasticity. *Int. J. Solids Structures* **21**, 121–132.

Ogden, R.W. 1986 Recent advances in the phenomenological theory of rubber elasticity. *Rubber Chem. Technol.* **59**, 361–383.

Ogden, R.W. 1987 On the stability of asymmetric deformations of a symmetrically-tensioned elastic sheet. *Int. J. Engng Sci.* **25**, 1305–1314.

Ogden, R.W. 1991 Nonlinear elasticity: incremental equations and bifurcation phenomena. In Ames, W.F. and Rogers, C., eds., *Nonlinear Equations in the Applied Sciences*. New York: Academic Press. 437–468.

Ogden, R.W. 1997 *Non-linear Elastic Deformations*. New York: Dover Publications.

Ogden, R.W. 2000 Elastic and pseudo-elastic instability and bifurcation. In *Material Instabilities in Elastic and Plastic Solids*, CISM Courses and Lectures No. 414 (ed. H. Petryk), 209–259.

Ogden, R.W. and Connor, P. 1995 On the stability of shear bands. In Parker, D.F. and England, A.H., eds., *Proceedings of the IUTAM & ISIMM Symposium on Anisotropy, Inhomogeneity and Nonlinearity in Solid Mechanics*. Dordrecht: Kluwer. 217–222.

Ogden, R.W. and Fu, Y.B. 1996 Nonlinear stability analysis of a pre-stressed elastic half-space. In *Contemporary Research in the Mechanics and Mathematics of Materials*, eds. R.C. Batra and M.F. Beatty, pp. 164–175. Barcelona: CIMNE.

Ogden, R.W. and Roxburgh, D.G. 1993 The effect of pre-stress on the vibration and stability of elastic plates. *Int. J. Engng Sci.* **30**, 1611–1639.

Ogden, R.W. and Sotiropoulos, D.A. 1997 The effect of pre-stress on the propagation and reflection of plane waves in incompressible elastic solids. *IMA J. Appl. Math.* **59**, 95–121.

Ogden, R.W. and Sotiropoulos, D.A. 1998 Reflection of plane waves from the boundary of a pre-stressed compressible elastic half-space. *IMA J. Appl. Math.* **61**, 61–90.

Rajagopal, K.R., Massoudi, M. and Wineman, A.S. 1989 Exact solutions in nonlinear dynamics. *Arch. Mech.* **41**, 779–784.

Reddy, B.D. 1982 Surface instabilities on an equibiaxially stretched half-space. *Math. Proc. Camb. Phil. Soc.* **91**, 491–501.

Reddy, B.D. 1983 The occurrence of surface instabilities and shear bands in plane-strain deformation of an elastic half-space. *Quart. J. Mech. Appl. Math.* **32**, 265–271.

Rivlin, R.S. 1948a Large elastic deformations of isotropic materials, I. Fundamental concepts. *Phil. Trans. R. Soc. Lond.* **A240**, 459–490.

Rivlin, R.S. 1948b Large elastic deformations of isotropic materials, IV. Further developments in the general theory. *Phil. Trans. R. Soc. Lond.* **A241**, 379–397.

Rivlin, R.S. 1949a Large elastic deformations of isotropic materials, V. The problem of flexure. *Proc. R. Soc. Lond.* **A195**, 463–473.

Rivlin, R.S. 1949b Large elastic deformations of isotropic materials, VI. Further results in the theory of torsion, shear and flexure. *Phil. Trans. R. Soc. Lond.* **A242**, 173–195.

Rivlin, R.S. 1974 Stability of pure homogeneous deformations of an elastic cube under dead loading. *Quart. Appl. Math.* **32**, 265–271.

Rogers, T.G. 1984a Finite deformation and stress in ideal fibre-reinforced materials. In Spencer, A.J.M., ed., *Continuum Theory of the Mechanics of Fibre-reinforced Composites*, CISM Courses and Lectures No. 282. Wien: Springer Verlag. 33–71.

Rogers, T.G. 1984b Problems for helically wound cylinders. In Spencer, A.J.M., ed., *Continuum Theory of the Mechanics of Fibre-reinforced Composites*, CISM Courses and Lectures No. 282. Wien: Springer Verlag. 147–178.

Rosakis, P. 1990 Ellipticity and deformations with discontinuous deformation gradients in finite elastostatics. *Arch. Ration. Mech. Anal.* 109, 1–37.

Roxburgh, D.G. and Ogden R.W. 1994 Stability and vibration of pre-stressed compressible elastic plates. *Int. J. Engng Sci.* 32, 427–454.

Sawyers, K.N. 1976 Stability of an elastic cube under dead loading: two equal forces. *Int. J. Non-Linear Mech.* 11, 11–23.

Sawyers, K.N. 1977 Material stability and bifurcation in finite elasticity. In *Finite Elasticity*, Applied Mechanics Symposia Series Vol. 27, ed. Rivlin, R.S. American Society of Mechanical Engineers, pp. 103–123.

Sawyers, K.N. and Rivlin, R.S. 1974 Bifurcation conditions for a thick elastic plate under thrust. *Int. J. Solids Structures* 10, 483–501.

Sawyers, K.N. and Rivlin, R.S. 1982 Stability of a thick elastic plate under thrust. *J. Elasticity* 12, 101–125.

Seth, B.R. 1964 Generalized strain measure with application to physical problems. In Reiner, M. and Abir, D., eds., *Second-Order Effects in Elasticity, Plasticity and Fluid Dynamics*. Oxford: Pergamon Press. 162–172.

Sewell, M.J. 1967 On configuration dependent loading. *Arch. Ration. Mech. Anal.* 23, 327–351.

Simpson and Spector 1983 On copositive matrices and strong ellipticity for isotropic elastic materials. *Arch. Ration. Mech. Anal.* 84, 55–68.

Spencer, A.J.M. 1970 The static theory of finite elasticity. *J. Inst. Maths Applics* 6, 164–200.

Spencer, A.J.M. 1972 *Deformations of Fibre-reinforced Materials*. Oxford: University Press.

Spencer, A.J.M. 1984 Constitutive theory for strongly anisotropic solids. In Spencer, A.J.M., ed., *Continuum Theory of the Mechanics of Fibre-reinforced Composites*, CISM Courses and Lectures No. 282. Wien: Springer Verlag. 1–32.

Truesdell, C.A. and Noll, W. 1965 In Flügge, S., ed., *The Nonlinear Field Theories of Mechanics: Handbuch der Physik Vol. III/3*. Berlin: Springer.

Usmani, S.A. and Beatty, M.F. 1974 On the surface instability of a highly elastic half-space. *J. Elasticity* 4, 249–263.

Valanis, K.C. and Landel, R.F. 1967 The strain-energy function of a hyperelastic material in terms of the extension ratios. *J. Appl. Phys.* 38, 2997–3002.

Varga, O.H. 1966 *Stress-Strain Behavior of Elastic Materials*. New York: Interscience.

Wang, C.-C. and Truesdell, C.A. 1973 *Introduction to Rational Elasticity*. Leyden: Noordhoff.

Wang, Y. and Aron, M. 1996 A reformulation of the strong ellipticity conditions for unconstrained hyperelastic media. *J. Elasticity* 44, 89–96.

Zee, L. and Sternberg, E. 1983 Ordinary and strong ellipticity in the equilibrium theory of incompressible hyperelastic solids. *Arch. Ration. Mech. Anal.* 83, 53–90.

2

Hyperelastic Bell materials: retrospection, experiment, theory

Millard F. Beatty

Department of Engineering Mechanics
University of Nebraska-Lincoln
Lincoln, NE 68588-0526, U.S.A.
Email: mb53810@mail.navix.net

This chapter is an overview of a theory of a class of nonlinear elastic materials for which the deformation is subject to an internal material constraint described in experiments by James F. Bell on the finite plastic deformation of a variety of annealed metals. Research by Bell and his associates published since about 1979 is reviewed, and Bell's empirically deduced rules and laboratory data are compared with analytical results obtained within the context of nonlinear elasticity theory. First, Bell's empirical characterization of the constrained response of polycrystalline annealed metals in finite plastic strain is sketched. A few kinematical consequences of Bell's constraint, an outline of the constitutive theory developed to characterize the isotropic, nonlinearly elastic response of Bell materials, and theoretical results that lead to Bell's empirical parabolic laws within the structure of isotropic, elastic and hyperelastic Bell constrained materials are presented. The study concludes with discussion of Bell's empirically based incremental theory of plasticity.

2.1 Introduction

It is common in technical writing to begin with a sketch of related research assembled to set the stage for the work ahead. But I'm not going to follow the usual path. There is more to this account than just its technical side - teachers and students, colleagues and associates, family and friends, places and events, life and death - the ingredients of the human side of the story. A reader who feels no interest in this sort of personal, anecdotal retrospection, however, will find immediate relief and surely suffer no loss in skipping ahead to Section 2.3 where Bell's important experiments and his internal material constraint are introduced. We'll return to this shortly.

On 26 October 1995, I had the distinct honor to present at the Johns Hopkins University the first of the *James F. Bell Lectures in Continuum Mechan-*

ics. The lecture on this special occasion focused on Bell's grand experimental achievements in finite strain plasticity and their connection with new theoretical results obtained by Hayes and me on a class of constrained, nonlinearly elastic materials that we named Bell materials. I don't recall precisely my introductory remarks guided by a few notes and spoken as reflections of the past came to mind, including a showing of a few photographs of Bell in his laboratory. My initial thoughts drifted backward in time to my early formative years at Hopkins and to my association with Bell and others‡.

2.2 Retrospection

Four decades earlier on 29 September 1955, I entered the Johns Hopkins University as a nontraditional (married) freshman student in the Department of Mechanical Engineering. This adventure in learning followed several years of employment and self-study after graduation from the Baltimore Polytechnic Institute in February 1950. In the year prior to my admission to Hopkins, I was employed by the Glen L. Martin Company, nowadays known as Lockheed Martin. I was called a design engineer. Actually, I was an advanced design draftsman assigned to the wind tunnel testing group. Driven by curiosity and unanswered questions on technical matters encountered in self-studies on mathematics and mechanical engineering design, I was highly motivated upon entering Hopkins with the idea that I wanted to become a better design engineer. But the influence of my Hopkins experience in just a couple of years at this great research institution changed my target in ways I had never dreamed of.

The time passed quickly. In my senior year in September 1958, in courses in Solid Mechanics and Experimental Solid Mechanics, I met for the first time my teacher and senior research advisor Professor James Frederick Bell. Bell was a charming, warm, and friendly personality, a dynamic lecturer well liked by his students. His exuberance and enthusiasm for teaching all aspects of

---

‡ The reflections that follow are a reconstructed, much expanded chronicle that mirrors my original presentation only in general details. A similar lecture that helped restore my fading memory of this event was presented on 28 April 1997 at the University of California-Berkeley. My preparation for the talk motivated the completion of this overview of a small fragment of Bell's research involving his loading constraint, a review encouraged long ago by Professor Akhtar Khan, a former student of Bell's who attended my presentation at Baltimore.

In the afternoon at Berkeley, I had the great pleasure to discuss with four graduate students - Eveline Baesu, Arthur Brown, Deepak Nath, and Ping Wong - some topics in elastic stability and some central ideas in Bell's research on torsion and extension of thin-walled tubes. In recognition of their broad interests in various aspects of stability, plasticity, and continuum mechanics, typical of Bell's enduring interests in teaching and research, this article was prepared mainly with these and other students in mind.

solid mechanics and his concurrent excitement with his research studies were infectious. His boundless energy radiated in all directions. Most certainly I was infected and began work in his laboratory studying the propagation of stress waves in aluminum bars having a high temperature gradient extending to the molten state. I had complete freedom to do what I liked. Bell furnished the materials and equipment; the path of exploration, however, was directed by my own curiosity. I graduated in June 1959 leaving the study incomplete, and returned to the Martin Company as an engineer.

But the Hopkins fever raged within me, and I was drawn back to graduate school in September 1959 to extend my previous experimental investigations to the study of waves in polymers, principally nylon. Several engineering departments at Hopkins were undergoing a reorganization to form the Department of Mechanics, eventually my new home department. I was awarded a University (tuition) Fellowship that I was most grateful to have received. With the birth of my son Scott in February 1958, however, my family responsibilities had increased; so my decision to return to Hopkins hinged on what outside employment I might arrange. Besides becoming a full time graduate student enrolled for 12 difficult credit hours, I accepted full-time employment as an Instructor of Engineering Physics at Loyola College, responsible for 12 hours of instruction in the fall semester, 4 more in the spring, and for grading all my students' papers.

In spite of this superabundant activity, I continued my stress wave experiments. So far as I was concerned, in my mind Bell was still my research mentor; however, I recall no formal procedure to actually fix this role. During my inaugural graduate year, I met for the first time in a course in Nonlinear Elasticity another great teacher and remarkable theoretician, Professor Jerald La Verne Ericksen, whose lectures I found especially stimulating. I would go often to discuss with him and review the lecture notes that I had rewritten and worked out in detail. I suppose my unusual interest in the subject eventually led Ericksen to offer me partial support from his research grant†, an invitation I accepted without hesitation. It was easy to surrender my heavy teaching load at Loyola for a much reduced teaching load as Instructor of Mechanics at Johns Hopkins coupled with a research assistantship and the unusual opportunity to work with Ericksen. Thereafter, I abandoned my novice experimental effort with stress waves; but my interest in Bell's courses never waned.

During 1960-1961, a year celebrated by the birth of my daughter Ann in

---

† Bell, according to his wife Perra, remembered this differently. Ericksen asked Bell if he knew of any developing theoreticians in solid mechanics; and Bell recommended me. I've never asked Ericksen about this; but I'll wager that my enthusiasm for learning nonlinear elasticity theory must have helped as well.

July 1961, I attended two other courses offered by Bell in Nonlinear Mechanics and Vibrations. And I enjoyed another course by Ericksen in Tensor Analysis during 1961-1962, a year of unusual opportunity to interact with Ericksen's postdoctoral associate Dr. Michael A. Hayes, a former research assistant to Professor Ronald S. Rivlin, a truly brilliant scholar and a warm and cheerful colleague who would become a lifelong professional friend and future collaborator.

In the same period, the Truesdell and Toupin (1960) article on *The Classical Field Theories* appeared; and in the fall of 1961 Truesdell† arrived at Hopkins. Thereafter, Professors Clifford Truesdell and Walter Noll (1965) were collaborating in writing the *Nonlinear Field Theories of Mechanics*, a draft version of which Ericksen suggested I should read. Looking backward, I can say certainly that this period at Hopkins was the most intellectually exhilarating experience of my academic career. It was a truly exciting time. I attended classes and lectures by Professors Truesdell, Noll, Coleman, Dillon, Phillips and other well-known celebrities of the day, including Serrin, Toupin, and Rivlin. Eventually, with Ericksen's guidance, my dissertation on elastic stability theory evolved (see Beatty 1965). Although Bell and Truesdell, a helpful reader and valued critic of my work, sat on the final examination committee at my dissertation defense on October 14, 1964, I had no other research interaction with Bell. Actually, I left Johns Hopkins in August 1963 to pursue regular employment at the University of Delaware as an Acting Assistant Professor of Mechanical Engineering, a temporary rank awarded me until my doctorate was officially conferred on November 4, 1964. I was then a bona fide Assistant Professor doing what I had done before. Though subsequently I often noticed Bell's work reported in the literature, I lost all contact with it.

There is much to add of the years that followed, too much to burden these pages; but a few facts are relevant. Professor Oscar Dillon in 1967 was Chairman at the University of Kentucky, and he invited me to join his department as Associate Professor of Engineering Mechanics. Troubled by disquieting administrative turmoil between my department chair, the college dean and other administrators at Delaware, I felt that Dillon's offer could not have come at a more opportune time. Though not without some hesitation over leaving a fresh opportunity to interact with my esteemed colleague and friend Professor Jerzey L. Nowinski, a renowned scholar and expert in boundary value problems in finite elasticity, I accepted the position. In April the following year my wife Nadine delivered our daughter Laura, the last of our children, a joyous occasion preceded and dimmed by the nearly simultaneous death of my father

† Truesdell joined the Hopkins faculty in 1960, while concluding at Bologna the final stages of his treatise with Toupin during a sabbatical leave from Indiana University.

at Baltimore. The next several years brought renewed academic stimulation
as Dillon's new mechanics group matured. While on sabbatical leave from my
position at Kentucky during the fall semester of 1973, I had the good fortune
to spend six months at Hopkins as Visiting Research Scientist. Besides the
refreshing experience of traditional research seminars in continuum mechanics run by Ericksen, Truesdell, and Bell, I found special interest in Dr. Ingo
Müller's work in the statistical mechanics of rubber elasticity. I don't recall,
however, the specific nature of Bell's research activity at that time.

Ericksen moved in 1982 to the University of Minnesota. He and others at
Minnesota obtained funding for a series of workshops to be held during the
1984-1985 *Program in Continuum Physics and Partial Differential Equations*
conducted within the Institute for Mathematics and Its Applications. In orga-
nizing this yearlong affair, Ericksen offered me a senior fellow appointment to
participate in residence throughout the year. My visit in Minnesota from August 1984 though June 1985 was a most memorable, productive, and rewarding
period, a stimulating year standing in memory second only to my graduate days
at Hopkins. I presented research at two of the workshops, attended all of the
others, and I had great fun in presenting a talk on engineering applications of
mathematics to a variety of advanced high school students enrolled in special
University courses. Toward the end of the year in May and June 1985, Bell and
Hayes also were participants at the Institute. It was during this period that I
was drawn again to Bell's work.

At the workshop on *Dynamical Problems in Continuum Physics* held June
3-7, 1985, I recall Bell's exciting lecture (see Bell 1987) in which he identified
an invariant kinematical constraint that occurred during loading in a variety
of experiments on annealed, polycrystalline metals subjected to substantial
plastic strains. Hayes and I were seated together in the packed auditorium.
During the seminar, as Bell raced through description of his experiments and
results, it occurred to us that we might take a look at this purely kinematical
constraint within the context of finite elasticity. Our interest stemmed more
from the standpoint of curiosity than expectation that any results we might
find could possibly relate to Bell's tests within the different context of the
plastic deformation of metals. At lunch that day, we met with Bell to discuss
in some detail his loading constraint and related experiments; subsequently,
we felt our idea might be the start of "a new ball game". Thus, on that
spring day in Minnesota, our research project on Bell constrained materials was
born, and within a short time later some preliminary results were produced.
Unfortunately, not long afterward we departed the Institute; I returned to the
University of Kentucky, Hayes to University College Dublin.

In the ten years that followed, we published several papers of which three (see Beatty and Hayes 1992a,b; 1995) constitute the main body of the theory describing the nonlinear elastic response of a class of materials subject to large deformations and characterized by the internal constraint tr $\mathbf{V} = 3$, where $\mathbf{V}$ is the left Cauchy-Green stretch tensor.† We named this constraint after Bell (1983a, 1985, 1989) who discovered it in countless experiments on various annealed metals studied in the different context of finite strain plasticity. The constraint, however, is strictly kinematical; so we decided to study its implications in the simpler context of finite elasticity theory, as mentioned previously. The results, however, when examined in the context of Bell's experiments were truly remarkable. We were able to produce many formulae derived in Bell's countless earlier experiments. Based on Bell's experimentally motivated work function, we investigated a special class of isotropic, hyperelastic materials for which we derived precisely Bell's parabolic law for uniaxial loading, precisely Bell's parabolic law for pure shear, and his fundamental invariant rule connecting the deviatoric stress intensity to the corresponding deviatoric strain intensity exactly as reported in his experiments. Nevertheless, we make no pretence that our results have any direct bearing on the plastic deformation of annealed metals in Bell's tests. And I say again that Bell's studies deal with finite strain plasticity theory, whereas the studies by Hayes and me focus strictly on nonlinear elastic deformations of isotropic solids characterized by a similar work function; and no immediate association with plasticity is implied. Therefore, our results should be viewed in this spirit. It is nonetheless remarkable that our theoretical results predict and concur accurately with the great body of Bell's experimental data.

The objectives of this essay are to review Bell's experimental results on his internal constraint and to examine the Bell constraint in the context of results derived from the Beatty-Hayes theory of isotropic, hyperelastic Bell constrained materials. The principal ideas needed from Bell's experiments will be sketched, the Beatty-Hayes theory will be outlined, some easy theoretical results will be illustrated, and the connection with results found in tests by Bell will be demonstrated. The review concludes with comparison of Bell's experimental data relating the deviatoric stress intensity with the corresponding deviatoric strain intensity. Every one of Bell's former students and research associates will surely recognize this rule, it is the foundation upon which all of Bell's own analysis of his experiments is developed in the sense of incremental plasticity theory discussed briefly at the end.

I proudly dedicate this essay to the memory of my esteemed teacher and friend Professor James Frederick Bell, whose careful, thorough experiments and

---

† Henceforward, unless explicitly stated otherwise, we shall adopt in this chapter notation previously defined in Chapter 1.

enduring scholarship on the quasi-static response of annealed metals in finite plastic strain are outlined below.

2.3 Experiment

In this section, Bell's kinematical constraint is introduced and compared with the incompressibility condition. It is shown that there are no incompressible Bell materials, yet both constraints reduce for infinitesimal strains to the familiar null dilatation condition for an incompressible material. A stress tensor used by Bell (1983a, 1985) to characterize his analytical and experimental results is introduced in equations derived for the modified 2-dimensional Bridgman compression experiment. These equations are used to demonstrate the validity of the constraint in comparison with test data. The Bell constraint is consistent with these data, while the incompressibility condition is not, except in the limit of infinitesimal strains from the genesis state defined later on. Other relevant experimental results establishing the isotropic character of the materials are noted. These primary experimental results are then employed as principal ingredients in the development of a general theory of nonlinearly elastic Bell constrained materials.

2.3.1 Bell's constraint

In a great variety of quasi-static tests on different kinds of annealed, polycrystalline solids, including aluminum, brass, copper, and mild steel, Bell (1985, 1989, 1996a) discovered that for the finite strain measure $\mathbf{e} \equiv \mathbf{V} - \mathbf{1}$ the constraint tr $\mathbf{e} = 0$ holds for all loading processes† relative to the undeformed, nat-

---

† Bell (1995b, 1996a) shows that for all loading processes in homogeneous strain relative to the undeformed, natural state, his primary internal constraint holds to within a certain upper bound on the deviatoric strain intensity $\Gamma = \sqrt{\operatorname{tr} \mathbf{e}^2} \leq 0.33$, an *averaged* experimental value that depends on the loading path and is subject to variation. Beyond this value of generalized strain in a tension-torsion experiment, in the course of the plastic deformation there is the sudden appearance of a highly localized inhomogeneous deformation that generally proceeds immediately to failure in either shear banding, necking, or gross buckling collapse of the specimen. The upper bound is attained when the measured change of volume per unit initial undeformed volume decreases below 4-5%. At this point in Bell's (1995b) stress controlled test, the specimen may suddenly spin to failure through one or more revolutions, leaving visible traces of shear banding, or necking, or a combination of these. The volume is then reported to increase toward its initial value before it decreases again with increasing plastic strains. Upon unloading, the material volume usually returns to its pre-deformation value.

In a private communication dated 23 August 1993, Bell reported the bound $\Gamma \simeq 28\%$, for example; and he expressed the belief that the domain on which the constraint (3.1) applies includes essentially the whole area of continuum plasticity. Beyond this bound, however, for many kinds of loading histories, he envisioned the problem as one of a metallurgical study of extrusion, rolling, necking, shear banding, and so on.

ural state of the material. We recall that $\mathbf{V}$ is the left Cauchy-Green stretch tensor (see Beatty 1996, Ogden 2001, Truesdell and Noll 1965). Thus, alternatively, Bell's loading constraint is

$$i_1 = \mathrm{tr}\, \mathbf{V} = 3, \text{ or } I_e \equiv I_1(\mathbf{e}) = \mathrm{tr}\, \mathbf{e} = 0. \qquad (3.1)$$

In the paper concluded just prior to his death in 1995, Bell (1996b) describes this internal constraint as a manifestation of the local, orientation dependent, inhomogeneous stress-strain response in the single crystal itself, an inhomogeneity that increases with deformation. On the macroscopic scale, however, the measurable inhomogeneous response on the small scale becomes statistically homogeneous, very nearly, and is reflected in the macroscopic constraint (3.1). In the polycrystalline solid, the principal difference between the finite strain response of the free single crystal and that of the same crystal imbedded in the aggregate is the presence of boundary constraints imposed by adjacent neighbors. Bell thus perceived (3.1) as the continuum manifestation of an internal constraint due to homogenization of local deformation inhomogeneities in the crystal and boundary interactions with adjacent neighbors in the polycrystalline aggregate. It is emphasized in Bell's studies that (3.1) is relaxed during unloading, for then the material volume increases toward its initial value in a different way. Bell is quite specific in stating that "the internal constraint tr e =0 does not apply during unloading"; it applies only to all increasing loading conditions relative to the undeformed, natural state. Some typical experimental data will be described farther on.

In the principal basis (3.1) becomes, respectively,

$$i_1 = \lambda_1 + \lambda_2 + \lambda_3 = 3, \quad I_e = e_1 + e_2 + e_3 = 0, \qquad (3.2)$$

in which $\lambda_k = 1 + e_k$ are the three principal stretch components, and hence e_k may be identified as the three principal engineering strains. We note for future use that in a plane strain deformation for which $e_2 = \lambda_2 - 1 = 0$, the constraint (3.2) reduces to

$$\lambda_1 + \lambda_3 = 2, \quad e_1 = -e_3. \qquad (3.3)$$

Hence, for a plane strain, only one of the principal components is independent.

Notice that tr e =0 implies that in a linearized approximation for infinitesimal engineering strains ε_k the dilatation tr $\varepsilon = 0$, which is the well-known infinitesimal condition for incompressibility. On the contrary, however, (3.1) is not an incompressibility condition for any sort of finite strain.

2.3.2 Incompressibility condition

A deformation $\mathbf{F}$ is isochoric if and only if

$$J \equiv \det \mathbf{F} = \det \mathbf{V} = \det(\mathbf{1} + \mathbf{e}) = 1. \tag{3.4}$$

Therefore, the common assumption in classical plasticity that the material is incompressible requires that (3.4) hold for all deformations of the material from the undeformed, natural state κ_0. It is easy to see, however, that the constraint (3.4) is independent of the reference configuration. Indeed, suppose an incompressible material is subjected to a deformation $\mathbf{F}_0$ from κ_0 followed by another deformation $\mathbf{F}_1$ from the primary deformed state κ_1, so that the total deformation from κ_0 to the present configuration κ_2 is $\mathbf{F}_2 = \mathbf{F}_1\mathbf{F}_0$. Then because (3.4) holds for both $\mathbf{F}_0$ and $\mathbf{F}_2$, it follows that $\det \mathbf{F}_1 = 1$ as well. Notice that (3.1) does not share this invariant configuration property valid for all loading and unloading conditions.

In terms of principal components, (3.4) requires

$$\lambda_1\lambda_2\lambda_3 = 1, \text{ or } (1 + e_1)(1 + e_2)(1 + e_3) = 1, \tag{3.5}$$

for all deformations from κ_0. Thus, for plane strain with $e_2 = \lambda_2 - 1 = 0$, for example, we have

$$\lambda_1 = \frac{1}{\lambda_3}, \text{ or } e_1 = -\frac{e_3}{1 + e_3}. \tag{3.6}$$

Thus, clearly the Bell constraint in (3.3) cannot hold: $e_1 \neq -e_3$, that is, $\text{tr}\,\mathbf{e} \neq 0$. In consequence, one sees that there are no incompressible Bell constrained materials for which both (3.1) and (3.4) hold; they share a common solution set only in the undeformed state.

On the other hand, to the first order in the infinitesimal engineering strain components ε_k, (3.5) delivers $\text{tr}\,\varepsilon = 0$, the familiar infinitesimal condition for incompressibility. We recall that the Bell constraint yields the same linearized, null dilatation condition for infinitesimal strains. Thus, every Bell constrained material behaves like an incompressible material in the neighborhood of the natural state, yet there are no incompressible Bell materials.

2.3.3 Bell's stress tensor

Bell (1983a, 1985) found it convenient to introduce the stress tensor τ defined by

$$\tau = J\mathbf{V}^{-1}\sigma = \mathbf{R}\pi^T. \tag{3.7}$$

We recall that $J \equiv i_3 = \det \mathbf{V}$ is a measure of the change of material volume, σ is the Cauchy stress tensor, and π is the 1st Piola-Kirchhoff stress tensor, also known as the engineering stress tensor. Beatty and Hayes (1992a) named τ the "Bell stress tensor". We observe from (3.7) that $\mathbf{S} = \pi^T = \mathbf{R}^T\tau$, where $\mathbf{S}$ is the nominal stress tensor in Ogden (2001); and hence in a pure homogeneous deformation, $\tau = \mathbf{S} = \pi^T$. In this case the principal Bell stress components τ_k are the principal engineering stress components, which are essentially the applied loads. We are now ready to consider Bell's test results on the loading constraint.

2.3.4 Modified Bridgman 2-dimensional compression experiment

Bell found the constraint $\mathrm{tr}\, \mathbf{e} = 0$ in countless quasi-static loading tests on annealed metals, but to improve upon and increase the accuracy of earlier measurements for determination of $\mathrm{tr}\, \mathbf{e} = 0$, he later (see Bell 1985) adopted a modified Bridgman two-dimensional, stress controlled compression experiment in which both principal strains e_1 and e_3 are measured with $e_2 = 0$ and $\tau_3 = 0$. Based upon his incremental plasticity equations, Bell (1983a, 1988) concluded that the ratio of principal Bell stresses is a constant: $\tau_2/\tau_1 = 1/2$; and hence his modified Bridgman experiment is a proportional loading test. Bell continues and shows that when $\mathrm{tr}\, \mathbf{e} = 0$ in the Bridgman test, so that $e_1 = -e_3$, the following theoretical relation must hold:

$$\tau_1^2 = -2^{3/2}\beta^2 e_3. \tag{3.8}$$

Hence, if the Bell constraint is valid, a plot of τ_1^2 versus $|e_3|$ should be linear and provide the constant slope value for the particular material to fix β.

On the other hand, if the material is incompressible, the response equation calculated for the incompressibility condition (3.6) must be given by

$$\tau_1^2 = -2^{3/2}\beta^2 \frac{e_3}{1 + e_3}. \tag{3.9}$$

Notice that (3.9) reduces to (3.8) for infinitesimal strains for which $e_3 = \varepsilon_3$ and terms of order ε_3^2 are neglected. Otherwise, the response (3.9) is not linear in $|e_3|$.

Bell's (1985) experimental data for annealed aluminum 1100-0 are shown in Figure 1 for the modified Bridgman 2-dimensional compression test. The stress in (3.8) for Bell's constraint $\mathrm{tr}\, \mathbf{e} = 0$ is compared with the stress (3.9) for the incompressibility constraint $\det(1 + \mathbf{e}) = 0$. The data show clearly that Bell's condition $\mathrm{tr}\, \mathbf{e} = 0$ is in accord with the physics characterized by decreasing material volume. Notice, however, that the data for the two models are essentially indistinguishable for small strains for which $\mathrm{tr}\, \varepsilon = 0$ describes both constraints. See Bell's (1979, 1980, 1996a) for additional discussion and historical review

Millard F. Beatty

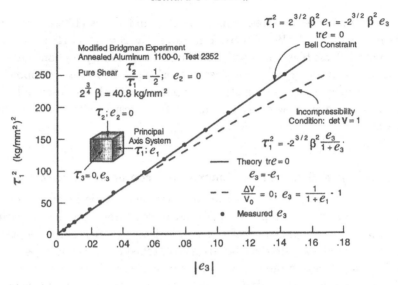

Fig. 1. Plot of theoretical relations for τ_1^2 versus $|e_3|$ comparing the Bell and in-compressibility constraints against measured values from Bell's modified Bridgman compression experiment for annealed aluminum. Both constraints agree closely for $|e_3| < 0.04$; but only the Bell constraint is consonant with data over the entire range of plastic strain shown here. This figure is adapted from Bell (1985).

on measurement of volume changes in plastic deformation and his analysis of Bridgman's data demonstrating a bound on the range of applicability of (3.2) mentioned in the footnote on page 64. This effect is illustrated in Figure 2 in Bell (1996a), which shows an upper limit on $|e_3| = 20\%$ for this pure shear test. This is followed by an increase in material volume with increasing strain as noted before.

2.3.5 Isotropic material response

Bell further established the characteristics of the material response relative to an undistorted reference configuration. This undeformed state is specific for annealed polycrystalline solids; it is identified with the undistorted state of the free single crystal. Hence, according to Bell, this undeformed "genesis" reference configuration for the fully annealed solid, regardless of how many times or the manner in which the specimen may have been loaded, unloaded, and reloaded, is the single reference configuration with respect to which all components of stress and strain are referred under all circumstances.

Relative to this reference state of annealed metals Bell (1995b) and Bell and

Baesu (1996) demonstrated for a great variety of experimental data and in different ways certain characteristics of the material response. The coaxiality of stress and stretch, so that $\sigma V = V\sigma$ holds for all deformations from the genesis reference configuration along loading paths of arbitrary composition and direction, is demonstrated, for example. Therefore, we know from Bell's studies, further confirmed in these detailed experimental data, that the material response for Bell's annealed metals is isotropic relative to its genesis reference configuration.

2.3.6 Basis for the theory and commencement of the project

Some additional experimental results will be introduced later. For the present, however, to develop a general theory of nonlinearly elastic Bell materials, we require only Bell's primary experimental results on material isotropy and his internal material constraint. Thus, if we do not concern ourselves with the inelastic material response exhibited in unloading, we may consider the response to be nonlinearly elastic relative to an undistorted, "genesis" state with respect to which the material is isotropic and satisfies the Bell constraint for all subsequent deformations. With this model of constrained, isotropic nonlinear elastic response in mind, we might hope to characterize the essential aspects of Bell's model. This was what Hayes and I proposed in 1985 following Bell's lecture on his experiments. And in further conversation with Bell at lunch that day, we discussed at greater length details of his results and our thoughts about using the constraint in the different context of finite elasticity. Our project on development of a theory of elastic Bell constrained materials was thus born. This is essentially the plan that we explored over the several years that followed. Throughout this effort, we have received many helpful suggestions and marvelous encouragement from Ericksen, for which we are especially grateful.

I want to emphasize that this was not to be a theory of finite strain plasticity, so we were not looking to model the detailed response observed in Bell's many experiments. Besides, Bell (1979, 1980, 1983a) had already developed an incremental theory that appeared to capture just about everything anyway; and when it did not do this in the same way for certain kinds of annealed solids, including mild steel (see Bell 1983b), he found new directions of explanation of other observed phenomena in description of quantum states† of

---

† Annealed mild steel, low purity aluminum and zinc, unlike annealed, high purity copper and aluminum, for example, demonstrate a second order transition structure that complicates characterization of the material response. Nevertheless, Bell's incremental law continues to hold piecewise throughout each transition stage. This is illustrated by Bell (1983b) for proportional and nonproportional dead loading of thin-walled tubes of annealed mild steel in combined states of tension and torsion and for similar solid bars under

material response. We have made no attempt to model these other kinds of effects. Nevertheless, you will see later on that the results achieved are indeed remarkable when viewed in the light of Bell's quasi-static test results.

2.4 Theory

Recalling Bell's experiments that identify two properties of his annealed metals - the internal constraint and isotropic response relative to a genesis reference configuration - we now focus on development of the theoretical model introduced by Beatty and Hayes (1992a,b; 1995) to characterize a Bell material modeled as an isotropic, internally constrained nonlinear elastic solid. We'll first examine some kinematical implications of the Bell constraint. Some homogeneous deformations possible in a Bell material, and others that are not, are illustrated; and some principal kinematical results are presented, including comparison with incompressible materials. The Poisson kinematical function for a Bell material is compared with the function for an incompressible material. Afterwards, the constitutive equation for our Bell model is introduced. It is then shown that this equation yields results for uniaxial loading and for pure shear that include relations of the kind found by Bell in corresponding experiments. The remarkable nature of these results will be demonstrated in our subsequent discussion of the theory of hyperelastic Bell materials in Section 2.5. Discussion of Bell's notable experimental result on the negligible rigid rotation of the principal stress-stretch axes in a finite torsion and extension of a thin-walled tube concludes our review of analytical results for the general class of isotropic Bell materials. We begin with some easy geometrical results that derive from the constraint.

2.4.1 *Kinematics: geometry of the constraint and the invariant plane*

The Bell constraint has a simple geometrical interpretation (see Beatty and Hayes 1993) easily visualized in Figure 2 illustrating the pure homogeneous deformation of a unit cube. The Bell constraint (3.2) shows that the perimeter of a unit cube is preserved in every pure homogeneous deformation with principal stretches λ_k. This geometrical result is strictly kinematical and is independent

tension and compression. Among the experts, Bell's characterization of quantum states of material response reflected in the quantized variability of the material parameter β (see (3.8)) with continuous gross plastic strain is controversial. We'll find no need to discuss this here.

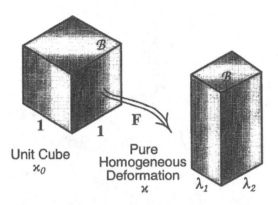

Fig. 2. Geometrical interpretation of the Bell constraint showing the invariance of the perimeter of the unit cube under a pure homogeneous deformation: $4\lambda_1 + 4\lambda_2 + 4\lambda_3 = 12$.

of any constitutive description of the material. Some additional geometrical results follow.

The constraint function

$$f(\lambda_1, \lambda_2, \lambda_3) \equiv \lambda_1 + \lambda_2 + \lambda_3 - 3 = 0 \qquad (4.1)$$

defines an invariant plane in λ-space. The principal stretches in any deformation whatsoever are restricted to values on this surface; and hence points on this surface may be described by the stretch vector $\lambda = (\lambda_1, \lambda_2, \lambda_3)$. Since $\lambda_k > 0$, the plane is contained in the first coordinate octant shown in Figure 3, and bounded by three lines $\lambda_i + \lambda_j = 3$, $i \neq j = 1, 2, 3$, that form the equiangular triangular base of an equilateral tetrahedron of edge length 3. The centroid of the triangle is at the undeformed state where the stretch vector $\lambda = (1, 1, 1)$. Every deformation path begins from this point and lies in the invariant plane. The binormal unit vector $\mathbf{b} = (1, 1, 1)/\sqrt{3}$ to any deformation trajectory traced by the stretch vector λ in the invariant triangle is also normal to the invariant plane. Thus, the Bell constraint (3.2) expresses the invariance of the orthogonal projection of the stretch vector λ upon $\mathbf{b}$, that is, $\lambda \cdot \mathbf{b} = \sqrt{3}$. Notice that the invariant plane of Bell's constraint is an octahedral plane! Some special deformation trajectories are illustrated next.

2.4.2 Equibiaxial stretch, plane stretch, and shear with contraction

The geometry in the invariant plane reveals easily that some familiar kinds of finite homogeneous deformations are kinematically possible in every Bell

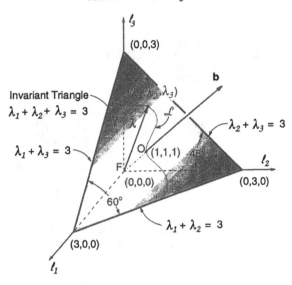

Fig. 3. Invariant octahedral plane of the Bell constraint.

material, while others are not, except possibly in a small neighborhood of the undeformed state.

The geometry in Figure 4 shows, for example, a line of equibiaxial stretch defined by

$$\lambda_1 = \lambda_2, \quad 2\lambda_1 + \lambda_3 = 3; \tag{4.2}$$

and another line of plane stretch defined by

$$\lambda_3 = 1, \quad \lambda_1 + \lambda_2 = 2. \tag{4.3}$$

Both deformations are separately possible in every Bell material. Clearly, these deformation paths have only the undeformed state in common; therefore, *an equibiaxial, plane stretch of a Bell constrained material is not possible.*

The trajectory of a simple shear defined by $\lambda_1\lambda_2 = 1$, $\lambda_3 = 1$ in λ-space is also shown in Figure 4. This path, however, is not in the invariant plane, except in a small neighborhood of the undeformed state at the centroid. Hence, *a simple shear deformation is not possible in any sort of Bell material.*[†]

[†] One consequence of this fact, pointed out by Bell (1996b), is that Taylor's simple shear hypothesis (see Bell 1995a) for describing the response of cubic single crystals is inadmissible and thus begs reassessment. Some ideas and facts Bell considered relevant to this are collected in his article (1996b).

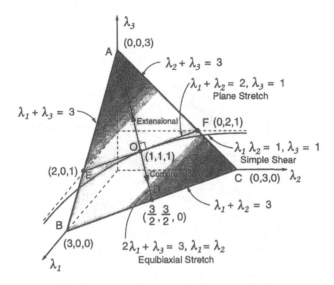

Fig. 4. Geometrical description of an equibiaxial stretch, a plane stretch, and a simple shear. Note that an equibiaxial plane stretch and a simple shear are impossible in any Bell constrained material.

On the other hand, *a generalized shear with normal contraction, defined by*

$$x = X + KY, \quad y = Y, \quad z = \mu Z, \tag{4.4}$$

is kinematically possible in every Bell constrained material. The deformation path of a shear with contraction is diagrammed in Figure 5. It is seen for this case that the deformation path $\lambda_1 \lambda_2 = 1$, $\lambda_3 = \mu$ lies in the invariant plane. Note that a simple shear requires $\mu = 1$, possible only in a neighborhood of the undeformed state of a Bell material.

We recall that a simple shear is an isochoric deformation. It may be seen that *no isochoric deformations whatever are possible in any Bell constrained material, except approximately in the vicinity of the undeformed state.* Recall that an isochoric deformation requires that the principal stretches are situated on the hyperbolic surface defined by (3.5); but this surface shares only the centroid with the invariant plane (4.1). In consequence, as noted earlier, *there are no incompressible, Bell constrained materials.*

On the contrary, *superimposed, incremental isochoric deformations may be possible in a Bell material.* It is shown by Beatty and Hayes (1995) that the increment i_3^* in the third principal invariant $i_3 = \det \mathbf{V}$ is given by $i_3^* = i_3 \mathrm{tr}\,\mathbf{H}$, in which $\mathbf{H} = \partial \mathbf{u}/\partial \mathbf{x}$ is the usual infinitesimal superimposed displacement gradient relative to the underlying state κ_1 of finite strain. As usual, $\mathrm{tr}\,\mathbf{H}$ is the

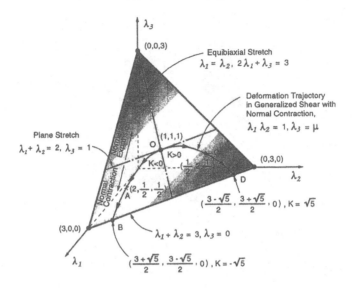

Fig. 5. Deformation trajectory of a general shear of amount K with normal stretch $\lambda_3 = \mu$ shown in the invariant plane. Only transverse contraction with $\mu < 1$ is possible.

incremental change in the material element of volume per unit element of volume in κ_1. In consequence, the incremental Bell constraint given by $\mathrm{tr}(\mathbf{HV}) = 0$ does not preclude an isochoric incremental deformation for which $\mathrm{tr}\,\mathbf{H} = 0$ relative to any configuration κ_1. In fact, when κ_1 is the undeformed state for which $\mathbf{V} = 1$, it is seen that an incremental isochoric deformation in the vicinity of the undeformed state is possible. This result is consistent with our previous observations on the nature of a simple shear; and it is consonant with Bell's experimental data presented in Figure 1. We'll see next that this property is also concordant with the familiar value of Poisson's ratio for infinitesimal isochoric deformations.

2.4.3 Poisson's kinematic function

A simple extension is an equibiaxial deformation for which $\lambda_1 = \lambda_2$ and $\lambda_3 = \lambda$, a homogeneous deformation kinematically possible in every Bell material. We recall that the Poisson kinematic function $\nu(\lambda)$ is defined as the ratio of the lateral contraction $-e_1 = -e_2$ to the longitudinal extension e_3 in a simple extension experiment (see Beatty and Stalnaker 1986). Thus,

$$\nu(\lambda) = -\frac{e_1}{e_3} = \frac{1 - \lambda_1(\lambda)}{\lambda - 1}. \tag{4.5}$$

For a Bell constrained material in simple extension, (3.2) gives $2e_1 = -e_3$, or $\lambda_1 = (3 - \lambda)/2$. *Therefore, every Bell constrained material in simple extension has the constant valued Poisson function,*

$$\nu(\lambda) \equiv \nu_o = \frac{1}{2}, \tag{4.6}$$

a value usually attributed to an incompressible material.

For an incompressible material, however, we recall the constraint (3.5). Hence, in a simple extension $\lambda_1 = \lambda_2 = 1/\sqrt{\lambda}$. Thus, by (4.5), *the Poisson function for an incompressible material is given by*

$$\nu(\lambda) = \frac{1}{\lambda + \sqrt{\lambda}}. \tag{4.7}$$

This is a monotone decreasing function of the extensional stretch λ; its value at $\lambda = 1$ is *the classical value of Poisson's kinematic function for an incompressible material in simple extension:*

$$\nu(1) \equiv \nu_o = \frac{1}{2}. \tag{4.8}$$

Notice that for simple extension of an incompressible material, ν has the value $\nu_o = 1/2$ only in the undeformed state, whereas for a Bell material $\nu = \nu_o = 1/2$ is constant for all values of λ. We thus see again that a Bell material appears to behave like an incompressible material in the neighborhood of its undeformed state, as demonstrated by test data in Figure 1, yet there are no incompressible Bell materials.

The foregoing properties of a Bell material are purely kinematical properties that derive from the Bell constraint. We next introduce the material response of a Bell material modeled as an isotropic, internally constrained nonlinear elastic solid.

2.4.4 Constitutive equation for general elastic Bell materials

Bell established that the material response from the "genesis" state κ_0 of his annealed polycrystalline solid is isotropic and subject to the constraint $\text{tr}\,\mathbf{V} = 3$ for all loading deformations from κ_0. Therefore, introducing a workless constraint reaction stress $\mathbf{N}$ corresponding to the Bell constraint, it turns out that

$$\mathbf{N} = \hat{p}(\mathbf{x})\mathbf{V}, \tag{4.9}$$

where $\hat{p}(\mathbf{x})$ is an arbitrary, undetermined, scalar constraint reaction parameter (see Beatty and Hayes 1992a). The total Cauchy stress is then given by

$$\boldsymbol{\sigma} = \mathbf{N} + \boldsymbol{\sigma}_{\mathbf{E}}, \tag{4.10}$$

in which the extra stress $\boldsymbol{\sigma}_{\mathbf{E}}$ is an isotropic tensor-valued function of $\mathbf{V}$ determined by

$$\boldsymbol{\sigma}_{\mathbf{E}} = \omega_0 \mathbf{1} + \omega_1 \mathbf{V} + \omega_2 \mathbf{V}^2. \tag{4.11}$$

The material response functions ω_Γ, $\Gamma = 0, 1, 2$, are isotropic scalar-valued functions of the indicated principal invariants of $\mathbf{V}$:

$$\omega_\Gamma = \omega_\Gamma(i_2, i_3). \tag{4.12}$$

With (3.1) in mind, we recall (see Ogden 2001)

$$i_2 = \frac{1}{2}\left(9 - \operatorname{tr} \mathbf{V}^2\right), \quad i_3 = \det \mathbf{V}. \tag{4.13}$$

Using (4.9) and (4.11) in (4.10) and introducing a modified constraint parameter p in an obvious way, we obtain *the general constitutive equation for an isotropic, nonlinearly elastic Bell constrained material*:

$$\boldsymbol{\sigma} = p\mathbf{V} + \omega_0 \mathbf{1} + \omega_2 \mathbf{V}^2. \tag{4.14}$$

Of course, other representations are possible (see Beatty and Hayes 1992a). Another form of (4.14) in terms of the Bell stress $\boldsymbol{\tau}$ is introduced later.

The equilibrium equation in the absence of body force requires $\operatorname{div} \boldsymbol{\sigma} = \mathbf{0}$. From now on, we'll ignore body forces. Then, for every kinematically admissible homogeneous deformation, the equilibrium equation is trivially satisfied for every isotropic Bell constrained material (4.14), if and only if the parameter $p(\mathbf{x}) = $ constant The constant p may be chosen to satisfy an assigned boundary condition, for example. We'll encounter this later on.

2.4.5 Restrictions on the response functions

The response functions, though general, are not entirely arbitrary. Besides being sufficiently smooth differentiable functions, the material response functions are required to satisfy the following inequalities, called the A-inequalities (see Beatty-Hayes 1992a), for all deformations relative to the undeformed state κ_0:

$$\omega_0 \leq 0, \quad \omega_2 > 0. \tag{4.15}$$

The (*ad hoc*) A-inequalities are modeled after the so-called empirical inequalities adopted for both unconstrained compressible and incompressible isotropic materials (see Truesdell and Noll 1965, Beatty 1996). No theoretical foundation for (4.15) is known; however, we'll see farther on that the A-inequalities share some empirical basis in Bell's experiments. These inequalities impose restrictions on the mechanical response to assure sensible material behavior of the theoretical model; and they are useful in assessing and checking on the physical interpretation of analytical results derived for every Bell constrained material. Two simple examples illustrate the idea.

Example 4.1. It is natural to expect that an elastic material will return to its undeformed state when the stress is removed. The A-inequalities assure this behavior for every Bell material.

To see this, let's suppose that $\sigma = 0$. This is a trivial equilibrium state. Then, in the principal basis, equation (4.14) provides

$$p\lambda_k + \omega_0 + \omega_2\lambda_k^2 = 0, \quad k = 1, 2, 3, \tag{4.16}$$

which yield

$$(\lambda_k - \lambda_j)[-\omega_0 + \lambda_k\lambda_j\omega_2] = 0, \quad j \neq k = 1, 2, 3. \tag{4.17}$$

Thus, if the A-inequalities (4.15) hold, (4.17) requires $\lambda_k = \lambda_j$; and the Bell constraint (3.2) shows that $\lambda_1 = \lambda_2 = \lambda_3 = 1$. Thus, the strain in an isotropic Bell material vanishes with the stress when the A-inequalities hold. $\square$

Example 4.2. It is also reasonable to expect that a real Bell material will elongate under tension, contract under compression. The A-inequalities guarantee this simple extension/compression response for every Bell material.

Consider a simple extension (4.2) with $\lambda_3 = \lambda$ under an uniaxial stress $\sigma_{33} = \sigma$ so that (4.14) yields

$$\sigma = p\lambda + \omega_0 + \lambda^2\omega_2, \tag{4.18}$$

$$0 = p\lambda_1 + \omega_0 + \lambda_1^2\omega_2. \tag{4.19}$$

Notice that (4.19) is the null lateral traction boundary condition for the homogeneous simple extension. The constant p may be chosen to satisfy this, as remarked earlier. Eliminating the parameter p from (4.18), we find

$$\sigma = \frac{\lambda - \lambda_1}{\lambda_1}[-\omega_0 + \lambda\lambda_1\omega_2]. \tag{4.20}$$

For a Bell constrained material, (4.2) gives $2\lambda_1 = 3 - \lambda$. Thus, the uniaxial stress in the simple extension is determined by

$$\sigma = \frac{3(\lambda - 1)}{3 - \lambda}[-\omega_0 + \frac{\lambda}{2}(3 - \lambda)\omega_2]. \tag{4.21}$$

Clearly, $0 < \lambda < 3$; therefore, in an isotropic Bell constrained material for which the A-inequalities hold, the stress is in the direction of the stretch. That is, when (4.15) hold and $1 < \lambda < 3$, then $\sigma > 0$ and tension accompanies the extension; but when $0 < \lambda < 1$, $\sigma < 0$ and compression accompanies the contraction. Thus, the A-inequalities support our intuitive perceptions of the response in a simple uniaxial stretch. $\square$

The ordered forces inequalities and other inequalities similar to the Baker-Ericksen inequalities are introduced in Beatty and Hayes (1992a) in terms of both the Cauchy and Bell stresses (see also Truesdell and Noll 1965). These further relations will not be needed here. There are many additional and nontrivial situations illustrated throughout the Beatty and Hayes articles (1992a,b;1995) in which these various inequalities are helpful in describing the physical response of Bell constrained materials. In consequence of such results, we may employ the various inequalities with confidence in their predictive value, especially in situations where conclusions drawn from their introduction may be less apparent.

2.4.6 Comparison of theoretical results with Bell's relations

It is shown next that our general constitutive equation (4.14) for nonlinearly elastic Bell constrained materials, recast in terms of the Bell stress, yields results for uniaxial loading and for pure shear that include relations of the kind demonstrated in corresponding experiments by Bell. To demonstrate this, we must choose the response functions appropriately. The extraordinary nature of these results will be demonstrated more naturally and exactly in our subsequent development of the theory of hyperelastic Bell materials in Section 2.5.

We first recast our constitutive equation (4.14) in terms of the Bell stress (3.7) to obtain

$$\tau = q\mathbf{1} + \Omega_1 \mathbf{V} + \Omega_{-1} \mathbf{V}^{-1}, \qquad (4.22)$$

in which $q(\mathbf{x}) \equiv i_3 p(\mathbf{x})$ is another undetermined parameter. The new response functions $\Omega_\Gamma(i_2, i_3)$ must satisfy the A-inequalities (4.15), so that

$$\Omega_1(i_2, i_3) \equiv i_3 \omega_2 > 0, \quad \Omega_{-1}(i_2, i_3) \equiv i_3 \omega_0 \leq 0 \qquad (4.23)$$

hold for all kinematically admissible deformations from the natural, undeformed state. In terms of principal components, (4.22) may be written as

$$\tau_k = q + \Omega_1 \lambda_k + \Omega_{-1} \lambda_k^{-1}, \qquad (4.24)$$

wherein it is helpful to recall $e_k = \lambda_k - 1$ for the principal engineering strains.

The equilibrium equation in terms of τ may be readily obtained from (3.7) (see Beatty and Hayes 1992a). In the absence of body force, this equation is trivially satisfied in every kinematically admissible homogeneous deformation of every isotropic elastic Bell material (4.22), if and only if the parameter $q(\mathbf{x}) = $ constant. We now turn to application of (4.22) in describing uniaxial loading of a rod and pure shear of a block.

2.4.6.1 Bell's parabolic law for uniaxial loading

Let's consider a rod subject to a simple extension by an uniaxial load $\tau_3 = \tau$ with corresponding principal engineering strain $e_3 = e = \lambda_3 - 1$ so that the Bell constraint (3.2) is satisfied with $e_1 = e_2 = -e$. The principal invariants (4.13) are functions of e alone; and hence $\Omega_\Gamma(i_2, i_3) = \tilde{\Omega}_\Gamma(e)$. Eliminating the constant q from the component equations, we obtain by (4.24) the uniaxial stress-strain relation

$$\tau = \frac{3}{2}e\Omega(e), \qquad (4.25)$$

an exact theoretical result in which the uniaxial response function is defined by

$$\Omega(e) \equiv \tilde{\Omega}_1(e) - \frac{2\tilde{\Omega}_{-1}(e)}{(1+e)(2-e)}. \qquad (4.26)$$

Notice that because $0 < \lambda_3 < 3$, we have $-1 < e < 2$.

Turning to experiments reported by Bell (1985), we find that *the result* (4.25) *for isotropic, nonlinearly elastic Bell materials includes as a special case Bell's parabolic law for uniaxial loading; namely,*

$$\tau = \frac{3}{2}\gamma(\text{sgn } e) \left|e\right|^{1/2}, \qquad (4.27)$$

in which γ is a positive material constant and for tension $e > 0$, compression $e < 0$. We obtain this result from our general relation (4.25) by choosing the uniaxial response function

$$\Omega(e) \equiv \gamma \left|e\right|^{-1/2}. \qquad (4.28)$$

We'll return to (4.28) in a different context in Section 2.5.

2.4.6.2 Bell's parabolic law for pure shear

Consider a rectangular block having cross sectional dimensions $2a$ and $2b$ in the XY-plane, and height $2c$ along its central Z-axis. A pure shear of the block, shown earlier in Figure 1, is characterized by

$$e_2 = \lambda_2 - 1 = 0, \quad \tau_3 = 0 \text{ on ends } Z = \pm c. \qquad (4.29)$$

This is a plane strain problem for which the Bell constraint reduces to (3.3) for $0 < \lambda_k < 2$. The principal Bell stresses provided by (4.24) are

$$\tau_1 = q + \lambda_1 \Omega_1 + \lambda_1^{-1} \Omega_{-1}, \tag{4.30}$$

$$\tau_2 = q + \Omega_1 + \Omega_{-1}, \tag{4.31}$$

$$\tau_3 = q + \lambda_3 \Omega_1 + \lambda_3^{-1} \Omega_{-1}, \tag{4.32}$$

in which $\Omega_\Gamma(i_2, i_3) = \Omega_\Gamma(3 - e_1^2, 1 - e_1^2) \equiv \hat{\Omega}_\Gamma(|e_1|)$ for $-1 < e_1 = \lambda_1 - 1 < 1$. The boundary condition (4.29) requires that $\tau_3 = 0$. Therefore, eliminating the constant q with the use of (4.32), we find the principal Bell stresses

$$\tau_1 = 2(1 - \lambda_3)(\hat{\Omega}_1 - \frac{1}{\lambda_1 \lambda_3} \hat{\Omega}_{-1}), \tag{4.33}$$

$$\tau_2 = (1 - \lambda_3)(\hat{\Omega}_1 - \frac{1}{\lambda_3} \hat{\Omega}_{-1}). \tag{4.34}$$

Recalling the A-inequalities (4.23), we thus note that

$$\frac{\tau_1 - \tau_2}{\lambda_1 - 1} = \frac{\tau_1 - \tau_2}{1 - \lambda_3} = \hat{\Omega}_1 - \frac{1}{\lambda_1} \hat{\Omega}_{-1} > 0. \tag{4.35}$$

If the plane ends $X = \pm a$ are under tensile loading $\tau_1 > 0$, which implies that $\lambda_1 > 1 > \lambda_3$, then (4.34) shows also that $\tau_2 > 0$. Hence, the faces $Y = \pm b$ are under tension too, and $\tau_1 > \tau_2 > 0$. Otherwise, $\lambda_1 < 1 < \lambda_3$ and both faces are in compression with $\tau_1 < \tau_2 < 0$. Note that (4.33) shows that $\tau_1 = 0$ on one face implies by (4.34) that $\tau_2 = 0$ on the other. In any case, $|\tau_1| > |\tau_2|$, and we obtain the ratio

$$\frac{\tau_2}{\tau_1} = \frac{1}{2} \left(\frac{\hat{\Omega}_1 - \frac{1}{\lambda_3} \hat{\Omega}_{-1}}{\hat{\Omega}_1 - \frac{1}{\lambda_1 \lambda_3} \hat{\Omega}_{-1}} \right) < 1. \tag{4.36}$$

Now let's recall Bell's modified Bridgman experiment on pure shear in which the stress ratio is controlled so that $\tau_2/\tau_1 = 1/2$. We see from (4.36) that this is possible for our isotropic, nonlinearly elastic Bell material model, if and only if $\hat{\Omega}_{-1}(|e_1|) = 0$ for all pure shear deformations. This is consistent with the A-inequality $\hat{\Omega}_{-1} \leq 0$ for all deformations from the undeformed state. Accepting this, we have from our theory in (4.33) and (4.34) for a pure shear deformation

$$\tau_1 = 2\tau_2 = 2e_1 \hat{\Omega}_1(|e_1|), \quad \hat{\Omega}_{-1}(|e_1|) = 0, \quad \forall e_1 \in (-1, 1). \tag{4.37}$$

We now recall Bell's (1985) empirical result that in a pure shear of the block

$$\tau_1 = 2^{3/4} \beta (\operatorname{sgn} e_1) |e_1|^{1/2} \equiv 2\alpha (\operatorname{sgn} e_1) |e_1|^{1/2}, \tag{4.38}$$

valid for both tension, $e_1 > 0$, or compression, $e_1 < 0$. Here we introduce the material constant $\alpha \equiv 2^{-1/4}\beta > 0$ for future convenience. Thus, *the result (4.37) for isotropic, nonlinearly elastic Bell materials includes as a special case Bell's parabolic law for pure shear*. Clearly, Bell's result (4.38) follows from our general relation (4.37) by our choosing the pure shear response function

$$\hat{\Omega}_1(|e_1|) = \alpha\,|e_1|^{-1/2}. \tag{4.39}$$

Although the pure shear result (4.39) appears the same as the uniaxial result (4.28), they actually differ in their response parameters.

It is seen that Bell's experimental data obtained in pure shear appear to support our A-inequalities. The result that $\Omega_{-1} = 0$ in every pure shear need not hold in all deformations of a Bell constrained material. The fact that it does vanish in this test provides support for study of the class of Bell materials for which $\Omega_{-1} = i_3\omega_0 \equiv 0$. The constitutive equation (4.14), or (4.22), for this special class of Bell materials has the respective reduced form

$$\boldsymbol{\sigma} = p\mathbf{V} + \omega_2\mathbf{V}^2, \text{ or } \boldsymbol{\tau} = q\mathbf{1} + \Omega_1\mathbf{V}. \tag{4.40}$$

We conclude our review of some theoretical results for general isotropic, nonlinearly elastic Bell constrained materials with discussion of Bell's notable experimental observation of the negligible rigid rotation of principal stress-stretch axes in the finite twist and extension of a thin-walled tube.

2.4.7 Finite twist and axial stretch of a thin-walled tube

For finite torsion-extension experiments in which thin-walled, cylindrical tubes of annealed copper, mild steel, and aluminum were severely twisted and extended in proportional and nonproportional loading†, Bell (1989, 1990b) reports the remarkable empirical result that the rigid rotation of the principal

---

† Bell mentioned in an earlier paper (1983a, p. 165) that for nonproportional loading along arbitrary stress paths in finite torsion-tension experiments, *very large* rigid rotations may occur, citing specifically an angle of 38° to 40° rigid rotation of principal axes for test 2211 by Bell and Khan (1980) describing an annealed copper tube extended and twisted through 297°. One finds a parallel remark in Bell (1985, p. 7). It is further noted in Bell (1983a) that even higher values had been obtained in some of the other tests for arbitrary loading paths described therein. In 1989, however, we see for the first time detailed calculations based on correct kinematics, suggested by Ericksen, demonstrating that rigid body rotation of principal stretch axes in the gross twist of thin-walled tubes is quite small compared with the total angle of twist. In the light of Bell's (1989, 1990b) subsequent work for both proportional and nonproportional loading, however, it appears to me that as the kinematics developed more clearly in Bell's mind, he realized a misinterpretation introduced in earlier calculations, calculations that, to my knowledge, were not used in any data analysis and were not published in any detail for examination. For test 2211, Bell (1989, 1990b) subsequently records in detail a considerably smaller rigid body rotation of 7.52° in response to a maximum angle of twist of 297°. Bell (1990b, p. 310) now recognizes the earlier anomaly; but I don't understand his brief explanation of this matter. Based

axes of $\mathbf{V}$ is negligible, so that to a close approximation the rotation $\mathbf{R} \simeq \mathbf{1}$. In one example, the angle of twist was 347° with an axial extensional strain of 24%; yet the rigid rotation was only 6.91°, minuscule in the context of the gross deformation of the tube, in fact less than 2% of the total twist. More recently, Bell (1995b) and Bell and Baesu (1996) examine a variety of experimental data that demonstrate in every instance that rigid rotation effects along arbitrary loading paths are negligible. The result that $\mathbf{R} \simeq \mathbf{1}$, very nearly, is a purely kinematical result essential to establishing the invariance of Bell's (1990b) incremental theory under superimposed rigid rotations. Here I review some analytical consequences of Bell's observations and describe their connection with his data.

The Beatty-Hayes (1992a) analysis of the finite twist and stretch of a thin-walled tube is too complex to exhibit in a few lines, so let's focus on some principal results. Accounting for the null traction conditions on the cylindrical surfaces of the tube, Beatty and Hayes derive a relation connecting the rotation ϕ with the angle of twist $\theta = L\psi$ and the axial stretch δ of a tube of initial length L, namely,

$$\phi = \frac{2r_m\psi}{3+\delta} \ll 1, \qquad (4.41)$$

where r_m is the mean radius of the deformed tube and ψ is the angle of twist per unit length. They thus deduce Bell's remarkable experimental result that in finite torsion and extension of thin-walled cylindrical tubes the rigid rotation of the coaxial principal axes of $\mathbf{V}$, $\boldsymbol{\sigma}$, and $\boldsymbol{\tau}$ must be very small, and hence to a close approximation $\mathbf{R} \simeq \mathbf{1}$. Their result for the thin-walled tube approximation, taking into account the experimental circumstances described by Bell (1990b), exhibits excellent agreement with averaged values obtained from 85 tests based on measurements of annealed aluminum, copper, and mild steel tubes at their ultimate state of deformation prior to failure under different proportional and nonproportional loading conditions.

It is also shown in Beatty and Hayes (1992a) and in Bell (1996a) that the volume of a Bell constrained material must always decrease with increasing deformation. In the case of the thin-walled tubes in "simple" torsion, i.e. with no axial loading, Bell (1989, 1990b) shows in all experiments that the inner and outer tube diameters remain unchanged, regardless of the magnitude of the angle of twist. Accordingly, we thus conclude that the volume can decrease only if the tube shortens under the applied torque. This decreasing volume

on well-defined calculations for data in Bell (1990b), however; I'm inclined to ignore his earlier, intuitive misperception. The test 2211 having the greatest angle of twist with large extension is discussed in further detail in (1995b) as regards the negligible character of rigid rotations in this test.

effect during loading is observed in all of Bell's (1990b) simple torsion tests for tubes of annealed copper and mild steel. We recall, however, Bell's upper bound on application of the constraint (3.1) fixed by an *averaged* experimental value of the deviatoric strain intensity $\Gamma = \sqrt{\operatorname{tr} e^2} \simeq 0.33$; therefore, according to Bell, subsequent deformations of thin-walled tubes under this new condition may increase or decrease volume, or permit isochoric deformation comparable to simple shear. In any event, the change of volume in Bell's (1995b) tests is usually recovered during unloading, at least very nearly. Obviously, the Beatty and Hayes analysis does not admit volume changes due to creep and it does not include possible volume changes due to porosity or other effects, it admits only decreasing volume with increasing deformation concordant with (3.1). It is emphasized that all of the foregoing results are purely kinematical results independent of the constitutive properties of the material.

Bell also found a certain universal relation connecting the engineering stress with the angle of rotation. In addition to this rule, Beatty and Hayes derive other universal relations consistent with the approximations and independent of the particular material response functions, including the following *universal rule for the finite twist and stretch of a thin-walled tube of undeformed mean radius* R_m, *with null lateral tractions and subjected to an applied torque* M:

$$\frac{R_m^2 N}{\tau(\delta)} = \frac{3(\delta - 1)(\delta + 3)}{\delta^2 (3 - \delta)^2}, \tag{4.42}$$

a rule relating the torsional modulus $\tau(\delta) \equiv M/(\psi/\delta)$ *to the axial force* N *and the stretch* δ *in the same way for every isotropic, elastic Bell constrained material.* The right and left-hand terms of this universal relation may be calculated independently for each of Bell's 85 tests. Beatty and Hayes (1992a) report that the averaged data for the two sides of (4.42) differ by only 2.2%, well within experimental variation.

Beatty and Hayes also derive the universal relation

$$\frac{\pi \langle \theta Z \rangle}{\pi \langle z Z \rangle} = \frac{\sigma_{\theta z}}{\sigma_{zz}} = \frac{\delta r_m \psi}{\delta^2 - \alpha^2} = \frac{\delta \tan \phi}{\delta - \alpha} \tag{4.43}$$

connecting the ratios of physical engineering stress components $\pi \langle jK \rangle$, physical Cauchy stress components σ_{jk}, the radial stretch α, and other kinematical quantities noted earlier. Of course, everything is not perfect. Beatty and Hayes found that consistency of (4.43) with Bell's data is poor. While the kinematical relations in (4.43) agree very closely with Bell's data in all of the tests, the comparison with the stress ratios often is poor and sometimes varies considerably from one test to another.

The gross variance reported in Beatty and Hayes (1992a) prompted Bell to

Millard F. Beatty

look again at his original data. He reasoned that during loading nonlinear elasticity and nonlinear plasticity are indistinguishable only for proportional loading; and on this basis he concluded that the primary difference between finite elasticity and plasticity is that experiment reveals the latter to be very path dependent, a distinction that is unimportant in describing the kinematics of a deformed state. Therefore, for plasticity, according to Bell, proportional loading tests are the only tests to which the *stress ratios* in (4.43) can apply. He then demonstrated† that among the 85 tests averaged in Beatty and Hayes (1992a), in all 26 proportional loading cases for annealed copper, iron, and aluminum, his experimental data are in close agreement with the universal rule (4.43), while those for nonproportional loading paths are not. Some further extraordinary analytical results connected with Bell's experiments are presented below.

2.5 Isotropic, hyperelastic Bell materials

An isotropic, hyperelastic Bell material is characterized by an isotropic, scalar-valued function $W(i_2, i_3)$, the strain energy per unit volume in κ_0. The mechanical energy principle requires

$$\dot{W} = J \mathrm{tr}(\sigma \mathbf{D}) \tag{5.1}$$

(see Beatty 1996, Ogden 2001). Here $\mathbf{D} = \mathrm{Sym}(\partial \dot{\mathbf{x}}(\mathbf{x}, t)/\partial \mathbf{x})$ is the stretching tensor and a superimposed dot denotes the material time derivative.

The energy function determines σ only to within an arbitrary workless stress. We recall the workless constraint function $\gamma(\mathbf{V}) \equiv i_1 - 3 = 0$ and expand the energy rate $\dot{W}(i_2, i_3)$ to obtain from (5.1)

$$[\sigma - \frac{\partial W}{\partial i_3}\mathbf{1} + \frac{1}{i_3}\frac{\partial W}{\partial i_2}\mathbf{V}^2] \cdot \mathbf{D} = 0 \quad \forall \mathbf{D} = \mathbf{D}^T \tag{5.2}$$

† In all comparisons with Bell's data in Beatty and Hayes (1992a), the only place where stress ratios arise is in (4.43), a fact noted in a letter of 8 July 1992 in which Bell first mentioned the improved correlation of his data with (4.43) for proportional loading cases. In a subsequent letter of 23 August 1993 to Hayes and me, copied to Ericksen, Bell noted that 28 of the 85 tests were listed as proportional loading cases. Rechecking his original data, however, Bell found that test 2143 for copper was actually a nonproportional test, and that the copper test 1987 had a laboratory notation of a problem with the torsion loading system. For the remaining 26 proportional loading tests, Bell notes the average engineering stress ratio $\pi \langle \theta Z \rangle / \pi \langle z Z \rangle$ is 0.605, while the average kinematical ratio $(\delta \tan \phi)/(\delta - \alpha)$ is 0.597, a difference of only 1.3%. For ten aluminum tests, Bell reports the respective ratios as 0.457 and 0.463, also a difference of only 1.3%. He reports for eleven copper tests the respective ratios 0.870 and 0.861, a difference of 1.0%, while for five iron specimens with average ratios 0.319 and 0.325, the difference is 1.8%. In all, Bell's experimental results for proportional loading stand in strong agreement with the universal rule (4.43).

subject to the condition

$$\dot{\gamma}(\mathbf{V}) = \operatorname{tr}\dot{\mathbf{V}} = \mathbf{V} \cdot \mathbf{D} = 0. \tag{5.3}$$

It thus follows that the constitutive equation for an isotropic, hyperelastic Bell material is given by the rule

$$\boldsymbol{\sigma} = p\mathbf{V} + \omega_0\mathbf{1} + \omega_2\mathbf{V}^2, \tag{5.4}$$

which is the same as (4.14), or alternatively by our previous rule (4.22),

$$\boldsymbol{\tau} = q\mathbf{1} + \Omega_1\mathbf{V} + \Omega_{-1}\mathbf{V}^{-1}. \tag{5.5}$$

The difference, however, is that here the response functions are determined explicitly in terms of the strain energy function. Recalling the A-inequalities (4.15), or equivalently (4.23), we have

$$\Omega_1 = i_3\omega_2 \equiv -\frac{\partial W}{\partial i_2} > 0, \quad \Omega_{-1} = i_3\omega_0 \equiv i_3\frac{\partial W}{\partial i_3} \leq 0. \tag{5.6}$$

The response functions also must satisfy the following compatibility relations

$$\frac{\partial \omega_0}{\partial i_2} + \frac{\partial}{\partial i_3}(i_3\omega_2) = 0, \quad \frac{\partial \Omega_{-1}}{\partial i_2} + i_3\frac{\partial \Omega_1}{\partial i_3} = 0. \tag{5.7}$$

2.5.1 Bell's law

Let's consider the special class of hyperelastic Bell materials described by the constitutive equation (4.40) for which $\Omega_{-1} = i_3\omega_0 = 0$. Then (5.6) shows that $W = W(i_2)$, and hence (5.7) yields

$$\Omega_1(i_2) = i_3\omega_2(i_2, i_3) = -\frac{dW(i_2)}{di_2}. \tag{5.8}$$

Bell (1985) reports that all of his trials are consistent with the specific work function defined by

$$\hat{W}(I_2(\mathbf{e})) = \frac{2}{3}\beta(-2I_2(\mathbf{e}))^{3/4} = \frac{2}{3}\beta[2(3 - i_2)]^{3/4} = W(i_2), \tag{5.9}$$

where β is a material constant; and, in view of (3.1),

$$-I_2(\mathbf{e}) = \frac{1}{2}\operatorname{tr}\mathbf{e}^2 = 3 - i_2 > 0. \tag{5.10}$$

Using (5.9) in (5.8), the response function for this class of hyperelastic Bell materials, all primary annealed metals in Bell's tests, is given explicitly by

$$\Omega_1 = \beta(-2I_2(\mathbf{e}))^{-1/4} = \beta[2(3 - i_2)]^{-1/4} = i_3\omega_2. \tag{5.11}$$

We thus derive a special constitutive law for hyperelastic Bell constrained materials in the class (4.40):

Bell's Law

$$\tau = q\mathbf{1} + \beta \mathbf{V}[2(3 - i_2)]^{-1/4} = \hat{q}\mathbf{1} + \beta \mathbf{e}(-2I_2(\mathbf{e}))^{-1/4}. \qquad (5.12)$$

Beatty and Hayes (1992a) so named this result in consequence of its development from Bell's work function.

Clearly, the first of (5.12) exhibits an evident singularity at $\mathbf{V} = 1$ where $i_2 = 3$; but this can be removed by introducing $q = \hat{q} - i_2\omega_2$ to obtain the second of (5.12), provided $\hat{q}$ remains well-behaved. We now turn to some applications of Bell's law for hyperelastic materials.

2.5.2 Applications of Bell's law

We'll now revisit the uniaxial loading and pure shear problems discussed earlier for general nonlinearly elastic Bell materials. With use of Bell's law, however, our results will be rendered more explicit and precise. Afterwards, we'll investigate a remarkable relation connecting the deviatoric stress intensity to the deviatoric strain intensity and then express the strain energy function in terms of these quantities.

2.5.2.1 Uniaxial loading of a rod

First, let's recall for a simple extension that $e_3 = e$, $e_1 = e_2 = -\frac{1}{2}e$, where we've introduced $(3.2)_2$. Then, by (5.10),

$$I_2(\mathbf{e}) = -\frac{3}{4}e^2. \qquad (5.13)$$

We thus obtain by (5.11) the response function defined earlier in (4.28), explicitly,

$$\Omega_1 = \Omega(e) = \gamma |e|^{-1/2} \text{ with } \gamma \equiv \beta \left(\frac{3}{2}\right)^{-1/4}; \qquad (5.14)$$

and we find that the uniaxial Bell stress obtained from (5.12) is given by (4.25):

$$\tau = \frac{3}{2}\gamma(\text{sgn } e) |e|^{1/2}. \qquad (5.15)$$

This is precisely Bell's parabolic law (4.27) for uniaxial loading and includes in (5.14) exactly the material parameter γ described in Bell's experiments.

2.5.2.2 Pure shear of a block

For the pure shear of a rectangular block, let's recall $e_2 = 0$, and $e_1 = -e_3$, which satisfies the Bell constraint $(3.2)_2$. Hence, (5.10) becomes

$$I_2(\mathbf{e}) = -e_1^2; \qquad (5.16)$$

and (5.11) yields the response function defined earlier in (4.39), explicitly

$$\Omega_1 = \hat{\Omega}(|e_1|) = \alpha |e_1|^{-1/2} \text{ with } \alpha \equiv 2^{-1/4}\beta, \tag{5.17}$$

which differs from the response parameter in (5.14). The normal Bell stress delivered by (5.12) is given by (4.37):

$$\tau_1 = 2\alpha(\operatorname{sgn} e_1) |e_1|^{1/2}. \tag{5.18}$$

We thus deduce exactly Bell's parabolic law (4.38) for pure shear including precisely in (5.17) the material parameter α characterized in Bell's work.

2.5.3 Bell's invariant parabolic law

In view of the Bell constraint (3.1), we observe that the strain tensor e is a deviatoric strain tensor. So, let's introduce Bell's deviatoric stress s defined by

$$s \equiv \tau - \frac{1}{3}(\operatorname{tr} \tau)\mathbf{1} \tag{5.19}$$

for which $\operatorname{tr} s = 0$. Then with the aid of Bell's law in $(5.12)_2$ and use of the Bell constraint to obtain $\operatorname{tr} \tau = 3\hat{q}$, we see by (5.12) and (5.19) that Bell's deviatoric stress,

$$s = \tau - \hat{q}\mathbf{1} = \beta e(-2I_2(e))^{-1/4}, \tag{5.20}$$

is the extra stress in Bell's law.

We next define the invariants T and Γ, the respective deviatoric stress and strain intensities,

$$T \equiv \sqrt{-2I_2(s)} = \sqrt{\operatorname{tr} s^2}, \tag{5.21}$$

$$\Gamma \equiv \sqrt{-2I_2(e)} = \sqrt{\operatorname{tr} e^2}. \tag{5.22}$$

Use of $(5.22)_1$ in (5.20) yields $s = \beta\Gamma^{-1/2}e$, and hence

$$s^2 = \beta^2\Gamma^{-1}e^2. \tag{5.23}$$

Finally, with the aid of $(5.21)_2$ and $(5.22)_2$, we derive from (5.23), in exact terms, *Bell's invariant parabolic law relating the deviatoric stress intensity to the corresponding strain intensity.*

Bell's Invariant Parabolic Law

$$T^2 = \beta^2\Gamma. \tag{5.24}$$

Now let's return to Bell's work function in (5.9) and introduce $(5.22)_1$ to obtain $\hat{W} = \frac{2}{3}\beta\Gamma^{3/2}$. Then use of (5.24) yields *Bell's invariant fundamental rule for the specific work function:*

$$\hat{W} = \frac{2}{3}T\Gamma. \tag{5.25}$$

2.5.4 Return to the laboratory

In accordance with (5.24), the graph of Bell's invariant constitutive equation for T^2 versus Γ is a straight line with slope β^2 characteristic of the particular Bell material. Now let's return to Bell's laboratory results. Graphs of a variety of experimental data by Bell in proportional dead loading tests for which various stress ratios τ_2/τ_1 of the principal Bell stresses are controlled are shown in Figure 6 for annealed copper. The parabolicity described by (5.24) is evident; and Bell reports the constant slope $\beta = 37\,\mathrm{kg/mm^2}$ for this material. Note that all graphs for the different tests demonstrate the same slope. Similar results are obtained by Bell and Khan (1980) and Bell (1983b, 1990b) in the finite tension and torsion of thin-walled, annealed tubes of mild steel, copper, and aluminum subjected to nonproportional dead loading along arbitrarily assigned loading paths, and precisely the same response consistent with (5.24) is demonstrated within the context of Bell's incremental theory of plasticity for which $d\Gamma = 2TdT/\beta^2$.

I'm certain that every one of Bell's former students and research associates will surely recognize this rule - it is the foundation upon which all of Bell's own analysis of his experiments, all consistent with the internal constraint tr $\mathbf{e} = 0$, is developed in the sense of incremental plasticity theory. It is surely most surprising that the theory by Beatty and Hayes (1992a,b, 1995), developed to characterize the nonlinear elastic deformation of an isotropic, hyperelastic Bell material, yields analytical results that agree precisely with the great body of experimental results described by Bell within the distinct framework of an incremental theory of plasticity in which the constraint tr $\mathbf{e} = 0$ holds for all loading conditions relative to the genesis state of annealed polycrystalline solids.

Finally, it is most important to recognize the major contributions of Bell and Ericksen to the foundations of the theory developed by Beatty and Hayes. The basis for the general elastic and hyperelastic Beatty-Hayes models was actually laid long ago in the incremental theory developed empirically by Bell (1987) as sketched in his Minnesota lecture in June 1985, a memorable presentation that motivated the Beatty-Hayes investigation of the general theory for elastic bodies. Moreover, as emphasized many times by Bell and sketched in Bell

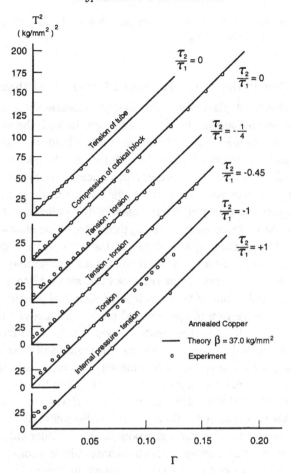

Fig. 6. Plots of T^2 versus Γ for a variety of Bell's test data in which various principal Bell stress ratios τ_2/τ_1 are controlled for specimens of annealed copper. All are straight lines having the same slope $\beta = 37$ kg/mm^2 for the material constant. This figure is adapted from Bell (1985).

(1983a), the hyperelastic theory for Bell's law was first outlined in private communication by Ericksen in terms of the Bell stress and the special empirical strain energy function w defined by† $w = k\left(\operatorname{tr} e^2\right)^{3/4} = k\Gamma^{3/2}$, where k is a material constant. With $k = \frac{2}{3}\beta$ and w replaced by $\hat{W}$, we see that this is our energy equation introduced in (5.9) for Bell's law, and ultimately written

† This expression corrects a minor typographical error in equation (40) in Bell (1983a).

as (5.25). Thus, the profound contributions due to both Bell and Ericksen pervade all matters presented here.

2.6 Concluding remarks: Bell's incremental theory

This is not a theory of plasticity. The theory presented here neglects concepts important to the foundation of such a theory. Indeed, the example 4.1 in Section 2.4.5 shows clearly that if the A-inequalities hold, an elastic Bell constrained material will always return to its natural, undeformed state when the stress is removed. Hence, under these circumstances, permanent set is impossible. Moreover, we have not introduced a yield stress or any sort of criterion that specifies when the material "yields plastically" under a critical value of shear stress, or shear-strain energy, or anything else. The elastic and plastic components of strain are not separated, in fact the latter are not identified; and other characteristics of a theory of finite strain plasticity that may include work hardening, strain rate dependence or the lack of it, strain history dependence, and so on, are ignored. None of these typical attributes of plasticity theory are introduced in the description of the material model and analysis here.

On the other hand, because the Beatty-Hayes theory is structured on two principal features of Bell's experiments - his constraint and isotropic material response from a genesis state - there are some aspects common to Bell's investigations of plasticity theory that have found their way into the structure studied here. We may recall in the classical theory of plasticity, for example, that the yielding of metals is considered little affected by all-around hydrostatic stress, so it is common to remove this stress component and consider only a deviatoric stress-strain relation. This concept arises here naturally as a consequence of the Bell constraint, which from the experiments shown in Figure 1, appears to be a plausible finite strain substitute for the incompressibility condition adopted within the structure of the classical theory, wherein plastic deformation is considered isochoric for all deformations and only elastic volume changes generally are considered. The Bell constraint is consistent with the incompressibility hypothesis for sufficiently small strains, yet, excepting these, it disallows all isochoric strains during the loading process and says nothing about the unloading process, plainly unimportant in elasticity theory. Of course, as Bell (1987) emphasizes, it is generally recognized that the constitutive equations for loading and unloading can be radically different at large strains, a primary difference being the characteristic permanent set deformation; and so under loading conditions that ignore yield surfaces, strain hardening, and other common aspects of a theory of plasticity, it is conceivable that a theoretical model of nonlinear elastic loading response may shed light on our understanding of the complex behavior of ordered solids in substantial states of strain. Beatty and Hayes provide a model that appears to capture several of the principal features of response demonstrated in Bell's experiments† and reported in his data. Moreover, the Bell constraint is prescribed only during loading, but not for arbitrarily large strains. In fact, Bell (1979, 1996a) reports a bound on its application of about‡ 33% in the strain intensity Γ. At larger strain intensity, other effects creep into the complex picture of the response of ordered solids; and the Beatty and Hayes model does not consider these effects.

One is inclined naturally to question (5.15) describing uniaxial loading of a material having an infinite slope at the undeformed state, hence an infinite Young's modulus. In the classical history of plasticity theory, however, Hill (1950, p.38) reminds us that the Lévy-Mises equations, for example, were

---

† In all, Bell (1983a) compiled data from more than 3000 stress controlled tests comprised of 30 kinds of experiments on 50 different ordered solids, all for initial loading from the undeformed genesis state of the fully annealed polycrystalline solid to states of substantial strains for which the incremental stress intensity $dT \geq 0$. It is noted that the undeformed genesis configuration for the single crystal components of the annealed polycrystalline solid coincides with the undeformed reference configuration of the free single crystal. Loading, unloading, reloading, and cyclic loading effects along stress paths of arbitrary proportional *and* nonproportional loading are discussed in Bell (1983a); and the occurrence of two yield surfaces is described.

All yield surfaces are of the Von Mises type (see Hill 1950) for which yielding occurs when the stress intensity T reaches a critical value. The elastic limit characterizes the first, inner yield surface at which the critical stress intensity is $T_Y = \sqrt{\text{tr } s_Y^2}$, say; and in accordance with Bell's (1983a) incremental theory, $d\Gamma = 2T dT/\beta^2$, and therefore (5.24), holds only for $T \geq T_Y$. The second, outer yield surface for annealed metals is determined by the subsequent maximum stress T_m to which the material is subjected. When the specimen is unloaded and reloaded, the material remembers the maximum previous stress T_{m0} to which it was subjected. Independent of the details of the intervening stress path history, when upon reloading T reaches the critical stress intensity T_{m0}, the new Von Mises yield surface, the deformation returns to and continues in the totally plastic region as though no unloading had occurred at all; and, according to Bell (1980, 1983a), the parabolicity described by (5.24) continues anew at this outer yield surface of the maximum previous stress. The same process is repeated upon unloading and reloading again to a different maximum previous stress T_{m1} to form a new outer yield surface, a process that may be repeated over and over to ultimate failure as described earlier. The description of unloading effects in the fully plastic domain and existence of the current outer yield surface of maximum previous strain intensity is provided by Moon (1973) for tension and torsion of thin-walled tubes of annealed polycrystalline aluminum subjected to proportional and nonproportional loading. Clearly, there is much here to sort out by analysis.

It is interesting that the property of the material's memory of its maximum previously strained state is also typical of elastomers for which the phenomenon is commonly known as the Mullins effect (see Johnson and Beatty 1993, Beatty 1999, Beatty and Krishnaswamy 2000). It is also observed that some elastomers when subjected to finite strain suffer permanent deformation upon release of loads. I have a sample of a short cylinder of carbon filled rubber that more than 30 years ago was severely compressed and still retains its permanently distorted, slightly barreled cylindrical surfaces and slightly concave end faces induced by platen friction.

‡ Bell reports this value variably between 28-38%. The average of these values, $\Gamma = 33\%$, is used throughout this presentation.

strictly applicable only to a material model for which Young's modulus must be regarded as infinitely large and for which the natural state is essentially rigid. Thus, in Hill's words, the Lévy-Mises equations are strictly applicable only to a *fictitious* material in which the elastic strains are zero. Now, according to Bell, the real origin for the parabolicity in (5.24), (5.25), and hence also in the result (5.15) is the "elastic limit", the inner yield surface. In fact, in Bell's incremental theory described by $2TdT = \beta^2 d\Gamma$, each of the aforementioned relations holds only for $T \geq T_Y$, the stress intensity at the inner yield surface. Moreover, we recall that Bell's (1979, 1980) continuum theory describes only loading effects into the totally plastic region bounded by the outer yield surface, thus avoiding the familiar complications of unloading and reloading (see Moon 1973). For a stress intensity increment $dT \geq 0$, and for loading paths of arbitrary composition and direction, in a great variety of tests on fifty ordered solids, high strength metal alloys, and pure metals, Bell reports the stress-strain ratios $de = sd\lambda$ for the incremental deviatoric strain tensor de and the deviatoric stress tensor s, thus recognizing the classical St. Venant hypothesis of plasticity theory that the principal axes of the strain increments de and the stress s coincide. Of course, here the tensors e and s are not those common to the classical formulation, and $d\lambda$ is an incremental parameter of proportionality. Bell (1985, 1987) informs us that the ratios $de_{11}/s_{11} = de_{12}/s_{12} = \cdots = d\lambda$ include in approximation the familiar Lévy-Mises relations for infinitesimal plastic strains. These empirical ratios are the incremental ingredients that led Bell to his invariant parabolic law (5.24) and corresponding work function (5.25).

In sum, the following empirically based constitutive equations† for quasi-static loading of fully annealed metals from their genesis state of the single crystal, with $dT \geq 0$, form the foundation of Bell's (1979, 1980, 1990b) incremental theory of plasticity for the fully plastic domain on which $T \geq T_Y$, the elastic limit yield stress intensity:

$$d\Gamma = 2\beta^{-2}TdT, \text{ for } T \geq T_Y, \tag{6.1}$$

$$de = 2\beta^{-2}sdT \text{ for } T \geq T_Y, \tag{6.2}$$

$$\text{tr } e = 0 \text{ only during loading } dT \geq 0, \tag{6.3}$$

$$\dot{w} \equiv \text{tr}(s\dot{e}) = T\dot{\Gamma}, \tag{6.4}$$

† The constitutive equations for high strength metal alloys are modified versions of the incremental equations (6.1) and (6.2) that account for a small intermediate region of plastic deformation beyond the inner yield surface and through which the material must pass in reaching the totally plastic region at an outer yield surface. According to Bell, this small intermediate region does not exist for fully annealed metals. In any case, experiments for both proportional and non-proportional loading paths show that the response in all regions is parabolic. See Bell (1979, 1980) for analytical details and supporting experimental data.

in which s is given by (5.19). We recall that Bell's incremental theory applies only in the loading domain of finite homogeneous plastic strains for which $\Gamma \leq 1/3$, approximately, an upper bound on the application of (6.3) that depends on the loading path and characterizes the mode of ultimate collapse of the specimen (see Bell 1995b). Moreover, it is shown in Bell (1990b) that these incremental constitutive equations are invariant under superimposed rigid body rotations and hence satisfy the principle of material objectivity subject to the condition $\mathbf{R} = \mathbf{1}$. This requirement, to a very good approximation, is demonstrated in data from 85 tests on torsion and tension of thin-walled tubes for which no unloading occurred along arbitrary paths of both proportional and nonproportional loading to states of maximum plastic deformation (see Bell 1995b, Bell and Baesu 1996). This is a strictly kinematical result independent of any sort of constitutive equation.‡ With the approximate condition $\mathbf{R} = \mathbf{1}$ appended to the incremental system (6.1)-(6.3), we have $\tau = \pi^T$, very nearly, from (3.7). Notice also that with $w = \hat{W}$, Bell's energy principle (6.4) follows from (5.24) and (5.25).

We observe also that by putting (6.1) and (6.2) together yields Bell's cartesian component form of the classical Lévy-Mises rule for finite strain plasticity:

$$\frac{de_{kl}}{s_{kl}} = 2\beta^{-2}dT = \frac{d\Gamma}{T} \equiv d\lambda, \; k,l = 1,2,3, \text{ no sum.} \qquad (6.5)$$

In the case of proportional loading the ratios of stress components are constants throughout the loading process; and in this instance (6.2) can be readily integrated to yield the constitutive equation $\mathbf{e} = \mathbf{e}_b + \beta^{-2}T\mathbf{s}$, where $\mathbf{e}_b$ is a constant tensor of integration.

These are the analytical tools discovered in the laboratory and believed by Bell to best describe the spectacular array of experimental data he assembled and interpreted to form a consistent incremental theory of plasticity, not everywhere, rather in the plastic regime of his principal interest, the gross plastic domain where the stress-strain curve changes to a parabolic form. For most annealed metals, this domain begins at the inner yield surface where Bell's parabola is joined to the essentially vertical line of the elastic stress-strain

‡ A similar situation is established in Bell (1990a) for the large deflection and plastic strain in an aluminum horizontal, cantilevered beam subjected to a vertical end load. Of course, the very large end displacement is accompanied by a maximum measured rigid body rotation (40° in Bell's test); however, the role of the rigid body rotation of principal axes is shown to be negligible in the high stress, large plastic strain region $\mathfrak{R}$ about 1/4 the length of the beam from the fixed end, a kinematical result that is independent of any sort of constitutive equation. In consequence, $\mathbf{R} = \mathbf{1}$, very nearly, and hence Bell's incremental theory may be used to characterize the plastic deformation of the beam in $\mathfrak{R}$. In fact, the data reveal a parabolic stress-strain response. Bell thus solves this problem to determine the stress and strain distribution in $\mathfrak{R}$. He finds that plane sections do not remain plane; rather, initially plane sections become parabolic in this theory.

curve for the comparatively small regime of infinitesimal elastic strains. So, perhaps one should not be too troubled that our model gives in (5.15) an infinite Young's modulus.

A great measure of the history of experimental mechanics and plasticity studies in particular, with countless references to original works, including descriptions of data and experimental apparatus, may be found in Bell's (1973) phenomenal encyclopedia article. Here one may find Bell's summary on works by himself and his associates up to 1973. In this article, I have focussed on discussion of papers by Bell and his associates published since 1979 and most directly related to the general study by Hayes and me on nonlinear elastic and hyperelastic Bell constrained materials. I'm not surprised that nonlinear elasticity theory can describe the gross deformation response of bodies under arbitrary kinds of loading. But I personally find it quite amazing that Hayes and I were able to deduce from an altogether different theoretical structure of constrained, isotropic nonlinear elasticity theory the many empirical results reviewed in Bell (1995b) and Bell and Baesu (1996), represented in equations derived by Bell on the basis on his incremental theory of finite strain plasticity. Unquestionably, there remain lots of interesting analytical puzzles to unravel and new applications to explore through Bell's studies, including some described by Bell (1996a, b) and left for others to pursue. If this review may excite the interest of researchers, especially unbiased young scholars like some I met at Berkeley, all far more knowledgeable of theories of finite strain plasticity than I, to search deeper into Bell's work with forbearance and the mindful purpose of fresh discovery, to formulate new theories, and to advance our knowledge of the plastic deformation of solids of all kinds, then this effort will have been rewarded beyond my expectations.

Acknowledgements

Acknowledgment: This work was partially funded by a Grant No. CMS-9634817 from the National Science Foundation. I thank Professors Jerald L. Ericksen and Michael A. Hayes for reading the original manuscript and providing helpful comments.

References

Beatty, M. F. 1965 Some static and dynamic implications of the general theory of elastic stability. *Arch. Rational Mech. Anal.* **19**, 167-188.

Beatty, M. F. 1996 Introduction to nonlinear elasticity. *Nonlinear Effects in Fluids and Solids* (eds. M. M. Carroll & M. A. Hayes) pp. 13-112. New York: Plenum Press.

Beatty, M. F. 1999 The Mullins effect in incompressible elastomers. In *Applied Mechanics in the Americas* Vol. 7, 1095-1098. (eds. P. B. Gonçalves, I. W. Jasiuk, D. Pamplona, C. R. Steele, H. I. Weber, and L. Bevilacqua) Proc. Sixth Pan-Am. Cong. Appl. Mech., Rio De Janeiro, Brazil 4-8 January 1999.

Beatty, M. F., and Hayes, M. A. 1992a Deformations of an elastic, internally constrained material. Part 1: Homogeneous deformations. *J. Elasticity* **29**, 1-84.

Beatty, M. F., and Hayes, M. A. 1992b Deformations of an elastic, internally constrained material. Part 2: Nonhomogeneous deformations. *Quart. J. Mech. Appl. Math.* **45**, 663-709.

Beatty, M. F., and Hayes, M. A. 1993 On Bell's constraint in finite elasticity. *Advances in Modern Continuum Mechanics.* International Conference in Memory of Antonio Signorini, Elba Island, July 6-11, 1991 (ed. G. Ferrarese) pp. 183-190. Bologna: Pitagora Editrice.

Beatty, M. F., and Hayes, M. A. 1995 Deformations of an elastic, internally constrained material. Part 3: Small superimposed deformations and waves. *Z. angew. Math. Phys.* **46**, S72-S106.

Beatty, M. F., and Krishnaswamy, S., 2000 A theory of stress-softening in incompressible isotropic materials. *J. Mech. Physics Solids* **48**, 1931-1965.

Beatty, M. F., and Stalnaker, D. O. 1986 The Poisson function of finite elasticity. *ASME J. Appl. Mech.* **53**, 807-813.

Bell, J. F. 1973 *The Experimental Foundation of Solid Mechanics.* Flügge's Handbuch der Physik, Vol. **VIa/1**. New York: Springer.

Bell, J. F. 1979 A physical basis for continuum theories of finite strain plasticity: Part I. *Arch. Rational Mech. Anal.* **70**, 319-338.

Bell, J. F. 1980 A physical basis for continuum theories of finite strain plasticity: Part II. *Arch. Rational Mech. Anal.* **76**, 1-24.

Bell, J. F., and Khan, A. S. 1980 Finite plastic strain in annealed copper during non-proportional loading. *Int. J. Solids Structures* **16**, 683-693.

Bell, J. F. 1983a Continuum plasticity at finite strain for stress paths of arbitrary composition and direction. *Arch. Rational Mech. Anal.* **84**, 139-170.

Bell, J. F. 1983b Finite plastic strain in annealed mild steel during proportional and non-proportional loading. *Int. J. Solids Structures* **19**, 857-872.

Bell, J. F. 1985 Contemporary perspectives in finite strain plasticity. *Int. J. Plasticity* **1**, 3-27.

Bell, J. F. 1987 A confluence of experiment and theory for waves of finite strain in the solid continuum. *Dynamical Problems in Continuum Physics* (eds. J. L. Bona, C. Dafermos, J. L. Ericksen & D. Kinderleher), IMA Volumes in Mathematics and Its Applications, Vol. 4, 89-130. New York: Springer.

Bell, J. F. 1988 Plane stress, plane strain, and pure shear at large finite strain. *Int. J. Plasticity* **4**, 127-148.

Bell, J. F. 1989 Experiments on the kinematics of large plastic strain in ordered solids. *Int. J. Solids Structures* **25**, 267-278.

Bell, J. F. 1990a Large deflection, rotation, and plastic strain in cantilevered beams. *Int. J. Engng. Sci.* **28**, 231-239.

Bell, J. F. 1990b Material objectivity in an experimentally based incremental theory of large finite plastic strain. *Int. J. Plasticity* **6**, 293-314.

Bell, J. F. 1995a A retrospect on the contributions of G. I. Taylor to the continuum physics of solids. *Experimental Mech.* **35**, 1-10.

Bell, J. F. 1995b Laboratory experiments on thin-walled tubes at large finite strain - symmetry, coaxiality, rigid body rotation, and the role of invariants, for the applied stress† $\sigma = \mathbf{R}\mathbf{T}_{\mathbf{R}}^{T}$, the Cauchy stress $\sigma^{*} = [III\mathbf{v}]^{-1}\mathbf{F}\mathbf{T}_{\mathbf{R}}^{T}$, and the left Cauchy-Green stretch tensor $\mathbf{V} = \mathbf{F}\mathbf{R}^{T}$. *Int. J. Plasticity* **11**, 119-144.

Bell, J. F. 1996a The decrease of volume during loading in finite plastic strain. *Meccanica* **31**, 461-472.

Bell, J. F. 1996b The kinematics of large plastic strain in cubic single crystals: A new look in the laboratory at G. I. Taylor's analysis of finite shear on face diagonals. *Contemporary Research in the Mechanics and Mathematics of Materials* (eds. R. C. Batra & M. F. Beatty). Barcelona: International Center for Numerical Methods in Engineering (CIMNE).

Bell, J. F., and Baesu, E. 1996 On the symmetry and coaxiality of pertinent stress and stretch tensors during non-proportional loading at finite plastic strain. *Acta Mech.* **115**, 1-14.

Hill, R. 1950 *The Mathematical Theory of Plasticity*. Oxford: Clarendon Press.

Johnson, M. A., and Beatty, M. F. 1993 A constitutive equation for the Mullins effect in stress controlled uniaxial extension experiments. *Continuum Mech. Thermodyn.* **5**, 83-115.

Moon, H. 1973 *Experimental Study of the Outer Yield Surface and of the Incremental Response Function in the Totally Plastic Region for Annealed Polycrystalline Aluminum*. Doctoral Dissertation, The Johns Hopkins University, Baltimore, Maryland.

Ogden, R. W. 2001 Elements of the theory of finite elasticity. Chapter 1 in *Nonlinear Elasticity: Theory and Applications* (eds. Y. B. Fu & R. W. Ogden). Cambridge: Cambridge University Press.

Truesdell, C., and Toupin, R. 1960 *The Classical Field Theories*. Flügge's Handbuch der Physik, Vol. **III/1**. Heidelberg: Springer.

Truesdell, C., and Noll, W. 1965 *The Non-Linear Field Theories of Mechanics*. Flügge's Handbuch der Physik, Vol. **III/3**. New York: Springer.

---

† Caution: Here the notation is that used by Bell, not that used in this text.

3

Universal results in finite elasticity

Giuseppe Saccomandi

Dipartimento di Ingegneria dell'Innovazione
Università degli Studi di Lecce, Italy
Email: giuseppe@ibm.isten.ing.unipg.it

A deformation or a motion is said to be a universal solution if it satisfies the balance equations with zero body force for all materials in a given class, and is supported in equilibrium by suitable surface tractions alone. On the other hand for a given deformation or motion, a local universal relation is an equation relating the stress components and the position vector which holds at any point of the body and which is the same for any material in a given class. Universal results of various kinds are fundamental aspects of the theory of finite elasticity and they are very useful in directing and warning experimentalists in their exploration of the constitutive properties of real materials. The aim of this chapter is to review universal solutions and local universal relations for isotropic nonlinearly elastic materials.

3.1 Introduction

One of the main problems encountered in the applications of the mechanics of continua is the complete and accurate determination of the constitutive relations necessary for the mathematical description of the behavior of real materials.

After the Second World War the enormous work on the foundations of the mechanics of continua, which began with the 1947 paper of Rivlin on the torsion of a rubber cylinder published in the Journal of Applied Physics, has allowed important insights on the above mentioned problem (Truesdell and Noll, 1965).

New achievements have been realized mainly by means of:

- a systematic and rigorous use of the restrictions imposed by thermodynamics;
- the derivation and use of the representation formulae, a consequence of the rigorous definitions of the different material behaviors and of the concept of material symmetry;
- the use of *a priori restrictions*, consequences of the requirement of mathematical well posedness for the models considered.

All these ideas have helped to focus the goal of experimental tests, which have to be the ultimate factor in determining the *right* constitutive equations.

We know that in the rheology laboratory a complicated situation exists because behind any experiment there is the need for a strong theoretical background. Indeed the notion of an experimentalist working alone measuring stresses and strains is too naive. A more realistic picture of what truly happens can be obtained from a recent overview by Leblanc (1998):

Basically any rheometry test consists in either applying a controlled rate of deformation and measuring the corresponding stress or vice versa. How the material flow defines the testing mode and, because stress and deformation rate are not directly measured, one needs equations to transform the readily measured quantities into the appropriate rheological parameters. It is worth mentioning here what hypotheses are set when developing those rheometrical equations. Whatever the testing mode, these hypotheses concern the type of flow, boundary conditions, thermal conditions and material properties. Furthermore, in order to give some meaning to measured quantities, one has to make an explicit hypothesis about the expected rheological behavior of the material.

Essentially any *rational* experimental test, which is not pure phenomenology†, needs a mathematical model to frame and interpret the data, but this is not sufficient. The experimentalist must be provided with key theoretical results that allows the *interpretation* of such a model.

First of all, as Muncaster (1979) points out, since it is not possible to have local (point) measurements both of the loads and of the deformations

the theorist must determine from the mathematical model which resultants, if any, have the principal influence upon the deformation.

Then, because measurements can be done only on the boundary of the test specimen, we must know *a priori* the kind of deformation that we want to reproduce experimentally in order to know what quantities can be effectively measured. This information

is to be supplied by the theorist through certain explicit solutions, either exact or approximate, predicted by the mathematical problem.

Lastly, we must be able to know the exact link between the various measured quantities in order

† Obviously the nature of the models of material behavior obtained using the continuum mechanics apparatus is phenomenological. The constitutive parameters or functions they contain have no physical meaning *a priori*. For this reason the *rationale* of the related experiment session is even more important than in the case of the models obtained using statistical mechanics.

to decide which values of the constitutive mappings or which quantities associated with them can be determined in the experiment, and then to provide the experimentalist with formulae expressing these in terms of quantities which he can measure.

Theoretical results which fit this program are the *universal* results.

A deformation, or a motion, which satisfies the balance equations with zero body force and, in equilibrium, is supported by suitable surface tractions alone, is called a **controllable solution**. A controllable solution which is the same for all materials in a given class is a **universal solution**.

Besides universal solutions, other kind of universal results, involving not only the strain, but also the stress exist. For a given deformation or motion, a **local universal relation** is an equation relating the stress components and the position vector which holds at any point of the body and which is the same for any material in a given class.

The importance of universal results is clearly pointed out by Beatty (1987a):

Universal deformations and universal results of various kinds are road signs posted to direct and to warn the experimenter in his exploration of the constitutive properties of real materials. If these signs are ignored, he can only wander aimlessly along an otherwise uncharted labyrinth in the vast realm of materials science. If they are thoughtfully evaluated, he will discover the rich rewards locked in the smaller domain of the science of highly elastic materials.

This paper is devoted to a selected review of the literature on universal results with the aim of a clarification of points two and three of the above quotes from Muncaster (1979). The first issue, which concerns Saint-Venant's principle, has been reviewed extensively by Horgan (1989, 1995b), and for this reason is not discussed here. We point out that here the main interest is in formal properties of the solutions of the balance equations. Thus, we are not directly involved with the fundamental problem of determining what restrictions should be placed on the stored-energy function of a hyperelastic material in order to ensure that physically reasonable and mathematically convenient results follow.

We point out that although Antman (1995), Beatty (1987a), Lurie (1990), Ogden (1984), Truesdell and Noll (1965), Wang and Truesdell (1973) and in general any textbook on continuum mechanics and finite elasticity have some discussion of universal results, to the best of our knowledge no reviews on this topic have ever been written. The interesting paper by Fosdick (1966) is about the related problem of the compatibility of the strain tensor and the brief review by Kafadar (1975) is about some universal motions in simple materials. A short review of universal relations in continuum mechanics may be found in Pucci and Saccomandi (1997).

Here we restrict our attention only to elastostatics. We do not consider the problem of universal motions and the problem of universal solutions for non-elastic theories, for the models of structural mechanics or for the case of coupled fields.

3.2 Isotropic unconstrained elasticity

Let us consider an isotropic elastic body and the corresponding representation formula

$$\sigma = \beta_0 \mathbf{I} + \beta_1 \mathbf{B} + \beta_{-1} \mathbf{B}^{-1} \tag{2.1}$$

for the Cauchy stress tensor, where $\mathbf{B} = \mathbf{FF}^T$ is the left Cauchy-Green deformation tensor and $\mathbf{F}$ the gradient of the deformation. The representation formula (2.1) *says* that all the real bodies that can be modelled as elastic and isotropic can be distinguished only on grounds of the response functions (or coefficients) β_i. Very little can be said, *a priori*, about these response functions. Requiring the well posedness of the model from the mechanical and mathematical point of view enables some restrictions to be obtained, but in any case the determination of the explicit functional form for the β_i's must be done by means of an experimental test (Ogden, 1986). In this section we show why universal results may help to achieve this goal.

Let us consider the homogeneous deformation of **simple shear**,

$$x_1 = X_1 + kX_2, \quad x_2 = X_2, \quad x_3 = X_3, \tag{2.2}$$

where k is an arbitrary parameter (see also Section 1.3.1.1 of Chapter 1).

A simple computation allows us to determine the physical components of $\mathbf{B}$ and of its inverse in the form

$$[B_{ij}] = \begin{pmatrix} 1+k^2 & k & 0 \\ k & 1 & 0 \\ 0 & 0 & 1 \end{pmatrix}, \quad [B_{ij}^{-1}] = \begin{pmatrix} 1 & -k & 0 \\ -k & 1+k^2 & 0 \\ 0 & 0 & 1 \end{pmatrix}, \tag{2.3}$$

so that the principal invariants of $\mathbf{B}$ are $I_1 = I_2 = 3 + k^2, I_3 = 1$.

Since the strain matrix has constant Cartesian components, $\mathrm{div}\,\sigma = \mathbf{0}$, i.e. the deformation is a solution of the balance equations in the absence of body forces. Moreover, because (2.2) is controllable for all the materials in the class given by (2.1) the deformation is universal.

The stress components for this deformation are

$$\sigma_{11} = \beta_0 + \beta_1(1+k^2) + \beta_{-1}, \quad \sigma_{12} = k(\beta_1 - \beta_{-1}), \quad \sigma_{13} = 0,$$

$$\sigma_{22} = \beta_0 + \beta_1 + \beta_{-1}(1+k^2), \quad \sigma_{23} = 0, \quad \sigma_{33} = \beta_0 + \beta_1 + \beta_{-1}, \tag{2.4}$$

and from these formulae it is possible to obtain

$$\beta_1 k^2 = \sigma_{11} - \sigma_{33}, \quad \beta_{-1} k^2 = \sigma_{22} - \sigma_{33},$$
$$\beta_0 k^2 = (2 + k^2)\sigma_{33} - (\sigma_{11} + \sigma_{22}).$$

This is an example of how it is possible to use a *universal solution* to design an experimental test to determine the β_i's. Obviously an experiment based on simple shear is too restrictive to determine completely the response functions, since this deformation allows exploration of what happens only along the line $I_3 = 1, I_1 = I_2$ in the space of invariants $I_1 > 0, I_2 > 0$, and $I_3 > 0$.†

From (2.4) it is possible to derive the relations

$$\sigma_{13} = \sigma_{23} = 0, \quad k\sigma_{12} = \sigma_{11} - \sigma_{22}. \qquad (2.5)$$

These relations are links between the stress components and the shear parameter k, which do not depend on the particular elastic isotropic material, and so they are universal relations. The first two are trivial expressions, but the third is a very interesting analytical expression of the Poynting effect, a peculiar behavior of nonlinear materials relating shear stress to normal stress differences.

The universal results just discussed are indications of the main topics of this chapter. These results were first reported by Rivlin, one of the pioneers of the modern theory of finite elasticity (see Rivlin, 1997, for the collected works). In many cases universal deformations are exact solutions in closed form of the balance equation. Thus, they are very useful as benchmarks for numerical solutions for more complex problems (Dragoni, 1996) or as prototypes for the study of complex physical mechanisms associated with dislocations (Nabarro, 1979) or cavitation phenomena (Horgan and Polignone, 1995).

The simple results reported above can be put in a very general framework using a theorem of Ericksen about universal solutions and a recent observation by Beatty about universal relations. These are the starting points for our investigations.

3.2.1 Ericksen's theorem and the coaxiality between stress and strain

In a fundamental paper by Ericksen (1955) it is shown that the homogeneous deformations are the only ones that are universal for hyperelastic, isotropic and

† How is possible to use general homogeneous deformations to determine completely the response functions can be found, for example, in Ogden (1984). We point out that the homogeneous deformations solve immediately the first and second point raised by Muncaster and reported in the Introduction. This is because the strain, and hence the stress, has the same value at each point of the body.

G. Saccomandi

unconstrained materials. Here we summarize the proof of this result by Shield (1971) with minor modifications (see also Ogden, 1984).

Let us introduce the nominal stress tensor (Section 1.2.4 of Chapter 1):

$$\mathbf{S} = \frac{\partial W}{\partial \mathbf{F}} = \sum_{i=1}^{3} \frac{\partial W}{\partial I_i} \frac{\partial I_i}{\partial \mathbf{F}}. \tag{2.6}$$

The problem is to find all the deformations such that $\mathrm{Div}\,\mathbf{S} = \mathbf{0}$, (here Div is the divergence with respect to the reference coordinates) and this for any choice of the strain energy density W.[†] We begin by choosing a strain-energy density linear with respect to the principal invariants ($W = \sum_i \mu_i I_i$, μ_i arbitrary real constants). Then, for $i = 1, 2, 3$, equation (2.6) gives

$$\mathrm{Div}\left(\frac{\partial I_i}{\partial \mathbf{F}}\right) = \mathbf{0}. \tag{2.7}$$

Next we consider the quadratic strain-energy density $W = \sum_i \widehat{\mu}_i I_i^2$, where the $\widehat{\mu}_i$ are arbitrary constants, and obtain, for $i = 1, 2, 3$,

$$\mathrm{Grad}\,I_i = \mathbf{0}. \tag{2.8}$$

Since $\partial I_1/\partial \mathbf{F} = 2\mathbf{F}$, from (2.7) we obtain $\mathrm{Div}\mathbf{F} = \mathbf{0}$ and using Cartesian coordinates (i.e. $\mathbf{F} = x_{i,\beta}$)

$$x_{i,\beta\beta} = 0. \tag{2.9}$$

In the case $i = 1$, equation (2.8) gives

$$x_{i,\alpha} x_{i,\alpha\beta} = 0, \tag{2.10}$$

and by differentiation of this last relation with respect to X_β, using (2.9), we obtain

$$x_{i,\alpha\beta}\, x_{i,\alpha\beta} = 0.$$

Since this is a sum of squares we obtain

$$x_{i,\alpha\beta} = 0,$$

the desired result: the only universal solutions are the homogeneous deformations.

On the other hand, the universal relations (2.5) can be deduced from an important property of isotropic materials: the coaxiality of stress and strain, i.e.

$$\boldsymbol{\sigma}\mathbf{B} = \mathbf{B}\boldsymbol{\sigma}. \tag{2.11}$$

[†] For the strain energy viewed as a function of I_1, I_2, I_3, we use the symbol W, whereas for the strain energy viewed as a function of $\lambda_1, \lambda_2, \lambda_3$, we use the symbol Σ.

The relationship between this property and the universal relations has been discussed extensively by Beatty (1987b).

In the treatise on the theory of elasticity by Wang and Truesdell (1973) we find the statement:

formulae of this kind [i.e. such as (2.11)], *which are called universal relations, are most important since they reflect a type of material symmetry directly in terms of stress and deformation. As we shall see later, the universal relation (3.21)* [i.e. (2.11)] *follows from applying to the case of simple shear the fact that for an isotropic elastic point a principal basis of* **B** *is a principal basis of* $\boldsymbol{\sigma}$.

In this affirmation we can discover for the first time the connection between coaxiality and universal relations when we consider isotropic materials[†].

3.2.2 Non-universal solutions

Universal solutions are not the only kind of possible solutions for the balance equations. The search for exact and analytical solutions which are not universal is a current topic of research and the main tool is always the semi-inverse method. For example, let us consider the radial deformation (Horgan, 1995a)

$$r = r(R), \quad \theta = \Theta, \quad z = Z, \tag{2.12}$$

where $dr/dR > 0$. The deformation gradient referred to cylindrical polar coordinates has components given by

$$[F_{ij}] = \operatorname{diag}(dr/dR, r/R, 1)$$

and the principal stretches are easily obtained as $\lambda_1 = dr/dR, \lambda_2 = r/R$ and $\lambda_3 = 1$. The balance equation $\operatorname{div}\boldsymbol{\sigma} = \mathbf{0}$, is reduced to a scalar ordinary differential equation in the unknown $r(R)$, namely

$$\frac{d}{dR}\left(R\frac{\partial\Sigma}{\partial\lambda_1}\right) - \frac{\partial\Sigma}{\partial\lambda_2} = 0. \tag{2.13}$$

Given a particular strain-energy function, this is a single ordinary differential equation in the single unknown $r = r(R)$. It is obvious that for some choices of Σ it is possible that (2.13) has no solution. For example, choosing $\Sigma = \lambda_2$ equation (2.13) becomes $1 = 0$, or, choosing $\Sigma = \lambda_2^2$, it becomes $\lambda_2 = r/R = 0$, and again this equation has no meaningful solutions because, for $r = 0$, the tensor **F** is not positive definite.

---

† The relationship between material symmetry and universal relations is a little bit more complex than the one naively reported in the book of Wang and Truesdell (1973). We know that only in the case of hyperelasticity is it possible to read material symmetry using universal relations, as has been shown in Vianello (1996) for isotropic elasticity and in Saccomandi and Vianello (1997) for transversely hemitropic elastic materials.

In general, if we discard such pathological cases, (2.13) admits a solution‡ for any reasonable choice of the function Σ. This solution depends on the choice of Σ and for this reason it is not useful in an experimental test, because the geometry of the deformation depends on the quantities we have to measure. For example, let us consider§

$$\Sigma = \alpha\lambda_1^2 + \beta\lambda_2, \tag{2.14}$$

where α and β are constant constitutive parameters to be determined by experiments. For (2.14), equation (2.13) reduces to

$$2\alpha\frac{d}{dR}\left(R\frac{dr}{dR}\right) - \beta = 0, \tag{2.15}$$

which has solution

$$r(R) = \frac{\beta}{2\alpha}R + C_1 + C_2\ln R, \tag{2.16}$$

where the C_i are integration constants. Now (2.16) is a family of deformations and for any material in the class defined by (2.14) there is always a solution of (2.15) in this family, but the geometry of the deformation is completely known only if we know the exact value of the constitutive parameters. Thus, (2.16) is **controllable**, but not **universal**.

It may happen that for a particular subclass of isotropic materials equation (2.13) admits a solution which does not depend on the response functions although the equation itself does. This is the case for the harmonic materials defined (for plane deformations with $\lambda_3 = 1$) by the strain-energy function

$$\Sigma = f(\lambda_1 + \lambda_2) + c\lambda_1\lambda_2 - c, \tag{2.17}$$

where f is a nonlinear constitutive function and c a constitutive parameter. In this case, for every choice of f and c, the solution of (2.13) is given by

$$r(R) = AR + \frac{B}{R}, \tag{2.18}$$

where A and B are constants of integration (Horgan, 1995b). This is an example of a **relatively universal solution**† (relative to the subclass of harmonic materials).

‡ Here we are not speaking of solutions of boundary-value problems, but of formal solutions of equation (2.13).

§ This and the above examples are based on strain-energy functions not compatible with isotropy.

† For these solutions Professor Rivlin in 1977 during a Euromech colloquium in Darmstadt proposed the term *quasi-controllable deformations*. In our opinion controllability and universality, as shown in (2.16), are two different ideas and for this reason we consider the term used by Rivlin to be misleading.

In the search for solutions of the balance equations by means of a semi-inverse method we are not always so lucky as in the previous example. It may happen that the balance equations reduce to an overdetermined set of differential equations which are not always compatible. This is the case for the deformation of *antiplane shear,* defined by

$$x_1 = X_1, \qquad x_2 = X_2, \qquad x_3 = X_3 + u(X_1, X_2); \qquad (2.19)$$

see Horgan (1995a) for a review. The physical components of **B** are given by

$$[B_{ij}] = \begin{pmatrix} 1 & 0 & u_1 \\ 0 & 1 & u_2 \\ u_1 & u_2 & 1+\kappa^2 \end{pmatrix},$$

and the principal invariants are $I_1 = I_2 = 3 + \kappa^2, I_3 = 1$, where $\kappa = |\nabla u|$.

It is easy to check that the balance equations now form a system of three differential equations in the unknown $u(X_1, X_2)$, i.e.

$$\left(2 \frac{\partial \overline{W}}{\partial I_2} u_\alpha u_\beta \right)_{,\beta} - \left[\frac{\partial \overline{W}}{\partial I_3} + \frac{\partial \overline{W}}{\partial I_1} + 2(2 + \kappa^2) \frac{\partial \overline{W}}{\partial I_2} \right]_{,\alpha} = 0, \qquad (2.20)$$

$$\left[\left(\frac{\partial \overline{W}}{\partial I_1} + \frac{\partial \overline{W}}{\partial I_2} \right) u_\beta \right]_{,\beta} = 0,$$

where the subscripts run from 1 to 2 (1 stands for differentiation with respect X_1 and 2 with respect X_2) and

$$\frac{\partial \overline{W}}{\partial I_i} = \frac{\partial W}{\partial I_i} \bigg|_{I_1 = I_2 = 3 + \kappa^2, \, I_3 = 1}.$$

As shown by Knowles, the overdetermined differential system (2.20) is compatible only for particular choices of the strain-energy density (Horgan, 1995a).

It is interesting to point out what happens to the universal relations that we can derive from the coaxiality between stress and strain when we consider these non-universal solutions. Obviously, the relation $\sigma \mathbf{B} = \mathbf{B}\sigma$ is always in force and it does not matter whether the strain corresponds to a universal or a non-universal solution. The problem is that, given a non-universal solution, the eigenvectors of the strain and the corresponding stress are the same, but the eigenbasis may change from material to material. Then, the three scalar relations that we can derive from the coaxiality condition, for a given strain, may depend on the response coefficients. For example, in the case of the deformation of antiplane shear, the three non-zero scalar components of the relation $\sigma\mathbf{B} = \mathbf{B}\sigma$ can be written as

$$
\begin{aligned}
u_2(\sigma_{33} - \sigma_{22}) &= \kappa^2\sigma_{23} + u_1\sigma_{12}, \\
u_1(\sigma_{33} - \sigma_{11}) &= \kappa^2\sigma_{13} + u_2\sigma_{12}, \\
u_1\sigma_{23} &= u_2\sigma_{13},
\end{aligned}
\tag{2.21}
$$

which depend on the response coefficients because, in general, $u\,(X_1, X_2)$ does. If we consider the family of strain-energy functions (Horgan, 1995a)

$$
W = h(I_1) + \frac{dh}{dI_1}f(I_3) + g(I_3),
$$

where f, g and h are arbitrary functions such that

$$
f(1) = 0, \quad f'(1) = -1, \quad g(1) = 0, \quad g'(1) = 0, \quad h(3) = 0,
$$

then equations (2.20) are compatible and reduce to

$$
\left(\frac{dh}{dI_1}\bigg|_{I_1=3+\kappa^2} u_\beta \right)_{,\beta} = 0.
\tag{2.22}
$$

In the case where $h(I_1)$ is a linear function of I_1 this last differential equation becomes the Laplace equation and the relations in (2.21) are then universal in a strict sense. Otherwise, $u(X_1, X_2)$ and its derivatives depend on the response functions and the relations (2.21) are not strictly the same for every member in the given class of materials.

We observe from (2.21) that it is always possible to find a universal relation for any compressible material that may sustain an antiplane shear deformation. The universal relation is nonlinear and reads

$$
\sigma_{23}\sigma_{13}(\sigma_{11} - \sigma_{22}) = \sigma_{12}(\sigma_{13}^2 - \sigma_{23}^2).
\tag{2.23}
$$

Pucci and Saccomandi (1997) introduced a powerful method based on algebraic geometry, more efficient than (2.11), to generate the complete set of universal relations between universal solutions for a large class of theories of interest in continuum mechanics. Previously, Hayes and Knops (1966), using linear algebra, presented a method of finding all the universal relations for isotropic elasticity corresponding to universal solutions.

3.3 Universal solutions for isotropic constrained elasticity

The theorem of Ericksen solves completely the issue of universal solutions for unconstrained isotropic elasticity, but the situation is different and more intriguing when the material is internally constrained. Although at first sight the theory for internally constrained materials may appear as a particular case of the unconstrained theory, this is not at all the case and many new and interesting results may be derived.

According to the principle of determinism (Section 1.2.3.4, Chapter 1) the stress in this case splits in two parts: the reactive stress and the extra stress

$$\sigma = \sigma^R + \sigma^E.$$

To have a representation formula for the reactive stress we have to characterize as ideal the internal constraints we are considering, i.e. we require that in any motion satisfying the constraint the reaction stress does no work. Obviously, the explicit determination of σ^R may be very complex (Podio-Guidugli, 1990).

Here, we will restrict our attention only to isotropic constraints, i.e. constraints that are represented analytically by equations such as

$$\gamma(I_1 - 3, I_2 - 3, I_3 - 1) = 0, \tag{3.1}$$

where γ is smooth function of the principal invariants with a simple root at zero (i.e., where $I_1 = I_2 = 3$, and $I_3 = 1$). The constraint (3.1) is the most general available in the framework of isotropy†.

The constraint of incompressibility $I_3 - 1 = 0$, the constraint of Ericksen (1986), $I_1 - 3 = 0$, and the constraint of Bell (1996) which is usually expressed as $\mathrm{tr}\,\mathbf{V} = 3$ but can also be recast as

$$I_1^2 - 18I_1 - 4I_2 - 24\sqrt{I_3} + 81 = 0,$$

are all examples of isotropic constraints.

For isotropic constraints the formula for the reactive stress is

$$\sigma^R = -\pi_0 \mathbf{F} \partial_{\mathbf{C}} \gamma \mathbf{F}^{\mathrm{T}}, \tag{3.2}$$

where $\pi_0(\mathbf{x})$ is the indeterminate reaction scalar, to be determined by the balance equations.

By introducing (3.1) and (2.1) in (3.2) the constitutive equation becomes

$$\sigma = (\beta_0 - \pi_0 \Delta)\,\mathbf{I} + \left(\beta_1 - \pi_0 \frac{\partial \gamma}{\partial I_1}\right)\mathbf{B} + \left(\beta_{-1} + \pi_0 I_3 \frac{\partial \gamma}{\partial I_2}\right)\mathbf{B}^{-1}, \tag{3.3}$$

which we rewrite as

$$\sigma = -\pi \mathbb{S} + \delta_1 \mathbf{B} + \delta_2 \mathbf{B}^{-1}, \tag{3.4}$$

† Podio-Guidugli and Vianello (1989) show that two isotropic constraints acting on the same material are equivalent to the rigidity constraint. This result is true for constraints describing manifolds in the set of all symmetric positive definite tensors; otherwise, a counterexample to this assertion is given by

$$I_1 - I_2 = 0, \quad I_3 = 1.$$

where‡

$$\Delta = I_2 \frac{\partial \gamma}{\partial I_2} + I_3 \frac{\partial \gamma}{\partial I_3}, \quad \mathbb{S} = \mathbf{I} + a\mathbf{B} + b\mathbf{B}^{-1},$$

$$\pi = -(\beta_0 - \pi_0 \Delta), \quad \delta_1 = \beta_1 - \frac{\beta_0 \gamma_1}{\Delta}, \quad \delta_2 = \beta_{-1} + \frac{\beta_0 I_3 \gamma_2}{\Delta}, \tag{3.5}$$

and $a = \gamma_1 \Delta^{-1}, b = -I_3 \gamma_2 \Delta^{-1}$.

We point out that

$$\mathbb{S}\mathbf{B} = \mathbf{B}\mathbb{S}$$

so that the tensors $\mathbb{S}$ and $\mathbf{B}$ are coaxial. Moreover, if $\mathbf{B}$ has a double eigenvalue the same must be true for $\mathbb{S}$. The tensor $\mathbb{S}$ can also be singular, but here we will not investigate this possibility†.

It is useful to use (3.1) to introduce two scalar functions $M(\mathbf{x})$ and $N(\mathbf{x})$ such that $I_i = I_i(M, N)$. The balance equations in the absence of body forces can then be rewritten as

$$\begin{aligned}
\mathbb{S} \nabla \pi &= \delta_1 \nabla \cdot \mathbf{B} + \delta_2 \nabla \cdot \mathbf{B}^{-1} - \pi \nabla \cdot \mathbb{S} \\
&+ \mathbf{B}(\delta_{1,M} \nabla M + \delta_{1,N} \nabla N) \\
&+ \mathbf{B}^{-1}(\delta_{2,M} \nabla M + \delta_{2,N} \nabla N),
\end{aligned} \tag{3.6}$$

where we have used ∇ for grad, $\nabla\cdot$ for div and $\nabla\times$ for curl. Since we have the existence of $\mathbb{S}^{-1}$, the gradient of the reaction scalar is given by

$$\begin{aligned}
\nabla \pi &= \delta_1 \mathbb{S}^{-1} \nabla \cdot \mathbf{B} + \delta_2 \mathbb{S}^{-1} \nabla \cdot \mathbf{B}^{-1} - \pi \mathbb{S}^{-1} \nabla \cdot \mathbb{S} \\
&+ (\delta_{1,M} \mathbb{S}^{-1} \mathbf{B} + \delta_{2,M} \mathbb{S}^{-1} \mathbf{B}^{-1}) \nabla M \\
&+ (\delta_{1,N} \mathbb{S}^{-1} \mathbf{B} + \delta_{2,N} \mathbb{S}^{-1} \mathbf{B}^{-1}) \nabla N,
\end{aligned} \tag{3.7}$$

and taking the curl of this last equation we obtain the compatibility equation for the existence of the scalar π:

$$\begin{aligned}
\pi \nabla \times \iota &= \delta_{1,MM}(\nabla M \times \mathbf{m}_1) + \delta_{1,NN}(\nabla N \times \mathbf{n}_1) \\
&+ \delta_{2,MM}(\nabla M \times \mathbf{m}_2) + \delta_{2,NN}(\nabla N \times \mathbf{n}_2) \\
&+ \delta_{1,MN}(\nabla N \times \mathbf{m}_1 + \nabla M \times \mathbf{n}_1) \\
&+ \delta_{2,MN}(\nabla N \times \mathbf{m}_2 + \nabla M \times \mathbf{n}_2) \\
&+ \delta_{1,M}(\nabla M \times \mathbf{v} + \nabla \times \mathbf{m}_1 - \iota \times \mathbf{m}_1) \\
&+ \delta_{1,N}(\nabla N \times \mathbf{v} + \nabla \times \mathbf{n}_1 - \iota \times \mathbf{n}_1) \\
&+ \delta_{2,M}(\nabla M \times \mathbf{w} + \nabla \times \mathbf{m}_2 - \iota \times \mathbf{m}_2) \\
&+ \delta_{2,N}(\nabla N \times \mathbf{w} + \nabla \times \mathbf{n}_2 - \iota \times \mathbf{n}_2) \\
&+ \delta_1(\nabla \times \mathbf{v} - \iota \times \mathbf{v}) + \delta_2(\nabla \times \mathbf{w} - \iota \times \mathbf{w}),
\end{aligned} \tag{3.8}$$

‡ Here we assume $\Delta \neq 0$. The case $\Delta = 0$ is of interest since it contains the Ericksen materials that have been examined in Pucci and Saccomandi (1998).

† The case where the tensor $\mathbb{S}$ is singular has been studied in detail in Pucci and Saccomandi (1998).

with the following meanings for the various new symbols:

$$\iota = \mathbb{S}^{-1}\nabla \cdot \mathbb{S}, \quad \mathbf{v} = \mathbb{S}^{-1}\nabla \cdot \mathbf{B}, \quad \mathbf{w} = \mathbb{S}^{-1}\nabla \cdot \mathbf{B}^{-1}, \quad \mathbf{m}_1 = \mathbb{S}^{-1}\mathbf{B}\nabla M,$$

$$\mathbf{m}_2 = \mathbb{S}^{-1}\mathbf{B}^{-1}\nabla M, \quad \mathbf{n}_1 = \mathbb{S}^{-1}\mathbf{B}\nabla N, \quad \mathbf{n}_2 = \mathbb{S}^{-1}\mathbf{B}^{-1}\nabla N.$$

The relation (3.8) can be satisfied in two ways †.

   (i) We require that the system (3.8) be complete for any choice of π; then, $\nabla \times \iota$ and all the coefficients of the response coefficients δ_i and their derivatives must be zero.

  (ii) We require that (3.8) define a particular π. If we denote by $\mathbf{\Psi}$ the right-hand side of (3.8), then

$$\pi = \frac{\mathbf{\Psi} \cdot \nabla \times \iota}{\nabla \times \iota \cdot \nabla \times \iota},$$

and

$$\mathbf{\Psi} \times \nabla \times \iota = \mathbf{0}. \tag{3.9}$$

In the first case, to determine the reaction scalar we have to integrate equation (3.7) and then we determine π up to an arbitrary constant; in the second case, the reaction scalar is determined directly and it does not depend on arbitrary elements. This difference may be important in the use of universal solutions to solve boundary-value problems (in the spirit of the semi-inverse method).

Moreover, the second possibility is completely missing if $\nabla \cdot \mathbb{S}$ is zero, and it can be shown that this is the case if and only if $a = b = 0 \Rightarrow \gamma_1 = \gamma_2 = 0$ and $\mathbb{S} = \mathbf{I}$, i.e. if the material is incompressible. This fact suggests a classification of the isotropic constraints on the basis of the eigenvalues of the tensor $\mathbb{S}$.

If $\mathbb{S}$ is spherical (incompressibility) then we have an exceptional behavior, which is the one we have just described.

If the tensor $\mathbb{S}$ has a double eigenvalue (and this for all the admissible deformations) it is still possible to characterize the family of corresponding constraints. Also, in this situation we have a strange behavior and this is because a small number of independent relations comprise the compatibility condition. Some technical details about this case may be found in Pucci and Saccomandi (1998).

If the tensor $\mathbb{S}$ has distinct eigenvalues, i.e. we are considering the standard case, the following analysis can be applied.

---

† The theory of overdetermined systems is an important topic in many fields of mathematics and physics (symmetries of differential equations, conservation laws, Hamiltonian systems, for example). The main mathematical tool for investigating this problem is the Cartan-Kahler theorem. A simple introduction to these topics can be found in Pucci and Saccomandi (1992).

Case I: $\nabla \times \mathbb{S}^{-1} \nabla \cdot \mathbb{S} = \mathbf{0}$

If we require directly the completeness of (3.7), then

$$\nabla \times \iota = \mathbf{0}, \qquad (3.10)$$

and

$$
\begin{aligned}
&\nabla \times \mathbf{v} - \iota \times \mathbf{v} = \mathbf{0}, && \nabla \times \mathbf{w} - \iota \times \mathbf{w} = \mathbf{0}, \\
&\nabla N \times \mathbf{m}_1 + \nabla M \times \mathbf{n}_1 = \mathbf{0}, && \nabla N \times \mathbf{m}_2 + \nabla M \times \mathbf{n}_2 = \mathbf{0}, \\
&\nabla M \times \mathbf{m}_1 = \mathbf{0}, && \nabla N \times \mathbf{n}_1 = \mathbf{0}, \\
&\nabla M \times \mathbf{m}_2 = \mathbf{0}, && \nabla N \times \mathbf{n}_2 = \mathbf{0}, \\
&\nabla M \times \mathbf{v} + \nabla \times \mathbf{m}_1 - \sigma \times \mathbf{m}_1 = \mathbf{0}, && \nabla N \times \mathbf{v} + \nabla \times \mathbf{n}_1 - \sigma \times \mathbf{n}_1 = \mathbf{0}, \\
&\nabla M \times \mathbf{w} + \nabla \times \mathbf{m}_2 - \sigma \times \mathbf{m}_2 = \mathbf{0}, && \nabla N \times \mathbf{w} + \nabla \times \mathbf{n}_2 - \sigma \times \mathbf{n}_2 = \mathbf{0}.
\end{aligned}
$$

$$(3.11)$$

This is an overdetermined system in the tensorial unknown $\mathbf{B}$, to which we have to append the restriction that the strain solution of this system is effectively a metric. This is done by requiring that the Riemann-Christoffel tensor based on $\mathbf{C} \,(:= \mathbf{F}^{\mathrm{T}}\mathbf{F})$ must be zero, i.e.

$$R_{skrl} = \Gamma_{kls,r} - \Gamma_{krs,l} - (\Gamma^p_{kl}\Gamma_{srp} - \Gamma^p_{kr}\Gamma_{slp}) = 0, \qquad (3.12)$$

where the subscripts run over $1, 2, 3$ and the notation

$$\Gamma_{kls} = \frac{1}{2}(C_{ks,l} + C_{ls,k} - C_{kl,s}), \qquad \Gamma^p_{kl} = C^{-1}_{pm}\Gamma_{klm},$$

has been used †.

The system (3.10), (3.12) for incompressible materials was first given, and partially solved, in Ericksen (1954).

In the case of a generic isotropic constraint the complete solution of this system was given for the first time in Pucci and Saccomandi (1998) (see also Martins and Duda, 1998), but only for essentially plane deformations of the form

$$x_1 = x_1(X_1, X_2), \quad x_2 = x_2(X_1, X_2), \quad x_3 = \lambda X_3. \qquad (3.13)$$

For general three-dimensional deformations, equations (3.10)-(3.12) have been examined in Pucci and Saccomandi (1999), where in the case of non-constant strain invariants the problem has been solved completely. In Pucci and Saccomandi (1999) it is also established that, when the invariants are not both constant, the surfaces $I_1 = $ constant, and this for all the constraints, can only be parallel planes, concentric cylinders or spheres. Then, in any constrained

---

† The problem of the compatibility of strains is reviewed in Fosdick (1966). General compatibility conditions for the left Cauchy-Green tensor are very hard to manage and obtain (Acharya, 1999).

material, only four families of universal deformations with non-constant strain invariants are possible. Obviously, although a geometrical correspondence exists between the various constrained materials, the forms of the functions describing such families are different from constraint to constraint.

The general solution for the case of constant invariants is still an open problem even in the incompressible case, but, as will be pointed out in the next subsection for non-spherical constraints, this case has to be formulated in a more general context. In fact, the only inhomogeneous solution we know for equation (3.10) when the invariants are constants is the Singh and Pipkin (1965) deformation, which, in terms of polar coordinates in both the deformed and reference configurations, is given by

$$r = \sqrt{AR}, \quad \theta = D\ln(BR) + C\Theta, \quad z = FZ, \tag{3.14}$$

where the parameters A, B, C, D and F have to be chosen such that the constraint is satisfied (Pucci and Saccomandi, 1996b).

Case II $\nabla \times \mathbb{S}^{-1}\nabla \cdot \mathbb{S} \neq 0$

Introducing (3.9) in (3.7) we obtain a cumbersome expression that must be satisfied for any choice of the response coefficients. When $\nabla M \neq 0$ and $\nabla N \neq 0$, this is possible if and only if

$$
\begin{aligned}
&\nabla N \times \mathbf{m}_1 + \nabla M \times \mathbf{n}_1 = \mathbf{0}, && \nabla N \times \mathbf{m}_2 + \nabla M \times \mathbf{n}_2 = \mathbf{0}, \\
&\nabla M \times \mathbf{m}_1 = \mathbf{0}, && \nabla N \times \mathbf{n}_1 = \mathbf{0}, \\
&\nabla M \times \mathbf{m}_2 = \mathbf{0}, && \nabla N \times \mathbf{n}_2 = \mathbf{0}, \\
&\nabla M \times \mathbf{v} + \nabla \times \mathbf{m}_1 - \boldsymbol{\iota} \times \mathbf{m}_1 = \mathbf{0}, && \nabla N \times \mathbf{v} + \nabla \times \mathbf{n}_1 - \boldsymbol{\iota} \times \mathbf{n}_1 = \mathbf{0}, \\
&\nabla M \times \mathbf{w} + \nabla \times \mathbf{m}_2 - \boldsymbol{\iota} \times \mathbf{m}_2 = \mathbf{0}, && \nabla N \times \mathbf{w} + \nabla \times \mathbf{n}_2 - \boldsymbol{\iota} \times \mathbf{n}_2 = \mathbf{0}.
\end{aligned}
\tag{3.15}
$$

Moreover, setting

$$
\Omega_1 = \frac{(\nabla \times \mathbf{v} - \boldsymbol{\iota} \times \mathbf{v}) \cdot \nabla \times \boldsymbol{\iota}}{\nabla \times \boldsymbol{\iota} \cdot \nabla \times \boldsymbol{\iota}}, \quad
\Omega_2 = \frac{(\nabla \times \mathbf{w} - \boldsymbol{\iota} \times \mathbf{w}) \cdot \nabla \times \boldsymbol{\iota}}{\nabla \times \boldsymbol{\iota} \cdot \nabla \times \boldsymbol{\iota}},
$$

and noting that

$$\pi = \delta_1 \Omega_1 + \delta_2 \Omega_2,$$

we must have

$$
\begin{aligned}
&\Omega_1 \nabla M = \mathbb{S}^{-1} \mathbf{B} \nabla M, && \Omega_1 \nabla N = \mathbb{S}^{-1} \mathbf{B} \nabla N, \\
&\Omega_2 \nabla M = \mathbb{S}^{-1} \mathbf{B}^{-1} \nabla M, && \Omega_2 \nabla N = \mathbb{S}^{-1} \mathbf{B}^{-1} \nabla N,
\end{aligned}
\tag{3.16}
$$

and

$$\nabla\Omega_1 = -\mathbb{S}^{-1}\nabla\cdot\mathbb{S}\Omega_1 + \mathbb{S}^{-1}\nabla\cdot\mathbf{B},$$

$$\nabla\Omega_2 = -\mathbb{S}^{-1}\nabla\cdot\mathbb{S}\Omega_2 + \mathbb{S}^{-1}\nabla\cdot\mathbf{B}^{-1}. \tag{3.17}$$

From (3.16) we find that ∇M and ∇N are eigenvectors of $\mathbb{S}^{-1}\mathbf{B}$ and $\mathbb{S}^{-1}\mathbf{B}^{-1}$ with eigenvalues Ω_1 and Ω_2. Moreover, in $(3.5)_1$, $a = a(M,N)$, $b = b(N,M)$, and ∇a and ∇b are eigenvectors of $\mathbb{S}^{-1}\mathbf{B}$ and $\mathbb{S}^{-1}\mathbf{B}^{-1}$ with the same eigenvalues. Considering these facts, the definition of ι and $(3.5)_1$, we have

$$\iota = \Omega_1\nabla a + \Omega_2\nabla b + a\mathbb{S}^{-1}\nabla\cdot\mathbf{B} + b\mathbb{S}^{-1}\nabla\cdot\mathbf{B}^{-1}.$$

Using (3.17) we obtain

$$\iota(1 - a\Omega_1 - b\Omega_2) = -\nabla(1 - a\Omega_1 - b\Omega_2),$$

and since we have, by hypothesis, $\nabla\times\iota\neq 0$, taking the curl of this last expression leads to

$$(1 - a\Omega_1 - b\Omega_2) = 0,$$

in contradiction of the relations (3.16), which may be rewritten as

$$b(\Omega_2\mathbf{B}\nabla M - \Omega_1\mathbf{B}^{-1}\nabla M) = \Omega_1\nabla M,$$

$$-a(\Omega_2\mathbf{B}\nabla M - \Omega_1\mathbf{B}^{-1}\nabla M) = \Omega_2\nabla M.$$

Thus, we require

$$a\Omega_1 + b\Omega_2 = 0.$$

This means that we must now have

$$\nabla M = \nabla N = \mathbf{0},$$

i.e. the invariants must be constants.

This situation allows us to simplify equations (3.8) to

$$\pi\nabla\times\iota = \delta_1(\nabla\times\mathbf{v} - \iota\times\mathbf{v}) + \delta_2(\nabla\times\mathbf{w} - \iota\times\mathbf{w}), \tag{3.18}$$

whereas (3.12) become

$$\Gamma^q_{km}\Gamma^m_{qr} - \Gamma^m_{kr,m} = 0. \tag{3.19}$$

The general solution of (3.18) and (3.19) is still an open problem, but an interesting class of solutions is obtained when the strain tensor is such that

$$\nabla\cdot\mathbf{B}\times\nabla\cdot\mathbf{B}^{-1} = 0. \tag{3.20}$$

In this case, as a consequence of (3.20), the vectors $\mathbb{S}^{-1}\nabla \cdot \mathbf{B}$, $\mathbb{S}^{-1}\nabla \cdot \mathbf{B}^{-1}$ and $\mathbb{S}^{-1}\nabla \cdot \mathbb{S}$ are parallel and (3.18) then admits the solution

$$\pi = k_1\delta_1 + k_2\delta_2, \tag{3.21}$$

where the k_i are *ad hoc* constants†.

Equation (3.20) is satisfied by all the essentially plane solutions because $\mathbf{B}$ can be rewritten as

$$\mathbf{B} = B_{\alpha\beta}\mathbf{e}_\alpha \otimes \mathbf{e}_\beta + \lambda^2\mathbf{e}_3 \otimes \mathbf{e}_3, \tag{3.22}$$

where the Greek subscripts range over 1 and 2, and $B_{\alpha\beta} = x_{\alpha,\Delta}x_{\beta,\Delta}$. Then, introducing the tensor

$$\widehat{\mathbf{B}} = B_{\alpha\beta}\mathbf{e}_\alpha \otimes \mathbf{e}_\beta,$$

the principal invariants are

$$I_1 = \widehat{I}_1 + \lambda^2, I_2 = \widehat{I}_2 + \lambda^2\widehat{I}_1, I_3 = \lambda^2\widehat{I}_1,$$

and the Cayley-Hamilton theorem gives $\widehat{\mathbf{B}} - \widehat{I}_1\widehat{\mathbf{I}} + \widehat{I}_2\widehat{\mathbf{B}}^{-1} = \mathbf{0}$. Then, if the deformation is such that $\widehat{I}_1$ and $\widehat{I}_2$ are constant then from the Cayley-Hamilton formula we recover the relation (3.20).

This is a surprising result: all the essentially plane deformations with constant invariants are admissible for any isotropically constrained material, but with no spherical reaction stress tensor.

There are also three-dimensional solutions that satisfy (3.20). For example, let us consider

$$r = \nu R, \quad \theta = \beta\Theta, \quad z = \gamma R + \eta Z, \tag{3.23}$$

which represents the torsion of a circular cone into another circular cone with different apex. The physical components of $\mathbf{B}$ are

$$[B_{ij}] = \begin{pmatrix} \nu^2 & 0 & \gamma\nu \\ 0 & \beta^2\nu^2 & 0 \\ \gamma\nu & 0 & \gamma^2 + \nu^2 \end{pmatrix},$$

and (3.20) is then satisfied if the various constants satisfy the relation

$$\left(\beta^2 - 1\right)\left(\eta^2 - \beta^2\nu^2\right) + \beta^2\gamma^2 = 0.$$

This deformation was first reported by Fosdick (1966), but it has never been recognized as a deformation admissible for some material. In the case of the

---

† If a solution is in correspondence with a pressure field such as (3.21) then (3.20) must be satisfied.

constraint of Bell (3.23) it is a solution of the balance equations and of the constraint equation if

$$\eta = \beta\,(3 - 2\beta\nu)\,, \gamma^2 = (1 - \beta^2)\,[(2\beta - 1)\nu - 3]\,[(2\beta + 1)\nu - 3]\,.$$

Another deformation for which (3.20) is satisfied is given by

$$x_1 = AX_1 + \sin\lambda X_2, \quad x_2 = DX_2, \quad x_3 = AX_3 - \cos\lambda X_2, \tag{3.24}$$

where A, D and λ constants. This deformation was first reported in a dynamic context by Carroll (1967) and studied by Beatty and Hayes (1992b) in the framework of the Bell constraint. Moreover, this deformation was shown to be universal for all isotropically constrained materials (apart from incompressibility) by Pucci and Saccomandi (1996b).

In summary, the situation for isotropically constrained materials is as follows.

- In the case of non-uniform invariants of the deformation, all the possible universal solutions have been determined up to the integration of a separable first-order ordinary differential equation. In this situation there is an *equivalence* among all the constraints.
- When the invariants are constants we do not know the general solution of (3.18). The only solutions we know are the class of generalized plane deformations and the families (3.23) and (3.24).

3.4 Details for essentially plane deformations

In the case of essentially plane deformations the general constraint may be rewritten as

$$\gamma(I_1 - 3, I_2 - 3, I_3 - 1) = \widehat{\gamma}(\widehat{I}_1 - 2, \widehat{I}_2 - 1, \lambda^2 - 1) = 0,$$

where λ has to be considered as a parameter and the balance equations are

$$\mathrm{div}\widehat{\sigma} = 0,$$

where div is the *in-plane* divergence.

The system (3.10) and (3.11) is simplified to three equations and (3.12) simplifies to only one equation. Indeed, using the Cayley-Hamilton theorem, the in-plane Cauchy stress becomes[†]

$$\widehat{\sigma} = \left[\widehat{\delta}_2\widehat{I}_2^{-1}\widehat{I}_1 - \pi(\widehat{b}\widehat{I}_2^{-1}\widehat{I}_1 + 1)\right]\widehat{\mathbf{I}} + \left[\widehat{\delta}_1 - \pi(\widehat{a} - \widehat{b}\widehat{I}_2^{-1}) - \widehat{\delta}_2\widehat{I}_2^{-1}\right]\widehat{\mathbf{B}}. \tag{4.1}$$

[†] We use a hat ^ to indicate that the various quantities are considered for the subclass of essentially plane deformations.

This representation formula can be rewritten in the compact form

$$\hat{\sigma} = q\mathbb{Z} + \delta\hat{\mathbf{I}}, \qquad (4.2)$$

where

$$q = \delta_1 - \pi(a - b\hat{I}_2^{-1}) - \delta_2\hat{I}_2^{-1}, \quad \mathbb{Z} = \hat{\mathbf{B}} + \nu\hat{\mathbf{I}},$$

and ν is computed by introducing q into (4.1).

The balance equations reduce to

$$\mathbb{Z}\nabla q + q\nabla \cdot \mathbb{Z} + \nabla\delta = \mathbf{0},$$

so it is clear that if the invariants of the strain are both constants then the choice $q \doteqdot 0$ satisfies the balance equations since $\nabla\delta = \mathbf{0}$. This means that all the essentially plane deformations with constant invariants are solutions of the balance equations. Obviously, this is true for all constrained materials that admit the representation formula (4.2)†.

Requiring the existence of $\mathbb{Z}^{-1}$, we have

$$\nabla q + q\mathbb{Z}^{-1}\nabla \cdot \mathbb{Z} + \mathbb{Z}^{-1}\nabla\delta = \mathbf{0},$$

and the compatibility equation is

$$-\mathbb{Z}^{-1}\nabla\delta \times \mathbb{Z}^{-1}\nabla \cdot \mathbb{Z} + q\nabla \times \left(\mathbb{Z}^{-1}\nabla \cdot \mathbb{Z}\right) + \nabla \times \left(\mathbb{Z}^{-1}\nabla\delta\right) = \mathbf{0}.$$

We require that this compatibility condition must be valid for any choice of δ and q, and, since

$$\frac{\partial\hat{\gamma}}{\partial\hat{I}_1}\nabla\hat{I}_1 + \frac{\partial\hat{\gamma}}{\partial\hat{I}_2}\nabla\hat{I}_2 = 0$$

is in force, this is possible if and only if

$$\nabla \times \left(\mathbb{Z}^{-1}\nabla \cdot \mathbb{Z}\right) = \mathbf{0}, \quad \mathbb{Z}^{-1}\nabla\hat{I}_1 \times \nabla\hat{I}_1 = \mathbf{0},$$

$$\nabla \times \left(\mathbb{Z}^{-1}\nabla\hat{I}_1\right) + \left(\mathbb{Z}^{-1}\nabla \cdot \mathbb{Z}\right) \times \mathbb{Z}^{-1}\nabla\hat{I}_1 = \mathbf{0}. \qquad (4.3)$$

From (4.3)$_2$ we have that $\nabla\hat{I}_1$ is an eigenvector of $\mathbb{Z}$, and hence it is also an eigenvector of $\hat{\mathbf{B}}$. Since the eigenvalues are functions of the invariants, equation (4.3)$_3$ reduces to

$$\left(\mathbb{Z}^{-1}\nabla \cdot \mathbb{Z}\right) \times \nabla\hat{I}_1 = 0 \quad \Rightarrow \quad \mathbb{Z}^{-1}\nabla \cdot \mathbb{Z} = \mu_Z\nabla\hat{I}_1,$$

or, in terms of $\hat{\mathbf{B}}$, to

$$\nabla \cdot \hat{\mathbf{B}} = \mu_B\nabla\hat{I}_1.$$

† We point out that incompressible materials do not admit this representation formula.

From (4.3)$_1$ we have $\nabla \times (\mu_Z \nabla \widehat{I_1}) = \nabla \mu_Z \times \nabla \widehat{I_1} = 0$, and both μ_Z and μ_B must be functions of $\widehat{I_1}$.

Rewriting $\widehat{\mathbf{B}}$ as

$$\widehat{\mathbf{B}} = \lambda_1^2 \mathbf{a}_1 \otimes \mathbf{a}_1 + \lambda_2^2 \mathbf{a}_2 \otimes \mathbf{a}_2 = \left(\lambda_1^2 - \lambda_2^2\right) \mathbf{a}_1 \otimes \mathbf{a}_1 + \lambda_2^2 \widehat{\mathbf{I}}, \tag{4.4}$$

we see that when $\widehat{I_1}$ is not constant we have

$$\mathbf{a}_1 = \frac{\nabla \widehat{I_1}}{\sqrt{\nabla \widehat{I_1} \cdot \nabla \widehat{I_1}}}, \tag{4.5}$$

and the eigenvalues λ_i are functions of $\widehat{I_1}$.

If $\lambda_1 \neq \lambda_2$, it follows from (4.4) and (4.5) that

$$\nabla \cdot \widehat{\mathbf{B}} = \mu_B \nabla \widehat{I_1} \quad \Rightarrow \quad \nabla \cdot \left(\frac{\nabla \widehat{I_1} \otimes \nabla \widehat{I_1}}{\nabla \widehat{I_1} \cdot \nabla \widehat{I_1}}\right) = F(\widehat{I_1}) \nabla \widehat{I_1}, \tag{4.6}$$

where $F(\widehat{I_1})$ is a certain function of its argument†.

Moreover,

$$\mathrm{tr}\,(\nabla \otimes \mathbf{a}_1) = \overline{F}(\widehat{I_1}). \tag{4.7}$$

Indeed, because $\mathbf{a}_1$ is a unit vector

$$(\nabla \otimes \mathbf{a}_1)\,\mathbf{a}_1 = 0, \tag{4.8}$$

and (4.6) may be rewritten as

$$\mathrm{tr}\,(\nabla \otimes \mathbf{a}_1) = F(\widehat{I_1})\sqrt{\nabla \widehat{I_1} \cdot \nabla \widehat{I_1}}. \tag{4.9}$$

On the other hand $\nabla(\nabla \widehat{I_1} \cdot \nabla \widehat{I_1})$ is parallel to $\nabla \widehat{I_1}$, and this is possible only if $\sqrt{\nabla \widehat{I_1} \cdot \nabla \widehat{I_1}} = H(\widehat{I_1})$, from which (4.7) follows.

A little bit of differential geometry allows us to complete the proof. For a plane curve $U(x_1, x_2) = 0$, the curvature is given by

$$k = \frac{U_{11} U_2^2 - 2U_{12} U_1 U_2 + U_{22} U_1^2}{\sqrt{(U_1^2 + U_2^2)^3}},$$

and then $\mathrm{tr}\,(\nabla \otimes \mathbf{a}_1)$ is the curvature of the curves $\widehat{I_1} = $ constant, which, from (4.9), we know to be constant. We introduce the curve $x_1 = x_1(s), x_2 = x_2(s)$ such that

$$\frac{dx_i}{ds} = a_{1i}.$$

---

† If $\lambda_1 = \lambda_2$, the only possible deformations are homogeneous.

The absolute derivative of this relation is

$$a_{1i,j}\frac{dx_j}{ds} = a_{1i,j}a_{1j},$$

and by (4.8) the characteristic curves of the vector field $\nabla\widehat{I}_1$ must be straight lines. Then, the curves $\widehat{I}_1$ = constant are either parallel straight lines or concentric circles. This information allows us to close completely the problem.

The in-plane tensor $\widehat{\mathbf{B}}$ has components

$$\left[\widehat{B}_{\alpha\beta}\right] = \frac{1}{2}\begin{pmatrix} \widehat{I}_1 + h\cos\phi & h\sin\phi \\ h\sin\phi & \widehat{I}_1 - h\cos\phi \end{pmatrix},$$

where $h = \sqrt{\widehat{I}_1^2 - 4\widehat{I}_2}$, and its eigenvectors are

$$\mathbf{a}_1 = \cos(\phi/2)\,\mathbf{e}_1 + \sin(\phi/2)\,\mathbf{e}_2, \quad \mathbf{a}_2 = -\sin(\phi/2)\,\mathbf{e}_1 + \cos(\phi/2)\,\mathbf{e}_2.$$

The compatibility condition can then be written, as was first recognized by Fosdick and Schuler (1969), in a very simple and compact complex form.

Skipping the details of the general case (to be found in Pucci and Saccomandi, 1996a and 1998), here we consider only the case when the lines $\widehat{I}_1$ = constant are concentric circles.

Choosing $\widehat{I}_1 = \widehat{I}_1(r)$ and $\phi = 2\theta$, so that $\mathbf{a}_1 = \cos\theta\,\mathbf{e}_1 + \sin\theta\,\mathbf{e}_2$, the compatibility condition becomes

$$\frac{d}{dr}\left[rM\frac{d\left(\widehat{I}_1 + h\right)}{dr} + 2r\left(\widehat{I}_1 + h\right)\frac{dM}{dr} + 2M\left(\widehat{I}_1 + h\right)\right] = 0,$$

where $M = 2(\widehat{I}_1^2 - h^2)^{-1/2}$. By integrating this last equation, we obtain

$$2(\widehat{I}_1 - h) - r\frac{d(\widehat{I}_1 - h)}{dr} = 2K_0(\widehat{I}_1 + h)^{-1/2}(\widehat{I}_1 - h)^{3/2}, \qquad (4.10)$$

or

$$\frac{d}{dr}\left(\frac{r^2}{\widehat{B}_{22}}\right) = 2K_0 r(\widehat{B}_{11}\widehat{B}_{22})^{-1/2}. \qquad (4.11)$$

When $K_0 = 0$, equation (4.11) implies that $r^2/\widehat{B}_{22} = \overline{K}_0^2$, and this suggests that we introduce Cartesian coordinates in the reference configuration so that

$$\left(\frac{\partial X_1}{\partial r}\right)^2 + \frac{1}{r^2}\left(\frac{\partial X_1}{\partial\theta}\right)^2 = \frac{1}{\widehat{B}_{11}},$$

$$\frac{\partial X_1}{\partial r}\frac{\partial X_2}{\partial r} + \frac{1}{r^2}\frac{\partial X_1}{\partial\theta}\frac{\partial X_2}{\partial\theta} = 0,$$

$$\left(\frac{\partial X_2}{\partial r}\right)^2 + \frac{1}{r^2}\left(\frac{\partial X_2}{\partial \theta}\right)^2 = \frac{1}{\widehat{B}_{22}}.$$

A solution in the form

$$X_1 = X_1(r), \quad X_2 = X_2(\theta) = \overline{K}_0\theta + K_2 \tag{4.12}$$

is possible if

$$\frac{\partial X_1}{\partial r} = \sqrt{\frac{1}{\widehat{B}_{11}}} = \sqrt{\frac{2}{I_1 + h}}.$$

We point out that the constraint can be read also as a relationship, $\Omega(\widehat{I}_1, h) = 0$ say, between $\widehat{I}_1$ and h instead that between $\widehat{I}_1$ and $\widehat{I}_2$. Then

$$\frac{\partial \Omega}{\partial \widehat{I}_1}\frac{d\widehat{I}_1}{dr} + \frac{\partial \Omega}{\partial h}\frac{dh}{dr} = 0. \tag{4.13}$$

This deformation represents the case of the bending and stretching of a rectangular parallelepiped into a cylindrical annular sector. We can imagine the rectangular block bounded by three pairs of parallel planes $X_1 = 0$, $X_1 = T$; $X_2 = 0$, $X_2 = L$; $X_3 = 0$, $X_3 = H$. In the current configuration we introduce cylindrical coordinates; then, the deformation is given by the formulae

$$r = f(X_1), \qquad \theta = CX_2, \qquad x_3 = FX_3,$$

wherein the positive-valued function $f(X_1)$ is different for different constraints, as the following examples show.

- For incompressible materials ($\det \mathbf{F} = 1$)

$$f(X_1) = \sqrt{2AX_1};$$

- for Bell materials ($\mathrm{tr}\mathbf{V} = 3$)

$$f(X_1) = \frac{(3-F)L}{C} + A\exp(-\frac{C}{L}X_1);$$

- for Ericksen materials ($\mathrm{tr}\mathbf{B} = 3$)

$$f(X_1) = \frac{L}{C}\sqrt{3-F^2}\sin\left(\frac{C}{L}X_1 + A\right).$$

The rectangular block is deformed into a cylindrical sector of height $h = FH$, central angle $\alpha = CL$, inner and outer cylindrical surfaces

$$r_I = r(0), \quad r_0 = r(T),$$

and the thickness is given by $t = r(T) - r(0)$.

The positive definite nature of the matrix $\mathbf{V}$, whose physical components are obtained as

$$[V_{ij}] = \operatorname{diag}\left(\frac{df}{dX_1}, Cr, F\right)$$

requires $df/dX_1 > 0$.

Now let us simplify the situation by taking $C = 1$ and using the set

$$X_1 = X'T, \quad X_2 = Y'L, \quad Z = X_3'H, \quad r = r'T, \quad x_3 = x_3'H,$$

of dimensionless variables, which allows us to rewrite, by a suitable renaming of A,

$$
\begin{aligned}
r' &= t\sqrt{2X'}, \\
r' &= \delta^{-1} + \frac{t}{\exp(-\delta) - 1}\exp(-\delta X'), \\
r' &= \delta^{-1}\sqrt{3 - F^2}\sin(\delta X' + \Lambda)
\end{aligned}
$$

respectively. Here $\delta = T/L$ is a form parameter and

$$\Lambda = 2\arctan\frac{\sqrt{3 - F^2}(\cos\delta - 1) \pm \sqrt{2(3 - F^2)(1 - \cos\delta) - t^2}}{t + \sqrt{3 - F^2}\sin\delta}.$$

Moreover, the parameter A has been eliminated by considering as prescribed the thickness of the cylindrical sector.

The different behavior for the three constraints is striking; the incompressible solution depends only on the thickness of the sector, the Bell material solution on the thickness and the form parameter, and the Ericksen material solution also on the axial stretch.

We point out that the existence of the solution is not always guaranteed for the Ericksen material and that the non-uniqueness is fictitious since either the $+$ sign or the $-$ sign must be chosen in Λ.

A comparative analysis of the different constraints based on homogeneous deformations has been published by Beatty and Hayes (1992a).

3.5 Incompressible materials

Rivlin, in studying the Mooney-Rivlin strain-energy density

$$W = \frac{1}{2}[\alpha(I_1 - 3) + \beta(I_2 - 3)],$$

discovered the existence of four families of *controllable deformations*† that are solutions of the equilibrium equations for any value of the constitutive parameters α and β. Then, he discovered that the same families are universal solutions for the whole class of incompressible isotropic hyperelastic materials.

Ericksen (1954) began to examine the problem of the complete determination of the universal solutions for this class of materials. By a general and direct method he was able to check that it was possible to enlarge the families of Rivlin, but he was not able to give a complete answer to this problem. Today this definitive answer is still lacking †.

The four families found by Rivlin and Ericksen are today so well known that they have a standard numeration and denomination:

- **Family 1**: Bending, stretching and shearing of a rectangular block

$$r = \sqrt{2AX}, \qquad \theta = BY, \qquad z = \frac{Z}{AB} - BCY.$$

- **Family 2**: Straightening, stretching and shearing of a sector of a hollow cylinder

$$x = \frac{1}{2}AB^2R^2, \qquad y = \frac{\Theta}{AB}, \qquad z = \frac{Z}{B} + \frac{C\Theta}{AB}.$$

- **Family 3**: Inflation, bending, torsion, extension and shearing of an annular wedge

$$r = \sqrt{AR^2 + B}, \qquad \theta = C\Theta + DZ, \qquad z = E\Theta + FZ,$$

with $A(CF - DE) = 1$.

- **Family 4**: Inflation or eversion of a sector of a spherical shell

$$r = (\pm R^3 + A)^{1/3}, \qquad \theta = \pm\Theta, \qquad \phi = \Phi.$$

Here A, B, C, D, E, F are constants; moreover, the isochoric homogeneous deformations are obviously universal and they are known as **Family 0**. As already pointed out, these solution have an *analogue* for any isotropically constrained materials although the functional form of the deformation is different from constraint to constraint ‡.

These deformations are interesting not only because they can be useful in the design of tests, but also for understanding the mechanics of many problems.

---

† It is very interesting to read about this point in the autobiographical postscript contained in Rivlin (1997).

† In Wang and Truesdell (1973), page 293, it is written that this problem
... *remains a major unsolved problem of the theory of elasticity.*

‡ A similar explicit list of deformations is available for the Bell constraint (Beatty and Hayes, 1992a, b), the Ericksen material (Pucci and Saccomandi, 1996a) and materials constrained such that $I_2 = 3$ (Bosi and Salvatori, 1996).

For example, Families 3 and 4 have been used to study cavitation phenomena in elastic materials (Horgan and Polignone, 1995) and to give some nonlinear examples of elastic dislocations (Nabarro, 1979).

The proof of Ericksen was not complete in two respects:

(i) when two principal stretches are equal and at least one of the principal invariants is not constant;
(ii) when all the principal invariants are constants.

Point (i) was resolved completely by Marris and Shiau (1970), who showed that if two principal stretches are equal then the universal deformations are homogeneous or enclosed in Family 2.

With regard to the second point the final answer is still lacking. In the paper of Ericksen this possibility is ignored and he only reported the two relations that the strain tensor has to satisfy to be a solution of the equilibrium equations. In the first edition of Truesdell and Noll (1965), page 209, they conjecture that the only solution with constant invariants must be homogeneous. However, they did not observe that the deformation

$$r = \sqrt{A}R, \qquad \theta = C\Theta, \qquad z = FZ,$$

enclosed in Family 3 is not homogeneous, but has constant invariants, as was shown by Fosdick (1966). Further development on this problem is contained in the work on universal solutions for the elastic dielectric by Singh and Pipkin (1965). As a by-product of this research they discovered the **Family 5,** which is given by formula (3.14) with $ACF = 1$, a universal deformation with all the invariants constant and not contained in the list of Ericksen.

Today we know that the solution of Ericksen's problem is complete for the following classes of deformations:
generalized plane deformations (Fosdick and Schuler, 1969; Müller, 1970):

$$x_1 = x_1(X_1, X_2), \qquad x_2 = x_2(X_1, X_2), \qquad x_3 = x_3(X_3);$$

radially symmetric deformations (Fosdick, 1971):

$$r = r(R, Z), \quad \theta = \Theta, \quad z = z(R, Z);$$

generalized cylindrical deformations (Müller, 1970):

$$r = r(R), \qquad \theta = f_1(\Theta) + g_1(Z), \qquad z = f_2(\Theta) + g_2(Z);$$

generalized spherical deformations (Müller, 1970):

$$r = r(R), \qquad \theta = \Theta, \qquad \phi = \phi(\Phi);$$

and for antiplane shear deformations (Knowles, 1979).

Moreover, Kafadar (1972) and Marris (1975) have established some general results about the possible existence of constant invariant deformations. To give a flavor of these results we have to introduce some definitions from classical differential geometry and first of all the *abnormality* of a vector†. Let e_i ($i = 1, 2, 3$) be three orthogonal unit vectors; the curl components $\pi_{ab} = e_b \cdot \text{curl} e_a$ are the abnormalities when $a = b$, and the curvature components when $a \neq b$. Kafadar (1972) showed that there is no possibility of new solutions with constant invariants if the eigenvectors of the strain are constant or if at least one of the abnormalities of these eigenvectors is zero. Moreover, the results of Kafadar imply that the principal stretches may be distinct. Otherwise, we recover the Singh and Pipkin deformation or the homogeneous deformations. On the other hand, Marris (1975) showed that no new solution exists if at least two of the three abnormalities are constants.

All these studies are on the negative side of the problem. They show that no new solutions exists under some conditions, and they give only vague information on existence and the way to construct new solutions. In Marris (1982) and in Adeleke (1984) the authors change the approach. In Marris (1982) it is shown that if a new solution exists then it must satisfy a certain polynomial link between the abnormalities, the curvature and the eigenvalues of the strain. This result is very important since for the first time we come back to the true nature of the problem, which is a compatibility problem. Starting with (3.7) we have in this case that

$$\nabla \pi = \delta_1 \nabla \cdot \mathbf{B} + \delta_2 \nabla \cdot \mathbf{B}^{-1},$$

and we have to solve the system given by

$$\nabla \times \nabla \cdot \mathbf{B} = \mathbf{0}, \quad \nabla \times \nabla \cdot \mathbf{B}^{-1} = \mathbf{0},$$

and the relation

$$\Gamma_{km}^q \Gamma_{qr}^m - \Gamma_{kr,m}^m = 0.$$

Then, we take into account that the deformation gradients may be written as

$$\mathbf{F} = \frac{d\mathbf{x}}{d\mathbf{X}} = \lambda_1 e_1 \otimes \mathbf{E}_1 + \lambda_2 e_2 \otimes \mathbf{E}_2 + \lambda_3 e_3 \otimes \mathbf{E}_3,$$

$$\mathbf{F}^{-1} = \frac{d\mathbf{X}}{d\mathbf{x}} = \lambda_1^{-1} \mathbf{E}_1 \otimes e_1 + \lambda_2^{-1} \mathbf{E}_2 \otimes e_2 + \lambda_3^{-1} \mathbf{E}_3 \otimes e_3,$$

where the λ_i are constants and $\lambda_1 \lambda_2 \lambda_3 = 1$.

We write 15 independent scalar relations involving the derivatives of the components of the strain and we have to set a compatibility problem. The

† The abnormalities are introduced through the derivatives along an eigenbasis.

direct computations are huge since we have to work using necessary conditions and then we have to raise the order of the derivation. In any case, in the end we obtain a full set of necessary compatibility conditions. Marris was able to find, using some geometry and a lot of algebra, one of these conditions. Obtaining all these conditions is a fundamental (although not final) step toward the complete solution and in principle all the manipulations can be done automatically by a symbolic software package.

In Adeleke (1984) the problem is attacked in a constructive way using a representation formula for the deformation with constant strain invariants. Under the assumption of analyticity he shows that the set of universal deformations with distinct principal stretches is at most a 19-parameter set to within rigid rotations, and then, by means of iterative schemes, he showed that if there is a point within the domain of analyticity where the first and second order derivatives of the strain are zero then the solution is homogeneous.

3.6 The universal manifold

The determination of the universal relations for a wide class of constitutive equations can be done using a simple method presented in Pucci and Saccomandi (1997). Here we adapt the general method to the case of elastic isotropic materials, as in Pucci and Saccomandi (1996c).

Let us consider a universal deformation and compute the corresponding deformation tensor $\mathbf{B}$. We choose a point $\mathbf{X}$ and then express the constitutive equation (2.1) in the space S, in which the coordinates[†]

$$\tau = \{\tau_1, \tau_2, \tau_3, \tau_4, \tau_5, \tau_6\}, \quad \beta = \{\beta_0, \beta_1, \beta_{-1}\},$$

have been introduced.

In this space (2.1) is a linear homogeneous manifold, $\mathcal{V}$ say, of dimension 3 and which is parametrized with β. The projection of this manifold in the subspace $S^*(\tau) \subset S$ is again a manifold, $\Pi(\mathcal{V})$ say, whose dimension k is given by the rank of the matrix

$$\begin{pmatrix} 1 & 0 & 0 & 1 & 0 & 1 \\ B_{11} & B_{12} & B_{13} & B_{22} & B_{23} & B_{33} \\ B_{11}^{-1} & B_{12}^{-1} & B_{13}^{-1} & B_{22}^{-1} & B_{23}^{-1} & B_{33}^{-1} \end{pmatrix}.$$

This last manifold is called the *universal manifold* since it does not depend on the response coefficients. Moreover, since (2.1) is linear and homogeneous it is

† Here, $\tau_1 = \sigma_{11}, \tau_2 = \sigma_{12}, \tau_3 = \sigma_{13}, \tau_4 = \sigma_{22}, \tau_5 = \sigma_{23}, \tau_6 = \sigma_{33}$.

always possible to give for $\Pi(\mathcal{V})$ the representation

$$H_q(\tau, \mathbf{X}) \equiv \sum_{p=1}^{6} \tau_p \phi_{pq}(\mathbf{X}) = 0,$$

where $q = 1, ..., 6 - k$, and all the equations are linear and homogeneous in τ. This idea allows all the universal relations to be found. Obviously, this method is more general than the coaxiality relation and it can be applied also to contexts other than elasticity.

We give two examples of applications of the universal manifold method: the special case of incompressible materials with a strain-energy function only of the first invariant I_1, and the case of constrained materials.

3.6.1 Universal relations when $W = W(I_1)$

In the literature several strain energies in the form $W = W(I_1)$ have been proposed to model the hyperelasticity of incompressible materials. In this case the Cauchy stress is given by

$$\sigma = -p\mathbf{I} + 2\frac{dW}{dI_1}\mathbf{B}, \tag{6.1}$$

and it is not surprising to discover a new fourth universal relation with respect to the general theory of elasticity. This is because the independent stress components are six and the response coefficients now are only two. The new universal relation cannot be recovered by considering the coaxiality condition. To describe the universal manifold it is necessary to compute the rank of the matrix

$$\begin{pmatrix} 1 & 0 & 0 & 1 & 0 & 1 \\ B_{11} & B_{12} & B_{13} & B_{22} & B_{23} & B_{33} \end{pmatrix},$$

which is at most two, and then to find the Euclidean imbedding of the linear manifold. This can be done by using the three usual universal relations arising from the coaxiality condition and the new one

$$(\sigma\mathbf{u} \times \mathbf{B}\mathbf{u}) \cdot \mathbf{u} = 0, \tag{6.2}$$

where $\mathbf{u}$ is an arbitrary vector.

In the case of a simple shear deformation this new universal relation is simply computed as $\sigma_{22} = \sigma_{33}$. This universal relation could be of practical significance because it provides a method of measuring quantitatively what it means to discard the β_{-1} constitutive function (see Section 3.2). A detailed discussion of the effects of discarding β_{-1} in the modelling of rubber can be found in Horgan and Saccomandi (1999).

3.6.2 Universal relations for constrained materials

In the case of constrained materials, since the scalar π is to be determined by the balance equations, it is usually represented in the form

$$\pi = \pi(\mathbf{x}, \delta_i, \partial\delta_i), \tag{6.3}$$

where the symbol $\partial\delta_i$ stands for the set of all the derivatives (of any order) of the response coefficients. The determination of universal relations, considering π as a response coefficient, is correct, but when a restriction of the form (6.3) holds, i.e.

$$\pi = b_1\delta_1 + b_2\delta_2, \tag{6.4}$$

the situation is obviously different.

Now, the Cauchy stress can be written as

$$\sigma = -\,(\,b_1\delta_1 + b_2\delta_2\,)\,\mathbb{S} + \delta_1\mathbf{B} + \delta_2\mathbf{B}^{-1},$$

i.e.

$$\sigma = \delta_1\,(\mathbf{B}-b_1\mathbb{S}) + \delta_2\,(\mathbf{B}^{-1} - b_2\mathbb{S}),$$

and we have to consider the rank of the 2×6 matrix whose entries in the first row are the independent components of the tensor $\mathbf{B} - b_1\mathbb{S}$ and in the second row the independent components of $\mathbf{B}^{-1} - b_2\mathbb{S}$. Then the rank is at most two and a new universal relation arises.

This observation was first made in Pucci and Saccomandi (1996b), where it was shown that when we have (6.4), a new universal relation peculiar to constrained materials arises. This new universal relation is a link between the eigenvalues of the stress and the strain and again cannot be recovered by the coaxiality condition. It is interesting to point out that for homogeneous deformations or deformations with constant invariants the associated reaction scalar is always as in (6.4).

For example, let us consider the superposition of a simple shear on a triaxial stretch, i.e.

$$x_1 = \lambda_1 X_1 + kX_2, \quad x_2 = \lambda_2 X_2, \quad x_3 = \lambda_3 X_3. \tag{6.5}$$

Now π assumes a constant value determined by the boundary conditions. The physical components of the stress tensor are

$$\sigma_{11} = -\pi\left[1 + a\left(\lambda_1^2 + k^2\right) + \frac{b}{\lambda_1^2}\right] + \delta_1\left(\lambda_1^2 + k^2\right) + \frac{\delta_2}{\lambda_1^2},$$

$$\sigma_{22} = -\pi\left[1 + a\lambda_2^2 + b\left(\frac{\lambda_1^2 + k^2}{\lambda_2^2\lambda_1^2}\right)\right] + \delta_1\lambda_2^2 + \delta_2\left(\frac{\lambda_1^2 + k^2}{\lambda_2^2\lambda_1^2}\right),$$

$$\sigma_{12} = -\pi \left[k \left(a\lambda_2 - \frac{b}{\lambda_2 \lambda_1^2} \right) \right] + k \left(\delta_1 \lambda_2 - \frac{\delta_2}{\lambda_2 \lambda_1^2} \right),$$

$$\sigma_{13} = \sigma_{23} = 0, \quad \sigma_{33} = -\frac{\pi}{\lambda_3^2}. \tag{6.6}$$

Using the coaxiality between stress and strain it is possible to obtain

$$k\lambda_2 (\sigma_{11} - \sigma_{22}) = \left(\lambda_1^2 - \lambda_2^2 + k^2 \right) \sigma_{12}, \tag{6.7}$$

and the trivial universal relations $\sigma_{13} = \sigma_{23} = 0$.

The new universal relation can be obtained by considering that once the boundary conditions are fixed π must be as in (6.4). For example, if we choose π such that $\sigma_{12} = 0$, we have

$$\pi = \frac{\delta_1 \lambda_2^2 \lambda_1 - \delta_2}{a \lambda_2^2 \lambda_1 - b},$$

and this is always possible if the reaction tensor stress is not spherical. Then, the new universal relation is† $\sigma_{11} = \sigma_{22}$. This is a very interesting fact. Indeed, for incompressible materials, the stress can have a double eigenvalue if and only if the strain has a double eigenvalue, but this is not the case for general isotropically constrained materials. The Cauchy stress of constrained materials is considered in a short note by Hayes and Saccomandi (2000) in which the general result in linear algebra and many details are given.

It is also interesting to note that the use of the fourth universal relation peculiar to constrained materials is effective in many experimental results. Indeed, in many experiments with incompressible rubber based on a triaxial stretch deformation, the pressure is chosen so that $\sigma_{33} = 0$. This is the new universal relation, which is peculiar to constrained materials, and which cannot be obtained from coaxiality (see, for example, page 363 of Ogden, 1986).

3.7 Relative-universal deformations

The search for controllable deformations in unconstrained elasticity has been, in some sense, *delayed* for many years because of the Ericksen result on universal deformations. Only in recent times has there been a revival of interest in these deformations. These new researches have been catalyzed by two important papers by Isherwood and Ogden (1977) and by Currie and Hayes (1981).

In Isherwood and Ogden (1977) a method for plane strain problems has been proposed based on complex variables which has been very effective in obtaining the direct solution to several boundary-value problems for particular

† From the coaxiality of stress and strain now we have only trivial relations.

classes of materials. These solutions usually are not relative-universal but only controllable deformations.

The paper by Currie and Hayes (1981) has been a landmark in the search for relatively universal solutions. The title of this paper is misleading since it contains the word non-universal, but here for the first time it was shown, by means of simple and direct examples, that it is possible to build classes of constitutive equations for which new controllable solutions can be found. The authors use as examples a variable shear deformation

$$x_1 = X_1 + f(X_2, X_3), \quad x_2 = X_2, \quad x_3 = X_3,$$

where $f(X_2, X_3)$ is a harmonic function, a uniaxial extension

$$x_1 = f(X_1), \quad x_2 = \frac{X_2}{df/dX_1}, \quad x_3 = \frac{X_3}{df/dX_1},$$

and the torsional deformation

$$r = R, \quad \theta = \Theta + DZ, \quad z = Z.$$

Since the publication of these papers various authors have presented examples of controllable deformations for particular compressible and incompressible elastic materials and some of these solutions are relatively-universal. Fundamental results in this direction has been obtained by Carroll, Beatty, Horgan, Rajagopal and many others, and are reviewed in the chapters by J.M. Hill and C.O. Horgan in this volume.

There are two typical problems arising in the search for relatively universal solutions.

- The first one is the following: given a deformation, is it possible to find the general form of the strain energy for which this is a controllable solution?
- The second one is a sort of Ericksen problem in miniature: given a particular class of models is it possible to find all the corresponding relatively universal solutions?

Both problems are very difficult to solve and generally only partial results are available.

We begin by studying the first problem, recalling the radial deformation considered in Section 3.2.2, i.e.

$$r = r(R), \quad \theta = \Theta, \quad z = Z.$$

Here, the question is what is the functional form of the most general (elastic, isotropic) strain-energy density for which the deformation

$$r(R) = AR + \frac{B}{R} \tag{7.1}$$

is a solution of the balance equations in the absence of body forces (see, e.g., Aron, 1994).

Because

$$\lambda_1 = A - \frac{B}{R^2}, \quad \lambda_2 = A + \frac{B}{R^2}, \tag{7.2}$$

and

$$\lambda_1 + \lambda_2 = 2A, \quad \lambda_2 - \lambda_1 = 2\frac{B}{R^2} = Rr'', \tag{7.3}$$

going back to equation (2.13) it is possible to write

$$R\left[r''\Sigma_{11} + R\left(\frac{r}{R}\right)'\Sigma_{12}\right] = \Sigma_2 - \Sigma_1, \tag{7.4}$$

and, using (7.3),

$$(\lambda_1 - \lambda_2)(\Sigma_{11} - \Sigma_{12}) = \Sigma_2 - \Sigma_1. \tag{7.5}$$

Equation (7.5) can be integrated once with respect to λ_1 to obtain the first-order linear partial differential equation

$$\Sigma_1 - \Sigma_2 = \frac{d^2 H(\lambda_2)}{d\lambda_2^2}(\lambda_1 - \lambda_2),$$

where $H(\lambda_2)$ is an arbitrary function. The general integral of this equation is given by

$$\Sigma = \widehat{H}(\lambda_1 + \lambda_2) + 2H - \frac{dH}{d\lambda_2}(\lambda_1 - \lambda_2), \tag{7.6}$$

where $\widehat{H}(\lambda_1 + \lambda_2)$ is a new arbitrary function. Imposing on (7.6) the symmetry requirement of isotropy, it may be shown that only the subclass of (7.6), given by

$$\Sigma = \widehat{H}(\lambda_1 + \lambda_2) + k\lambda_1\lambda_2, \tag{7.7}$$

is an admissible strain energy, where k is an arbitrary parameter.

The strain energy (7.7) *seems* to be the most general isotropic strain energy for which the radial deformation (7.1) is a universal deformation. Now, it is important to point out that our computations have been done not in the whole positive cone of (λ_1, λ_2)-space, but in a submanifold of this cone, namely $\lambda_1 + \lambda_2 = 2A$. This is because we have used (7.3) in deriving (7.5) from (7.4).

The problem of recovering the most general strain energy in (λ_1, λ_2)-space from information on a submanifold is obviously very hard and, generally, it is not possible to solve it completely and explicitly. Indeed, (7.1) is a universal deformation also for the materials with strain energy given by

$$\Sigma = \widehat{H}(\lambda_1 + \lambda_2) + \widetilde{H}(\lambda_1 + \lambda_2)\lambda_1\lambda_2, \tag{7.8}$$

where $\widetilde{H}\left(\lambda_1 + \lambda_2\right)$ is an arbitrary function such that $\widetilde{H'}\left(2A\right) = 0$. Aron (1994), using the fact that, for a range of values of the parameter A, $\lambda_1 + \lambda_2 = 2A$ fills the entire positive cone of (λ_1, λ_2)−space, was able to show that (7.8) is the complete answer to our question. The result of Aron is rather special; in general, it is not possible to recover the general form of the strain energy.

The indeterminacy in solving these kinds of problems is subtle. To the best of our knowledge, the suggestion that the restriction to the manifold be used to enlarge the results of this sort of computation is exploited only in Currie and Hayes (1981) and in Jiang and Ogden (1998, 2000).

The problem arising from the second question is an Ericksen problem with a changed point of view. We are searching for universal solutions in a class which is no longer a material symmetry class. Also, this approach is very difficult and the best results that we know have been given by Carroll (1995) for special classes of unconstrained materials. This is also a problem of interest in the case of constrained elasticity. For example, if we consider a neo-Hookean material, because the Cauchy stress tensor is given by

$$\sigma = -p\mathbf{I} + \mu\mathbf{B},$$

where μ is constant, all the solutions of the balance equations are given by those strain tensors such that

$$\nabla \times \nabla \cdot \mathbf{B} = \mathbf{0}.$$

Then all possible deformations are independent of the constitutive parameter μ and are therefore relatively-universal in the *neo-Hookean world*. This is a trivial example, but more general examples are very difficult to find. Let us consider more general incompressible materials with a strain-energy function $W = W(I_1)$. Now, the universal deformations are characterized by the three relations

$$\nabla \times \nabla \cdot \mathbf{B} = \mathbf{0}, \quad \nabla I_1 \times \mathbf{B}\nabla I_1 = \mathbf{0}, \quad \nabla I_1 \times \nabla \cdot \mathbf{B} + \nabla \times \mathbf{B}\nabla I_1 = \mathbf{0}$$

and the compatibility equations for the strain†. In this case all the information with respect to $\mathbf{B}^{-1}$ is missing. It is possible to show that, if new universal solutions exist they must be characterized by the fact that ∇I_1 is not parallel to ∇I_2 (then, for example, new solutions cannot be plane deformations), but we do not know of any examples of new deformations.

In the framework of unconstrained isotropic materials we have only to record

---

† In a brief note by Parry (1979) it is pointed out that it is not necessary to use all the equations of the overdetermined system (3.11) to deduce the Ericksen result. It is possible that if we pick a restricted class of isotropic materials, we still cannot find new universal deformations.

the partial results of Carroll (1995) concerning materials of class I, II and III, i.e.

$$W = f(i_1) + c_2(i_2 - 3) + c_3(i_3 - 1),$$

$$W = c_1 (i_1 - 3) + g(i_2) + c_3(i_3 - 1),$$

$$W = c_1 (i_1 - 3) + c_2(i_2 - 3) + h(i_3),$$

respectively where the c_i are constants and f, g, h are functions of the principal invariants i_1, i_2, i_3 of the stretch tensor $\mathbf{V}$ in the polar decomposition of $\mathbf{F} = \mathbf{VR}$.

A general result obtained by Carroll is that universal deformations for these materials must be such that $\mathrm{Div}\,\mathbf{R} = 0$. The explicit examples of universal solutions obtained by this approach are, however, just the radial deformations previously found by other methods (see Horgan, 1995b, and references cited therein).

3.8 Concluding remarks

Considerable progress has been made in universal results in finite isotropically constrained elasticity. About universal solutions there is one open question remaining: the question of the existence of further deformations with constant invariants.

Section 3.7 shows that the idea of relative-universal solutions poses difficulties in finding new exact solutions of the equations in unconstrained nonlinear elasticity. We also have to cope with inhomogeneous deformations whose geometry depends, in some sense, on the response coefficients.

Another approach considered by Carroll uses the concept of a universal state of stress, i.e. a state that can exist, in absence of body forces, in every material of the class whose stress range allows it (Carroll, 1973a, b). The results are that the only states of stress universal for unconstrained materials are homogeneous and for incompressible materials must have uniform deviatoric stress invariants. Thus, there are no analogues to the four families of finite deformations with non-uniform stretch invariants already listed.

The problem of Ericksen for anisotropic materials is not truly different from the one analyzed here, and this is because an isotropic material is also a special case of an anisotropic material. In this case, the problem is only to consider the universal solutions for the isotropic case and to check by direct computation when they are also universal solutions for the anisotropic class of materials under investigation. This has been done for some classes of anisotropy by

Ericksen and Rivlin (1954) and by Huilgol (1966). This means that here the results are also quite complete.

On the other hand the problem of Ericksen, when we consider the case of *anisotropic constraints,* is still an obscure matter and many questions are waiting for a rigorous and definitive answer. To the best of our knowledge the only anisotropic constraints investigated are inextensibility and area preservation. A brief survey of results obtained in this field is presented in Beatty and Saccomandi (2000), in which the universal manifold method is used to determine the complete set of universal relations.

The universal manifold method is completely general and can be used to compute the complete set of universal relations for any anisotropy class and also when anisotropic constraints are considered (Pucci and Saccomandi, 1997).

Acknowledgements

It has been a very great pleasure to collaborate with Professor Edvige Pucci on universal results. I thank her for unfailing help, support and very many kindnesses. I am also much indebted to Professors Mike Hayes, Cornelius Horgan and Ray Ogden for stimulating discussions and for their criticisms of the manuscript. My thanks also go to Yibin Fu and Ray Ogden for inviting me to contribute to this volume.

References

Acharya, A. 1999 On compatibility conditions for the left Cauchy-Green deformation field in three dimensions. *J. Elasticity* **56**, 95-105.

Antman, S.S. 1995 *Nonlinear Problems in Elasticity.* New York: Springer-Verlag.

Adeleke, S.A. 1984 On Ericksen's problem in elasticity. *Mech. Res. Comm.* **11**, 21-27.

Aron, M. 1994 On a class of plane deformations of compressible nonlinearly elastic solids. *IMA J. Appl. Math.* **52**, 289-296.

Beatty, M.F. 1978 General solutions in the equilibrium theory of inextensible elastic materials. *Acta Mechanica* **29**, 119-126.

Beatty, M.F. 1987a Topics in finite elasticity: Hyperelasticity of rubber, elastomers, and biological tissues - with examples. *Appl. Mech. Rev.* **40**, 1699-1733. (Reprinted with minor modifications as "Introduction to nonlinear elasticity". In *Nonlinear Effects in Fluids and Solids,* eds. M.M. Carroll and M.A. Hayes, 1996. New York: Plenum Press.)

Beatty, M.F. 1987b A class of universal relations in isotropic elasticity. *J. Elasticity* **17**, 113-121.

Beatty, M.F. and Hayes, M.A. 1992a Deformations of an elastic, internally constrained material, part 1: homogeneous deformations. *J. Elasticity* **29**, 1-84.

Beatty, M.F. and Hayes, M.A. 1992b Deformations of an elastic, internally constrained material, part 2: nonhomogeneous deformations. *Q. J. Mech. Appl. Math.* **45**, 663-709.

Beatty, M.F. and Saccomandi, G. 2000 Universal relations for fiber reinforced materials. *J. Elasticity*, to appear.

Bell, J.F. 1996 The decrease of volume during loading and coaxiality in finite plastic strain. *Meccanica* **31**, 461-472.

Bosi, M. and Salvatori, M.C. 1996 Some inhomogeneous deformations for a special class of isotropic constrained materials. *Rend. Mat. Appl.* **16**, 689-713.

Carroll, M.M. 1973a Controllable state of stress for compressible elastic solids. *J. Elasticity* **3**, 57-61.

Carroll, M.M. 1973b Controllable states of stress for incompressible elastic solids. *J. Elasticity* **3**, 147-153.

Carroll, M.M. 1995 On obtaining closed form solutions for compressible elastic materials. *ZAMP* **46**, S126-S145.

Currie, P.K. and Hayes, M. 1981 On non-universal finite elastic deformations. In *Finite Elasticity*, eds. D. Carlson and R.T. Shield. The Hague: Martinus Nijhoff Publishers, 143-150.

Dragoni, E. 1996 The radial compaction of a hyperelastic tube as a benchmark in compressible finite elasticity. *Int. J. Non-Linear Mech.* **41**, 483-493.

Ericksen, J.L. 1954 Deformations possible in every isotropic, incompressible, perfectly elastic body. *ZAMP* **5**, 466-488.

Ericksen, J.L. 1955 Deformations possible in every isotropic compressible, perfectly elastic material. *J. Math. Phys.* **34**, 126-128.

Ericksen, J.L 1977 Semi-inverse methods in finite elasticity theory. In *Finite Elasticity* AMD 27, ed. R.S. Rivlin, 11-21.

Ericksen, J.L. 1986 Constitutive theory for some constrained elastic crystals. *Int. J. Solids Structures* **22**, 951-964.

Ericksen, J.L. and Rivlin R.S 1954 Large elastic deformation of homogeneous anisotropic materials. *J. Rational Mech. Anal.* **3** , 281-301.

Fosdick, R.L. 1966 Remarks on compatibility. In *Modern Developments in the Mechanics of Continua*, 109-127. New York: Academic Press.

Fosdick, R.L. 1971 Statically possible radially symmetric deformations in isotropic incompressible elastic solids. *ZAMP* **22**, 590-607.

Fosdick, R.L. and Schuler, K.W. 1969 On Ericksen's problem for plane deformations with uniform transverse stretch. *Int. J. Engng Sci.* **7**, 217-233.

Green, A.E. and Adkins, J.E. 1960 *Large Elastic Deformations*. Oxford: Clarendon Press.

Hayes, M.A and Knops, R.J. 1966 On universal relations in elasticity theory. *ZAMP* **17**, 636-639.

Hayes, M.A. and Saccomandi, G. 2000 The Cauchy stress tensor for a material subject to an isotropic internal constraint. *J. Engng Math.*, **37**, 85-92.

Huilgol, R.R. 1966 A finite deformation possible in transversely isotropic materials. *ZAMP* **17**, 787-788.

Horgan, C.O. 1989 Recent developments concerning Saint-Venant's principle: an update. *Appl. Mech. Rev.* **42**, 295-303.

Horgan, C.O. 1995a Antiplane-shear deformations in linear and nonlinear solid mechanics. *SIAM Review* **37**, 53-81.

Horgan, C. O. 1995b On axisymmetric solutions for compressible nonlinearly elastic solids. *ZAMP* **46**, S107-S125.

Horgan, C.O. 1996 Recent developments concerning Saint-Venant's principle: a second update. *Appl. Mech. Rev.* **49**, S101-S111.

Horgan, C.O. and Polignone D.A. 1995 Cavitation in nonlinearly elastic solids: A review. *Appl. Mech. Rev.* **48**, 471-485.

Horgan, C.O. and Saccomandi, G. 1999 Simple torsion of isotropic hyperelastic incompressible materials with limiting chain extensibility. *J. Elasticity* **56**, 159-170.

Isherwood, D.A. and Ogden, R.W. 1977 Finite plane strain problems for compressible elastic solids: General solution and volume changes. *Rheol. Acta* **16**, 113-122.

Jiang, X. and Ogden, R.W. 1998 On azimuthal shear of a circular cylindrical tube of compressible material. *Q. J. Mech. Appl. Math.* **51**, 143-158.

Jiang, X. and Ogden, R.W. 2000 Some new solutions for the axial shear of a circular tube of compressible material. *Int. J. Non-Linear Mech.* **35**, 361-369.

Kafadar, C.B. 1972 On Ericksen's problem. *Arch. Rat. Mech. Anal.* **47**, 15-27.

Kafadar, C.B 1975 Exact solutions in fluids and solids. In *Continuum Physics*, ed. A.C. Eringen. New York: Academic Press, 407-448.

Knowles, J.K. 1979 Universal states of finite anti-plane shear, Ericksen's problem in miniature. *Am. Math. Monthly* **86**, 109-113.

Leblanc, J.L 1998 What's new in rheometry and vulcametry of rubber materials. *Kautschuk Gummi Kunststoffe* **51**, 19-27.

Lurie, A.I. 1990 *Nonlinear Theory of Elasticity.* Amsterdam: North-Holland.

Marris, A.W. 1975 Universal deformations in incompressible isotropic elastic materials. *J. Elasticity* **5**, 111-128.

Marris, A.W. 1982 Two new theorems on Ericksen's problem. *Arch. Rat. Mech. Anal.* **79**, 131-173.

Marris, A.W. and Shiau, J.F. 1970 Universal deformations in isotropic incompressible hyperelastic materials when the deformation tensor has equal proper values. *Arch. Rat. Mech. Anal.* **36**, 135-160.

Martins, L.C. and Duda, F.P. 1998 Constrained elastic bodies and universal solutions in the class of plane deformations with uniform transverse stretch. *Math. Mech. Solids* **3**, 91-106.

Martins, L.C. and Duda, F.P. 1997 Plane deformations with constant strain invariants, Programa de engenharia mecanaica relatorio interno 7/87 Universidade Federaldo Rio de Janeiro.

Müller, W. C. 1970 Some further results on the Ericksen problem for deformations with constant strain invariants. *ZAMP* **21**, 633-636.

Muncaster, R.G. 1979 Saint Venant problems in nonlinear elasticity: a study of cross-section. In *Nonlinear Analysis and Mechanics: Heriot-Watt symposium*, ed. R.J. Knops. London: Pitman, 17-75.

Nabarro, F.R.N. 1979 Nonlinear elastic problems. In *Dislocations in Solids*, ed. B.K.D Gairola. Amsterdam: North Holland.

Ogden R.W. 1984 *Non-linear Elastic Deformations.* Chichester: Ellis Horwood. Reprinted 1997 New York: Dover Publications.

Ogden R.W. 1986 Recent advances in the phenomenological theory of rubber elasticity. *Rubber Chem. Technol.* **59**, 361-383.

Parry, G. P. 1979 Corollaries of Ericksen's theorems on the deformations possible in every isotropic hyperelastic body. *Arch. Mech.* **31**, 757-760.

Podio-Guidugli, P. 1990 Constrained elasticity. *Rend. Mat. Lincei* **s.9**, 341-350.

Podio-Guidugli, P. and Vianello, M. 1989 Constraint manifolds for isotropic solids. *Arch. Rat. Mech. Anal.* **105**, 105-121

Pucci, E. and Saccomandi, G. 1992 On the weak symmetry groups of partial differential equations. *J. Math. Anal. Appl.* **163**, 588-598.

Pucci, E. and Saccomandi, G. 1996a Some universal solutions for totally inextensible isotropic constrained materials. *Q.J. Mech. Appl. Math.* **49**, 147-162.

Pucci, E. and Saccomandi, G. 1996b Universal relations in constrained elasticity, *Math. Mech. Solids* **1**, 207-217.

Pucci, E. and Saccomandi, G. 1996c Universal relations in finite elasticity. In *Contemporary Research in the Mechanics and Mathematics of Materials*, eds R.C. Batra and M.F. Beatty. Barcelona: CIMNE, 176-184.

Pucci, E. and Saccomandi, G. 1997 On universal relations in continuum mechanics. *Cont. Mech. Thermodyn.* **9**, 61-72.

Pucci, E. and Saccomandi, G. 1998 Universal generalized plane deformations in constrained elasticity. *Math. Mech. Solids* **3**, 201-216.

Pucci, E. and Saccomandi, G. 1999 Universal solutions in constrained simple materials. *Int. J. Nonlinear Mech.* **34**, 469-484.

Rivlin, R. S. 1997 *Collected Papers*, eds G. I. Barenblatt and D. D. Joseph. New York: Springer-Verlag.

Saccomandi, G. and Vianello, M. 1997 A universal relation characterizing hemitropic hyperelastic materials. *Math. Mech. Solids* **2**, 181-188.

Shield, R. T. 1971 Deformations possible in every compressible isotropic perfectly elastic material. *J. Elasticity* **1**, 145-161.

Singh, M and Pipkin, A.C. 1965 Note on Ericksen's problem. *ZAMP* **16**, 706-709.

Truesdell, C. and Noll, W. 1965. In Flügge, S., ed., *The Nonlinear Field Theories of Mechanics: Handbuch der Physik III/3*. Berlin: Springer-Verlag.

Truesdell, C. 1956 Das Ungelöste Hauptproblem der Endlichen Elastizitätstheorie. *ZAMP* **36**, 97-103. (Translated with the title "The main unsolved problem in finite elasticity theory". In 1965 *Continuum Mechanics III*. New York: Gordon and Breach).

Vianello, M. 1996 Optimization of the strain energy and coaxiality between stress and strain in finite elasticity. *J. Elasticity* **44**, 193-202.

Wang, C.C. and Truesdell C. 1973 *Introduction to Rational Elasticity*. Leyden: Noordhoff International Publishing.

Wineman, A.S. and McKenna, G.B. 1996 Determination of the strain energy density function for compressible isotropic nonlinear elastic solids by torsion-normal force experiments. In *Nonlinear effects in fluids and solids*, eds. M.M. Carroll and M. Hayes. New York: Plenum Press.

4

Equilibrium solutions for compressible nonlinearly elastic materials

Cornelius O. Horgan

Department of Civil Engineering
University of Virginia
Charlottesville, VA 22903, U.S.A.
Email: coh8p@virginia.edu

For homogeneous isotropic *incompressible* nonlinearly elastic solids in equilibrium, the simplified kinematics arising from the constraint of no volume change has facilitated the analytic solution of a wide variety of boundary-value problems. For *compressible* materials, the situation is quite different. Firstly, the absence of the isochoric constraint leads to more complicated kinematics. Secondly, since the only controllable deformations are the homogeneous deformations, the discussion of inhomogeneous deformations has to be confined to a particular strain-energy function or class of strain-energy functions. Nevertheless, in recent years, substantial progress has been made in the development of analytic forms for the deformation and in the solution of boundary value problems. The purpose of this Chapter is to review some of these recent developments.

4.1 Introduction

For homogeneous isotropic *incompressible* materials in equilibrium, the simplified kinematics arising from the constraint of no volume change has facilitated the analytic solution of a wide variety of boundary-value problems, see, e.g., Ogden (1982, 1984), Antman (1995), and Chapter 1 of the present volume. Most well-known among these are the *controllable* or *universal* deformations, namely, those inhomogeneous deformations which are independent of material properties and thus can be sustained in *all* incompressible materials in the absence of body forces. For homogeneous isotropic *compressible* materials, Ericksen (1955) established that the only controllable deformations are *homogeneous* deformations. Thus, *inhomogeneous* deformations for compressible materials necessarily have to be discussed in the context of a particular strain-energy function or class of strain-energy functions. Even with this restriction, the establishment of analytic closed-form solutions of the governing nonlinear differential equations (even without the additional difficulty of satisfying boundary conditions in the solution of boundary-value problems), has been a formidable task. Nevertheless, in recent years substantial progress has been made in the development of analytic forms for the deformation and in the solution of boundary-value problems. The purpose of this chapter is to review some of these recent developments. We confine our attention primarily to progress which has been made since the publication of the book by Ogden (1984), where references to earlier work may be found. Emphasis is placed on research areas where the present author has made some contributions.

We begin, in Sections 4.2-4.6, with a discussion of problems involving spherically and cylindrically symmetric deformations. Since the governing equations reduce, in this case, to a single second-order nonlinear ordinary differential equation, this is the framework within which the most rapid progress has been made to date. Two analytic techniques are described in Sections 4.3 and 4.4, leading to consideration of first-order differential equations. Illustrative examples for specific classes of strain-energies are provided in Sections 4.5 and 4.6. In Section 4.5, the strain-energies are functions of the principal invariants of the stretch tensor and include the well-known *harmonic* material model due to John (1960), while Section 4.6 considers the special Blatz-Ko material (Blatz-Ko 1962). Some other equilibrium solutions and work on controllable deformations are briefly described in Section 4.7. In Section 4.8, we consider the generalized Blatz-Ko material and its various specializations, and describe some boundary-value problems that have been solved for this experimentally-based material model. The Blatz-Ko strain-energy has the feature that the associated equilibrium equations (a system of second-order quasilinear partial differential equations) can lose ellipticity at sufficiently large stretches. This issue and its ramifications for some boundary-value problems are discussed in Section 4.9. Finally, in Section 4.10, we briefly describe some recent work on *isochoric* deformations for compressible materials. Such deformations include, for example, pure torsion, anti-plane shear, pure axial shear, and pure azimuthal shear.

4.2 Axisymmetric deformations of homogeneous isotropic compressible elastic solids

Consider spherically symmetric deformations of an elastic sphere. The deformation, which takes the point with spherical polar coordinates (R, Θ, Φ) in the undeformed region to the point (r, θ, ϕ) in the deformed region, has the form

$$r = r(R), \quad \theta = \Theta, \quad \phi = \Phi; \quad \frac{dr}{dR} > 0, \tag{2.1}$$

where $r(R)$ is to be determined. The polar components of the deformation gradient tensor associated with (2.1) are given by

$$\mathbf{F} = \text{diag}\left(\dot{r}(R), \frac{r(R)}{R}, \frac{r(R)}{R}\right) \tag{2.2}$$

while the principal stretches are

$$\lambda_1 = \dot{r}(R), \quad \lambda_2 = \lambda_3 = \frac{r(R)}{R}. \tag{2.3}$$

In (2.2), (2.3), the superposed dot denotes differentiation with respect to R.

As discussed in Chapter 1, the strain-energy density per unit undeformed volume for isotropic elastic compressible materials is given by

$$W = W(I_1, I_2, I_3), \tag{2.4}$$

where I_1, I_2, I_3 are the principal invariants of the deformation tensor $\mathbf{C} = \mathbf{F}^T\mathbf{F}$. We have

$$I_1 = \lambda_1^2 + \lambda_2^2 + \lambda_3^2, \quad I_2 = \lambda_1^2\lambda_2^2 + \lambda_2^2\lambda_3^2 + \lambda_3^2\lambda_1^2, \quad I_3 = \lambda_1^2\lambda_2^2\lambda_3^2. \tag{2.5}$$

In view of (2.4), (2.5), W can also be regarded as a function of $\lambda_1, \lambda_2, \lambda_3$ so that

$$\dot{W} = W(\lambda_1, \lambda_2, \lambda_3). \tag{2.6}$$

For simplicity here, we use the same symbol W for the different functions arising in (2.4), (2.6). The principal components of the Cauchy stress σ are given by

$$\sigma_i = \frac{\lambda_i}{\lambda_1\lambda_2\lambda_3} \frac{\partial W}{\partial \lambda_i} \quad \text{(no sum on } i\text{)}. \tag{2.7}$$

It is customary in finite elasticity to normalize W so that $W = 0$, $\sigma = 0$ in the undeformed state. This will impose some minor restrictions on $W(3, 3, 1)$ (or $W(1, 1, 1)$) and on its derivatives. We shall not record these here as no direct use is made of them in this Chapter.

For the radially symmetric deformation (2.1), the equilibrium equations in the absence of body forces $\text{div } \sigma = 0$ can be shown to reduce to the single equation

$$\frac{d}{dR}\left[R^2 \frac{\partial W}{\partial \lambda_1}\right] - 2R\frac{\partial W}{\partial \lambda_2} = 0, \tag{2.8}$$

where W is evaluated at the principal stretches (2.3). By virtue of (2.3) and

(2.6), we see that equation (2.8) is a *second-order* nonlinear ordinary differential equation for $r(R)$. Thus boundary-value problems for spherically symmetric deformations of isotropic compressible bodies can be formulated in terms of solutions of (2.8), subject to appropriate boundary conditions. We note that equation (2.8), in expanded form, may be found in Ogden (1984), p. 247. As remarked there, solutions to (2.8) will, in general, require a numerical treatment. However, for certain special forms for W, exact analytical solutions may be obtained. This issue will be our chief concern in Sections 4.3-4.6.

The two-dimensional analog of the foregoing considerations arises in plane strain axisymmetric deformations of a cylinder. Thus points with plane polar coordinates (R, Θ) are mapped to points (r, θ) in the deformed region where

$$r = r(R), \quad \theta = \Theta; \quad \frac{dr}{dR} > 0. \tag{2.9}$$

Using analogous notation to that already introduced, we have

$$\lambda_1 = \dot{r}(R), \quad \lambda_2 = \frac{r(R)}{R}, \tag{2.10}$$

while the two invariants of concern are

$$I_1 = \lambda_1^2 + \lambda_2^2, \quad I_2 = \lambda_1^2 \lambda_2^2. \tag{2.11}$$

The analog of (2.8) is easily shown to be

$$\frac{d}{dR}\left[R\frac{\partial W}{\partial \lambda_1}\right] - \frac{\partial W}{\partial \lambda_2} = 0, \tag{2.12}$$

where now $W = W(\lambda_1, \lambda_2)$.

4.3 Transformation to a pair of first-order differential equations

It was shown in Horgan (1989) how equations (2.8), (2.12) may each be transformed to a pair of *first-order* differential equations. The substitution

$$t(R) = R\frac{\dot{r}(R)}{r(R)} \quad \left(= \frac{\lambda_1}{\lambda_2}\right) \tag{3.1}$$

is introduced in both the spherically symmetric and cylindrically symmetric problems in turn. Then, on using the chain rule, it may be readily verified that equations (2.8) and (2.12) may both be written in the form

$$R\dot{t} = k(\lambda_2 W_{11})^{-1} \{W_2 - W_1 + \lambda_2(1-t)W_{12}\} + t(1-t), \tag{3.2}$$

where the notation

$$W_\alpha = \frac{\partial W}{\partial \lambda_\alpha}, \quad W_{\alpha\beta} = \frac{\partial^2 W}{\partial \lambda_\alpha \partial \lambda_\beta}, \quad (\alpha, \beta = 1, 2), \tag{3.3}$$

has been introduced and the constant k is given by

$$k = \begin{cases} 2 & \text{(spherically symmetric),} \\ 1 & \text{(cylindrically symmetric).} \end{cases} \tag{3.4}$$

If W is such that the right hand side of (3.2) can be expressed in terms of R, $t(R)$, then equation (3.2) can be regarded as a first-order ordinary differential equation for $t(R)$ and the relation (3.1) as a first-order ordinary differential equation for $r(R)$. For special forms for W, it was shown in Horgan (1989), (1995a) that (3.1), (3.2) lead to a parametric solution to equations (2.8), (2.12) respectively, with t as a parameter. In some cases, the parameter t can be eliminated to yield an explicit closed form expression for $r(R)$.

4.4 An alternative transformation

Using (2.3), it was shown in Hill (1993) and Horgan (1995a) that the equilibrium equation (2.8) governing radially symmetric deformations is invariant under a stretching transformation. Thus, (2.8) can always be reduced to an ordinary differential equation of *first-order* for $p = p(u)$ by the sequence of transformations

$$R = e^s, \quad u = \frac{r(R)}{R}, \quad p = \frac{du}{ds}, \tag{4.1}$$

the first of these being the usual Euler transformation. On using the first two of (4.1) we rewrite (2.8) as

$$\frac{d}{ds}\left(e^{2s}\frac{\partial W}{\partial \lambda_1}\right) - 2e^{2s}\frac{\partial W}{\partial \lambda_2} = 0 \tag{4.2}$$

where, by virtue of (2.3), (4.1) we have

$$\lambda_1 = \dot{u}(s) + u(s), \quad \lambda_2 = u(s). \tag{4.3}$$

Equation (4.2) is a *second-order* non-linear ordinary differential equation for $u(s)$, which may be written in autonomous form as

$$\frac{d}{ds}\left(\frac{\partial W}{\partial \lambda_1}\right) + 2\left(\frac{\partial W}{\partial \lambda_1} - \frac{\partial W}{\partial \lambda_2}\right) = 0. \tag{4.4}$$

Finally, on using the third of (4.1), we may rewrite (4.4) as the *first-order* ordinary differential equation

$$p\frac{d}{du}\left(\frac{\partial W}{\partial \lambda_1}\right) + 2\left(\frac{\partial W}{\partial \lambda_1} - \frac{\partial W}{\partial \lambda_2}\right) = 0 \tag{4.5}$$

for $p(u)$, where

$$\lambda_1 = p(u) + u, \quad \lambda_2 = u. \tag{4.6}$$

It has been shown in Horgan (1995a) that it is of interest to consider all three equivalent forms of the equilibrium equation, namely (2.8) directly for $r(R)$, *or* (4.4) for $u(s)$ and then obtain $r(R)$ on using (4.1) *or* (4.5) for $p(u)$, with $u(s)$ then obtained from (4.1) and then $r(R)$ from (4.1). It is also worth observing that for spherically symmetric deformations, the result of Ericksen (1955) that all compressible nonlinearly elastic materials must admit the homogeneous deformation $r = cR$ (c constant) may be readily verified in all three formulations (see Horgan 1995a).

Analogous results hold for the plane strain problem. Thus on using (4.1) in (2.12) we find that the counterpart of (4.4) is

$$\frac{d}{ds}\left(\frac{\partial W}{\partial \lambda_1}\right) + \left(\frac{\partial W}{\partial \lambda_1} - \frac{\partial W}{\partial \lambda_2}\right) = 0, \qquad (4.7)$$

where

$$\lambda_1 = \dot{u}(s) + u(s), \quad \lambda_2 = u(s). \qquad (4.8)$$

On using (4.1), this may be rewritten as

$$p\frac{d}{du}\left(\frac{\partial W}{\partial \lambda_1}\right) + \left(\frac{\partial W}{\partial \lambda_1} - \frac{\partial W}{\partial \lambda_2}\right) = 0, \qquad (4.9)$$

where

$$\lambda_1 = p(u) + u, \quad \lambda_2 = u. \qquad (4.10)$$

4.5 Spherically and cylindrically symmetric deformations for special classes of compressible materials

In previous investigations, six classes of compressible materials have received much attention. Following Carroll (1988), (1991a, b), Murphy (1992) we list these as:

$$
\begin{array}{lll}
\text{I} & W = f(i_1) + c_2(i_2 - 3) + c_3(i_3 - 1), & f''(i_1) \neq 0, \\
\text{II} & W = c_1(i_1 - 3) + g(i_2) + c_3(i_3 - 1), & g''(i_2) \neq 0, \\
\text{III} & W = c_1(i_1 - 3) + c_2(i_2 - 3) + h(i_3), & h''(i_3) \neq 0, \\
\text{IV} & W = c_1 i_1 i_2 + c_2 i_1 + c_3 i_2 + c_4 i_3 + c_5, & c_1 \neq 0, \\
\text{V} & W = c_1 i_2 i_3 + c_2 i_1 + c_3 i_2 + c_4 i_3 + c_5, & c_1 \neq 0, \\
\text{VI} & W = c_1 i_1 i_3 + c_2 i_1 + c_3 i_2 + c_4 i_3 + c_5, & c_1 \neq 0.
\end{array}
\qquad (5.1)
$$

In (5.1), f, g, and h are arbitrary functions of their indicated arguments, where i_1, i_2, i_3 are the principal invariants of the stretch tensor and so

$$i_1 = \lambda_1 + \lambda_2 + \lambda_3, \quad i_2 = \lambda_1\lambda_2 + \lambda_2\lambda_3 + \lambda_3\lambda_1, \quad i_3 = \lambda_1\lambda_2\lambda_3. \qquad (5.2)$$

Also, $c_1, \ldots, c_5$ are constants, which differ from class to class. The invariants (5.2) are related to those of (2.5) as follows:

$$I_1 = i_1^2 - 2i_2, \quad I_2 = i_1^2 - 2i_1 i_3, \quad I_3 = i_3^2. \tag{5.3}$$

Materials of Classes I-III were extensively investigated by Carroll in (1988, 1991a, b). Class I materials are the *harmonic* materials introduced by John (1960). Class III materials are called *generalized Varga materials* (Horgan 1992); for $c_2 = 0$ they are known as Varga materials (Haughton 1987, Horgan 1989) since they may be viewed as a generalization, to include the effect of compressibility, of an incompressible material model proposed by Varga. The class IV-VI materials were introduced by Murphy (1992). The constants $c_1, \ldots, c_5$ and the constitutive functions f, g, h are usually subject to further restrictions to ensure physically realistic response but we shall not require consideration of these here. Explicit solutions for the deformation field $r(R)$ satisfying (2.8) have been obtained for all six classes of materials by several authors and by a variety of methods. In Horgan (1995a), we have examined equations (2.8), (4.4), (4.5) for each of the six cases and recorded the corresponding solution $r(R)$. These results are now summarized.

Class I: In this case, on using the hypothesis (5.1) that $f''(\cdot) \neq 0$, one finds that

$$r(R) = AR + \frac{B}{R^2}, \tag{5.4}$$

for arbitrary constants A and B. The solution (5.4), for the harmonic material I, was obtained independently by Abeyaratne and Horgan (1984), and by Ogden (1984). It was also obtained by Carroll (1988). All these authors observed that (2.8) implies that the first invariant i_1 is constant, from which (5.4) follows immediately. The deformation (5.4), being independent of the c_1, c_2, c_3 and the constitutive function f, is *controllable*, i.e. it is a possible equilibrium state (with no body forces) in all harmonic materials of the form I. It was shown in Horgan (1989) that (5.4) may also be derived from the *pair of first-order* differential equations (3.1), (3.2). The deformation (5.4) was employed in Abeyaratne and Horgan (1984) and in Ogden (1984) to obtain closed-form solutions for pressurized hollow spheres composed of harmonic materials. Application to micromechanical modeling of multiphase composites has been made recently by Aboudi and Arnold (2000).

Class II: Here, on using $g''(\cdot) \neq 0$, one finds that

$$r^2(R) = AR^2 + \frac{B}{R}. \tag{5.5}$$

The solution (5.5) was obtained directly from (2.8) by Carroll (1988). It can also be derived from the *pair* of first-order differential equations (3.1), (3.2) as was done for Class I materials. This controllable deformation was used by Murphy (1993) to treat the problems of inflation and eversion of hollow spheres of Class II materials, and was utilized by Aboudi and Arnold (2000) in their recent study of micromechanics of multiphase composites.

Class III: Here, on using $h''(\cdot) \neq 0$, one finds that

$$r^3(R) = A R^3 + B. \tag{5.6}$$

The solution (5.6) was obtained directly from (2.8) by Carroll (1988). For the case when $c_2 = 0$ in the definition of Class III in (5.1), this result was also obtained by Haughton (1987). It was shown by Horgan (1989) that (5.6) can also be derived from the pair of first-order equations (3.1), (3.2). The controllable deformation (5.6) for the generalized Varga material III was used by Horgan (1992) to illustrate the phenomenon of *cavitation* for compressible materials in a particularly tractable setting. (See Horgan and Polignone (1995) for a review of cavitation in nonlinear elasticity). Extensions of this work have been carried out by Murphy and Biwa (1997). The solution (5.6) has also been utilized by Aboudi and Arnold (2000) in their micromechanics analysis of composites undergoing finite deformation.

Class IV: It was shown by Hill (1993), on using (4.5), that

$$r^3(R) = \frac{(A + B R^3)^2}{R^3}, \tag{5.7}$$

where A and B are arbitrary constants. The solution (5.7) was also derived by Murphy (1992) using the methods involving the transformation (3.1).

Class V: This case was also treated by Hill (1993), using the formulation (4.5), to yield

$$r^5(R) = \frac{(A + B R^3)^2}{R}. \tag{5.8}$$

The solution (5.8) was also obtained by Murphy (1992) using the approach of Horgan (1989).

Class VI: It was shown by Horgan (1995) that the deformation field here is identical to that given for Class II by (5.5), namely,

$$r^2(R) = A R^2 + \frac{B}{R}. \tag{5.9}$$

This result was also obtained by Murphy (1992) by a different argument.

The foregoing solutions for Classes I-III have been generalized by Murphy

(1997a), who considers a strain-energy density with principal part given by

$$W = H(\alpha i_1 + \beta i_2 + \gamma i_3) + c_1 i_1 + c_2 i_2 + c_3 i_3, \qquad (5.10)$$

where α, β, γ are arbitrary constants. Solutions $r = r(R)$ for this material are given by Murphy (1997a) in the form

$$\frac{\gamma}{3} r^3 + \beta R r^2 + \alpha R^2 r = A R^3 + B, \qquad (5.11)$$

where A and B are arbitrary constants. When each of the constants α, β, γ is taken to be nonzero in turn, one recovers from (5.11) the solutions (5.4)–(5.6) respectively.

A solution of the form (5.11) with $\beta = 0$ and $A = 0$ also arises when a different strain-energy density is considered. Let

$$W = c_1(i_1 - 3) + c_2 \left(\frac{i_2}{i_3} - 3 \right), \qquad (5.12)$$

or equivalently,

$$W = c_1(\lambda_1 + \lambda_2 + \lambda_3 - 3) + c_2 \left(\frac{1}{\lambda_1} + \frac{1}{\lambda_2} + \frac{1}{\lambda_3} - 3 \right). \qquad (5.13)$$

This model may be viewed as another compressible Varga-type material since it reduces, in the incompressible limit, where $\lambda_1 \lambda_2 \lambda_3 = 1$, to a Varga material. On using the methods of Section 4.3 *or* Section 4.4, it may be shown that solutions $r = r(R)$ for the material (5.13) satisfy

$$r^3 + C R^2 r + D = 0, \qquad (5.14)$$

where C, D are arbitrary constants.

The plane strain analogs of the foregoing deformations are also described in Horgan (1995a). One can obtain these results by formally letting $\lambda_3 \to 0$ in the preceding so that $i_3 \to 0$ and we now have the two fundamental invariants

$$i_1 = \lambda_1 + \lambda_2, \quad i_2 = \lambda_1 \lambda_2, \qquad (5.15)$$

where λ_1, λ_2 are given by (2.10). We note that this deformation is *different* from the cylindrical inflation *with axial stretch* considered by Carroll (1988, 1991a, b), Hill (1993) and Murphy (1992). One may formally obtain the plane strain results by letting the axial stretch tend to zero.

Class I: Here

$$W = f(i_1) + c_2(i_2 - 1), \quad f''(i_1) \neq 0. \qquad (5.16)$$

This is the *harmonic* material proposed by John (1960) in plane strain. In this case, we find, on using $f''(\cdot) \neq 0$, that

$$r(R) = A R + \frac{B}{R}. \tag{5.17}$$

The plane strain deformation field (5.17) for harmonic materials was first obtained by Ogden and Isherwood (1978) using a complex variable approach. It was also derived directly from (2.12) in Ogden (1984) and independently by Jafari *et al.* (1984), and later on by Carroll (1988). In these references, the authors deduced from (2.12) that i_1 given by the first of (5.15) is constant, from which (5.17) follows. It was shown in Horgan (1989) that (5.17) may also be derived from the pair of first-order differential equations (3.1), (3.2). Solutions to boundary-value problems for hollow cylinders are described in Ogden and Isherwood (1978), Ogden (1984), and Jafari *et al.* (1984).

Class II: Here

$$W = c_1(i_1 - 2) + g(i_2), \quad g''(i_2) \neq 0. \tag{5.18}$$

In this case, we find, on using $g''(\cdot) \neq 0$, that

$$r^2(R) = A R^2 + B. \tag{5.19}$$

The solution (5.19) for plane strain deformations of the Class II material defined by (5.18) may be deduced from equation (5.13) of Carroll (1988) on letting the axial stretch tend to zero. It was also obtained in Horgan (1989) (see equation (47), p. 192) from the pair of first-order equations (3.1), (3.2). The deformation (5.19) was used by Horgan (1992) to investigate cavitation phenomena in plane strain (see also Steigmann 1992).

Class III: Since the plane strain elastic potentials W follow formally from (5.1) on letting $i_3 \to 0$, we see that the Class III materials are strictly *linear* in i_1 and i_2. As discussed by Carroll (1988), for example, strain-energies which are linear in the strain invariants i_1 and i_2 are not physically reasonable. We note here that it is the Class II material (5.18) which might be called a generalized Varga material (for plane strain) and *not* the Class III as in three dimensions.

Class IV: The plane strain analog of (5.1) for Class IV may be written as

$$W = c_1 i_1 i_2 + c_2(i_1 - 2) + c_3(i_2 - 1), \quad c_1 \neq 0. \tag{5.20}$$

In this case, it is shown in Horgan (1995a) that

$$r(R) = R^{\frac{1}{3}} \left(\frac{A}{R} + B R \right)^{\frac{2}{3}}, \tag{5.21}$$

where A, B are arbitrary constants. The plane strain solution in the form

(5.21) may also be derived on letting the axial stretch tend to zero in equation
(4.7) of Hill (1993).

We note that the plane strain analogs of Classes V and VI lead to materials
which are linear in i_1 and i_2 and so are not of physical interest (see remarks
above concerning Class III).

In the foregoing discussion, we have first specialized the subclass of compressible materials and then obtained explicit forms for the spherically and
cylindrically symmetric deformations corresponding to these subclasses. There
remains the converse question of whether the subclasses of materials are the
most general for which these deformations hold. This issue was addressed
by Carroll (1991b), Aron(1994), Murphy (1996), and recently by Martins and
Duda (1999) for the plane deformation (5.17).

4.6 The Blatz-Ko material

Based on their experimental investigations on the behavior of a compressible
foam rubber material, Blatz and Ko (1962) proposed the use of the strain-
energy function

$$
\begin{aligned}
2W &= \mu \left[\frac{I_2}{I_3} + 2(I_3)^{\frac{1}{2}} - 5 \right] \\
&= \mu \left[\lambda_1^{-2} + \lambda_2^{-2} + \lambda_3^{-2} + 2\lambda_1\lambda_2\lambda_3 - 5 \right] \\
&= \mu \left[\frac{i_2^2}{i_3^2} - 2\frac{i_1}{i_3} + 2i_3 - 5 \right],
\end{aligned}
\tag{6.1}
$$

where the constant $\mu > 0$ denotes the shear modulus of the material at infinitesimal deformations. The strain-energy (6.1), and generalizations thereof,
have been widely adopted in many investigations of finite deformations of compressible materials. (See Section 4.8 below for a discussion.)

For spherically symmetric deformations of the Blatz-Ko material (6.1), it
was shown by Chung *et al.* (1986), Horgan (1989, 1995a) that the method of
Section 4.3 leads to a parametric solution. One finds that

$$
r^6 = \frac{D\,(2t^2 + 2t + 5)\,d(t)}{(1-t)^2},
\tag{6.2}
$$

$$
R^{15} = \frac{C\,t^9\,d(t)}{(2t^2 + 2t + 5)^2\,(1-t)^5},
\tag{6.3}
$$

where C, D are arbitrary constants and the function $d(t)$ is given by

$$
d(t) = \exp\left\{ 2\tan^{-1}\left(\frac{2t+1}{3} \right) \right\} > 0.
\tag{6.4}
$$

Equations (6.2)-(6.4) provide a parametric solution where t is a parameter. This solution form was utilized by Chung *et al.* (1986) to obtain the deformation and stress fields in a pressurized hollow sphere composed of the material (6.1). (See also Beatty 1987, 1996 for an alternative treatment of this problem in the case of a thin spherical shell.) Applications to cavitation in compressible materials have been made by Horgan and Abeyaratne (1986).

The alternative solution method described in Section 4.4 was examined in Horgan (1995a). It was shown that

$$u^5(g+1)^3(2g^2 + 6g + 9)^{-3/2} = K \exp\left\{ \tan^{-1}\left(\frac{2g+3}{3} \right) \right\}, \qquad (6.5)$$

where K is a constant, and

$$g(u) = \frac{p(u)}{u}. \qquad (6.6)$$

Equation (6.5) is an implicit relation for the solution $p(u)$ of the first-order differential equation (4.5). This implicit relation is a first-order differential equation for $u(s)$ which does not, however, appear to be amenable to integration in order to determine $u(s)$ and thereby to find $r(R)$ from (4.1)$_2$. Thus, for spherically symmetric deformations of the Blatz-Ko material (6.1), *the method described in Section 4.3 leading to the parametric solution (6.2)-(6.4) seems preferable to the approach of Section 4.4.* Parametric solutions, using the method of Section 4.3, have also been obtained by Lei and Chang (1996) and Wang and Aron (1996) for some generalizations of the material model (6.1). Applications to the study of cavitation in compressible materials are described by Lei and Chang (1996).

The plane strain analog of (6.1) is given by

$$2W = \mu\left[\frac{I_1}{I_2} + 2(I_2)^{1/2} - 4 \right]$$
$$= \mu\left[\lambda_1^{-2} + \lambda_2^{-2} + +2\lambda_1\lambda_2 - 4 \right]. \qquad (6.7)$$

For the plane strain axisymmetric deformations (2.9) of the Blatz-Ko material (6.7), the parametric solution analogous to (6.2)-(6.4) was obtained by Abeyaratne and Horgan (1985) as

$$r^4 = \frac{D\,(t^2 + t + 4)\,h(t)}{(1-t)^2}, \qquad (6.8)$$

$$R^8 = \frac{C\,t^6\,h(t)}{(t^2 + t + 4)(1-t)^4}, \qquad (6.9)$$

where C, D are arbitrary constants and the function $h(t)$ is given by

$$h(t) = \exp\left\{\frac{6}{\sqrt{15}}\tan^{-1}\left(\frac{2t+1}{\sqrt{15}}\right)\right\} > 0. \qquad (6.10)$$

This solution scheme was used by Chung *et al.* (1986) in their study of a pressurized hollow cylinder composed of the material (6.7). This solution was also employed by Abeyaratne and Horgan (1985) to investigate shear localization and by Horgan and Abeyaratne (1986) in an analysis of cavitation for compressible materials.

4.7 Other equilibrium solutions

As one might expect, the spherically and cylindrically symmetric solutions described in Sections 4.2-4.6 are the most readily available analytical results obtained to date for compressible nonlinearly elastic materials in equilibrium and have been the most widely used in solving specific boundary-value problems. Similarly, the strain-energies most commonly employed are the Class I (harmonic), Class III (Varga), and the Blatz-Ko material (6.1). The latter model (and its generalizations) will be discussed in detail in the next Section. Some other problems that have been solved for the harmonic material are described by Ogden (1984), including the use of complex variable techniques for plane problems. Other more recent results, for example, are those of Wheeler (1985) on ellipsoidal cavities in a harmonic material, Carroll (1991a, b, 1995), Carroll and Horgan (1990) and Podio-Guidugli and Tomassetti (1999) on universal deformations, Carroll *et al.* (1994) on plane stress, Aron and Wang (1995) on flexure, Aron *et al.* (1998) on straightening of annular sectors, Haughton (1996), Haughton and Orr (1997) on eversion problems, Haughton (1998) on membrane theories, Hill and Arrigo (1996) on axially symmetric deformations and Murphy (1997b) on irrotational deformations. Other problems involving *isochoric* (volume preserving) deformations will be discussed in Section 4.10.

4.8 The generalized Blatz-Ko material and its various specializations

The strain-energy density (6.1) is a specialization of the general form

$$W = \frac{\mu}{2}f\left(I_1 - 1 - \frac{1}{\nu} + \frac{(1-2\nu)}{\nu}I_3^{-\nu/(1-2\nu)}\right)$$

$$+ \frac{\mu}{2}(1-f)\left(\frac{I_2}{I_3} - 1 - \frac{1}{\nu} + \frac{(1-2\nu)}{\nu}I_3^{\nu/(1-2\nu)}\right), \qquad (8.1)$$

where the constants μ, ν, f are such that $\mu > 0$, $0 \leq f \leq 1$, $0 < \nu < 1/2$. The constants μ, ν are the shear modulus and Poisson's ratio for infinitesimal deformations while f is related to the volume fraction of voids in the foam-rubber material modeled by (8.1). The material model (8.1) is known as the *generalized Blatz-Ko material* and was first proposed by Blatz and Ko (1962) as a model for foam rubber compressible materials undergoing large deformations.

Two special cases of (8.1) have received considerable attention in the literature, namely those for which $f = 0$ or 1, respectively. In the former case, (8.1) reduces to

$$\text{Case A } (f = 0): \quad W = \frac{\mu}{2}\left(\frac{I_2}{I_3} - 1 - \frac{1}{\nu} + \frac{(1 - 2\nu)}{\nu}I_3^{\nu/(1-2\nu)}\right), \quad (8.2)$$

while in the latter, one has

$$\text{Case B } (f = 1): \quad W = \frac{\mu}{2}\left(I_1 - 1 - \frac{1}{\nu} + \frac{(1 - 2\nu)}{\nu}I_3^{-\nu/(1-2\nu)}\right). \quad (8.3)$$

An extensive discussion of the model (8.1) and its special cases (8.2), (8.3) may be found in the review articles by Beatty (1987, 1996). As noted there, (8.2) characterizes the class of *foamed* polyurethane elastomers and (8.3) the class of *solid* polyurethane rubbers investigated in the experiments of Blatz and Ko (1962). A further specialization of (8.2) arises when it is matched with the particular foam rubber material used by Blatz and Ko (1962). It is shown by Blatz and Ko (1962) (see also the discussions in Knowles and Sternberg (1975) and Beatty (1987)) that, for this particular material, the value of Poisson's ratio ν can be taken as $\nu = 1/4$ and so (8.2) now becomes

$$W = \frac{\mu}{2}\left(\frac{I_2}{I_3} + 2I_3^{1/2} - 5\right), \quad (8.4)$$

which is often called the (special) *Blatz-Ko material* and coincides with (6.1) introduced earlier. For purposes of later discussion, we record here the further specialization of (8.3) when $\nu = 1/4$, namely

$$W = \frac{\mu}{2}\left[I_1 + 2I_3^{-1/2} - 5\right]. \quad (8.5)$$

We note that in the limit as $I_3 \to 1$, (8.1) yields

$$W \to \frac{\mu}{2}f(I_1 - 3) + \frac{\mu}{2}(1 - f)(I_2 - 3), \quad (8.6)$$

which has the form of the well-known Mooney-Rivlin strain-energy density for *incompressible* materials. Thus, as remarked by Beatty (1987), (8.1) may be viewed as a generalization to compressible materials of the Mooney-Rivlin incompressible material model (8.6). Similarly, the special case (8.3) may be viewed as a compressible generalization of the neo-Hookean incompressible material $2W = \mu(I_1 - 3)$.

The response of the material (8.1) to some *homogeneous* deformations is described in Knowles and Sternberg (1975) and Beatty (1987). In particular, for the special Blatz-Ko material (8.4), it is shown that the Cauchy stress response in simple extension approaches a *finite* asymptote for large values of the axial stretch while the nominal (or engineering) stress versus axial stretch relation is non-monotone. This rather special behavior and its implications for the solutions of crack problems was one of the motivations for the study of Knowles and Sternberg (1975) on the issue of *ellipticity* of the governing equilibrium partial differential equations for the special material (8.4). We discuss this aspect in more detail in Section 4.9.

The most widely used of the above models is (8.4), describing the special Blatz-Ko material. Both analytic and numerical solutions to a variety of boundary-value problems in finite elastostatics for compressible materials have been obtained for (8.4) and its two-dimensional counterpart. Even though (8.4) is indeed a very special compressible material model, such results provide considerable insight into the differences between compressible and incompressible material response to large deformations. The problems solved *analytically* for (8.4) include, for example, spherical inflation as described in Section 4.6, pure torsion of a circular cylinder (Beatty 1987, Carroll and Horgan 1990, Polignone and Horgan 1991), and analogs of the standard controllable deformations for incompressible materials (Carroll and Horgan 1990). The *pure torsion solution*, consisting of a deformation with no radial displacement, is remarkably simple (see the discussion in Section 4.10 below).

The special case (8.3), and its further specialization (8.5), has been less widely employed. As noted by Polignone and Horgan (1992), (8.3) has the form

$$W = \frac{\mu}{2} [I_1 - 3 + H(I_3)],\qquad(8.7)$$

where

$$H(I_3) = \frac{(1 - 2\nu)}{\nu} I_3^{-\nu/(1-2\nu)} + 2 - \frac{1}{\nu},\qquad(8.8)$$

and is called a *special Hadamard* material, since the general Hadamard material has the form

$$W = c_1(I_1 - 3) + c_2(I_2 - 3) + H(I_3),\qquad(8.9)$$

where c_1, c_2 are constants and $H(I_3)$ is an arbitrary function.

Several other specific examples of materials of the form (8.7) are given by Polignone and Horgan (1992). Materials of the form (8.7), with H arbitrary, are sometimes called *compressible neo-Hookean materials*. The response of the special model (8.5) to some *homogeneous* deformations is described by Bolzon and Vitaliani (1993). While it is not explicitly carried out there, one can readily verify that, in contrast to the situation for (8.4), the Cauchy stress response in simple extension for (8.5) does *not* reach an asymptote for large values of the axial stress and the nominal stress versus axial stretch relation is monotone. (See, e.g., Figures 1 and 3 in Brockman 1986, which also discusses the use of (8.3) in finite element analysis). Further properties of the material (8.5) are described in Bolzon (1993) and Bolzon and Vitaliani (1993).

For the class of materials (8.7) with H arbitrary, and so in particular for (8.3) and (8.5), it is shown by Polignone and Horgan (1992) that *axisymmetric anti-plane shear* deformations are possible, and explicit solutions for the displacement and stress are given. (See also Jiang and Beatty 1995, Beatty and Jiang 1996, Jiang and Ogden 2000). On the other hand, it is shown by Polignone and Horgan (1991) that *pure torsional deformations* of a circular cylinder (i.e. a deformation with *zero* radial displacement) are *never* possible for (8.7), in contrast to the situation for (8.2) and thus (8.4), for which such deformations are explicitly determined. For the special form (8.3) of (8.7) existence and properties of smooth radial solutions of the torsion problem are discussed by Paullet and Polignone (1996). The existence of such solutions is consistent with a global ellipticity result for (8.3) established by Horgan (1996).

4.9 Ellipticity of the governing equilibrium equations

The system of second-order quasilinear partial differential equations for the displacement field $u_i(\mathbf{X})$ in a homogeneous compressible nonlinearly elastic solid can be written as (see, e.g., Knowles and Sternberg 1975, Rosakis 1990)

$$C_{ijkl}\left(1+\nabla\mathbf{u}\right)u_{k,lj}=0 \quad \text{on } R_0, \tag{9.1}$$

where

$$C_{ijkl}(\mathbf{F})=C_{klij}(\mathbf{F})=\frac{\partial^2 W(\mathbf{F})}{\partial F_{ij}\partial F_{kl}}, \quad \mathbf{F}=1+\nabla\mathbf{u}. \tag{9.2}$$

Here $\mathbf{X}$ denotes the material coordinates in the reference configuration R_0 and the usual Cartesian tensor notation is employed. The equations (9.1) are said to be *elliptic* at the solution $\mathbf{u}$ and at the point P with position vector $\mathbf{X}$ if and only if

$$\det\left[C_{ijkl}(\mathbf{F}(\mathbf{X}))n_j n_l\right]\neq 0 \tag{9.3}$$

for every unit vector $\mathbf{n}$. In Horgan (1996), the following result was established:

Theorem: For the isotropic compressible material modeled by (8.1), the equations (9.1) are elliptic (globally) on R_0 when $f = 1$. Thus, ellipticity holds globally for the material (8.3). When $0 \leq f < 1$ in (8.1), ellipticity no longer holds at sufficiently large stretches.

The proof of this result was given by Horgan (1996) on using a general theorem due to Rosakis (1990). Since ellipticity holds globally for the material (8.3) modeling *solid* polyurethane rubber, one would anticipate no difficulty with loss of smoothness in the solution of boundary-value problems for this material. This was confirmed by the existence results of Paullet and Polignone (1996) for the torsion problem. In contrast, for all other values of f in $0 \leq f < 1$ for the generalized Blatz-Ko material (8.1), since loss of ellipticity can occur at sufficiently large stretches, a loss of smoothness can also be expected. As pointed out in Section 4.8, this was first investigated by Knowles and Sternberg (1975) for the special Blatz-Ko material (8.4). It was shown there that, for the material (8.4), equations (9.1) are locally elliptic at a solution if and only if the local principal stretches satisfy

$$2 - \sqrt{3} < \frac{\lambda_i}{\lambda_j} < 2 + \sqrt{3}, \ (i \neq j). \tag{9.4}$$

The relatively simple ellipticity conditions (9.4) have facilitated the use of the special model (8.4) to illustrate the physical interpretations and consequences of loss of ellipticity at sufficiently large deformations for several nonlinear elastostatic problems. In fact, the pioneering papers of Knowles and Sternberg (1975, 1977) on ellipticity for this material were motivated by the study of *elastostatic shocks*, involving discontinuities in displacement gradients, near crack tips (see, e.g., Knowles and Sternberg 1978, Rosakis and Jiang 1993 and the references cited therein). The explicit conditions (9.4) were applied in another physical context by Horgan and Polignone (1995), namely that of *torsion* of compressible nonlinearly elastic circular cylinders. As was pointed out in Section 4.8, an earlier study of this problem by Polignone and Horgan (1991) shows that *pure torsion*, i.e. a deformation with zero radial displacement, is possible for the generalized Blatz-Ko material (8.1) if and only if $f = 0$. Thus (8.2) and its further specialization (8.4) admits a pure torsional deformation. The associated stress distribution, found explicitly by Polignone and Horgan (1991), is remarkably simple. It was shown by Horgan and Polignone (1995) that ellipticity is lost for the pure torsion solution for the material (8.4) when the prescribed torque reaches a critical value. For values of the twisting moment greater than this critical value, there is an axial core of the cylinder on which ellipticity holds, surrounded by an annular region where loss of ellipticity has occurred (see Figure 1 of Horgan and Polignone 1995). The results have suggestive interpretations analogous to those associated with elastic-plastic torsion.

Another specific problem in which loss of ellipticity has interesting consequences is that of azimuthal (or circular) shear of a hollow cylinder (Polignone and Horgan 1994, Simmonds and Warne 1992, Wineman and Waldron 1995, Paullet *et al.* 1998). The inner surface of the tube is fixed and its outer surface is subjected to a uniform circumferential shear. This problem was studied by Simmonds and Warne (1992) for the special Blatz-Ko material (8.4). Numerical solutions exhibit a kink (discontinuity in slope) in the moment-rotation relation at a critical applied moment. Computational difficulties encountered by Simmonds and Warne (1992) in extending the solution beyond this critical moment are due to the loss of ellipticity. Existence and non-existence results for this problem were established by Paullet *et al.* (1998). This problem was also investigated by Wineman and Waldron (1995) for the generalized Blatz-Ko material (8.1). For the special case $f = 1$, $\nu = 1/4$, that is, for (8.5), no numerical difficulties were encountered. This is explained by the theorem quoted earlier since ellipticity holds globally for the material (8.1) when $f = 1$. Combined azimuthal shear and torsion for the material (8.1) was recently investigated numerically by Zidi (2000).

As described in Section 4.6, spherical inflation of an internally pressurized hollow sphere composed of the special material (8.4) was investigated by Chung *et al.* (1986). An explicit analytic solution to this problem was obtained, and conditions for the initiation of a localized shear bifurcation found. It was shown that when the ratio of the outer undeformed radius to the inner undeformed radius is less than a critical value, the shear bifurcation occurs *before* the pressure maximum is attained while when this ratio is smaller than the critical value, the converse is true.

Most of the results just discussed have their analogs for two-dimensional *plane deformations*. For general isotropic compressible elastic materials, necessary and sufficient conditions for ellipticity of the two-dimensional version of (8.1) were obtained by Knowles and Sternberg (1975) and by Rosakis (1990). For the two-dimensional counterpart of (8.1), a theorem completely analogous to that quoted earlier can be established (Horgan 1996). In particular, for the two-dimensional special Blatz-Ko material (6.7) one recovers the result of Knowles and Sternberg (1975, 1977) that ellipticity holds if and only if the local principal stretches λ_1, λ_2 satisfy

$$2 - \sqrt{3} < \frac{\lambda_1}{\lambda_2} < 2 + \sqrt{3} \,. \tag{9.5}$$

The remarkably simple condition (9.5) was used in a number of studies to examine the consequences of loss of ellipticity in physical problems, see, e.g., Knowles and Sternberg (1978) for crack problems, Abeyaratne and Horgan (1985) for localization at cavity boundaries, Horgan and Abeyaratne (1986) for cavitation problems, Chung *et al.* (1986) for cylindrical inflation and Silling (1991) for creasing singularities. The condition (9.5) and its consequences were also discussed by Roxburgh and Ogden (1994) in the context of stability and vibration of pre-stressed compressible elastic plates.

4.10 Isochoric deformations

A comparatively recent research activity in compressible nonlinear elasticity is the investigation of classes of materials for which certain *isochoric* deformations are possible. One of the first such problems studied was the torsion problem for a compressible circular cylinder (Polignone and Horgan 1991). For general torsion of such a cylinder, one would expect a radial deformation. This is in contrast with the situation for an *incompressible* cylinder, where the only possible deformation is that of *pure torsion*, i.e. no extension occurs in the radial direction and the cross-section remains circular. In Polignone and Horgan (1991), the issue discussed was to determine the class of compressible materials for which the *isochoric* deformation describing *pure torsion* can occur. This question was originally raised by Currie and Hayes (1982) in their study of nonuniversal deformations. A necessary condition on the strain-energy density for pure torsion to be possible was obtained in Polignone and Horgan (1991), and examples of specific compressible materials satisfying this condition were given there. In particular, it was shown that the subclass of Blatz-Ko materials (8.2) can sustain pure torsion. This result was also obtained by Beatty (1987) and Carroll and Horgan (1990). The stresses arising in pure torsion of a circular cylinder composed of the material (8.2) were found explicitly by Polignone and Horgan (1991), and are remarkably simple. For the general Hadamard material (8.9), it was shown by Polignone and Horgan (1991) that pure torsion is possible provided that

$$2c_2 = c_1 \,. \tag{10.1}$$

This result was obtained in a different way by Currie and Hayes (1982).

In a subsequent paper (Polignone and Horgan 1992) an analogous issue was addressed for the deformation of *axisymmetric anti-plane shear* of a hollow circular cylinder. This is a special case of the general *anti-plane shear* deformation defined by

$$x_1 = X_1, \quad x_2 = X_2, \quad x_3 = X_3 + u(X_1, X_2), \tag{10.2}$$

on an arbitrary plane domain D (see, e.g., Horgan 1995b for a review of anti-plane shear). The deformation (10.2) is such that det $\mathbf{F} = 1$ and so is *isochoric*. Necessary and sufficient conditions on the strain-energy density W which ensure that a nontrivial state of anti-plane shear can be sustained in a *compressible* material were obtained by Knowles (1977) and Jiang and Knowles (1991). (See also Tsai and Rosakis 1994, Tsai and Fan 1999.) In particular, it was shown by Knowles (1977) that a Hadamard material (8.9) can sustain anti-plane shear if and only if

$$c_2 = 0. \tag{10.3}$$

For such special Hadamard materials, (i.e. materials of the form (8.7) with $H(I_3)$ arbitrary) the governing equation for anti-plane shear, which is a second-order quasi-linear partial differential equation, simplifies to Laplace's equation. There is still an axial normal stress which is strictly a nonlinear effect (see, e.g., Jiang and Knowles 1991, Horgan 1995b for details).

The special case of *axisymmetric anti-plane shear* is the deformation described by

$$r = R, \quad \theta = \Theta, \quad z = Z + w(R), \tag{10.4}$$

in cylindrical polar coordinates. This deformation is also called *pure axial shear* (see Jiang and Ogden 2000), and is the analog of *pure torsion* described earlier in this Section. Necessary conditions on the strain-energy W for (10.4) to be sustainable in a circular cylindrical tube were obtained in Polignone and Horgan (1992). The problem was further investigated by Jiang and Beatty (1995), Beatty and Jiang (1996), and Jiang and Ogden (2000). In particular, for the class of special Hadamard materials (8.7) with $H(I_3)$ arbitrary, it was shown by Polignone and Horgan (1992) that $w(R)$ in (10.4) has the form

$$w(R) = \frac{BT_0}{\mu} \ln \frac{R}{A}, \tag{10.5}$$

where A and B are the inner and outer radii of the tube and T_0 is the prescribed shear traction at the outer surface. The axial displacement (10.5) satisfies the axisymmetric form of Laplace's equation and so this result is consistent with that described earlier for the general anti-plane shear deformation (10.2). It was shown by Jiang and Beatty (1996) that when the *shear response function* is constant, the deformation is necessarily of the form (10.5). The converse of this was established by Jiang and Ogden (2000). New solutions, not of the form (10.5), were also obtained by Jiang and Ogden (2000) for more general classes of strain-energies. It was also shown by Jiang and Ogden (2000) how these results may be used to generate solutions for axial shear for *incompressible* materials. An interesting connection between the pure axial shear problem and concentrated line load problems was discussed by Warne and Polignone (1996).

A third *isochoric* deformation for compressible materials that has been investigated in a similar fashion is that of *azimuthal shear* (or circular shear) of a cylindrical tube. Conditions under which a *pure azimuthal shear* i.e. a deformation of the form

$$r = R, \quad \theta = \Theta + g(R), \quad z = Z, \tag{10.6}$$

can be sustained have been examined by Haughton (1993), Polignone and Horgan (1994), Beatty and Jiang (1997), and Jiang and Ogden (1998). It was shown by Polignone and Horgan (1994) that for classes of compressible materials, the function $g(R)$ in (10.6) has the form

$$g(R) = C + \frac{D}{R^2}, \tag{10.7}$$

where C and D are constants to be determined from the boundary conditions. Beatty and Jiang (1997) established that (10.7) holds for all materials for which the shear response function is constant. (For other work on general (non-isochoric) azimuthal shear deformations, see, e.g., Ertepinar 1990, Carroll and Murphy 1993, Simmonds and Warne 1992, Haughton 1993, Wineman and Waldron 1995, Paullet *et al.* 1998, Jiang and Ogden 1998, and Zidi 2000).

The isochoric deformation consisting of the composition of the shearing deformations (10.4) and (10.6) is called *helical shear* and has been examined by Beatty and Jiang (1999).

Acknowledgements

This research was supported by the U.S. National Science Foundation under Grant DMS-96-22748, and by the U.S. Air Force Office of Scientific Research under Grant AFOSR-F49620-98-1-0443.

156 *Cornelius O. Horgan*

References

Abeyaratne, R. and Horgan, C. O. 1984 The pressurized hollow sphere problem in finite elastostatics for a class of compressible materials. *Int. J. Solids Struct.* **20**, 715-723.

Abeyaratne, R. and Horgan, C. O. 1985 Initiation of localized plane deformations at a circular cavity in an infinite compressible nonlinearly elastic medium. *J. of Elasticity* **15**, 243-256.

Aboudi, J. and Arnold, S. M. 2000 Micromechanical modeling of the finite deformation of thermoelastic multiphase composites. *Math. and Mech. of Solids* **5**, 75-100.

Antman, S. S. 1995 *Nonlinear Problems of Elasticity*, Springer, New York.

Aron, M. 1994 On a class of plane radial deformations of compressible nonlinearly elastic solids. *IMA J. of Appl. Math.* **52**, 289-296.

Aron, M. and Wang, Y. 1995 Remarks concerning the flexure of a compressible nonlinearly elastic block. *J. of Elasticity* **40**, 99-106.

Aron, M., Christopher, C. and Wang, Y. 1998 On the straightening of compressible, nonlinearly elastic, annular cylindrical sectors. *Math. and Mech. of Solids* **3**, 131-145.

Beatty, M. F. 1987 Topics in finite elasticity: hyperelasticity of rubber, elastomers, and biological tissues - with examples. *Appl. Mech. Reviews* **40**, 1699-1734.

Beatty, M. F. 1996 Introduction to nonlinear elasticity, in *Nonlinear Effects in Fluids and Solids* (ed. by M. M. Carroll and M. A. Hayes), pp. 13-112, Plenum Press, New York.

Beatty, M. F. and Jiang, Q. 1996 Compressible, isotropic hyperelastic materials capable of sustaining axisymmetric, antiplane shear deformations, in *Contemporary Research in the Mechanics and Mathematics of Materials* (ed. by R. C. Batra and M. F. Beatty), pp. 133-144. International Center for Numerical Methods in Engineering (CIMNE), Barcelona.

Beatty, M. F. and Jiang, Q. 1997 On compressible materials capable of sustaining axisymmetric shear deformations. Part 2: Rotational shear of isotropic hyperelastic materials. *Q. J. Mech. Appl. Math.* **50**, 211-237.

Beatty, M. F. and Jiang, Q. 1999 On compressible materials capable of sustaining axisymmetric shear deformations. Part 3: Helical shear of isotropic hyperelastic materials. *Quart. Appl. Math.*, **57**, 681-698.

Blatz, P. J. and Ko, W. L. 1962 Application of finite elasticity to the deformation of rubbery materials. *Trans. Soc. Rheol.* **6**, 223-251.

Bolzon, G. 1993 On a class of constitutive models for highly deforming compressible materials. *Archive of Applied Mechanics* **63**, 296-300.

Bolzon, G. and Vitaliani, G. 1993 The Blatz-Ko material model and homogenization. *Archive of Applied Mechanics* **63**, 228-241.

Brockman, R. A. 1986 On the use of the Blatz-Ko constitutive model in nonlinear finite element analysis. *Computers and Structures* **24**, 607-611.

Carroll, M. M. 1988 Finite strain solutions in compressible isotropic elasticity. *J. of Elasticity* **20**, 65-92.

Carroll, M. M. 1991a Controllable deformations for special classes of compressible elastic solids. *Stability and Applied Analysis of Continuous Media* **1**, 309-323.

Carroll, M. M. 1991b Controllable deformations in compressible finite elasticity. *Stability and Applied Analysis of Continuous Media* **1**, 373-384.

Carroll, M. M. 1995 On obtaining closed form solutions for compressible nonlinearly elastic materials. *Zeit. Angew. Math Phys.* **46**, S126-S145.

Carroll, M. M. and Horgan, C. O. 1990 Finite strain solutions for a compressible elastic solid. *Quart. Appl. Math.* **48**, 767-780

Carroll, M. M. and Murphy, J. G. 1993 Azimuthal shearing of special compressible materials. *Proc. R. Irish Academy* **93A**, 209-230.

Carroll, M. M., Murphy, J. G. and Rooney, F. J. 1994 Plane stress problems for compressible materials. *Int. J. Solids Struct.* **36**, 1597-1607.

Chung, D. T., Horgan, C. O., and Abeyaratne, R. 1986 The finite deformation of internally pressurized hollow cylinders and spheres for a class of compressible elastic materials. *Int. J. Solids Struct.* **22**, 1557-1570.

Currie, P. K. and Hayes, M. 1982 On non-universal finite elastic deformations. In *Finite Elasticity*, Proceedings of IUTAM Symposium (Carlson, D. E. and Shield, R. T., editors), pp. 143-150, Martinus Nijhoff, The Hague.

Ericksen, J. L. 1955 Deformations possible in every compressible isotropic perfectly elastic material. *J. Math. Physics* **34**, 126-128.

Ertepinar, A. 1990 On the finite circumferential shearing of compressible hyperelastic tubes. *Int. J. Eng. Sc.* **28**, 889-896.

Haughton, D. M. 1987 Inflation of thick-walled compressible elastic spherical shells. *IMA J. Appl. Math.* **39**, 259-272.

Haughton, D. M. 1993 Circular shearing of compressible elastic cylinders. *Quart. J. Mech. Appl. Math.* **46**, 471-486.

Haughton, D. M. 1996 Further results for the eversion of highly compressible elastic cylinders. *Math. and Mech. of Solids* **1** 355-367.

Haughton, D. M. 1998 Exact solutions for elastic membrane disks. *Math. and Mech. of Solids* **3**, 393-410.

Haughton, D. M. and Orr, A. 1997 On the eversion of compressible elastic cylinders. *Int. J. Solids Struct.* **34**, 1893-1914.

Hill, J. M. 1993 Cylindrical and spherical inflation in compressible finite elasticity. *IMA J. Appl. Math.* **50**, 195-201.

Hill, J. M. and Arrigo, D. J. 1996 On axially symmetric deformations of perfectly elastic compressible materials. *Q. J. Mech. Appl. Math.* **49**, 19-28.

Horgan, C. O. 1989 Some remarks on axisymmetric solutions in finite elastostatics for compressible materials. *Proc. Royal Irish Academy* **89A**, 185-193.

Horgan, C. O. 1992 Void nucleation and growth for compressible non-linearly elastic materials: an example. *Int. J. Solids Struct.* **29**, 279-291.

Horgan, C. O. 1995a On axisymmetric solutions for compressible nonlinearly elastic solids. *Zeit. Angew. Math. Phys.* **46**, S107-S125.

Horgan, C. O. 1995b Anti-plane shear deformations in linear and nonlinear solid mechanics. *SIAM Review* **37**, 53-81.

Horgan, C. O. 1996 Remarks on ellipticity for the generalized Blatz-Ko constitutive model for a compressible nonlinearly elastic solid. *J. of Elasticity* **42**, 165-176.

Horgan, C. O. and Abeyaratne, R. 1986 A bifurcation problem for a compressible nonlinearly elastic medium: growth of a micro-void. *J. of Elasticity* **16**, 189-200.

Horgan, C. O. and Polignone, D. A. 1995 A note on the pure torsion of a circular cylinder for a compressible nonlinearly elastic material with nonconvex strain energy. *J. of Elasticity* **37**, 167-178.

Horgan, C. O. and Polignone, D. A. 1995 Cavitation in nonlinearly elastic solids: a review. *Appl. Mech. Revs.* **48**, 471-485.

Jafari, A.H., Abeyaratne, R., and Horgan, C.O. 1984 The finite deformation of a pressurized circular tube for a class of compressible materials. *Zeit. Angew. Math. Phys.* **35**, 227-246.

Jiang, Q. and Beatty, M. F. 1995 On compressible materials capable of sustaining axisymmetric shear deformations. Part 1: Anti-plane shear of isotropic hyperelastic materials. *J. of Elasticity* **39**, 75-95.

Jiang, Q. and Knowles, J. K. 1991 A class of compressible materials capable of sustaining finite anti-plane shear. *J. of Elasticity* **25**, 193-201.

Jiang, X. and Ogden, R. W. 1998 On azimuthal shear of a circular cylindrical tube of compressible elastic material. *Q. J. Mech. Appl. Math.* **51**, 143-158.

Jiang, X. and Ogden, R. W. 2000 Some new solutions for the axial shear of a circular cylindrical tube of compressible elastic material. *Int. J. of Nonlinear Mechanics* **35**, 361-369.

John, F. 1960 Plane strain problems for a perfectly elastic material of harmonic type. *Comm. Pure Appl. Math.* **13**, 239-296.

Knowles, J. K. 1977 A note on anti-plane shear for compressible materials in finite elastostatics. *J. of the Australian Math. Soc. (Ser. B)* **20**, 1-7.

Knowles, J. K. and Sternberg, E. 1975 On the ellipticity of the equations of nonlinear elastostatics for a special material. *J. of Elasticity* **5**, 341-361.

Knowles, J. K. and Sternberg, E. 1977 On the failure of ellipticity of the equations for finite elastostatic plane strain. *Arch. Rational Mech. and Anal.* **63**, 321-336.

Knowles, J. K. and Sternberg, E. 1978 On the failure of ellipticity and the emergence of discontinuous deformation gradients in plane finite elastostatics. *J. of Elasticity* **8**, 329-379.

Lei, H. C. and Chang, H. W. 1996 Void formation and growth in a class of compressible solids. *J. of Engineering Mathematics* **30**, 693-706.

Martins, L. C. and Duda, F. P. 1999 Maximal classes of stored energies compatible with cylindrical inflations. *J. of Elasticity* **53**, 189-198.

Murphy, J. G. 1992 Some new closed-form solutions describing spherical inflation in compressible finite elasticity. *IMA J. Appl. Math.* **48**, 305-316.

Murphy, J. G. 1993 Inflation and eversion of spherical shells of a special compressible material. *J. of Elasticity* **30**, 251-276.

Murphy, J. G. 1996 A family of solutions describing plane strain cylindrical inflation in finite compressible elasticity. *J. of Elasticity* **45**, 1-11.

Murphy, J. G. 1997a A family of solutions describing spherical inflation in finite compressible elasticity. *Q. J. Mech. Appl. Math.* **50**, 35-45.

Murphy, J. G. 1997b Irrotational deformations in finite compressible elasticity. *Math. and Mech. of Solids* **3**, 491-502.

Murphy, J. G. and Biwa, S. 1997 Non-monotonic cavity growth in finite compressible elasticity. *Int. J. Solid Struct.* **34**, 3859-3872.

Ogden, R. W. 1982 Elastic deformations of rubberlike solids, in *Mechanics of Solids* (R. Hill 60th Anniversary Volume, ed. by H. G. Hopkins and M. J. Sewell), pp. 499-537. Pergamon Press, New York.

Ogden, R. W. 1984 *Nonlinear Elastic Deformations*. Wiley, New York. (Reprinted by Dover, New York, 1997).

Ogden, R. W. and Isherwood, D. A. 1978 Solutions of some finite plane-strain problems for compressible elastic solids. *Q. J. Mech. Appl. Math.* **31**, 219-249.

Paullet, J. E. and Polignone, D. A. 1996 Existence and a priori bounds for the finite torsion solution for a class of general Blatz-Ko materials. *Math. and Mech. of Solids* **1**, 315-326.

Paullet, J. E., Polignone Warne, D. A. and Warne, P. G. 1998 Existence and uniqueness of azimuthal shear solutions in compressible isotropic nonlinear elasticity. *Math. and Mech. of Solids* **3**, 53-69.

Podio-Guidugli, P. and Tomassetti, G. 1999 Universal deformations for a class of compressible isotropic hyperelastic materials. *J. of Elasticity* **52**, 159-166.

Polignone, D. A. and Horgan, C. O. 1991 Pure torsion of compressible nonlinearly elastic circular cylinders. *Quart. Appl. Math.* **49**, 591-607.
Polignone, D. A. and Horgan, C. O. 1992 Axisymmetric finite anti-plane shear of compressible nonlinearly elastic circular tubes. *Quart. Appl. Math.* **50**, 323-341.
Polignone, D. A. and Horgan, C. O. 1994 Pure azimuthal shear of compressible nonlinearly elastic circular tubes. *Quart. Appl. Math.* **52**, 113-131.
Rajagopal, K. R. and Carroll, M. M. 1992 Inhomogeneous deformations of nonlinearly elastic wedges. *Int. J. Solids Struct.* **29**, 735-744.
Rosakis, P. 1990 Ellipticity and deformations with discontinuous gradients in finite elastostatics. *Arch. for Rational Mech. and Anal.* **109**, 1-37.
Rosakis, P. and Jiang, Q. 1993 Deformations with discontinuous gradients in finite elastostatics. *J. of Elasticity* **33**, 233-257.
Roxburgh, D. G. and Ogden, R. W. 1994 Stability and vibration of pre-stressed compressible elastic plates. *Int. J. Eng. Sc.* **32**, 427-454.
Silling, S. A. 1991 Creasing singularities in compressible elastic materials. *J. of Applied Mechanics* **58**, 70-74.
Simmonds, J. G. and Warne, P. 1992 Azimuthal shear of compressible, or incompressible, nonlinearly elastic polar-orthotropic tubes of infinite extent. *Int. J. Non-Linear Mechanics* **27**, 447-464.
Steigmann, D. J. 1992 Cavitation in elastic membranes: an example. *J. of Elasticity* **28**, 277-287.
Tao, L. and Rajagopal, K. R. 1990 On an inhomogeneous deformation of an isotropic compressible elastic material. *Arch. Mech. Stos.* **42**, 729-736.
Tsai, H. and Rosakis, P. 1994 On anisotropic compressible materials that can sustain dynamic antiplane shear. *J. of Elasticity* **35**, 213-222.
Tsai, H. and Fan, X. 1999 Anti-plane shear deformations in compressible transversely isotropic materials. *J. of Elasticity* **54**, 73-88.
Wang, Y. and Aron, M. 1996 Radial deformations of cylindrical and spherical shells composed of a generalized Blatz-Ko material. *Zeit. Angew. Math. Mech.* **76**, 535-536.
Warne, P. G. and Polignone, D. A. 1996 Solution for an infinite compressible nonlinearly elastic body under a line load. *Quart. Appl. Math.* **54**, 317-326.
Wheeler, L. T. 1985 Finite deformations of a harmonic elastic medium containing an ellipsoidal cavity. *Int. J. of Solids and Structures* **21**, 799-804.
Wineman, A. S. and Waldron Jr., W. K. 1995 Normal stress effects induced during circular shear of a compressible nonlinear elastic cylinder. *Int. J. Non-Linear Mechanics* **30**, 323-339.
Zidi, M. 2000 Azimuthal shearing and torsion of a compressible hyperelastic and pre-stressed tube. *Int. J. Non-Linear Mechanics* **35**, 201-209.

5

Exact integrals and solutions for finite deformations of the incompressible Varga elastic materials

J.M. Hill

School of Mathematics and Applied Statistics
University of Wollongong
Wollongong, NSW, Australia
Email: Jim_Hill@uow.edu.au

The purpose of this chapter is to focus on a variety of exact results applying to the perfectly elastic incompressible Varga materials. For these materials it is shown that the governing equations for plane strain, plane stress and axially symmetric deformations, admit certain first order integrals, which together with the constraint of incompressibility, give rise to various second order problems. These second order problems are much easier to solve than the full fourth order systems, and indeed some of these lower order problems admit elegant general solutions. Accordingly, the Varga elastic materials give rise to numerous exact deformations, which include the controllable deformations known to apply to all perfectly elastic incompressible materials. However, in addition to the standard deformations, there are many exact solutions for which the corresponding physical problem is not immediately apparent. Indeed, many of the simple exact solutions display unusual and unexpected behaviour, which possibly reflects non-physical behaviour of the Varga elastic materials for extremely large strains. Alternatively, these exact results may well mirror the full consequences of nonlinear theory. This chapter summarizes a number of recent developments.

5.1 Introduction

Natural and synthetic rubbers can accurately be modelled as homogeneous, isotropic, incompressible and hyperelastic materials, and which are sometimes referred to as perfectly elastic materials. The governing partial differential equations tend to be highly nonlinear and as a consequence the determination of exact analytical solutions is not a trivial matter. Such materials are characterised by a strain–energy function $W(\lambda_1, \lambda_2, \lambda_3) \geq 0$ which is a symmetric function of the three principal stretches λ_i $(i = 1, 2, 3)$, where $\lambda_1 \lambda_2 \lambda_3 = 1$ for incompressibility. In adopting a particular strain–energy function there is a "trade-off" between correlation with experimental data and the prospect of mathematical tractability. In other words, a more complicated strain–energy function involving several parameters may well accommodate available experimental data, but may prohibit the determination of simple mathematical solutions. Simple exact results are obviously important if they happen to also represent the solution of a practical problem, or if they can be utilized as a bench mark for a full numerical scheme.

However, there is another reason for their importance, which is not always recognised. Mathematical models of physical phenomena tend to embody more than the strict axioms which underly the model. In our case these axioms are homogeneity, isotropy, incompressibility, the assumption that a strain–energy function exists, as well as all the usual hypotheses underlying rational continuum mechanics. Now any exact consequences of the theory represent a manifestation of these combined hypotheses, and therefore can be viewed as an "outcome" or the "end product" of the mathematical model. Further, any shortcomings or non-physical behaviour of an exact result would tend to infer the inadequacy of one or more of the underlying axioms of the model, except to say that nonlinear models are frequently unpredictable and counter-intuitive.

In finite elasticity the majority of known exact results have been determined by a semi-inverse procedure with a particular simple physical problem in mind. The exact results examined in this chapter were not determined in this manner. Here we provide a synopsis of a number of recent developments which involve exact consequences of the so-called Varga elastic materials. Our results apply solely to these special materials and many of them describe unexpected and even bizarre physical behaviour. For the particular class of perfectly elastic incompressible materials, they represent the only exact results of any generality which are presently known. Their existence indicates that the subject of finite elasticity admits considerable research potential and that there is still much that is not properly understood concerning nonlinearity and nonlinear phenomena.

The Varga strain–energy function is given by

$$W = 2\mu(\lambda_1 + \lambda_2 + \lambda_3 - 3), \tag{1.1}$$

where μ coincides with the usual infinitesimal shear modulus. This strain–energy function was originally introduced by Varga (1966) for natural rubber vulcanizates and later by Dickie and Smith (1971) for styrene-butadiene vul-

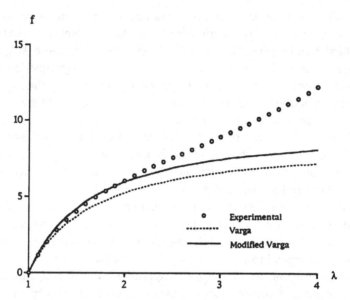

Fig. 1. Variation of the force f per unstrained area, with extension λ in simple tension for the Varga and modified Varga materials compared with typical experimental data.

canizates. The modified Varga material first proposed by Hill and Arrigo (1995) has strain–energy function

$$W = \alpha_1(\lambda_1 + \lambda_2 + \lambda_3 - 3) + \alpha_2\left(\frac{1}{\lambda_1} + \frac{1}{\lambda_2} + \frac{1}{\lambda_3} - 3\right), \qquad (1.2)$$

where α_1 and α_2 are two material parameters such that $\alpha_1 + \alpha_2 = 2\mu$. This form of the strain–energy function considerably enhances the experimental validity of (1.1), which strictly speaking is only valid for deformations with maximum principal stretch $\lambda_{\max} \leq 2$, whereas (1.2) applies for $\lambda_{\max} \leq 3$. Typical behaviour of these materials in simple tension and equibiaxial tension is indicated in Figures 1 and 2 respectively, where we have taken $\alpha_1 = 8.749$, $\alpha_2 = 0.576$. The points shown in the figures are calculated from the strain–energy function proposed by Ogden (1972) which provides a very accurate model of the experimental data of Rivlin and Saunders (1951).

The exact solutions summarized here can be traced back to a result first given by Holden (1968). For plane equilibrium deformations of perfectly elastic materials,

$$x = x(X, Y), \qquad y = y(X, Y), \qquad z = Z, \qquad (1.3)$$

where (X, Y, Z) and (x, y, z) denote respectively material and spatial rectangu-

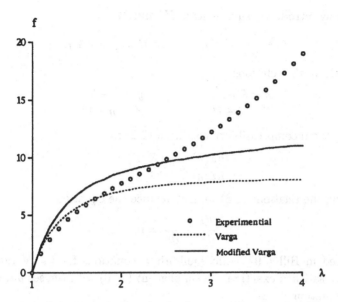

Fig. 2. Variation of the force f per unstrained area, with extension λ in equibiaxial tension for the Varga and modified Varga materials compared with typical experimental data.

lar Cartesian coordinates, Holden (1968) derived the exact parametric solution

$$X + iY = \omega - f(\bar{\omega}), \qquad x + iy = \omega + f(\bar{\omega}), \qquad (1.4)$$

where $\omega = \xi + i\eta$ is the complex parameter, $\bar{\omega}$ is the complex conjugate and $f(\omega)$ denotes any analytical function of ω. Actually Holden (1968) also imposed the condition $|f'(\omega)| < 1$, but this turns out not to be necessary. Holden's general solution applies to the particular elastic material which for the plane deformation (1.3) has strain–energy function $W^*(\zeta)$ defined parametrically by

$$W^*(\zeta) = 4\mu \cosh \zeta, \qquad I = 2 \cosh 2\zeta, \qquad (1.5)$$

where I denotes the strain invariant which is given by (2.5). However, Holden did not make the important identification that the material defined by (1.5) actually coincides with that proposed by Varga (1966), namely (1.1) with $\lambda_3 = 1$. Hill (1972) (see page 203) first made this identification and subsequently Hill (1975) gave an independent derivation of (1.4) from a variational approach. From (1.4) the "mean" coordinates (ξ, η) are given by

$$\xi = \frac{1}{2}(x + X), \qquad \eta = \frac{1}{2}(y + Y), \qquad (1.6)$$

while we may introduce displacements $2U$ and $2V$ by

$$U = \frac{1}{2}(x - X), \qquad V = \frac{1}{2}(y - Y), \qquad (1.7)$$

so that we have the relations

$$\begin{aligned} x &= \xi + U, & y &= \eta + V, \\ X &= \xi - U, & Y &= \eta - V. \end{aligned} \qquad (1.8)$$

On writing the incompressibility condition (2.2) as

$$\frac{\partial(x, y)}{\partial(\xi, \eta)} = \frac{\partial(X, Y)}{\partial(\xi, \eta)}, \qquad (1.9)$$

and utilising the relations (1.8) we may deduce the equation

$$\frac{\partial U}{\partial \xi} + \frac{\partial V}{\partial \eta} = 0. \qquad (1.10)$$

As described in Hill (1975) the equilibrium equations for the Varga elastic material are satisfied exactly if we supplement (1.10) with the second Cauchy-Riemann equation

$$\frac{\partial U}{\partial \eta} - \frac{\partial V}{\partial \xi} = 0, \qquad (1.11)$$

so that we have

$$U - iV = f(\xi + i\eta), \qquad (1.12)$$

where f is an arbitrary analytical function, which constitutes the first general solution for a perfectly elastic material. Although Holden's general solution is extremely simple and elegant, it is clearly not in a form suitable for direct use because the (ξ, η) variables involve both the undeformed and deformed co-ordinates. However, Hill (1975) utilized this result to deduce some extremely elegant closed expressions for small deformations superimposed upon the symmetrical expansion of a cylindrical tube, and which are mentioned briefly in Section 5.7 of this chapter. The initial large deformation (see equation (7.3)) is obtained by taking

$$f(\xi + i\eta) = \frac{1}{4}K(\xi + i\eta)^{-1}, \qquad (1.13)$$

where K is the constant appearing in (7.3), and the expressions deduced in Hill (1975) for the small superimposed deformation, still constitute the only known exact solutions of these complicated nonhomogeneous but linear ordinary differential equations. Subsequently Hill (1990) also exploited Holden's result in the form of (1.10) and (1.11) for the plane similarity deformation

$$r = Rf(a\Theta + b\log R), \qquad \theta = \Theta + g(a\Theta + b\log R), \qquad z = Z, \qquad (1.14)$$

where (R, Θ, Z) and (r, θ, z) denote respectively, material and spatial cylindrical polar coordinates and a and b designate arbitrary real constants. After a good deal of laborious calculations we may ultimately deduce the following parametric representation of (1.14) for the Varga elastic material, thus

$$R = 2^{1/2} \rho e^{-\beta\tau/2} (\cosh \beta\tau - \cos \alpha\tau)^{1/2}, \quad \Theta = \psi + \tan^{-1}\left(\frac{\sin \alpha\tau}{e^{\beta\tau} - \cos \alpha\tau}\right),$$

$$r = 2^{1/2} \rho e^{-\beta\tau/2} (\cosh \beta\tau + \cos \alpha\tau)^{1/2}, \quad \theta = \psi - \tan^{-1}\left(\frac{\sin \alpha\tau}{e^{\beta\tau} + \cos \alpha\tau}\right),$$

$$(1.15)$$

where the two parameters ρ and τ are defined as follows,

$$\xi + i\eta = \rho e^{i\psi}, \qquad \tau = a\psi + b \log \rho, \qquad (1.16)$$

and the constants α and β are defined by

$$\alpha = \frac{-2a}{(a^2 + b^2)}, \qquad \beta = \frac{-2b}{(a^2 + b^2)}. \qquad (1.17)$$

Thus, although Holden's result has received limited application, its usefulness is restricted, because it is couched in terms of the (ξ, η) variables and consequently gives rise to parametric solutions such as (1.15), which although explicit, generate little insight because they involve obscure variables such as the parameters ρ and τ. The recent developments described here originate from the important recognition that equation (1.11) is in fact equivalent to the condition $\partial x/\partial Y = \partial y/\partial X$. This can be most easily seen by writing (1.11) as

$$\frac{\partial(U, \xi)}{\partial(X, Y)} + \frac{\partial(V, \eta)}{\partial(X, Y)} = 0, \qquad (1.18)$$

and simplifying by means of (1.6) and (1.7). This identification first given by Hill and Arrigo (1995) means that the Cauchy-Riemann pair (1.10) and (1.11) becomes

$$\frac{\partial(x, y)}{\partial(X, Y)} = 1, \qquad \frac{\partial x}{\partial Y} = \frac{\partial y}{\partial X}, \qquad (1.19)$$

which although nonlinear, they are much more useful, because they involve the material and spatial coordinates directly. As a consequence any exact solutions of this system are more amenable to physical interpretation. The above represents the historical development of the results to be described in this chapter.

Roughly speaking the incompressible Varga elastic materials are analogous to the "harmonic" compressible elastic materials originally introduced by John (1960) and which posess strain–energy functions of the form

$$W(\lambda_1, \lambda_2, \lambda_3) = 2\mu\{h(\lambda_1 + \lambda_2 + \lambda_3) - \lambda_1\lambda_2\lambda_3\}, \qquad (1.20)$$

where again μ corresponds to the linear elastic shear modulus and the function h is assumed to satisfy $h(3) = h'(3) = 1$ and $h''(3) > 0$. John (1960) shows that plane strain deformations of these compressible elastic materials reduce to a pair of Cauchy-Riemann equations, while Hill and Arrigo (1996a) give the corresponding result for axially symmetric deformations. John's result has been exploited to deduce the solution of a variety of plane strain problems in finite elasticity (see Hill and Arrigo 1996a for further references). However, the results presented in this chapter have not been developed to the same extent and they admit considerable scope for further research.

In the following section we summarize the general equations for plane strain and axially symmetric deformations. For these special situations general integrals for the Varga materials have been identified. No doubt there are other special circumstances for which corresponding general integrals may be determined, such as that for thin plane elastic sheets given by Arrigo and Hill (1996) and that applying to the three dimensional deformation (8.16) examined in Hill and Arrigo (1999b), both of which are included in the final section of this chapter. However, corresponding results for fully three dimensional deformations have not yet been established. In the subsequent two sections respectively we detail the known integrals and some exact solutions for the two special situations of plane strain and axially symmetric deformations. Sections 5.5 and 5.6 deal with further specific examples of exact solutions for plane strain and axially symmetric deformations. In Section 5.7 of the chapter we describe recent results for small deformations superimposed upon an initial large deformation which is assumed to be a known solution of one of the integrals. As previously mentioned the final section of the chapter details some further general integrals.

5.2 General equations for plane strain and axially symmetric deformations

5.2.1 Plane strain deformations

In terms of material and spatial rectangular Cartesian coordinates (X, Y, Z) and (x, y, z) respectively, the plane deformation

$$x = x(X, Y), \qquad y = y(X, Y), \qquad z = Z, \tag{2.1}$$

of an incompressible material satisfies the condition

$$\frac{\partial(x, y)}{\partial(X, Y)} = x_X y_Y - x_Y y_X = 1, \tag{2.2}$$

where subscripts denote partial derivatives. For static deformations of homogeneous isotropic hyperelastic materials, the equilibrium equations are (see Hill

1973a)

$$p_x = \phi\nabla^2 x + \phi_X x_X + \phi_Y x_Y,$$
$$p_y = \phi\nabla^2 y + \phi_X y_X + \phi_Y y_Y, \tag{2.3}$$

where ∇^2 denotes the usual Laplacian in rectangular Cartesian coordinates. The response function $\phi(I)$ is defined in terms of the strain–energy $W(I)$ by

$$\phi = 2\frac{dW}{dI}, \tag{2.4}$$

and invariant I of the deformation tensor is given by

$$I = x_X^2 + x_Y^2 + y_X^2 + y_Y^2. \tag{2.5}$$

The pressure function p in the equilibrium equations arises as a Lagrangian multiplier corresponding to the constraint of incompressibility.

In terms of material and spatial cylindrical polar coordinates (R, Θ, Z) and (r, θ, z) respectively, the plane strain deformation with a uniform transverse stretch α^{-1} takes the form

$$r = r(R, \Theta), \qquad \theta = \theta(R, \Theta), \qquad z = \frac{Z}{\alpha}, \tag{2.6}$$

where α is a constant. For an incompressible material (2.6) satisfies the condition

$$r_R\theta_\Theta - r_\Theta\theta_R = \frac{\alpha R}{r}. \tag{2.7}$$

In Hill (1973a) the equilibrium equations are shown to become

$$p_r = \phi\left\{\nabla^2 r - r\left(\theta_R^2 + \frac{\theta_\Theta^2}{R^2}\right)\right\} + \phi_R r_R + \frac{\phi_\Theta r_\Theta}{R^2},$$
$$\frac{p_\theta}{r^2} = \phi\left\{\nabla^2\theta + \frac{2}{r}\left(r_R\theta_R + \frac{r_\Theta\theta_\Theta}{R^2}\right)\right\} + \phi_R\theta_R + \frac{\phi_\Theta\theta_\Theta}{R^2}, \tag{2.8}$$

where ∇^2 is the usual two-dimensional Laplacian, that is

$$\nabla^2 = \frac{\partial^2}{\partial R^2} + \frac{1}{R}\frac{\partial}{\partial R} + \frac{1}{R^2}\frac{\partial^2}{\partial\Theta^2}, \tag{2.9}$$

where p is the pressure function and ϕ and ψ are given respectively by

$$\phi = 2\frac{\partial W}{\partial I}, \qquad \psi = 2\frac{\partial W}{\partial\alpha}, \tag{2.10}$$

where $W(I, \alpha^2)$ is the strain–energy function of the material and I is given by

$$I = r_R^2 + \frac{r_\Theta^2}{R^2} + r^2\left(\theta_R^2 + \frac{\theta_\Theta^2}{R^2}\right). \tag{2.11}$$

As shown in Hill (1976b) the resultant force F^* acting on the semi-circle

$(-\pi/2 \le \Theta \le \pi/2)$ in the direction $\Theta = 0$, which is given originally by the circular cylinder $R = $ constant, is given by

$$F^* = \frac{L}{\alpha} \int_{-\pi/2}^{\pi/2} [-p(r\sin\theta)_\Theta + \alpha R\phi(r\cos\theta)_R]\, d\Theta, \qquad (2.12)$$

where L is the original length of the cylinder.

5.2.2 Axially symmetric deformations

In terms of material and spatial cylindrical polar coordinates (R, Θ, Z) and (r, θ, z) respectively, the axially symmetric deformation

$$r = r(R, Z), \qquad \theta = \Theta, \qquad z = z(R, Z), \qquad (2.13)$$

of an incompressible material satisfies the condition

$$\frac{\partial(r, z)}{\partial(R, Z)} = \frac{R}{r}. \qquad (2.14)$$

For static deformations of homogeneous isotropic hyperelastic materials, the equilibrium equations are (see, for example, Hill 1973b)

$$p_r = \phi_1 \nabla^2 r + \phi_{1R} r_R + \phi_{1Z} r_Z - \frac{r}{R^2}\phi_2,$$

$$p_z = \phi_1 \nabla^2 z + \phi_{1R} z_R + \phi_{1Z} z_Z, \qquad (2.15)$$

where

$$\nabla^2 = \frac{\partial^2}{\partial R^2} + \frac{1}{R}\frac{\partial}{\partial R} + \frac{\partial^2}{\partial Z^2}.$$

The response functions $\phi_1(I, \lambda)$ and $\phi_2(I, \lambda)$ are defined in terms of the strain–energy $W(I, \lambda)$ by

$$\phi_1 = 2\frac{\partial W}{\partial I}, \qquad \phi_2 = 2\frac{\partial W}{\partial \lambda^2}, \qquad (2.16)$$

where the invariant I of the deformation tensor and λ are given respectively by

$$I = r_R^2 + r_Z^2 + z_R^2 + z_Z^2, \qquad \lambda = r/R, \qquad (2.17)$$

and noting that in terms of the principal stretches I and λ are given by

$$I = \lambda_1^2 + \lambda_2^2, \qquad \lambda = \lambda_3. \qquad (2.18)$$

In terms of material and spatial spherical polar coordinates (R, Θ, Φ) and (r, θ, ϕ) respectively, the axially symmetric deformation

$$r = r(R, \Theta), \qquad \theta = \theta(R, \Theta), \qquad \phi = \Phi, \qquad (2.19)$$

satisfies the incompressibility condition

$$r_R\theta_\Theta - r_\Theta\theta_R = \frac{R^2 \sin\Theta}{r^2 \sin\theta},\tag{2.20}$$

where as usual subscripts denote partial derivatives. The equilibrium equations can be shown to become

$$p_r = \phi_1\left\{\nabla^2 r - r\left(\theta_R^2 + \frac{\theta_\Theta^2}{R^2}\right)\right\} + \phi_{1R}r_R + \frac{\phi_{1\Theta}r_\Theta}{R^2} - \phi_2\frac{r\sin^2\theta}{R^2\sin^2\Theta},$$

$$\frac{p_\theta}{r^2} = \phi_1\left\{\nabla^2\theta + \frac{2}{r}\left(r_R\theta_R + \frac{r_\Theta\theta_\Theta}{R^2}\right)\right\} + \phi_{1R}\theta_R + \frac{\phi_{1\Theta}\theta_\Theta}{R^2} - \phi_2\frac{\sin\theta\cos\theta}{R^2\sin^2\Theta},$$
$$\tag{2.21}$$

where p is the pressure function and ∇^2 is the Laplacian given by

$$\nabla^2 = \frac{\partial^2}{\partial R^2} + \frac{2}{R}\frac{\partial}{\partial R} + \frac{1}{R^2}\frac{\partial^2}{\partial\Theta^2} + \frac{\cot\Theta}{R^2}\frac{\partial}{\partial\Theta},\tag{2.22}$$

and the response functions ϕ_1 and ϕ_2 are given by

$$\phi_1 = 2\left\{\frac{\partial W}{\partial I_1} + \lambda^2\frac{\partial W}{\partial I_2}\right\}, \qquad \phi_2 = 2\left\{\frac{\partial W}{\partial I_1} + \left(I - \frac{1}{\lambda^4}\right)\frac{\partial W}{\partial I_2}\right\},\tag{2.23}$$

where $W(I_1, I_2)$ is the strain–energy function of the material. Further, I_1 and I_2 are the first two invariants of the inverse Cauchy deformation tensor which are given by

$$I_1 = I + \lambda^2, \qquad I_2 = \lambda^2 I + \frac{1}{\lambda^2},\tag{2.24}$$

where I and λ are defined by

$$I = r_R^2 + \frac{r_\Theta^2}{R^2} + r^2\left(\theta_R^2 + \frac{\theta_\Theta^2}{R^2}\right), \qquad \lambda = \frac{r\sin\theta}{R\sin\Theta};\tag{2.25}$$

and in terms of I and λ the response functions ϕ_1 and ϕ_2 as given by (2.23) become (2.16). We note that for prescribed $\phi_1(I, \lambda)$ and $\phi_2(I, \lambda)$ we can determine the partial derivatives $\partial W/\partial I_1$ and $\partial W/\partial I_2$ from the relations (2.23) provided the Jacobian of (2.24) is finite.

Further, as described in Hill (1976c) the resultant force F^* acting in the conventional z-direction on a spherical cap of original radius R and subtended by an angle Θ_0, may be shown to be determined by the expression

$$F^* = 2\pi\int_0^{\Theta_0}[-pr\sin\theta(r\sin\theta)_\Theta + R^2\sin\Theta\phi_1(r\cos\theta)_R]\,d\Theta,\tag{2.26}$$

where p is the pressure function arising in the equilibrium equations (2.21) and ϕ_1 is the first response function defined by (2.23). The basic governing equations summarized here for plane strain and axially symmetric deformations form the basis of the results presented in subsequent sections.

5.3 Exact integrals and solutions for plane strain deformations

For the plane strain deformations (2.1) and (2.6) we have $I = \lambda_1^2 + \lambda_2^2$ and $\lambda_3 = \alpha^{-1}$ (assuming $\alpha > 0$ and noting $\alpha = 1$ for (2.1)) and therefore $\lambda_1 + \lambda_2 = (I + 2\alpha)^{1/2}$, from which we see that the strain–energy function (1.2) becomes

$$W(I, \alpha) = \alpha_1 \left((I + 2\alpha)^{1/2} + \frac{1}{\alpha} - 3 \right) + \alpha_2 \left(\frac{1}{\alpha}(I + 2\alpha)^{1/2} + \alpha - 3 \right). \quad (3.1)$$

From this equation and (2.4) we have

$$\phi(I, \alpha) = \left(\alpha_1 + \frac{\alpha_2}{\alpha} \right) (I + 2\alpha)^{-1/2}. \quad (3.2)$$

For simplicity, we first examine (2.1) for which $\alpha = 1$ and (3.2) becomes

$$\phi(I) = 2\mu(I + 2)^{-1/2}. \quad (3.3)$$

Now the identification that equation (1.11) is equivalent to $x_Y = y_X$ is important because it indicates that there is a more immediate derivation of the Holden result which follows from a reformulation of the equilibrium equations. Indeed (2.3) can be shown to become (see Hill and Arrigo 1995)

$$\begin{aligned}
(p + \phi)_x &= [\phi(x_X + y_Y)]_X + [\phi(x_Y - y_X)]_Y, \\
(p + \phi)_y &= [\phi(x_X + y_Y)]_Y - [\phi(x_Y - y_X)]_X.
\end{aligned} \quad (3.4)$$

Now, from (2.2) and (2.5) we observe that (2.5) becomes

$$I + 2 = (x_X + y_Y)^2 + (x_Y - y_X)^2, \quad (3.5)$$

and it can be seen that (3.4) is satisfied if

$$p = -\phi + p_0, \qquad x_Y = y_X, \qquad \phi(x_X + y_Y) = u_0, \quad (3.6)$$

where p_0 and u_0 denote arbitrary constants. This is certainly the case for the modified Varga strain-energy function (1.2) for which $u_0 = 2\mu$.

Further, if we introduce u and v defined by

$$u = \phi(x_X + y_Y), \qquad v = \phi(x_Y - y_X), \quad (3.7)$$

then (3.4) becomes

$$(p + \phi)_x = u_X + v_Y, \qquad (p + \phi)_y = u_Y - v_X. \quad (3.8)$$

But from (3.3) and (3.5) for the Varga material (1.2) we have

$$u^2 + v^2 = \phi^2 \left[(x_X + y_Y)^2 + (x_Y - y_X)^2 \right] = \phi^2(I + 2) = 4\mu^2, \quad (3.9)$$

and therefore there exists $\psi(X, Y)$ such that

$$u = 2\mu \cos \psi, \qquad v = 2\mu \sin \psi. \quad (3.10)$$

On substituting these expressions into (3.8) and requiring compatibility Hill and Arrigo (1996c), after considerable manipulation, deduce the expression

$$\psi(X,Y) = 2\tan^{-1}\left(\frac{aY+b}{aX+c}\right) + d, \tag{3.11}$$

where a, b, c and d denote arbitrary constants. Without loss of generality we may take $b = c = d = 0$ so that (3.11) becomes

$$\psi(X,Y) = 2\tan^{-1}(Y/X), \tag{3.12}$$

and from (3.7) and (3.10) we may deduce the following first integral of the equilibrium equations for the Varga material, namely

$$x_Y - y_X = \frac{2XY}{(X^2 - Y^2)}(x_X + y_Y), \tag{3.13}$$

which must be solved in conjunction with the incompressibility condition (2.2).

Now applying the Adkins reciprocal theorem (see Adkins 1958) the condition $x_Y = y_X$ remains unaltered whereas the condition (3.13) yields a further independent integral of the equilibrium equations. In summary, for the Varga material (1.2) we have established three independent integrals:

$$
\begin{aligned}
&\text{(I)} \qquad x_Y - y_X = 0,\\
&\text{(II)} \qquad x_Y - y_X = \frac{2XY}{(X^2 - Y^2)}(x_X + y_Y),\\
&\text{(III)} \qquad x_Y - y_X = \frac{-2xy}{(x^2 - y^2)}(x_X + y_Y),
\end{aligned}
\tag{3.14}
$$

which together with the incompressibility condition (2.2) constitute three second order systems all of which admit linearization.

For (I) we introduce $\omega(X,Y)$ such that $x = \omega_X$ and $y = \omega_Y$ and (2.2) becomes $\omega_{XX}\omega_{YY} - \omega_{XY}^2 = 1$, which is a Monge-Ampère equation which may be solved parametrically by introducing $\Psi(s,t)$ such that

$$X = t, \qquad Y = \Psi_s, \qquad \omega = s\Psi_s - \Psi. \tag{3.15}$$

The Monge-Ampère equation simplifies to give the harmonic equation

$$\Psi_{tt} + \Psi_{ss} = 0. \tag{3.16}$$

Notice that it is not immediately clear as to the connection between (3.16) and the Cauchy-Riemann pair (1.10) and (1.11). Both involve harmonic functions for the same problem, but use different variables.

In cylindrical polar coordinates the three integrals (3.14) become respectively

(I) $(Rr_R + r\theta_\Theta)\sin(\theta - \Theta) + (rR\theta_R - r_\Theta)\cos(\theta - \Theta) = 0,$

(II) $(Rr_R + r\theta_\Theta)\sin(\theta + \Theta) + (rR\theta_R - r_\Theta)\cos(\theta + \Theta) = 0,$ (3.17)

(III) $(Rr_R + r\theta_\Theta)\sin(\theta + \Theta) - (rR\theta_R - r_\Theta)\cos(\theta + \Theta) = 0,$

each of which must be solved in conjunction with the incompressibility condition (2.7) with $\alpha = 1$. Integral (II) admits a remarkable linearization as first shown by Hill and Arrigo (1996c). By a complicated sequence of transformations it is shown in the Appendix of that paper that the general solution for (II) is given by

$$r = \left(\omega^2 + \omega_\Theta^2\right)^{1/2}, \qquad \theta = -\Theta + \tan^{-1}(\omega_\Theta/\omega), \qquad R = \omega_\lambda,   \qquad (3.18)$$

where ω_λ is assumed positive and $\omega(\lambda, \Theta)$ satisfies the linear Helmholtz equation

$$\omega_{\lambda\lambda} + \omega_{\Theta\Theta} + \omega = 0,   \qquad (3.19)$$

where λ is a parameter which turns out to be essentially $(p + \phi)/4\mu$.

The integral (III) can also be linearized. This is formally achieved by rewriting (3.14)$_3$ as

$$\frac{\partial}{\partial X}\left(\frac{y}{x^2 + y^2}\right) + \frac{\partial}{\partial Y}\left(\frac{x}{x^2 + y^2}\right) = 0,$$

and introducing a potential $\Omega(X, Y)$ such that

$$\Omega_X = \frac{x}{(x^2 + y^2)}, \qquad \Omega_Y = \frac{-y}{(x^2 + y^2)}.$$

From the relations

$$x = \frac{\Omega_X}{(\Omega_X^2 + \Omega_Y^2)}, \qquad y = \frac{-\Omega_Y}{(\Omega_X^2 + \Omega_Y^2)},$$

and incompressibility we may deduce a Monge-Ampère equation which may be formally solved using the same procedure as that employed for (II). The details can be found in Hill and Arrigo (1996c).

Hill and Arrigo (1999a) show that the above three integrals can be generalized as

$$\frac{\partial x}{\partial Y} - \frac{\partial y}{\partial X} = \tan\psi\left(\frac{\partial x}{\partial X} + \frac{\partial y}{\partial Y}\right),   \qquad (3.20)$$

where $\psi(X, Y, x, y)$ is defined to be

$$\psi = 2\tan^{-1}\left(\frac{ax + by + cX + dY + e}{-bx + aY + dX - cY + f}\right) + k,   \qquad (3.21)$$

where a, b, c, d, e, f and k denote seven arbitrary real constants. We note that on performing the constant rotations of the material and spatial coordinates

$$\begin{pmatrix} X \\ Y \end{pmatrix} = \begin{pmatrix} \cos \Phi & \sin \Phi \\ -\sin \Phi & \cos \Phi \end{pmatrix} \begin{pmatrix} X_1 \\ Y_1 \end{pmatrix}, \quad \begin{pmatrix} x \\ y \end{pmatrix} = \begin{pmatrix} \cos \phi & \sin \phi \\ -\sin \phi & \cos \phi \end{pmatrix} \begin{pmatrix} x_1 \\ y_1 \end{pmatrix},$$
(3.22)

it is not difficult to show that (2.2) remains unchanged, while (3.20) has the same structure except that ψ is replaced by $\psi_1 = \psi - (\phi - \Phi)$. Now on examination of (3.21) we can always translate our coordinates such that the constants e and f are zero. Moreover, by an appropriate choice of Φ and ϕ in the rotations (3.22) we may eliminate two of the constants a, b, c and d. Thus, for example if we select

$$\tan \Phi = \frac{c}{d}, \qquad \tan \phi = -\frac{b}{a},$$

and choose the constant $k = \phi - \Phi$, then equation (3.21) becomes

$$\psi_1 = 2\tan^{-1}\left(\frac{(a^2 + b^2)^{1/2} x_1 + (c^2 + d^2)^{1/2} Y_1}{(a^2 + b^2)^{1/2} y_1 + (c^2 + d^2)^{1/2} X_1}\right),$$

and therefore we can if necessary assume that $\psi(X, Y, x, y)$ is given simply by

$$\psi = 2\tan^{-1}\left(\frac{ax + dY}{ay + dX}\right),$$
(3.23)

but there are other choices of Φ and ϕ which would allow other possibilities for constants to be set to zero. In addition, ψ given by (3.23) involves only one essential constant, namely either of the ratios d/a or a/d. However, it is sometimes convenient to leave the two constants a and d so that the special cases $a = 0, d \neq 0$ and $d = 0, a \neq 0$ recover the two integrals (II) and (III) respectively. In terms of cylindrical polar coordinates (3.20) and (3.21), with the three translational constants e, f and k set to zero, become

$$\frac{r_\Theta}{R} - r\theta_R = \tan(\psi + \theta - \Theta)\left(r_R + \frac{r}{R}\theta_\Theta\right),$$
(3.24)

$$\tan\left(\frac{\psi + \theta - \Theta}{2}\right) = \left(\frac{(ar + cR)\cos\left(\frac{\theta + \Theta}{2}\right) + (br + dR)\sin\left(\frac{\theta + \Theta}{2}\right)}{-(br - dR)\cos\left(\frac{\theta + \Theta}{2}\right) + (ar - cR)\sin\left(\frac{\theta + \Theta}{2}\right)}\right),$$
(3.25)

where subscripts denote partial derivatives and the curious equation (3.25) is most easily established from the left-hand side by expanding $\tan[\psi/2 + (\theta - \Theta)/2]$ and then using (3.21) to determine an expression for $\tan(\psi/2)$. In Beatty and Hill (2000) the angle ψ as defined by equation (3.20) is shown to be the local rigid body rotation angle.

Questions concerning the extent to which solutions of these integrals span the entire solution space are non-trivial. Hill and Cox (2000) attempt this question by examining the plane strain similarity deformation

$$r = Rf(\Theta), \qquad \theta = g(\Theta), \qquad z = Z, \tag{3.26}$$

which for the Varga material can be fully integrated, and the idea is to determine how solutions of the above reduced systems fit into the general solution. Even for this restricted deformation the situation is complicated and we refer the reader to the paper for further details. Further details on similarity deformations in finite elasticity can be found in Hill (1982, 1990).

Numerous specific solutions of the above integrals (I) and (II) can be found in Hill and Arrigo (1995, 1996c) and for the deformation (3.26) in Hill and Cox (2000). Here we merely indicate one or two illustrative examples. Hill and Arrigo (1995) show that the general solution of (2.2) and (3.14) such that y/x is a function of Y/X becomes

$$x = (bY + cX)\left[\frac{1}{ac - b^2} + \frac{c^*}{aY^2 + 2bXY + cX^2}\right]^{1/2},$$

$$y = (aY + bX)\left[\frac{1}{ac - b^2} + \frac{c^*}{aY^2 + 2bXY + cX^2}\right]^{1/2},$$

where again a, b, c, and c^* denote arbitrary real constants and for which we may verify the two key relations

$$\frac{y}{x} = \left(a\frac{Y}{X} + b\right)\bigg/\left(b\frac{Y}{X} + c\right),$$

$$ax^2 - 2bxy + cy^2 = aY^2 + 2bXY + cX^2 + (ac - b^2)c^*.$$

Thus, the deformation is such that material points move along radial lines in such a way that particles originally along one binary quadratic $aY^2 + 2bXY + cX^2 = $ constant are deformed to another binary quadratic $ax^2 - 2bxy + cy^2 = $ constant.

Hill and Arrigo (1996c) show that the corresponding solution of integral (II) (namely (2.7) with $\alpha = 1$ and $(3.17)_2$) is given by

$$r = \left\{\left[\frac{R^2}{C\cos^2(C_1\Theta + C_2)} + C_3\right]\left[1 + C\sin^2(C_1\Theta + C_2)\right]\right\}^{1/2},$$

$$\theta = -\Theta + \tan^{-1}\{C_1\tan(C_1\Theta + C_2)\}, \tag{3.27}$$

where C_1, C_2 and C_3 denote three arbitrary constants and $C = C_1^2 - 1$. It is of some considerable interest to show that these equations also emerge from the general solution (3.18) and (3.19) and the details can be found in the paper.

Hill and Cox (2000) give a number of simple explicit forms for (3.26) which are similar to (3.27) and are obtained from (2.7) with $\alpha = 1$ and (3.24) and (3.25) by taking special values of the constants a, b, c and d such as:

(i) $a = b = c = 0$ and $d \neq 0$.

$$f(\Theta) = \left[\frac{1 + C_1^2 \tan^2(C_1\Theta + C_2)}{C_1^2 - 1}\right]^{1/2}, \quad g(\Theta) = -\Theta + \tan^{-1}[C_1 \tan(C_1\Theta + C_2)],$$

where $C_1 \neq \pm 1$ and C_2 denote arbitrary constants and we note that the same deformation arises for the case $a = b = d = 0$ and $c \neq 0$.

(ii) $b = c = d = 0$ and $a \neq 0$.

$$f(\Theta) = \frac{(1 - C_1^2)^{1/2}}{C_1} \cos(\Theta + g), \quad g(\Theta) = C_1 \tan^{-1}[C_1 \tan(\Theta + g(\Theta)] + C_2,$$

and the same deformation arises for the case $a = c = d = 0$ and $b \neq 0$.

(iii) $a = b = 0$ and $c = d \neq 0$.

$$f(\Theta) = \left[\frac{1 - C_1^2 \cot^2(C_1\Theta + C_2)}{C_1^2 - 1}\right]^{1/2}, \quad g(\Theta) = -\Theta + \tan^{-1}\left[\frac{\tan(C_1\Theta + C_2)}{C_1}\right],$$

and this is well defined provided $C_1 \neq \pm 1$.

(iv) $c = d = 0$ and $a = b \neq 0$.

$$f(\Theta) = \frac{(1 - C_1^2)^{1/2}}{C_1} \sin(\Theta + g), \quad g(\Theta) = \frac{1}{C_1} \tan^{-1}[C_1 \tan(\Theta + g(\Theta))] + C_2,$$

and again this is well defined provided $C_1 \neq \pm 1$.

5.4 Exact integrals and solutions for axially symmetric deformations

For the axially symmetric deformation (2.13) we have from the relations (2.18) that the strain–energy function (1.2) becomes

$$W(I, \lambda) = \alpha_1 \left(\left(I + \frac{2}{\lambda}\right)^{1/2} + \lambda - 3\right) + \alpha_2 \left(\lambda \left(I + \frac{2}{\lambda}\right)^{1/2} + \frac{1}{\lambda} - 3\right), \quad (4.1)$$

and therefore from (2.16) the response functions $\phi_1(I,\lambda)$ and $\phi_2(I,\lambda)$ are given by

$$\phi_1(I,\lambda) = (\alpha_1 + \alpha_2\lambda)\left(I + \frac{2}{\lambda}\right)^{-1/2},$$

$$\phi_2(I,\lambda) = \frac{\alpha_1}{\lambda} + \frac{\alpha_2}{\lambda}\left(I + \frac{2}{\lambda}\right)^{1/2} - \frac{\alpha_2}{\lambda^3} - \frac{1}{\lambda^3}(\alpha_1 + \alpha_2\lambda)\left(I + \frac{2}{\lambda}\right)^{-1/2}.$$

$$(4.2)$$

Now on introducing u and v defined by

$$u = \phi_1(r_R + z_Z), \qquad v = \phi_1(r_Z - z_R), \tag{4.3}$$

we may reformulate the equilibrium equations (2.15) to give

$$\left(p + \frac{\phi_1}{\lambda}\right)_r = u_R + v_Z + \frac{u}{R} - \frac{r}{R^2}\left(\phi_2 + \frac{\phi_1}{\lambda^3}\right),$$

$$\left(p + \frac{\phi_1}{\lambda}\right)_z = u_Z - v_R - \frac{v}{R}, \tag{4.4}$$

and from (2.14) and (2.17) we have

$$I + \frac{2}{\lambda} = (r_R + z_Z)^2 + (r_Z - z_R)^2 = \frac{(u^2 + v^2)}{\phi_1^2}. \tag{4.5}$$

As described by Hill and Arrigo (1995) we may confirm that the obvious generalization of $(3.14)_1$, namely $r_Z = z_R$ provides an integral for the modified Varga strain–energy function (1.2) with

$$u = \alpha_1 + \alpha_2\lambda, \qquad v = 0, \qquad p = -\frac{(\phi_1 + \alpha_2)}{\lambda} + p_0, \tag{4.6}$$

where p_0 denotes an arbitrary constant.

For axially symmetric deformations, we find that the integral corresponding to $(3.14)_2$ applies only to the Varga strain–energy (1.1) and not to the more general form (1.2). In this case (4.2) becomes simply

$$\phi_1(I,\lambda) = 2\mu\left(I + \frac{2}{\lambda}\right)^{-1/2}, \qquad \phi_2(I,\lambda) = \frac{2\mu}{\lambda} - \frac{2\mu}{\lambda^3}\left(I + \frac{2}{\lambda}\right)^{-1/2}, \tag{4.7}$$

from which we may deduce from (4.5)

$$u^2 + v^2 = (2\mu)^2. \tag{4.8}$$

Accordingly, we may again introduce $\psi(R, Z)$ such that

$$u = 2\mu\cos\psi, \qquad v = 2\mu\sin\psi, \tag{4.9}$$

so that

$$r_Z - z_R = \tan\psi(r_R + z_Z). \tag{4.10}$$

Again after a laborious calculation which is described in Arrigo and Hill (1996) we may eventually deduce, modulo unessential translational constants,

$$\psi(R, Z) = -2\tan^{-1}(R/Z), \tag{4.11}$$

as providing an integral of the equilibrium equations (4.4) for the Varga strain-energy function (1.1). From the inverse results of Shield (1967), we find that the deformation which is inverse to one for the Varga material applies to the strain–energy function

$$W = 2\mu\left(\frac{1}{\lambda_1} + \frac{1}{\lambda_2} + \frac{1}{\lambda_2} - 3\right), \tag{4.12}$$

which has no physical or experimental basis and we do not pursue these results. In summary, for axially symmetric deformations of the Varga elastic material with strain–energy function (1.1) we have established the following two independent integrals:

$$
\begin{aligned}
&\text{(I)} \quad r_Z - z_R = 0, \\
&\text{(II)} \quad r_Z - z_R = \frac{2RZ}{(R^2 - Z^2)}(r_R + z_R),
\end{aligned} \tag{4.13}
$$

noting that (4.13)$_1$ also applies to the more general modified Varga strain-energy function (1.2). These integrals along with the incompressibility condition (2.14) constitute two nonlinear second order problems, but unlike the corresponding problems for plane deformations, they are not readily linearized.

For (I) we may introduce $\Omega(R, Z)$ such that $r = \Omega_R$ and $z = \Omega_Z$ and from (2.14) we obtain

$$\Omega_{RR}\Omega_{ZZ} - \Omega_{RZ}^2 = \frac{R}{\Omega_R}. \tag{4.14}$$

For (II) we may introduce new coordinates (α, β) defined by

$$\alpha = \frac{R}{(R^2 + Z^2)}, \qquad \beta = \frac{-Z}{(R^2 + Z^2)}, \tag{4.15}$$

so that equations (2.14) and (4.13)$_1$ become simply

$$\frac{\partial(r, z)}{\partial(\alpha, \beta)} = \frac{\alpha}{r(\alpha^2 + \beta^2)^3}, \qquad r_\beta = z_\alpha. \tag{4.16}$$

Thus, there exists a potential $\Omega(\alpha, \beta)$ such that $r = \Omega_\alpha$, $z = \Omega_\beta$, and the equation $(4.16)_1$ reduces to the single Monge-Ampère equation

$$\Omega_{\alpha\alpha}\Omega_{\beta\beta} - \Omega_{\alpha\beta}^2 = \frac{\alpha}{\Omega_\alpha(\alpha^2 + \beta^2)^3}. \tag{4.17}$$

From the equilibrium equation (4.4) we find that the pressure function p becomes

$$p + \phi_1/\lambda = 8\mu(\Omega - \alpha\Omega_\alpha - \beta\Omega_\beta) + p_0, \tag{4.18}$$

where p_0 denotes the constant of integration and λ denotes r/R. Although the linearization of both (4.14) and (4.17) is presently unknown, there are many known special solutions of these equations. Before examining some of these we first note that in terms of material and spatial spherical polar coordinates (R, Θ, Φ) and (r, θ, ϕ) respectively the integrals (I) and (II) become simply

(I) $(Rr_R + r\theta_\Theta)\sin(\theta - \Theta) + (rR\theta_R - r_\Theta)\cos(\theta - \Theta) = 0,$
(II) $(Rr_R + r\theta_\Theta)\sin(\theta + \Theta) + (rR\theta_R - r_\Theta)\cos(\theta + \Theta) = 0,$ (4.19)

which formally coincide with (I) and (II) of the previous section and each of which must be solved in conjunction with the incompressibility condition (2.20).

The following special solutions for (I) are derived in Hill and Arrigo (1995) while those for (II) are derived in Arrigo and Hill (1996). The coupled system (2.14) and $(4.13)_1$ admits the similarity solutions

$$r = R^{(2m+1)/3}F(\zeta), \qquad z = R^{(4-m)/3}G(\zeta), \tag{4.20}$$

where $\zeta = R^m/Z$ and m is an arbitrary parameter. In general the resulting ordinary differential equations need to be solved numerically, but for three special values of m analytical solutions may be determined.

(i) **m = −2.** In this case one noteworthy special solution is

$$r = \left(b\frac{Z}{R} - 2RZ^2\right)^{1/3}, \qquad z = \left(b\frac{R^2}{8Z^2} - \frac{R^4}{4Z}\right)^{1/3},$$

where b denotes an arbitrary constant. We note that this solution possesses singularities on both the Z axis and the plane $Z = 0$ and has the remarkable property that the inverse deformation obtained by the formal interchange of the material and spatial coordinates produces exactly the same deformation.

(ii) **m = 0.** In this case (4.20) becomes simply

$$r = R^{1/3}A(Z), \qquad z = R^{4/3}B(Z),$$

for certain functions $A(Z)$ and $B(Z)$. In particular $A(Z)$ is determined by the integral expression

$$\int_{\omega_0}^{A(Z)^{-3}} \frac{d\omega}{(c_0^3 - \omega^3)^{1/2}} = \pm 2\sqrt{2}(Z - Z_0),$$

where ω_0, c_0 and Z_0 all denote arbitrary real constants.

(iii) **m = 4.** In this case we may deduce

$$r = \left(aR^9 - 2RZ^2\right)^{1/3}, \qquad z = -\frac{1}{3}\int^{R^4/Z} \frac{d\xi}{(a\xi^3 - 2\xi)^{2/3}} + b,$$

where a and b denote arbitrary real constants.

For both (I) and (II) the solutions for which r/z is a function of R/Z are of particular interest. With $r/z = f(R/Z)$ we may deduce for (I) the following ordinary differential equation

$$\zeta f(\zeta f + 1)f'' + f'(1 - 2\zeta f)(\zeta f' - f) = 0, \qquad (4.21)$$

where here $\zeta = R/Z$. Although there appears to be no simple general solution of this equation, remarkably the transformation $f = g'/(2g - \zeta g')$ can be utilized to produce two integrals

$$(2gg'' - g'^2)g' = c_1\zeta, \qquad g'^2(\zeta g' - 2g) + c_1\zeta^2 = c_2 g^{3/2},$$

where c_1 and c_2 denote constants of integration. The second integral gives the general solution in terms of the roots of a cubic, after making the transformation $g = \zeta^{4/3}h$. Alternatively, the solution of (4.21) can be expressed as the root of a quadratic since in terms of f, the second integral is quadratic in f'. In both cases the final results are not particularly illuminating. However, one simple solution of (4.21) is $f(\zeta) = a\zeta$ where a denotes an arbitrary constant. This gives rise to the following exact solution

$$r = a^{1/3}R\frac{\left[(aR^2 + Z^2)^{3/2} + c\right]}{(aR^2 + Z^2)^{1/2}}, \qquad z = Z\frac{\left[(aR^2 + Z^2)^{3/2} + c\right]}{a^{2/3}(aR^2 + Z^2)^{1/2}}, \qquad (4.22)$$

where c denotes a further arbitrary constant. This solution constitutes a modest generalization of the well known controllable exact solution (namely $a = 1$) describing the uniform inflation of a thick walled spherical balloon. We note that this solution can also be obtained directly from the known controllable deformation from the following scaling transformations

$$r \to a^{1/6}r, \qquad z \to a^{2/3}z, \qquad R \to a^{1/2}R, \qquad Z \to Z,$$

because the reduced partial differential equations (2.14) and (4.13)$_1$ are left invariant by this transformation. We also note that a similar deformation to

(4.22) arises from the second integral on looking for solutions of (4.17) which are spherically symmetric. We obtain

$$r = R \frac{\left[C - (R^2 + Z^2)^{\frac{3}{2}}\right]^{\frac{1}{3}}}{(R^2 + Z^2)^{\frac{1}{2}}}, \qquad z = -Z \frac{\left[C - (R^2 + Z^2)^{\frac{3}{2}}\right]^{\frac{1}{3}}}{(R^2 + Z^2)^{\frac{1}{2}}}, \qquad (4.23)$$

where C is an arbitrary positive constant and the deformation is restricted to the sphere $R^2 + Z^2 \leq C^{2/3}$.

For solutions of (II) for which r/z is a function of R/Z, Arrigo and Hill (1996) deduce from (2.20) and $(4.19)_2$ the following deformation in spherical polar coordinates

$$r = \left(\frac{R^3 \sin \Theta}{g' \sin g} + h(\Theta) \right)^{1/3}, \qquad \theta = g(\Theta), \qquad \phi = \Phi, \qquad (4.24)$$

where $g(\Theta)$ and $h(\Theta)$ denote functions of Θ and primes here denote differentiation with respect to Θ. Now from $(4.19)_2$ we obtain

$$\frac{3 \sin \Theta}{g' \sin g}(g' + 1) \sin(g + \Theta) = \left(\frac{\sin \Theta}{g' \sin g} \right)' \cos(g + \Theta),$$

$$3hg' \sin(g + \Theta) = h' \cos(g + \Theta), \qquad (4.25)$$

the first of which readily integrates to give

$$g' = C_1 \frac{\sin \Theta \cos^3(g + \Theta)}{\sin g}, \qquad (4.26)$$

where C_1 denotes the constant of integration. Unlike the corresponding solution (3.27) for plane deformations we are unable to effect a further integration of these equations. However, we observe that (4.23) is a special case of the above with $C_1 = 1$ given by

$$r = (C - R^3)^{1/3}, \qquad \theta = \pi - \Theta,$$

which arises from (4.25) and (4.26) with $g(\Theta) = \pi - \Theta$ and $h(\Theta) = C$, a constant. In the following section we investigate further plane strain deformations for the Varga elastic material.

5.5 A further plane strain deformation

In this section we examine the deformation

$$r = \left(\alpha R^2 + f(\Theta) \right)^{1/2}, \qquad \theta = \Theta, \qquad z = Z/\alpha, \qquad (5.1)$$

and summarise the results given by Hill and Milan (1999) for the Varga elastic material, where α is a constant which may be positive or negative. For this deformation $\alpha > 0$ corresponds to inflation whereas $\alpha < 0$ corresponds to eversion, and there is an important distinction that we need to identify between these two cases. For both situations the principal stretches $(\lambda_1, \lambda_2, \lambda_3)$ are such that

$$I = \lambda_1^2 + \lambda_2^2, \qquad \lambda_3 = \frac{1}{|\alpha|}, \tag{5.2}$$

where the quantity I is defined by equation (2.11). From $(5.2)_2$ and the relation $\lambda_1\lambda_2\lambda_3 = 1$ we may deduce

$$I + 2|\alpha| = (\lambda_1 + \lambda_2)^2. \tag{5.3}$$

However, from (5.1) and the definition of I we have

$$I = 2\alpha + \frac{A(\Theta)}{r^2 R^2}, \tag{5.4}$$

where the function $A(\Theta)$ is defined by

$$A(\Theta) = f^2 + \frac{f'^2}{4}, \tag{5.5}$$

where primes here and throughout denote differentiation with respect to Θ. Thus from (5.3) and (5.4) we have

$$\lambda_1 + \lambda_2 = (I + 2|\alpha|)^{1/2} = \begin{cases} A(\Theta)^{1/2}/rR & \text{for } \alpha < 0, \\ (4|\alpha| + (A(\Theta)/r^2 R^2))^{1/2} & \text{for } \alpha > 0, \end{cases} \tag{5.6}$$

and we note that $\lambda_1 + \lambda_2$ has a simpler structure for $\alpha < 0$ than for $\alpha > 0$. Paradoxically, this in effect means that it is mathematically easier to find solutions for the supposedly physically more complicated problem of eversion ($\alpha < 0$) than it is for the problem of inflation ($\alpha > 0$).

Now for (5.1) the equilibrium equations (2.8) can be shown to become

$$(p + \alpha\phi)_R = \frac{\alpha\phi f''}{2r^2 R} - \frac{\alpha A\chi}{r^4 R} + \frac{\alpha f' A'}{2r^4 R^3}\frac{d\phi}{dI},$$

$$(p + \alpha\phi)_\Theta = \frac{\alpha\phi f'}{r^2} + \frac{\chi}{2r^2}\left(\frac{A'}{r^2} - \frac{Af'}{r^4}\right) - \frac{\alpha A'(2\alpha R^2 + f)}{r^4 R^2}\frac{d\phi}{dI}, \tag{5.7}$$

where the function $\chi(I)$ is defined by

$$\chi(I) = \phi + 2(I + 2\alpha)\frac{d\phi}{dI}.$$

For the Varga strain–energy function, with appropriate re-definition of the shear modulus μ, the response function $\phi(I)$ is given by

$$\phi(I, \alpha) = \frac{2\mu}{(I + 2|\alpha|)^{1/2}}. \tag{5.8}$$

For this response function $\chi(I) \equiv 0$ provided that $\alpha > 0$, and when $\alpha < 0$, we have

$$\chi(I) = \phi + 2(I + 2|\alpha| - 4|\alpha|)\frac{d\phi}{dI} = \frac{8\mu|\alpha|}{(I + 2|\alpha|)^{3/2}},$$

again indicating the differences in the two cases of inflation and eversion.

For $\alpha > 0$ the only solution is $f(\Theta) =$ constant. However, for $\alpha < 0$, we have from $(5.6)_1$ and (5.8)

$$\phi(I, \alpha) = \beta r R B(\Theta), \tag{5.9}$$

where β denotes 2μ and the function $B(\Theta) = A(\Theta)^{-1/2}$, where $A(\Theta)$ is defined by (5.5). In this case the equilibrium equations (5.7) eventually give

$$\frac{(Bf')''}{2} = 2(Bf)', \qquad \frac{f(Bf')''}{2} - \frac{f'(Bf')'}{4} = 3f(Bf)'. \tag{5.10}$$

On using the first equation in the second equation we have an identity which arises merely from the relation

$$B = \left(f^2 + \frac{f'^2}{4}\right)^{-\frac{1}{2}}, \tag{5.11}$$

that is

$$(Bf)(Bf)' + \frac{(Bf')}{2}\frac{(Bf')'}{2} = 0.$$

Now the·first equation of (5.10) integrates immediately to give

$$\frac{(Bf')'}{2} = 2Bf + c, \tag{5.12}$$

where c denotes the constant of integration. In order to integrate (5.12) further we exploit (5.11) and introduce ρ such that

$$Bf = \cos\rho, \qquad \frac{Bf'}{2} = \sin\rho, \tag{5.13}$$

and from (5.12) we obtain

$$\frac{d\rho}{d\Theta} = 2 + c\sec\rho. \tag{5.14}$$

If $c = 0$ then $\rho = 2\Theta + e$ where e is a constant and from (5.13) we may deduce

$$f(\Theta) = d\sec(2\Theta + e), \tag{5.15}$$

where d denotes a further arbitrary constant. If $c \neq 0$ then (5.14) may be integrated using the half-tangent substitution $t = \tan(\rho/2)$, thus

$$d\Theta = \frac{d\rho}{2 + c\sec\rho} = \frac{1}{2}\left\{d\rho - \frac{2\delta dt}{(\delta + 1) + (\delta - 1)t^2}\right\},$$

where δ denotes $c/2$. From this equation and assuming $\delta > 1$ we may deduce

$$\rho - \frac{2\delta}{(\delta^2 - 1)^{1/2}}\tan^{-1}\left\{\left(\frac{\delta - 1}{\delta + 1}\right)^{1/2}\tan\left(\frac{\rho}{2}\right)\right\} = 2\Theta + e, \tag{5.16}$$

and from (5.13) and (5.14) we have

$$f(\Theta) = \frac{d}{(\delta + \cos\rho)}, \tag{5.17}$$

where again d denotes a further constant of integration. Further, with $c = 2\delta$, the pressure function is given by

$$p - \alpha\phi = \frac{2\alpha\beta\delta}{\gamma^{1/2}}\sin^{-1}\left(\left(\frac{\gamma}{f}\right)^{1/2}R\right) + p_0,$$

where p_0 denotes an arbitrary constant and $\gamma = |\alpha|$.

In summary, for the Varga elastic materials there are no nontrivial $f(\Theta)$ other than $f(\Theta) = $ constant for the case $\alpha > 0$. However, for the case $\alpha < 0$ the general solution for $f(\Theta)$ may be determined in terms of three arbitrary constants δ, d and e and arising from the fact that the governing system $(5.10)_1$ and (5.11) is third order. At first sight this result is a little unexpected, but a similar situation also arises for the integrals (3.17). For example, for (I) and the deformation

$$r = \left(R^2 + f(\Theta)\right)^{1/2}, \qquad \theta = \Theta, \qquad z = Z,$$

the only possibility arises from $f'(\Theta) = 0$, whereas for the deformation

$$r = \left(-R^2 + f(\Theta)\right)^{1/2}, \qquad \theta = -\Theta, \qquad z = Z,$$

equation $(3.17)_1$ yields $f(\Theta) = d\sec 2\Theta$ which is essentially the solution (5.15).

Hill and Milan (2000) exploit the same solution for the deformation

$$r = (-\alpha R^2 + d\sec 2\Theta)^{1/2}, \qquad \theta = \pi - \Theta, \qquad z = Z/\alpha, \tag{5.18}$$

where in this context α denotes a positive constant for the problem of the plane strain bending of a sector of a circular cylindrical tube by a resultant force applied to two of the faces only. With the resultant forces and moments as shown in Figure 3, the constants α, d and p_0 are determined from $U = M = 0$, along with the condition that the resultant force over each cross-section is also

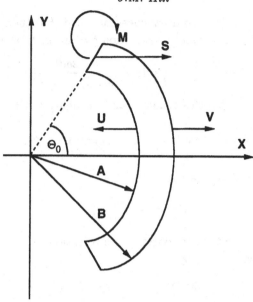

Fig. 3. Force and moment resultants referred to the undeformed body $(U = V + 2S)$.

zero. For this problem λ_{max} is less than the value 2 and therefore the strains involved here lie within the range for which the Varga material is physically meaningful. However, the numerical procedure to determine the constants α, d and p_0 is only meaningful provided $d \geq \alpha B^2$ which occurs only for $B/A \leq 1.325$. The general problem of determining all those strain-energy functions which admit non-constant $f(\Theta)$ is examined in Hill and Milan (2001).

5.6 A further axially symmetric deformation

In spherical polar coordinates, we examine the axially symmetric deformation

$$r = \left(-R^3 + f(\Theta)\right)^{1/3}, \qquad \theta = \pi - \Theta, \qquad \phi = \Phi, \qquad (6.1)$$

where $f(\Theta)$ is a function of Θ only and we summarize the results given by Hill (2000).

From (6.1) and the integral (I) as given by $(4.19)_1$, we may readily deduce

$$f' = 3f \tan 2\Theta, \qquad (6.2)$$

where primes here denote differentiation with respect to Θ. On integration this equation gives rise to the simple elegant solution

$$f(\Theta) = k(\cos 2\Theta)^{-3/2}, \qquad (6.3)$$

where k is a constant. This exact solution applies to the modified Varga strain–energy function (2.1) for which the pressure function is given explicitly by the expression (4.6)$_3$. Hill (2000) shows that for the Varga and modified Varga strain–energy functions, the only possible solutions of the form (6.1) are either $f(\Theta) = $ constant or (6.3). In addition, other than (6.3) for $n = 1$, it is shown that for the class of strain–energy functions for which the first response function $\phi_1(I, \lambda)$ is given by

$$\phi_1(I, \lambda) = \frac{2^n \mu}{(I + 2/\lambda)^{n/2}}, \qquad (6.4)$$

only the case $n = 2$ permits a nontrivial $f(\Theta)$ other than $f(\Theta) = $ constant. For $n = 2$, $f(\Theta)$ satisfies the highly non-linear ordinary differential equation

$$f\left\{ \frac{(gf')'}{3} + \cot\Theta\frac{gf'}{3} - 6gf \right\} = 3\log\left(\sin\Theta g^{1/2}\right) + C_1, \qquad (6.5)$$

where $g(\Theta)$ as a function of f and f' is defined by

$$g(\Theta) = \frac{1}{(f^2 + f'^2/9)}, \qquad (6.6)$$

and C_1 denotes a constant of integration. Equations (6.5) and (6.6) constitute a well-defined second order ordinary differential equation for the determination of $f(\Theta)$ and at least in principle, for this particular response function, we know there exists non-trivial $f(\Theta)$, perhaps involving three arbitrary constants. However, determination of a simple analytical expression appears to be difficult and further progress can only be achieved using a numerical solution procedure.

Returning to the modified Varga material and (6.1) with $f(\Theta) = k(\cos 2\Theta)^{-3/2}$ we note that the controllable deformation involved in turning a spherical shell inside out by means of a cut and due originally to Green and Shield (1950) is given by

$$r = (-R^3 + K)^{1/3}, \qquad \theta = \pi - \Theta, \qquad \phi = \Phi, \qquad (6.7)$$

where K is a constant. Namely, this deformation describes the eversion problem for a complete thick-walled spherical shell, which is everted by means of a cut. For other contributions to this problem, we refer the reader to Eringen (1962), (see page 182). For a portion of a spherical shell, a deformation of the form (6.1) is more likely to apply and the question arises as to whether we might utilise (6.3), for example, to describe the 'snap-buckling' of a spherical cap which is a familiar physical effect. For the majority of exact deformations which apply to particular finite elastic materials, it is not usually possible to satisfy stress boundary conditions in a pointwise sense, and at best only approximate or 'averaged' stress conditions on the boundary can be satisfied. In many instances such solutions lead to useful practical load-deflection relations (see, for example, Klingbeil and Shield 1966 and Hill and Lee 1989).

From equation (2.26) we may show, as described by Hill (2000), that the resultant force F^* acting in the z-direction, for the deformation given by (6.1) and (6.3), on a spherical cap of radius R and subtended by an angle Θ_0 is given by the expression

$$F^* = \pi R^2 \sin^2 \Theta_0 \left\{ \alpha_1 + 2\alpha_2\lambda - p_0\lambda^2 \right\}, \tag{6.8}$$

where λ is identified by

$$\lambda = \frac{r}{R} = \left(\frac{1}{\xi_0^3} - 1 \right)^{1/3}, \qquad \xi_0 = \frac{R}{k^{1/3}} \left(\cos 2\Theta_0 \right)^{1/2}. \tag{6.9}$$

We observe that α_1 and α_2 in (6.8) are the material constants arising in the modified Varga strain–energy function (6.2), and the limited experimental investigation given in Hill and Arrigo (1996b) indicates that both α_1 and α_2 are positive and that α_2/α_1 is approximately $1/15$. We also note that the resultant force F^* is quadratic in the principal stretch λ. Assuming for the time being that we may neglect the flat surface $\Theta = \Theta_0$ of the spherical cap which we take to be defined by

$$\{(R, \Theta, \Phi) : A \le R \le B; \quad 0 \le \Theta \le \Theta_0; \quad 0 \le \Phi \le 2\pi \},$$

where A and B here denote the inner and outer radii respectively, we can in principle determine the remaining unknown constants k and p_0 such that the condition F^* vanishing on the inner and outer spherical surfaces is satisfied, which would produce an approximate solution to the snap buckling problem. This implies that the values of λ at the inner and outer surfaces, namely λ_A and λ_B are determined as roots of

$$p_0\lambda^2 - 2\alpha_2\lambda - \alpha_1 = 0, \tag{6.10}$$

thus

$$\lambda_A = \frac{\alpha_2 + (\alpha_2^2 + \alpha_1 p_0)^{1/2}}{p_0}, \qquad \lambda_B = \frac{\alpha_2 - (\alpha_2^2 + \alpha_1 p_0)^{1/2}}{p_0}, \tag{6.11}$$

and therefore

$$\lambda_A + \lambda_B = \frac{2\alpha_2}{p_0}, \qquad \lambda_A\lambda_B = -\frac{\alpha_1}{p_0} \tag{6.12}$$

It is clear that for positive $\alpha_1, \alpha_2, \lambda_A$ and λ_B these conditions are not compatible and therefore at least for this particular material, the phenomenon of snap buckling of a spherical cap is not embodied in the deformation given by (6.1) and (6.3).

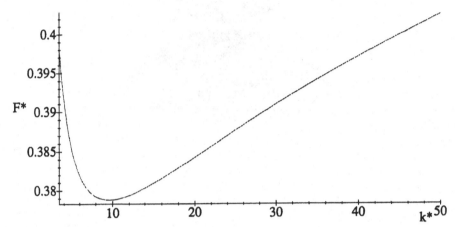

Fig. 4. Variation of $F/\pi\alpha_1 A^2$ with $k^* = k/A^3$ as given by (6.14) and (6.15) for $\Theta_0 = \pi/6$ and $B/A = 1.5$.

Alternatively, we can suppose that the resultant force F^* vanishes on one of the spherical surfaces and is prescribed on the other, thus

$$F^*(A) = 0, \qquad F^*(B) = F, \tag{6.13}$$

in which case we may deduce

$$F = \frac{\pi B^2}{\lambda_A^2} \sin^2 \Theta_0 (\lambda_A - \lambda_B) \{\alpha_1(\lambda_A + \lambda_B) + 2\alpha_2 \lambda_A \lambda_B\}, \tag{6.14}$$

noting that $\lambda_A > \lambda_B$ since $\lambda(R)$ is monotonically decreasing. Figure 4 shows the variation of $F/\pi\alpha_1 A^2$ with $k^* = k/A^3$ for $\Theta_0 = \pi/6$ and for the value of $B/A = 1.5$ assuming that $\alpha_2/\alpha_1 = 1/15$. It is clear from the figure that as noted previously F can never be zero and that for any given F there are two possible values of K. Further, in this context, λ_A and λ_B are given by

$$\lambda_A = \left(k^* (\cos 2\Theta_0)^{-3/2} - 1 \right)^{1/3}, \qquad \lambda_B = \left(\frac{k^* (\cos 2\Theta_0)^{-3/2}}{\delta^3} - 1 \right)^{1/3}. \tag{6.15}$$

where δ denotes B/A and the actual deformation of the outer surface is shown in Figure 5 for the case $B/A = 1.5$ and $k^* = 8.0$.

J.M. Hill

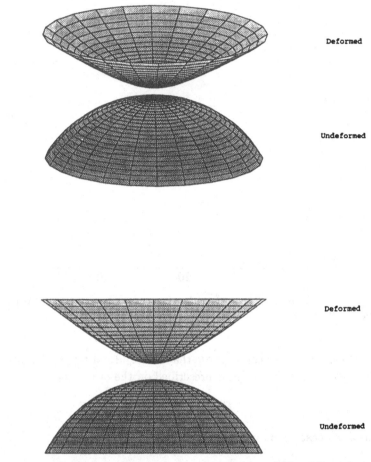

Fig. 5. Actual deformation of the outer surface for $B/A = 1.5$ and $k^* = 8.0$ as given by (6.1) and (6.3).

5.7 Small superimposed deformations

In this section we summarize some of the results given in Hill and Arrigo (1999c, d). A number of practical problems, including stability investigations, associated with finite elastic deformations, can be viewed as a "small" deformation which is superimposed upon a known "large" deformation, and which gives rise to a mathematical problem which is linear but highly non-homogeneous (see, for example, Green and Spencer (1958), Nowinski and Shahinpoor (1969), Wang and Ertepinar (1972) and Hill (1975, 1976a, b, c). Thus, for plane strain deformations in material and spatial rectangular Cartesian coordinates (X, Y, Z)

and (x, y, z) respectively, we examine

$$x = f(X, Y) + \epsilon U(X, Y), \qquad y = g(X, Y) + \epsilon V(X, Y), \qquad z = Z, \qquad (7.1)$$

where $f(X, Y)$ and $g(X, Y)$ are assumed to be a known large deformation and $\epsilon U(X, Y)$ and $\epsilon V(X, Y)$ represent the superimposed small deformation for which we suppose we can neglect all products and higher powers. Our purpose here is to identify the general structure for small superimposed deformations of the form (7.1), assuming only that the initial large deformation is a solution of one of the three first integrals (I), (II) or (III) defined by (3.14). Similarly, for axially symmetric deformations in material and spatial cylindrical polar co-ordinates (R, Θ, Z) and (r, θ, z) respectively, we examine deformations of the form

$$r = f(R, Z) + \epsilon U(R, Z), \qquad \theta = \Theta, \qquad z = g(R, Z) + \epsilon V(R, Z), \qquad (7.2)$$

where $f(R, Z)$ and $g(R, Z)$ are assumed to be a known large deformation and $\epsilon U(R, Z)$ and $\epsilon V(R, Z)$ represent the small deformation for which we suppose we can neglect all products and higher powers and we assume only that the initial large deformation is a known solution of either integral (I) or (II) as defined by (4.13).

In order to illustrate our results, we consider in material and spatial cylindrical polar coordinates (R, Θ, Z) and (r, θ, z) respectively, the plane strain deformation

$$r = (R^2 + K)^{1/2} + \epsilon u(R) \cos n\Theta, \qquad \theta = \Theta + \epsilon v(R) \sin n\Theta, \qquad z = Z, \quad (7.3)$$

where K is a constant, n denotes a positive integer and $u(R)$ and $v(R)$ denote functions of R only. Now Hill (1977b) shows that for the Varga strain-energy function (1.1), the function $u(R)$ satisfies the self-adjoint fourth order differential equation

$$\tilde{D}^2 \left\{ c(R) \tilde{D}^2 u \right\} = 0, \qquad (7.4)$$

where the second order differential operator $\tilde{D}^2$ and $c(R)$ are defined respectively by

$$\tilde{D}^2 u = \frac{d}{dR} \left\{ (R^2 + K)^{3/2} \frac{du}{dR} \right\} - (n^2 - 1)(R^2 + K)^{1/2} u,$$

$$c(R) = \frac{1}{(2R^2 + K)(R^2 + K)^{1/2}}, \qquad (7.5)$$

and we may exploit this identification to derive the exact solutions given by

Hill (1975). In cylindrical polars we may show from the constraint of incompressibility (2.7) with $\alpha = 1$ and the first integral (I) as given by $(3.17)_1$, that since the large deformation in (7.3) is a solution of these equations, we obtain the following two equations

$$\frac{du}{dR} + \frac{Ru}{(R^2 + K)} + \frac{nRv}{(R^2 + K)^{1/2}} = 0,$$

$$\frac{dv}{dR} + \frac{(2R^2 + K)v}{R(R^2 + K)} + \frac{nu}{R(R^2 + K)^{1/2}} = 0. \tag{7.6}$$

On eliminating $v(R)$ we find that $u(R)$ satisfies $\tilde{D}^2 u = 0$, where $\tilde{D}^2$ is precisely as defined above. Thus, we may utilize the first integrals to initially identify the relevant linear second order differential operator $\tilde{D}^2$ and then determine the structure of the fourth order equation such as (7.4).

A further illustration of the results given here is given by Hill and Arrigo (1996b) for small deformations superimposed upon the symmetrical expansion of a spherical shell. With material and spatial spherical polar coordinates (R, Θ, Φ) and (r, θ, ϕ) respectively, the authors examine a small superimposed axially symmetric deformation of the form

$$r = (R^3 + K)^{1/3} + \epsilon u(R) P_n(\cos \Theta),$$

$$\theta = \Theta + \epsilon v(R) \frac{d}{d\Theta} P_n(\cos \Theta), \quad \phi = \Phi, \tag{7.7}$$

where K is a constant and P_n denotes the Legendre function of degree n. It is shown that the small deformation is determined by solving the fourth order equation

$$\tilde{D}^2 \left\{ \frac{(\alpha_1 R + \alpha_2 \rho) \tilde{D}^2 u}{\rho^2 (2R^3 + K)} \right\} = 0, \tag{7.8}$$

where $\rho = (R^3 + K)^{1/3}$ and the second order differential operator $\tilde{D}^2$ is given by

$$\tilde{D}^2 u = \frac{d}{dR} \left\{ \frac{\rho^5}{R} \frac{du}{dR} \right\} - (n-1)(n+2)\rho^2 u. \tag{7.9}$$

Moreover, it is also shown that the operator $\tilde{D}^2$ is defined by examining small superimposed deformations of the form (7.7) as solutions of (2.20) and the first integral (I) defined by $(4.19)_1$. From these equations we find

$$\frac{d}{dR}(\rho^2 u) - n(n+1)\rho^2 v = 0, \quad \frac{d}{dR}(\rho R v) - u = 0, \tag{7.10}$$

so that on elimination of v we obtain $\tilde{D}^2 u = 0$, where $\tilde{D}^2$ is precisely as defined above. Thus we may exploit the first integrals to initially identify the relevant linear second order operator and subsequently determine the structure of the general fourth order equation such as (7.8). Our purpose here is to undertake this process for any solution of (2.20) and one of the integrals (I) or (II) as defined by (4.19).

We emphasize that the procedure of correctly factorizing a fourth order operator is by no means a trivial matter and the idea of identifying the operator from an examination of an easier lower order problem is an important one. Here we merely summarize the main results derived in Hill and Arrigo (1999c, d), where for plane strain and axially symmetric deformations these results are also given in terms of cylindrical and spherical polar coordinates respectively.

5.7.1 Initial deformation is a known solution of (2.2) and (3.14)₁

As described in Section 5.3 we introduce $\Omega(X,Y)$ such that $x = \Omega_X$ and $y = \Omega_Y$ and (2.2) becomes $\Omega_{XX}\Omega_{YY} - \Omega_{XY}^2 = 1$ and we suppose $\Omega(X,Y)$ is a known solution of this equation. In terms of Jacobians, $U(X,Y)$ and $V(X,Y)$ are given by

$$U = \frac{\partial(\Omega_X, \omega)}{\partial(X,Y)}, \qquad V = \frac{\partial(\Omega_Y, \omega)}{\partial(X,Y)}, \tag{7.11}$$

where $\omega(X,Y)$ is a second potential which satisfies the fourth order equation

$$D^2 \left\{ \frac{D^2\omega}{\nabla^2\Omega} \right\} = 0, \tag{7.12}$$

where the operators D^2 and ∇^2 are defined respectively by

$$D^2 = \Omega_{YY}\frac{\partial^2}{\partial X^2} - 2\Omega_{XY}\frac{\partial^2}{\partial X\partial Y} + \Omega_{XX}\frac{\partial^2}{\partial Y^2},$$
$$\nabla = \frac{\partial^2}{\partial X^2} + \frac{\partial^2}{\partial Y^2}. \tag{7.13}$$

5.7.2 Initial deformation is a known solution of (2.2) and (3.14)₂

As described in Hill and Arrigo (1996c) we may introduce new variables α and β defined by

$$\alpha = \frac{X}{(X^2+Y^2)}, \qquad \beta = \frac{-Y}{(X^2+Y^2)}, \tag{7.14}$$

and equations (2.2) and (3.14)₂ become respectively

$$\frac{\partial(x,y)}{\partial(\alpha,\beta)} = (\alpha^2+\beta^2)^{-2}, \qquad x_\beta = y_\alpha, \tag{7.15}$$

so that on introducing a potential $\Omega(\alpha, \beta)$ such that $x = \Omega_\alpha$ and $y = \Omega_\beta$, equation $(7.15)_1$ becomes

$$\Omega_{\alpha\alpha}\Omega_{\beta\beta} - \Omega_{\alpha\beta}^2 = (\alpha^2 + \beta^2)^{-2}. \tag{7.16}$$

Assuming that $\Omega(\alpha, \beta)$ is a solution of this equation we now look for a small upon large deformation of the form

$$x = \Omega_\alpha + \epsilon U(\alpha, \beta), \qquad y = \Omega_\beta + \epsilon V(\alpha, \beta), \qquad z = Z, \tag{7.17}$$

and we find that

$$
\begin{aligned}
U &= (\alpha^2 + \beta^2)^2(\Omega_{\alpha\alpha}\omega_\beta - \Omega_{\alpha\beta}\omega_\alpha) = (\alpha^2 + \beta^2)^2 \frac{\partial(\Omega_\alpha, \omega)}{\partial(\alpha, \beta)}, \\
V &= (\alpha^2 + \beta^2)^2(\Omega_{\alpha\beta}\omega_\beta - \Omega_{\beta\beta}\omega_\alpha) = (\alpha^2 + \beta^2)^2 \frac{\partial(\Omega_\beta, \omega)}{\partial(\alpha, \beta)},
\end{aligned} \tag{7.18}
$$

where $\omega(\alpha, \beta)$ is a second potential which satisfies the fourth order equation

$$D^2 \left\{ \frac{D^2\omega}{(\alpha^2 + \beta^2)\nabla_1^2\Omega} \right\} + 2D^2\omega = 0, \tag{7.19}$$

where here the operators D^2 and ∇_1^2 are defined respectively by

$$
\begin{aligned}
D^2 &= \frac{\partial}{\partial\alpha}\left((\alpha^2 + \beta^2)^2\Omega_{\beta\beta}\frac{\partial}{\partial\alpha}\right) - \frac{\partial}{\partial\alpha}\left((\alpha^2 + \beta^2)^2\Omega_{\alpha\beta}\frac{\partial}{\partial\beta}\right) \\
&\quad - \frac{\partial}{\partial\beta}\left((\alpha^2 + \beta^2)^2\Omega_{\alpha\beta}\frac{\partial}{\partial\alpha}\right) + \frac{\partial}{\partial\beta}\left((\alpha^2 + \beta^2)^2\Omega_{\alpha\alpha}\frac{\partial}{\partial\beta}\right),
\end{aligned} \tag{7.20}
$$

$$\nabla_1^2 = \frac{\partial^2}{\partial\alpha^2} + \frac{\partial^2}{\partial\beta^2}.$$

5.7.3 *Initial deformation is a known solution of (2.2) and (3.14)$_3$*

As described in Section 5.3 the first integral $(3.14)_3$ can be re-arranged to give the existence of a potential $\Omega(X, Y)$ such that

$$x = \frac{\Omega_X}{(\Omega_X^2 + \Omega_Y^2)}, \qquad y = \frac{-\Omega_Y}{(\Omega_X^2 + \Omega_Y^2)}, \tag{7.21}$$

and from (2.2) we find that $\Omega(X, Y)$ satisfies

$$\Omega_{XX}\Omega_{YY} - \Omega_{XY}^2 = (\Omega_X^2 + \Omega_Y^2)^2. \tag{7.22}$$

Further, we find that $U(X, Y)$ and $V(X, Y)$ are given by

$$U = \left(\frac{\Omega_X}{(\Omega_X^2 + \Omega_Y^2)}\right)_X \omega_Y - \left(\frac{\Omega_X}{(\Omega_X^2 + \Omega_Y^2)}\right)_Y \omega_X,$$

$$V = -\left(\frac{\Omega_Y}{(\Omega_X^2 + \Omega_Y^2)}\right)_X \omega_Y + \left(\frac{\Omega_Y}{(\Omega_X^2 + \Omega_Y^2)}\right)_Y \omega_X,$$

(7.23)

where $\omega(X, Y)$ satisfies the fourth order equation

$$D^2 \left\{ \frac{D^2 \omega}{(\Omega_X^2 + \Omega_Y^2)\nabla^2\Omega} \right\} - 2D^2\omega = 0,$$

(7.24)

where D^2 and ∇^2 are precisely as defined by (7.13).

5.7.4 Initial deformation is a known solution of (2.14) and $(4.13)_1$

As described in Section 5.4 we may introduce a potential $\Omega(R, Z)$ such that $r = \Omega_R$ and $z = \Omega_Z$, where $\Omega(R, Z)$ satisfies (4.14). From Hill and Arrigo (1999d) we find that $U(R, Z)$ and $V(R, Z)$ in (7.2) are given by

$$U = \frac{\Omega_{RR}\omega_Z - \Omega_{RZ}\omega_R}{R} = \frac{1}{R}\frac{\partial(\Omega_R, \omega)}{\partial(R, Z)}, \quad V = \frac{\Omega_{RZ}\omega_Z - \Omega_{ZZ}\omega_R}{R} = \frac{1}{R}\frac{\partial(\Omega_Z, \omega)}{\partial(R, Z)},$$

(7.25)

where $\omega(R, Z)$ is a second potential which satisfies

$$D^2 \left\{ \left(\frac{\alpha_1 R + \alpha_2 \Omega_R}{\Omega_{RR} + \Omega_{ZZ}}\right) D^2\omega \right\} = 0,$$

(7.26)

where the operator D^2 is defined by

$$D^2\omega = \frac{1}{R}\left\{\frac{\partial(\Omega_R, \omega_Z)}{\partial(R, Z)} - \frac{\partial(\Omega_Z, \omega_R)}{\partial(R, Z)} + \frac{1}{R}\frac{\partial(\Omega_Z, \omega)}{\partial(R, Z)}\right\},$$

(7.27)

or either of the alternative expressions

$$D^2 = \frac{1}{R}\left\{\Omega_{ZZ}\frac{\partial^2}{\partial R^2} - 2\Omega_{RZ}\frac{\partial^2}{\partial R\partial Z} + \Omega_{RR}\frac{\partial^2}{\partial Z^2} - \frac{\Omega_{ZZ}}{R}\frac{\partial}{\partial R} + \frac{\Omega_{RZ}}{R}\frac{\partial}{\partial Z}\right\},$$

$$D^2 = \frac{\partial}{\partial R}\left(\frac{\Omega_{ZZ}}{R}\frac{\partial}{\partial R}\right) - \frac{\partial}{\partial R}\left(\frac{\Omega_{RZ}}{R}\frac{\partial}{\partial Z}\right) - \frac{\partial}{\partial Z}\left(\frac{\Omega_{RZ}}{R}\frac{\partial}{\partial R}\right) + \frac{\partial}{\partial Z}\left(\frac{\Omega_{RR}}{R}\frac{\partial}{\partial Z}\right).$$

(7.28)

5.7.5 Initial deformation is a known solution of (2.14) and (4.13)₂

As described in Section 5.4 we may introduce new coordinates α and β defined by (4.15) so that there exists a potential $\Omega(\alpha, \beta)$ such that $r = \Omega_\alpha$ and $z = \Omega_\beta$, where $\Omega(\alpha, \beta)$ satisfies (4.17). We may show that on assuming a small on large deformation of the form

$$r = \Omega_\alpha + \epsilon U(\alpha, \beta), \qquad z = \Omega_\beta + \epsilon V(\alpha, \beta), \tag{7.29}$$

the functions $U(\alpha, \beta)$ and $V(\alpha, \beta)$ are given by

$$U = \frac{(\alpha^2 + \beta^2)^3}{\alpha} \frac{\partial(\Omega_\alpha, \omega)}{\partial(\alpha, \beta)}, \qquad V = \frac{(\alpha^2 + \beta^2)^3}{\alpha} \frac{\partial(\Omega_\beta, \omega)}{\partial(\alpha, \beta)}, \tag{7.30}$$

where $\omega(\alpha, \beta)$ satisfies the equation

$$D^2 \left\{ \frac{\alpha D^2 \omega}{(\alpha^2 + \beta^2)^2 \nabla_1^2 \Omega} \right\} + 4D^2 \omega = 0, \tag{7.31}$$

where D^2 and ∇_1^2 are defined respectively by

$$D^2 = \frac{\partial}{\partial \alpha} \left(\frac{(\alpha^2 + \beta^2)^3}{\alpha} \Omega_{\beta\beta} \frac{\partial}{\partial \alpha} \right) - \frac{\partial}{\partial \alpha} \left(\frac{(\alpha^2 + \beta^2)^3}{\alpha} \Omega_{\alpha\beta} \frac{\partial}{\partial \beta} \right)$$

$$- \frac{\partial}{\partial \beta} \left(\frac{(\alpha^2 + \beta^2)^3}{\alpha} \Omega_{\alpha\beta} \frac{\partial}{\partial \alpha} \right) + \frac{\partial}{\partial \beta} \left(\frac{(\alpha^2 + \beta^2)^3}{\alpha} \Omega_{\alpha\alpha} \frac{\partial}{\partial \beta} \right). \tag{7.32}$$

$$\nabla_1^2 = \frac{\partial^2}{\partial \alpha^2} + \frac{\partial^2}{\partial \beta^2}.$$

Here we have simply summarized the formal structural aspects of these small on large problems for the Varga materials. Specific examples and special cases are given by Hill and Arrigo (1999c, d).

5.8 Further general integrals

In this secton we note two further situations giving rise to integrals of the types noted previously for plane strain and axial symmetry The first arises for the plane stress or membrane theory of thin plane sheets of perfectly elastic materials as formulated by Green and Adkins (1960), while the second arises in an examination of the remaining, as yet unresolved, case for Ericksen's problem.

5.8.1 Plane stress theory of thin elastic sheets

Here we present a brief summary of results first derived by Arrigo and Hill (1996). The basic equations for the membrane theory of thin plane sheets of isotropic incompressible hyperelastic material, assuming uniform original thickness $2h_0$ are given by Hill (1977a). Using material and spatial rectangular Cartesian coordinates (X, Y, Z) and (x, y, z) respectively, the basic equations are as follows. The deformation is assumed to take the form

$$x = x(X, Y), \qquad y = y(X, Y), \qquad z = \Lambda(X, Y)^{-1} Z, \qquad (8.1)$$

so that the condition of incompressibility becomes

$$\frac{\partial(x, y)}{\partial(X, Y)} = \Lambda(X, Y). \qquad (8.2)$$

Directly from the general equations given in Hill (1977a) the equilibrium equations can be shown to become

$$\frac{\partial(\phi_2, y)}{\partial(X, Y)} + (\phi_1 x_X)_X + (\phi_1 x_Y)_Y = 0,$$
$$-\frac{\partial(\phi_2, x)}{\partial(X, Y)} + (\phi_1 y_X)_X + (\phi_1 y_Y)_Y = 0, \qquad (8.3)$$

where here $\phi_1(I, \Lambda)$ and $\phi_2(I, \Lambda)$ are two response functions defined by

$$\phi_1 = 2\frac{\partial W}{\partial I}, \qquad \phi_2 = \frac{\partial W}{\partial \Lambda}, \qquad (8.4)$$

and $W(\Lambda, I)$ is the strain–energy function and the invariant I is again defined by (2.5). We note that from (2.2) and (2.5) we have

$$I + 2\Lambda = (x_X + y_Y)^2 + (x_Y - y_X)^2. \qquad (8.5)$$

Now for the Varga elastic material with strain–energy function (1.1) we have

$$W = 2\mu \left[(I + 2\Lambda)^{\frac{1}{2}} + \Lambda^{-1} - 3 \right], \qquad (8.6)$$

since $I = \lambda_1^2 + \lambda_2^2$ and $\Lambda = \lambda_3^{-1} = \lambda_1 \lambda_2$, and therefore from (8.4) we obtain

$$\phi_1 = 2\mu(I + 2\Lambda)^{-\frac{1}{2}}, \quad \phi_2 = 2\mu \left[(I + 2\Lambda)^{-\frac{1}{2}} - \Lambda^{-2} \right], \qquad (8.7)$$

so that in particular

$$\phi_2 - \phi_1 = -2\mu\Lambda^{-2}. \qquad (8.8)$$

On introducing u, v and ψ precisely as given by (3.7) and (3.10), except ϕ is replaced by ϕ_1, Arrigo and Hill (1996) again deduce the two exact integrals:

$$
\begin{align}
\text{(I)} \qquad & x_Y - y_X = 0, \\
\text{(II)} \qquad & (x_Y - y_X) = \frac{2XY}{(X^2 - Y^2)}(x_X + y_Y),
\end{align}
\tag{8.9}
$$

corresponding respectively to the values $\psi = 0$ and $\psi = 2\tan^{-1}(Y/X)$. In the first case $\Lambda(X, Y)$ turns out to be a constant and the resulting Monge-Ampère equation can be linearized precisely as described in Section 5.3, equations (3.15)–(3.16). In the second case Arrigo and Hill (1996) introduce a potential $\Omega(X, Y)$ so that x, y and Λ are given by

$$
\begin{align}
x &= (Y^2 - X^2)\Omega_X - 2XY\Omega_Y, \\
y &= (Y^2 - X^2)\Omega_Y + 2XY\Omega_X, \\
\Lambda &= -(X\Omega_X + Y\Omega_Y + \Omega)^{-1}.
\end{align}
\tag{8.10}
$$

In terms of new variables α and β defined by

$$
\alpha = \frac{X}{(X^2 + Y^2)}, \qquad \beta = \frac{-Y}{(X^2 + Y^2)},
\tag{8.11}
$$

the Monge-Ampère equation resulting from substitution of (8.10) into (8.2) ultimately yields

$$
\Omega_{\alpha\alpha}\Omega_{\beta\beta} - \Omega_{\alpha\beta}^2 = (\alpha^2 + \beta^2)^{-2}(\alpha\Omega_\alpha + \beta\Omega_\beta - \Omega)^{-1}.
\tag{8.12}
$$

We note in terms of the new variables α and β equation $(8.9)_2$ becomes simply $x_\beta = y_\alpha$ and x and y are given by $x = \Omega_\alpha$ and $y = \Omega_\beta$. Equation (8.12) admits a remarkable linearization which in cylindrical polar coordinates (R, Θ, Z) and (r, θ, z), namely

$$
r = r(R, \Theta), \qquad \theta = \theta(R, \Theta), \qquad z = \lambda(R, \Theta)Z,
\tag{8.13}
$$

is given parametrically by

$$
r = (\omega^2 + \omega_\Theta^2)^{\frac{1}{2}}, \qquad \theta = -\Theta - \tan^{-1}(\omega_\Theta/\omega), \qquad R = \omega_\lambda,
\tag{8.14}
$$

where ω_λ is assumed positive and $\omega(\lambda, \Theta)$ denotes any solution of the "linear" equation

$$
\omega_{\lambda\lambda} + \lambda(\omega_{\Theta\Theta} + \omega) = 0.
\tag{8.15}
$$

As with the linearization (3.18) and (3.19) the parameterization used above employs the physical angle Θ and therefore the solutions are potentially of great importance for practical problems. In the present context the above linearization is particularly important since both parameters (λ and Θ) are physically meaningful.

5.8.2 Ericksen's problem for radially symmetric deformations

The determination of exact deformations which apply to all homogeneous hyperelastic materials is referred to as Ericksen's problem, and for this class of materials there remains only the case of constant strain invariants to be examined. Here we consider, in spherical polar coordinates, a radially symmetric deformation, for which the constancy of the strain invariants also arises as a natural part of the equilibrium conditions. In addition, we show that the remaining equilibrium equations admit two exact first integrals of the type derived here, which can be exploited to reduce the order of the problem. Ericksen's original problem (Ericksen 1954) consisted of determining all static deformations which can be produced in every homogeneous isotropic incompressible hyperelastic material under the sole action of surface tractions and we refer the reader to Beatty (1996) for an excellent review of the problem. Such deformations are referred to as either "controllable" or "universal". Hill and Arrigo (1999b) examine radially symmetric deformations in spherical polar coordinates (R, Θ, Φ) and (r, θ, ϕ),

$$r = R, \qquad \theta = \theta(\Theta, \Phi), \qquad \phi = \phi(\Theta, \Phi), \tag{8.16}$$

as potential constant strain invariant solutions of Ericksen's problem and show that for the deformation (8.16) the equilibrium conditions actually require the strain invariants to be constant, while the remaining equilibrium equations give rise to two exact integrals. Hill and Arrigo (1999b) show that the problem reduces to determining $\theta(\Theta, \Phi), \phi(\Theta, \Phi)$ and $q(\Theta, \Phi)$ such that

$$\frac{\partial(\theta, \phi)}{\partial(\Theta, \Phi)} = \frac{\sin \Theta}{\sin \theta}, \quad \left(\theta_\Theta + \frac{\sin \theta}{\sin \Theta}\phi_\Phi\right)^2 + \left(\frac{\theta_\Phi}{\sin \Theta} - \sin \theta \phi_\Theta\right)^2 = \alpha^2,$$

$$\frac{\partial q}{\partial \theta} = \left\{\nabla^2 \theta - \sin \theta \cos \theta \left(\phi_\Theta^2 + \frac{\phi_\Phi^2}{\sin^2 \Theta}\right)\right\}, \tag{8.17}$$

$$\frac{\partial q}{\partial \phi} = \sin^2 \theta \left\{\nabla^2 \phi + 2 \cot \theta \left(\theta_\Theta \phi_\Theta + \frac{\theta_\Phi \phi_\Phi}{\sin^2 \Theta}\right)\right\},$$

where $\alpha = (I + 2)^{1/2}$ is a constant. After a long and elaborate calculation Hill and Arrigo (1999b) show that these equilibrium equations with constant strain invariants admit the two exact first integrals,

$$\left(\frac{1}{\sin \Theta}\frac{\partial \theta}{\partial \Phi} - \sin \theta \frac{\partial \phi}{\partial \Theta}\right) = \tan \psi \left(\frac{\partial \theta}{\partial \Theta} + \frac{\sin \theta}{\sin \Theta}\frac{\partial \phi}{\partial \Phi}\right), \tag{8.18}$$

where $\delta = \pm 1$ and $\tan \psi$ is defined by

$$\tan \psi = \frac{\sinh(A + \delta a) \sin(B - \delta b)}{1 + \cosh(A + \delta a) \cos(B - \delta b)}. \tag{8.19}$$

We have introduced new material and spatial coordinates (A, B) and (a, b) respectively which are defined by

$$A = \log \tan(\Theta/2), \qquad B = \Phi, \qquad a = \log \tan(\theta/2), \qquad b = \phi, \qquad (8.20)$$

and we note that with these new variables $(8.17)_1$ and (8.18) become respectively

$$\frac{\partial a}{\partial A}\frac{\partial b}{\partial B} - \frac{\partial a}{\partial B}\frac{\partial b}{\partial A} = \frac{\cosh^2 a}{\cosh^2 A}, \qquad \frac{\partial a}{\partial B} - \frac{\partial b}{\partial A} = \tan\psi \left(\frac{\partial a}{\partial A} + \frac{\partial b}{\partial B} \right). \qquad (8.21)$$

We emphasize that not all deformations of the form (8.16) satisfy these integrals. However, every deformation which satisfies the integrals also satisfies the full equilibrium conditions and clearly the integrals involve a lower order problem.

In terms of complex coordinates $(Z, \bar{Z})$ defined by

$$Z = \frac{1}{2}(\Phi + i\log\tan(\Theta/2)), \qquad \bar{Z} = \frac{1}{2}(\Phi - i\log\tan(\Theta/2)), \qquad (8.22)$$

Hill and Arrigo (1999b) show that the problem of determining further solutions of Ericksen's problem reduces to the determination of a real function $P(Z, \bar{Z})$ which satisfies the two partial differential equations

$$\beta \frac{\partial^2 P}{\partial Z \partial \bar{Z}} + 1 + \frac{\partial P}{\partial Z}\frac{\partial P}{\partial \bar{Z}} + \tan(Z - \bar{Z})\left(\frac{\partial P}{\partial Z} - \frac{\partial P}{\partial \bar{Z}} \right) = 0,$$

$$\left| 1 + \left(\frac{\partial P}{\partial Z} \right)^2 - \frac{\partial^2 P}{\partial Z^2} \right| = (1 - \beta^2)^{1/2} \frac{\partial^2 P}{\partial Z \partial \bar{Z}}, \qquad (8.23)$$

where $\|$ denotes the modulus and β is a constant defined by $\beta = 2\delta/\alpha$, where $\delta = \pm 1$ and $\alpha = (I + 2)^{1/2}$, where I is the constant strain invariant such that $I_1 = I_2 = \alpha + 1$. However, the authors were unable to solve these equations. The two examples presented in this section indicate that there may well be other circumstances in finite elasticity which permit the derivation of exact integrals.

Acknowledgements

This work is supported by the Australian Research Council through the Large Grant Scheme and a Senior Research Fellowship.

References

Adkins, J.E. 1958 A reciprocal property of the finite plane strain equations. *J. Mech. Phys. Solids* **6**, 267-275.

Arrigo, D.J. and Hill, J.M. 1996 Transformations and equation reductions in finite elasticity II: Plane stress and axially symmetric deformations. *Math. Mech. Solids* 1, 177-192.

Beatty, M.F. 1996 Introduction to nonlinear elasticity. *Nonlinear effects in fluids and solids.* Eds. M.M. Carroll and M.A. Hayes, Plenum Press, New York, 13-112.

Beatty, M.F. and Hill, J.M. 2000 A note on some general integrals arising in plane finite elasticity. *Quart. J. Mech. Appl. Math.* 53, 421-428.

Dickie, R.A. and Smith T.L. 1971 Viscoelastic properties of rubber vulcanizates under large deformations in equal biaxial tension, pure shear and simple tension. *Trans. Soc. Rheol.* 36, 91-110.

Ericksen, J.L. 1954 Deformations possible in every isotropic incompressible perfectly elastic body. *Z. angew. Math. Phys.* 5, 466-486.

Eringen, A.C. 1962. *Nonlinear Theory of Continuous Media.* McGraw-Hill, New York.

Green, A.E. and Adkins, J.E. 1960 *Large Elastic Deformations and Non-linear Continuum Mechanics.* Oxford University Press.

Green, A.E. and Shield, R.T. 1950 Finite elastic deformation of incompressible isotropic bodies. *Proc. Roy. Soc. Lond.* A202, 407-419.

Green, A.E. and Spencer, A.J.M. 1958 The stability of a circular cylinder under finite extension and torsion. *J. Math. Phys.* 37, 316-338.

Hill, J.M. 1972 *Some partial solutions of finite elasticity.* PhD Thesis, University of Queensland, Australia.

Hill, J.M. 1973a Partial solutions of elasticity I - Plane deformations. *Z. angew. Math. Phys.* 24, 401-408.

Hill, J.M. 1973b Partial solutions of elasticity II - Axially symmetric deformations. *Z. angew. Math. Phys.* 24, 409-418.

Hill, J.M. 1973c Partial solutions of elasticity III - Three dimensional deformations. *Z. angew. Math. Phys.* 24, 609-618.

Hill, J.M. 1975 Buckling of long thick-walled circular cylindrical shells of isotropic incompressible hyperelastic materials under uniform external pressure. *J. Mech. Phys. Solids* 23, 99-112.

Hill, J.M. 1976a Critical pressures for the buckling of thick-walled spherical shells under uniform external pressure. *Quart. J. Mech. Appl. Math.* 29, 179-196.

Hill, J.M. 1976b Closed form solutions for small deformations superimposed upon the simultaneous inflation and extension of a cylindrical tube. *J. Elasticity* 6, 113-123.

Hill, J.M. 1976c Closed form solutions for small deformations superimposed upon the symmetrical expansion of a spherical shell. *J. Elasticity* 6, 125-136.

Hill, J.M. 1977a Radial deflections of thin precompressed cylindrical rubber bush mountings. *Int. J. Solids Structures* 13, 93-104.

Hill, J.M. 1977b Self-adjoint differential equations arising in finite elasticity for small superimposed deformations. *Int. J. Solids Structures* 13, 813-822.

Hill, J.M. 1982 On static similarity deformations for isotropic materials. *Quart. Appl. Math.* 40, 287-292.

Hill, J.M. 1990 Some integrable and exact similarity solutions for plane finite elastic deformations. *IMA J. Appl. Math.* 44, 111-126.

Hill, J.M. 2000 Finite elastic eversion of a thick-walled incompressible spherical cap. *J. Engng. Math.* 37, 93-109.

Hill, J.M. and Arrigo, D.J. 1995 New families of exact solutions for finitely deformed incompressible elastic materials. *IMA J. Appl. Math.* 54, 109-123.

Hill, J.M. and Arrigo, D.J. 1996a On axially symmetric deformations of perfectly elastic compressible materials. *Quart. J. Mech. Appl. Math.* **49**, 19-28.

Hill, J.M. and Arrigo, D.J. 1996b Small deformations superimposed upon the symmetrical expansion of a spherical shell, *Quart. J. Mech. Appl. Math.* **49**, 337-351.

Hill, J.M. and Arrigo, D.J. 1996c Transformations and equation reductions in finite elasticity I: Plane strain deformations. *Math. Mech. Solids* **1**, 155-175.

Hill, J.M. and Arrigo, D.J. 1999a Transformations and equation reductions in finite elasticity III: A general integral for plane strain deformations. *Math. Mech. Solids* **4**, 3-15.

Hill, J.M. and Arrigo, D.J. 1999b A note on Ericksen's problem for radially symmetric deformations. *Math. Mech. Solids* **4**, 395-405.

Hill, J.M. and Arrigo, D.J. 1999c On the general structure of small on large problems for elastic deformations of Varga materials II: Plane strain deformations, *J. Elasticity*, **54**, 193-212.

Hill, J.M. and Arrigo, D.J. 1999d On the general structure of small on large problems for elastic deformations of Varga materials II: Axially symmetric deformations, *J. Elasticity*, **54**, 213-227.

Hill, J.M. and Cox, G.M. 2000 Transformations and equation reductions in finite elasticity IV: Illustration of the general integral for a plane strain similarity deformation. *Math. Mech. Solids* **5**, 3-17.

Hill, J.M. and Lee, A.I. 1989 Large elastic compression of finite rectangular blocks of rubber. *Quart. J. Mech. Appl. Math.* **42**, 267-287.

Hill, J.M. and Milan, A.M. 1999 Finite elastic nonsymmetrical inflation and eversion of circular cylindrical rubber tubes. *Proc. Roy. Soc. Lond.* **A 455**, 1067-1082.

Hill, J.M. and Milan, A.M. 2000 Finite elastic plane strain bending of sectors of circular cylindrical sectors. *Int. J. Eng. Sci.* to appear.

Hill, J.M. and Milan, A.M. 2001 On a non-symmetrical plane strain deformation for hyperelastic materials. *IMA J. Appl. Math.*, to appear.

Holden, J.T. 1968 A class of exact solutions for finite plane strain deformations of a particular elastic material. *Appl. Sci. Res.* **19**, 171-181.

John, F. 1960 Plane strain problems for a perfectly elastic material of harmonic type. *Commun. Pure Appl. Math.* **13**, 239-296.

Klingbeil, W.W. and Shield, R.T. 1966 Large deformation analysis of bonded elastic mounts. *Z. angew. Math. Phys.* **17**, 281-305.

Nowinski, J.L. and Shahinpoor, M. 1969 Stability of an elastic circular tube of arbitrary wall thickness subjected to an external pressure. *Int. J. Nonlinear Mech.* **4**, 143-158.

Ogden, R.W. 1972 Large deformation isotropic elasticity - on the correlation of theory and experiment for incompressible rubberlike solids. *Proc. Roy. Soc. Lond.* **A 326**, 565-584.

Rivlin, R.S. and Saunders, D.W. 1951 Large elastic deformations of isotropic materials, Par VII. Experiments on the deformation of rubber. *Phil. Trans. Roy. Soc. Lond.* **A 243**, 251-288.

Shield, R.T. 1967 Inverse deformation results in finite elasticity. *Z. angew. Math. Phys.* **18**, 490-500.

Varga, O.H. 1966 *Stress-strain Behaviour of Elastic Materials.* New York, Interscience.

Wang, A.S.D. and Ertepinar, A. 1972 Stability and vibrations of elastic thick-walled cylindrical and spherical shells subjected to pressure. *Int. J. Nonlinear Mech.* **7**, 539-555.

6

Shear

Ph. Boulanger

Département de Mathématique
Université Libre de Bruxelles
Campus Plaine C.P.218/1
1050 Bruxelles, Belgium
Email: phboul@ulb.ac.be

M. Hayes

Department of Mathematical Physics
University College Dublin
Belfield, Dublin 4, Ireland
Email: Michael.Hayes@ucd.ie

In this chapter, we deal with the theory of finite strain in the context of non-linear elasticity. As a body is subjected to a finite deformation, the angle between a pair of material line elements through a typical point is changed. The change in angle is called the "shear" of this pair of material line elements. Here we consider the shear of all pairs of material line elements under arbitrary deformation. Two main problems are addressed and solved. The first is the determination of all "unsheared pairs", that is all pairs of material line elements which are unsheared in a given deformation. The second is the determination of those pairs of material line elements which suffer the maximum shear.

Also, triads of material line elements are considered. It is seen that, for an arbitrary finite deformation, there is an infinity of oblique triads which are unsheared in this deformation and it is seen how they are constructed from unsheared pairs.

Finally, for the sake of completeness, angles between intersecting material surfaces are considered. They are also changed as a result of the deformation. This change in angle is called the "planar shear" of a pair of material planar elements. A duality between the results for shear and for planar shear is exhibited.

6.1 Introduction

At a typical particle P in a body, material line elements are generally translated, rotated and stretched as a result of a deformation, so that angles between intersecting material lines are generally changed. A pair of material line elements subtending at angle Θ at P before deformation is said to be "sheared" if the angle becomes $\Theta + \gamma$ following the deformation. Then γ is called the *shear* of the pair of material line elements at P. Shear has long been a subject of interest in elasticity (Cauchy 1841, Tait 1859, Thomson and Tait 1879, Love 1927, Thomas 1961), fracture mechanics (Jaeger, 1969), and also rock mechanics and structural geology (see, for instance, Ramsay and Huber 1987). The shear of material line elements which are initially orthogonal, the *orthogonal shear* is considered, in the context of infinitesimal strain theory, in many textbooks on continuum mechanics and elasticity (see, for instance, Love 1927, Thomas 1961).

Fundamental in describing *finite* deformation is the right Cauchy-Green strain tensor $\mathbf{C}$ whose eigenvalues at P are denoted by λ_1^2, λ_2^2 and λ_3^2, the squares of the principal stretches and whose principal axes are along the orthogonal unit vectors $\mathbf{E}_{(1)}, \mathbf{E}_{(2)}$ and $\mathbf{E}_{(3)}$ (say). We assume throughout that $\lambda_3 > \lambda_2 > \lambda_1$. Associated with $\mathbf{C}$ is the spatial strain-ellipsoid (Truesdell and Toupin 1960), or $\mathbf{C}$-ellipsoid, $d\mathbf{X} \cdot \mathbf{C} d\mathbf{X} = K^2$, where K is infinitesimal.

Ericksen (quoted in Truesdell and Toupin 1960) has shown that the maximum orthogonal shear at P is $\gamma_G(\pi/2)$ given by

$$\sin \gamma_G(\pi/2) = \frac{\lambda_3^2 - \lambda_1^2}{\lambda_3^2 + \lambda_1^2}. \tag{1.1}$$

It occurs for that pair of orthogonal material elements which is coplanar with the principal $\mathbf{E}_{(1)}$ and $\mathbf{E}_{(3)}$ axes of the $\mathbf{C}$-ellipsoid corresponding to the largest and the least principal stretches and which lie along $(\mathbf{E}_{(1)} \pm \mathbf{E}_{(3)})/\sqrt{2}$, the bisectors of the angles between the principal axes.

It was shown by Cauchy that the material elements along the principal axes, $\mathbf{E}_{(1)}, \mathbf{E}_{(2)}$ and $\mathbf{E}_{(3)}$ of the $\mathbf{C}$- ellipsoid before deformation are mutually orthogonal after the deformation, so that none of the pairs of material elements along the principal axes of the $\mathbf{C}$-ellipsoid is sheared. Also, any pair of orthogonal material elements at P such that one element is along a principal axis of the $\mathbf{C}$-ellipsoid is unsheared. Further, it was shown by Hayes (1988) that any pair of material elements lying in either of the planes of the central circular sections of the $\mathbf{C}$-ellipsoid is unsheared in the deformation. This pair of planes is denoted by $\Sigma^\pm(\mathbf{C})$ – there is always a pair of such planes if the principal stretches λ_α are all different from one another; just one plane if two of the principal stretches are equal but different from the third and an infinity of such planes if all three principal stretches are equal to each other. This result of Hayes had previously been noted by Thomson and Tait (1879) for the special case of simple shear.

Here we consider the shear of all pairs of material line elements under arbitrary deformation. Two main problems are addressed and solved. The first is the determination of all pairs of material line elements which are unsheared. The second is the determination of those pairs of material line elements which suffer the maximum shear. It is shown that the maximum shear is not the maximum orthogonal shear given by Ericksen's formula (1.1), but that it occurs for a pair of material line elements subtending a certain acute angle.

Also, for the sake of completeness, a section is devoted to the change in angle between intersecting material surfaces. This change in angle is called *planar shear* of a pair of material planar elements. A duality between the results for shear and for planar shear is exhibited.

We first (Section 6.2) recall and derive some basic results from the theory of finite strain: the stretch of a material line element lying along the unit vector $\mathbf{N}$ is introduced, and the formula giving the angle after deformation of material line elements lying along $\mathbf{N}$ and $\mathbf{M}$ is recalled.

Then (Section 6.3) special properties of the planes of central circular section of the $\mathbf{C}$-ellipsoid at $\mathbf{X}$ are presented. All material elements in these planes are stretched by the same amount, λ_2. Also, no pair of material elements in any of these planes is sheared.

Then (Section 6.4), the problem of finding those pairs of line elements which are not sheared in the deformation is solved. We consider all the unsheared pairs of material elements in a given plane $\pi(\mathbf{S})$ with normal $\mathbf{S}$ passing through P. Provided $\pi(\mathbf{S})$ is not either $\Sigma^+(\mathbf{C})$ or $\Sigma^-(\mathbf{C})$, it is seen that corresponding to any material element in $\pi(\mathbf{S})$ there is in general a 'companion' material element such that the element and its companion are unsheared. There are, however, two elements for which the companion in $\pi(\mathbf{S})$ reduces to the element itself, so that they do not belong to any unsheared pair of this plane. We call their directions *limiting directions*. These are equally inclined with respect to the direction of least stretch in the plane $\pi(\mathbf{S})$. It is also shown that the product of the stretches of two unsheared material line elements in the plane $\pi(\mathbf{S})$ is equal to a constant, the product of the least and greatest stretch in this plane, which may be expressed in terms of $\mathbf{C}$ and $\mathbf{S}$ only.

Next (Section 6.5) we determine the maximum and minimum shear, i.e. the largest increase and the largest decrease of the angle between a pair of material line elements in a given plane $\pi(\mathbf{S})$ assuming that the angle Θ between them before deformation is given. The maximum (minimum) shear occurs for those material line elements which are equally inclined at the angle $\Theta/2$ with respect to the direction of least (greatest) stretch in $\pi(\mathbf{S})$. If the aspect ratio (ratio of major to minor semi-axes) of the elliptical section of the $\mathbf{C}$-ellipsoid by the plane $\pi(\mathbf{S})$ is ϵ (≥ 1), then the maximum shear $\gamma_M(\Theta)$ for given Θ is given by

$$\gamma_M(\Theta) = 2\tan^{-1}\left(\epsilon\tan\frac{\Theta}{2}\right) - \Theta . \qquad (1.2)$$

In the special case when $\Theta = \pi/2$, $\sin\gamma_M(\pi/2) = (\epsilon^2 - 1)/(\epsilon^2 + 1)$ which is a plane version of Ericksen's result (1.1) for maximum orthogonal shear.

Having considered the maximum shear for elements in $\pi(\mathbf{S})$ we first (Section 6.6) allow $\pi(\mathbf{S})$ to be arbitrary and determine the global maximum shear for given Θ. This global maximum occurs when $\pi(\mathbf{S})$ is the principal plane of the **C**-ellipsoid to which the axis of intermediate stretch is orthogonal − the $\mathbf{E}_{(1)} - \mathbf{E}_{(3)}$ plane. In particuler, when $\Theta = \pi/2$, we recover Ericksen's formule (1.1).

Then (Section 6.7), we allow Θ to be arbitrary and show that, in a given plane $\pi(\mathbf{S})$, the pair of material line elements which suffer the maximum shear is the pair of elements lying along the limiting directions in this plane $\pi(\mathbf{S})$. These subtend, before deformation, an acute angle Θ_L given by $\tan(\Theta_L/2) = \epsilon^{-1/2}$. After deformation, this angle is sheared into its supplement $\pi - \Theta_L$, so that the maximum shear γ^* (say) is $\gamma^* = \pi - 2\Theta_L$. Then, considering all possible planes, we conclude that the global maximum shear γ_G^* (say) occurs for material elements lying along the limiting directions in the $\mathbf{E}_{(1)} - \mathbf{E}_{(3)}$ plane of the **C**-ellipsoid at P. These subtend the angle Θ_G^* (say), and

$$\tan(\Theta_G^*/2) = (\lambda_1/\lambda_3)^{1/2} , \quad \gamma_G^* = \pi - 2\Theta_G^* . \qquad (1.3)$$

Then (Section 6.8), another point of view about unsheared pairs and limiting directions is presented. Indeed, from Section 6.4 we know that for a given material line element along the unit vector $\mathbf{N}$, there is, in general, in each plane passing through $\mathbf{N}$, a companion vector $\mathbf{M}$ such that line elements along $\mathbf{N}$ and $\mathbf{M}$ are unsheared. Here, we present a formula for $\mathbf{M}$ in terms of $\mathbf{N}$ and the unit normal $\mathbf{S}$ to the plane $\pi(\mathbf{S})$ passing through $\mathbf{N}$ ($\mathbf{N} \cdot \mathbf{S} = 0$). This formula was first derived by Joly (1905), using quaternions. Also, an explicit expression is obtained for the unit vectors along the limiting directions in the plane $\pi(\mathbf{S})$, assuming that an unsheared pair $(\mathbf{N}, \mathbf{M})$ in this plane is known.

Next (Section 6.9) we recall a classical result due to Cauchy that at every particle $\mathbf{X}$ in the body before deformation there exists a triad of mutually orthogonal infinitesimal material line elements whose members are also mutually orthogonal after deformation. When the principal stretches are all unequal, there is just one such orthogonal triad. However (Boulanger and Hayes in press), there are other oblique unsheared triads. It is seen how these are constructed.

Finally, we consider (Section 6.10) the change in the angle between a pair of material planar elements as a result of the deformation. The theory of planar shear runs parallel to the theory of shear of material line elements. It is seen that the results for planar shear may be read off from the previously derived results for shear of material line elements by making appropriate substitutions – replacing $\mathbf{C}$ by $\mathbf{C}^{-1}$, $\mathbf{B}$ by $\mathbf{B}^{-1}$, etc.

Notation: A hat (^) denotes a unit vector: e.g. $\hat{\mathbf{a}}$. Also vectors $\mathbf{N}$, $\mathbf{M}$, $\mathbf{P}$, ... in bold face capitals are unit vectors. The summation convention is used throughout: repeated Roman subscripts, i, j or a, b imply summation over those subscripts from 1 to 3. The summation convention does not apply to repeated Greek subscripts. Rectangular Cartesian coordinates are used throughout.

6.2 Basic equations

Let a body of material be subjected to the deformation

$$\mathbf{x} = \mathbf{x}(\mathbf{X}), \quad x_i = x_i(X_A), \tag{2.1}$$

in which the particle initially at $\mathbf{X}$ is displaced to $\mathbf{x}$. All quantities are referred to a fixed rectangular Cartesian coordinate system. A material line element $d\mathbf{X}$ at $\mathbf{X}$ is deformed into the element $d\mathbf{x}$, given by

$$d\mathbf{x} = \mathbf{F}d\mathbf{X}, \quad dx_i = F_{iA}dX_A, \tag{2.2}$$

where $\mathbf{F}$, the deformation gradient, evaluated at $\mathbf{X}$, is given by

$$\mathbf{F} = \partial\mathbf{x}/\partial\mathbf{X}, \quad F_{iA} = \partial x_i/\partial X_A. \tag{2.3}$$

The right Cauchy-Green strain tensor $\mathbf{C}$ and its inverse $\mathbf{C}^{-1}$, are given by

$$\mathbf{C} = \mathbf{F}^T\mathbf{F}, \qquad C_{AB} = \frac{\partial x_i}{\partial X_A}\frac{\partial x_i}{\partial X_B}, \tag{2.4}$$

$$\mathbf{C}^{-1} = \mathbf{F}^{-1}\mathbf{F}^{-T}, \qquad C_{AB}^{-1} = \frac{\partial X_A}{\partial x_i}\frac{\partial X_B}{\partial x_i}. \tag{2.5}$$

These tensors $\mathbf{C}$ and $\mathbf{C}^{-1}$ are symmetric positive definite. At $\mathbf{X}$, the infinitesimal ellipsoids

$$d\mathbf{X} \cdot \mathbf{C}d\mathbf{X} = K^2, \qquad C_{AB}dX_AdX_B = K^2,$$
$$d\mathbf{X} \cdot \mathbf{C}^{-1}d\mathbf{X} = L^2, \qquad C_{AB}^{-1}dX_AdX_B = L^2, \tag{2.6}$$

where K, L are infinitesimals, are called the $\mathbf{C}$-*ellipsoid* and the $\mathbf{C}^{-1}$-*ellipsoid* respectively.

The eigenvalues of $\mathbf{C}$ are denoted by λ_α^2 ($\alpha = 1, 2, 3$), and the corresponding mutually orthogonal unit eigenvectors by $\mathbf{E}_{(\alpha)}$ ($\alpha = 1, 2, 3$). Unless otherwise stated, it is assumed throughout that $\lambda_3 > \lambda_2 > \lambda_1$.

The left Cauchy-Green strain tensor $\mathbf{B}$ and its inverse $\mathbf{B}^{-1}$, are given by

$$\mathbf{B} = \mathbf{FF}^T, \qquad B_{ij} = \frac{\partial x_i}{\partial X_A} \frac{\partial x_j}{\partial X_A}, \qquad (2.7)$$

$$\mathbf{B}^{-1} = \mathbf{F}^{-T} \mathbf{F}^{-1}, \qquad B_{ij}^{-1} = \frac{\partial X_A}{\partial x_i} \frac{\partial X_A}{\partial x_j}. \qquad (2.8)$$

These tensors $\mathbf{B}$ and $\mathbf{B}^{-1}$ are also symmetric positive definite. At $\mathbf{x}$, the infinitesimal ellipsoids

$$dx \cdot \mathbf{B} dx = l^2, \qquad B_{ij} dx_i dx_j = l^2,$$
$$dx \cdot \mathbf{B}^{-1} dx = k^2, \qquad B_{ij}^{-1} dx_i dx_j = k^2, \qquad (2.9)$$

where k, l are infinitesimals, are called the $\mathbf{B}$-*ellipsoid* and the $\mathbf{B}^{-1}$-*ellipsoid* respectively.

Now

$$\mathbf{CE}_{(\alpha)} = \lambda_\alpha^2 \mathbf{E}_{(\alpha)}, \quad C_{AB} E_{(\alpha)B} = \lambda_\alpha^2 E_{(\alpha)A}, \quad \alpha = 1, 2, 3, \qquad (2.10)$$

and it follows that

$$\mathbf{BFE}_{(\alpha)} = \lambda_\alpha^2 \mathbf{FE}_{(\alpha)}, \quad B_{ij} F_{jA} E_{(\alpha)A} = \lambda_\alpha^2 F_{iA} E_{(\alpha)A}, \quad \alpha = 1, 2, 3 . \qquad (2.11)$$

Hence, material elements along $\mathbf{E}_{(\alpha)}$, that is along the principal axes of the C-ellipsoid at $\mathbf{X}$, are deformed into elements along $\mathbf{e}_{(\alpha)}$, defined by

$$\mathbf{e}_{(\alpha)} = \mathbf{FE}_{(\alpha)}, \qquad (2.12)$$

which are along the principal axes of the $\mathbf{B}^{-1}$-ellipsoid at $\mathbf{x}$. Thus, we have a result of Cauchy (1841): the principal axes of the spatial strain-ellipsoid at $\mathbf{X}$ are deformed into the principal axes of the material strain-ellipsoid, at $\mathbf{x}$.

Note that

$$\mathbf{e}_{(\alpha)} \cdot \mathbf{e}_{(\alpha)} = \mathbf{E}_{(\alpha)} \cdot \mathbf{CE}_{(\alpha)} = \lambda_\alpha^2 \mathbf{E}_{(\alpha)} \cdot \mathbf{E}_{(\alpha)} = \lambda_\alpha^2, \qquad (2.13)$$

so that, recalling that a 'hat' denotes a unit vector,

$$\hat{\mathbf{e}}_{(\alpha)} = \mathbf{FE}_{(\alpha)}/\lambda_\alpha . \qquad (2.14)$$

If $\mathbf{N}$ is a unit vector along a material line element of infinitesimal length L, before deformation, at $\mathbf{X}$, then, after the deformation the length of this element at $\mathbf{x}$ is $\lambda_{(\mathbf{N})} L$, where $\lambda_{(\mathbf{N})}$, called *the stretch along* $\mathbf{N}$, is given by

$$\lambda_{(\mathbf{N})} = |\mathbf{FN}| = (\mathbf{N} \cdot \mathbf{CN})^{1/2} = (C_{AB} N_A N_B)^{1/2}. \qquad (2.15)$$

Similarly, if $\hat{\mathbf{n}}$ is a unit vector along a material line element of infinitesimal length l, after deformation, at $\mathbf{x}$, then, before the deformation, this element had length $\lambda_{(\mathbf{n})}l$ at $\mathbf{X}$, where

$$\lambda_{(\mathbf{n})} = (\hat{\mathbf{n}} \cdot \mathbf{B}^{-1}\hat{\mathbf{n}})^{1/2}. \tag{2.16}$$

If $\mathbf{n} = \mathbf{FN}$, so that $\mathbf{n}$, along $\hat{\mathbf{n}}$, is the element into which $\mathbf{N}$ is deformed, then

$$\lambda_{(\mathbf{n})} = 1/\lambda_{(\mathbf{N})}. \tag{2.17}$$

In particular,

$$\lambda_1 \equiv \lambda_{(\mathbf{E}_1)} = 1/\lambda_{(\mathbf{e}_1)}, \quad \lambda_3 \equiv \lambda_{(\mathbf{E}_3)} = 1/\lambda_{(\mathbf{e}_3)}. \tag{2.18}$$

If $\mathbf{N}$ and $\mathbf{M}$ are unit vectors along two material elements at $\mathbf{X}$, and Θ is the angle between them $(0 < \Theta < \pi)$ then, as a result of the deformation, the angle between them becomes θ (say), so that the shear, $\gamma_{(\mathbf{M},\mathbf{N})}$, of the pair is given by $\theta = \Theta + \gamma_{(\mathbf{M},\mathbf{N})}$. It may be shown (Thomson and Tait 1879) that

$$\cos\theta = \cos(\Theta + \gamma_{(\mathbf{M},\mathbf{N})}) = (C_{AB}N_A M_B)/\{\lambda_{(\mathbf{N})}\lambda_{(\mathbf{M})}\},$$

$$\lambda_{(\mathbf{N})} = (C_{AB}N_A N_B)^{1/2}, \quad \lambda_{(\mathbf{M})} = (C_{AB}M_A M_B)^{1/2}. \tag{2.19}$$

It is clear that if $\mathbf{N}$ is along a principal axis of the C-ellipsoid and $\mathbf{M}$ is any material unit vector orthogonal to $\mathbf{N}$ so that it lies in a principal plane of the C-ellipsoid, then $C_{AB}N_A M_B = 0$ and thus $\theta = \pi/2$ and so the pair is unsheared.

Similarly, for two material elements along the unit vectors $\hat{\mathbf{n}}$ and $\hat{\mathbf{m}}$ at $\mathbf{x}$, the angle, Θ at $\mathbf{X}$, between these elements before deformation is given by

$$\cos\Theta = B_{ij}^{-1}\hat{n}_i \hat{m}_j/\{\lambda_{(\mathbf{n})}\lambda_{(\mathbf{m})}\}. \tag{2.20}$$

The unit normals to the planes of central circular section of the C-ellipsoid are $\mathcal{S}^{\pm}$ (see, for instance, Boulanger and Hayes 1993), given by

$$(\lambda_3^2 - \lambda_1^2)^{1/2}\mathcal{S}^{\pm} = (\lambda_2^2 - \lambda_1^2)^{1/2}\mathbf{E}_{(1)} \pm (\lambda_3^2 - \lambda_2^2)^{1/2}\mathbf{E}_{(3)}, \tag{2.21}$$

and the unit normals to the planes of central circular section of the $\mathbf{C}^{-1}$-ellipsoid are $\mathcal{N}^{\pm}$, given by

$$\lambda_2(\lambda_3^2 - \lambda_1^2)^{1/2}\mathcal{N}^{\pm} = \lambda_3(\lambda_2^2 - \lambda_1^2)^{1/2}\mathbf{E}_{(1)} \pm \lambda_1(\lambda_3^2 - \lambda_2^2)^{1/2}\mathbf{E}_{(3)}. \tag{2.22}$$

In terms of these vectors, the tensors $\mathbf{C}$ and $\mathbf{C}^{-1}$ may then be written as (see, for instance, Boulanger and Hayes 1993)

$$\mathbf{C} = \lambda_2^2\mathbf{1} - \frac{1}{2}(\lambda_3^2 - \lambda_1^2)(\mathcal{S}^+ \otimes \mathcal{S}^- + \mathcal{S}^- \otimes \mathcal{S}^+), \tag{2.23}$$

and

$$\mathbf{C}^{-1} = \lambda_2^{-2}\mathbf{1} + \frac{1}{2}(\lambda_1^{-2} - \lambda_3^{-2})(\boldsymbol{\mathcal{N}}^+ \otimes \boldsymbol{\mathcal{N}}^- + \boldsymbol{\mathcal{N}}^- \otimes \boldsymbol{\mathcal{N}}^+). \qquad (2.24)$$

The unit normals $\boldsymbol{\mathcal{N}}^+$ and $\boldsymbol{\mathcal{N}}^-$ are coplanar with the unit normals $\boldsymbol{\mathcal{S}}^+$ and $\boldsymbol{\mathcal{S}}^-$ and we have

$$\begin{aligned}
\boldsymbol{\mathcal{N}}^+ \cdot \boldsymbol{\mathcal{S}}^+ = \boldsymbol{\mathcal{N}}^- \cdot \boldsymbol{\mathcal{S}}^- &= (\lambda_2^2 + \lambda_1\lambda_3)/\lambda_2(\lambda_3 + \lambda_1), \\
\boldsymbol{\mathcal{N}}^+ \cdot \boldsymbol{\mathcal{S}}^- = \boldsymbol{\mathcal{N}}^- \cdot \boldsymbol{\mathcal{S}}^+ &= (\lambda_2^2 - \lambda_1\lambda_3)/\lambda_2(\lambda_3 - \lambda_1),
\end{aligned} \qquad (2.25)$$

and also

$$\boldsymbol{\mathcal{S}}^+ \cdot \boldsymbol{\mathcal{S}}^- = (2\lambda_2^2 - \lambda_1^2 - \lambda_3^2)/(\lambda_3^2 - \lambda_1^2), \quad \boldsymbol{\mathcal{N}}^+ \cdot \boldsymbol{\mathcal{N}}^- = (\lambda_1^2 + \lambda_3^2 - 2\lambda_2^{-2}\lambda_1^2\lambda_3^2)/(\lambda_3^2 - \lambda_1^2). \qquad (2.26)$$

In the case of special deformations such that $\lambda_2^2 = \lambda_1\lambda_3$, as happens, for example, with simple shear, we note that (2.21) and (2.22) become

$$(\lambda_3 + \lambda_1)^{1/2}\boldsymbol{\mathcal{S}}^\pm = \lambda_1^{1/2}\mathbf{E}_{(1)} \pm \lambda_3^{1/2}\mathbf{E}_{(3)}, \quad (\lambda_3 + \lambda_1)^{1/2}\boldsymbol{\mathcal{N}}^\pm = \lambda_3^{1/2}\mathbf{E}_{(1)} \pm \lambda_1^{1/2}\mathbf{E}_{(3)}, \qquad (2.27)$$

and then (2.25) and (2.26) reduce to

$$\boldsymbol{\mathcal{N}}^+ \cdot \boldsymbol{\mathcal{S}}^+ = \boldsymbol{\mathcal{N}}^- \cdot \boldsymbol{\mathcal{S}}^- = 2(\lambda_1\lambda_3)^{1/2}/(\lambda_3 + \lambda_1), \quad \boldsymbol{\mathcal{N}}^+ \cdot \boldsymbol{\mathcal{S}}^- = \boldsymbol{\mathcal{N}}^- \cdot \boldsymbol{\mathcal{S}}^+ = 0, \qquad (2.28)$$

and

$$\boldsymbol{\mathcal{N}}^+ \cdot \boldsymbol{\mathcal{N}}^- = -\boldsymbol{\mathcal{S}}^+ \cdot \boldsymbol{\mathcal{S}}^- = (\lambda_3 - \lambda_1)/(\lambda_3 + \lambda_1). \qquad (2.29)$$

Hence, in the case of these special deformations, $(\boldsymbol{\mathcal{N}}^+, \boldsymbol{\mathcal{N}}^-)$ and $(\boldsymbol{\mathcal{S}}^+, \boldsymbol{\mathcal{S}}^-)$ are along reciprocal sets of vectors in the $\mathbf{E}_{(1)} - \mathbf{E}_{(3)}$ principal plane. In this case, we say that $\mathbf{C}$ and $\mathbf{C}^{-1}$ have *reciprocal central circular sections*.

6.3 Properties of the planes of central circular section of the C-ellipsoid

The planes of central circular section of the C-ellipsoid at $\mathbf{X}$ are special in the theory of strain, for two reasons (Hayes 1988). Firstly, all material line elements in these planes, denoted by $\Sigma^\pm(\mathbf{C})$, with normals $\boldsymbol{\mathcal{S}}^\pm$ given by (2.21), are stretched by the same amount λ_2, the intermediate principal stretch. Secondly, all pairs of material line elements in $\Sigma^\pm(\mathbf{C})$ are unsheared. Conversely, they are the only planes with these properties: if all material line elements in a plane at $\mathbf{X}$ are stretched by the same amount, then that plane is one of $\Sigma^\pm(\mathbf{C})$; if all pairs of material line elements in a plane at $\mathbf{X}$ are unsheared in the deformation, then that plane is one of $\Sigma^\pm(\mathbf{C})$.

Finally, it is seen that if the pair of material line elements along $(\mathbf{N}, \mathbf{M})$ is unsheared and if also the stretches along $\mathbf{N}$ and $\mathbf{M}$ are equal, $\lambda_{(\mathbf{N})} = \lambda_{(\mathbf{M})}$, then $\mathbf{N}, \mathbf{M}$ lie in a plane of central circular section of the $\mathbf{C}$-ellipsoid (Boulanger and Hayes in press).

First we recall that if $\Phi_{ij}y_iy_j = 1$ is the equation of an ellipsoid, with intermediate principal semi-axis of length $1/\sqrt{\Phi_2}$, where Φ_2 is the intermediate eigenvalue of the positive definite matrix $(\mathbf{\Phi})$, then the planes of central circular section of the ellipsoid are given (Boulanger and Hayes 1993) by $\Phi_{ij}y_iy_j = \Phi_2 y_i y_i$. Then, we recall the identity (Hayes 1988)

$$d\mathbf{X} \cdot (\mathbf{C} - \lambda_2^2 \mathbf{1})d\mathbf{X} = \lambda_2^2 d\mathbf{x} \cdot (\lambda_2^{-2}\mathbf{1} - \mathbf{B}^{-1})d\mathbf{x}. \tag{3.1}$$

Now, because λ_2^2 is the intermediate eigenvalue of $\mathbf{C}$, and λ_2^{-2} the intermediate eigenvalue of $\mathbf{B}^{-1}$, it follows that the planes $\Sigma^\pm(\mathbf{C})$ of the central circular sections of the $\mathbf{C}$-ellipsoid at $\mathbf{X}$ are deformed into the planes $\Sigma^\pm(\mathbf{B}^{-1})$ of the central circular sections of the $\mathbf{B}^{-1}$-ellipsoid at $\mathbf{x}$.

It may be noted from (2.21) and (2.22) that

$$\mathbf{C}(\mathcal{S}^+ + \mathcal{S}^-) = \lambda_1^2(\mathcal{S}^+ + \mathcal{S}^-),$$
$$\mathbf{C}(\mathcal{S}^+ - \mathcal{S}^-) = \lambda_3^2(\mathcal{S}^+ - \mathcal{S}^-), \tag{3.2}$$
$$\mathbf{C}(\mathcal{S}^+ \times \mathcal{S}^-) = \lambda_2^2(\mathcal{S}^+ \times \mathcal{S}^-).$$

It follows that

$$\mathcal{S}^+ \cdot \mathbf{C}^{-1} \cdot \mathcal{S}^+ = \mathcal{S}^- \cdot \mathbf{C}^{-1} \cdot \mathcal{S}^- = \lambda_2^2/(\lambda_1^2\lambda_3^2). \tag{3.3}$$

Generally, if $\mathbf{H}$ is an eigenvector of $\mathbf{C} = \mathbf{F}^T\mathbf{F}$ with eigenvalue $\lambda^2 : \mathbf{CH} = \lambda^2\mathbf{H}$, then this equation may be written $\mathbf{B}^{-1}(\mathbf{F}^{-T}\mathbf{H}) = (1/\lambda^2)\mathbf{F}^{-T}\mathbf{H}$, so that $\mathbf{F}^{-T}\mathbf{H}$ is an eigenvector of $\mathbf{B}^{-1}$ with eigenvalue $1/\lambda^2$. Thus, the three equations in (3.3) may be rewritten as

$$\mathbf{B}^{-1}\{\mathbf{F}^{-T}\mathcal{S}^+ + \mathbf{F}^{-T}\mathcal{S}^-\} = \lambda_1^{-2}(\mathbf{F}^{-T}\mathcal{S}^+ + \mathbf{F}^{-T}\mathcal{S}^-),$$
$$\mathbf{B}^{-1}\{\mathbf{F}^{-T}\mathcal{S}^+ - \mathbf{F}^{-T}\mathcal{S}^-\} = \lambda_3^{-2}(\mathbf{F}^{-T}\mathcal{S}^+ - \mathbf{F}^{-T}\mathcal{S}^-), \tag{3.4}$$
$$\mathbf{B}^{-1}\{\mathbf{F}^{-T}(\mathcal{S}^+ \times \mathcal{S}^-)\} = \lambda_2^{-2}\mathbf{F}^{-T}(\mathcal{S}^+ \times \mathcal{S}^-).$$

On comparing (3.2) with (3.4) and using (3.3), it may be deduced (and checked directly) that

$$\mathbf{B}^{-1} = \lambda_2^{-2}\mathbf{1} + [(\lambda_3^2 - \lambda_1^2)/(2\lambda_2^2)] \left[\mathbf{F}^{-T}\mathcal{S}^+ \otimes \mathbf{F}^{-T}\mathcal{S}^- + \mathbf{F}^{-T}\mathcal{S}^- \otimes \mathbf{F}^{-T}\mathcal{S}^+\right]. \tag{3.5}$$

It follows that $\mathbf{F}^{-T}\mathcal{S}^+$ and $\mathbf{F}^{-T}\mathcal{S}^-$ are the normals to the planes of the central circular sections of the $\mathbf{B}^{-1}$- ellipsoid.

Another approach is to recall (Chapter 1) that if $d\mathbf{X}$ and $d\mathbf{Y}$ are two material line elements at $\mathbf{X}$, then the plane material element which they span has normal $\mathbf{N}$ (say) along $d\mathbf{X} \times d\mathbf{Y}$. After deformation these elements lie along $d\mathbf{x} =$

$\mathbf{F}d\mathbf{X}, dy = \mathbf{F}d\mathbf{Y}$ and the corresponding plane spanned by these has normal $\mathbf{n}$ (say) along $d\mathbf{x} \times d\mathbf{y} = \mathbf{F}d\mathbf{X} \times \mathbf{F}d\mathbf{Y} = (\det \mathbf{F})\mathbf{F}^{-T}(d\mathbf{X} \times d\mathbf{Y})$. Thus the normals transform as $\mathbf{n} = \alpha\mathbf{F}^{-T}\mathbf{N}$, where α is some multiplier. [The normals do not transform as material elements in general: $\mathbf{n} \times \mathbf{F}\mathbf{N} \neq \mathbf{0}$. The exceptions are when $\mathbf{N}$ is an eigenvector of $\mathbf{C}$. Then $\mathbf{C}\mathbf{N} = \lambda^2\mathbf{N}$ for some λ, which may be written $\mathbf{F}\mathbf{N} = \lambda^2\mathbf{F}^{-T}\mathbf{N}$, expressing the fact that it is only in this special case that the normal along $\mathbf{N}$ transforms in the same way as a material element along $\mathbf{N}$.] Hence the planar material elements with normals along $\mathcal{S}^+$ and $\mathcal{S}^-$ at $\mathbf{X}$ are deformed into planar material elements with normals along $\mathbf{F}^{-T}\mathcal{S}^+$ and $\mathbf{F}^{-T}\mathcal{S}^-$, respectively, at $\mathbf{x}$. But the normals to the planes of the central circular equations of the C-ellipsoid at $\mathbf{X}$ are along $\mathcal{S}^+$ and $\mathcal{S}^-$, and the normals to the planes of the central circular sections of the $\mathbf{B}^{-1}$-ellipsoid at $\mathbf{x}$ are along $\mathbf{F}^{-T}\mathcal{S}^+$ and $\mathbf{F}^{-T}\mathbf{S}^-$. Hence the planes of the central circular sections of the C-ellipsoid at $\mathbf{X}$ are deformed into the planes of the central circular sections of the $\mathbf{B}^{-1}$-ellipsoid at $\mathbf{x}$.

Finally, consider $d\mathbf{X} = \beta\mathbf{M}, d\mathbf{Y} = \delta\mathbf{N}$ (β, δ some scalars; $\mathbf{N}$, $\mathbf{M}$ unit vectors) which are any two material elements at $\mathbf{X}$ lying in a plane of a central circular section of the C-ellipsoid so that, e.g., $\mathcal{S}^+ \cdot \mathbf{M} = 0$, $\mathcal{S}^+ \cdot \mathbf{M} = 0$. Then, using (2.23), $\mathbf{N} \cdot \mathbf{C}\mathbf{N} = \mathbf{M} \cdot \mathbf{C}\mathbf{M} = \lambda_2^2$ so that these elements suffer the same stretch.

Conversely, suppose that at $\mathbf{X}$ there is a plane Γ such that all material line elements suffer the same stretch λ (say). Then, $\mathbf{N} \cdot \mathbf{C}\mathbf{N} = \lambda^2$ for all $\mathbf{N}$ in Γ. Let $\mathbf{D}$ and $\mathbf{E}$ be a pair of orthogonal unit vectors in Γ, and let $\mathbf{N} = \cos\Theta\mathbf{D} + \sin\Theta\mathbf{E}$. By hypothesis, $\mathbf{D} \cdot \mathbf{C}\mathbf{D} = \mathbf{E} \cdot \mathbf{C}\mathbf{E} = \lambda^2$, so that $\mathbf{N} \cdot \mathbf{C}\mathbf{N} = \lambda^2$ leads to $\cos\Theta\sin\Theta\mathbf{D} \cdot \mathbf{C}\mathbf{E} = 0$. It follows that the pair of orthogonal unit vectors $\mathbf{D}$ and $\mathbf{E}$ are conjugate with respect to the C-ellipsoid and the radii to this ellipsoid along $\mathbf{D}$ and $\mathbf{E}$ are equal in length. It follows that the plane Γ is either $\Sigma^+(\mathbf{C})$ or $\Sigma^-(\mathbf{C})$.

The shear, γ (say), of a pair of material line elements along $\mathbf{N}$ and $\mathbf{M}$ in $\Sigma^\pm(\mathbf{C})$ is given by

$$\cos(\Theta + \gamma) = (\mathbf{N} \cdot \mathbf{C}\mathbf{M})/\{\lambda_{(\mathbf{N})}\lambda_{(\mathbf{M})}\} = (\lambda_2^2\mathbf{N} \cdot \mathbf{M})/\lambda_2^2 = \mathbf{N} \cdot \mathbf{M} = \cos\Theta, \quad (3.6)$$

where Θ is the angle between $\mathbf{N}$ and $\mathbf{M}$. Thus these elements are unsheared. Finally, it is easily shown that if all material line elements in a certain plane at $\mathbf{X}$ are unsheared in the deformation then all stretches in that plane must be equal, so that the plane is a plane of central circular section of the C-ellipsoid at $\mathbf{X}$.

Conversely, suppose that at $\mathbf{X}$ there is a plane Γ such that no pair of material line element in Γ is sheared in the deformation. Let $\mathbf{D}$ and $\mathbf{E}$ be a pair of orthogonal unit vectors in Γ, and let $\mathbf{N} = \cos\Theta\mathbf{D} + \sin\Theta\mathbf{E}$, $\mathbf{M} = \cos(\Theta + \Psi)\mathbf{D} + \sin(\Theta + \Psi)\mathbf{E}$, be a pair of unit vectors subtending the acute angle Ψ

and lying along material line elements in Γ at $\mathbf{X}$. Because they are unsheared, they must satisfy

$$\lambda_{(\mathbf{N})}\lambda_{(\mathbf{M})}\mathbf{N} \cdot \mathbf{M} = \mathbf{N} \cdot \mathbf{CM}. \tag{3.7}$$

This is to hold for all Θ and Ψ, $0 \leq \Theta \leq 2\pi$, $0 < \Psi < \pi$. Let $\Psi = \pi/2$. then we have $\mathbf{N} \cdot \mathbf{CM} = 0$, so that (3.7) gives $\{\lambda_{(\mathbf{D})}^2 - \lambda_{(\mathbf{E})}^2\} \cos\theta \sin\Theta = \mathbf{D} \cdot \mathbf{CE} \cos 2\Theta$. Choosing $\Theta = \pi/4$ leads to $\lambda_{(\mathbf{D})}^2 = \lambda_{(\mathbf{E})}^2$, so that we must also have $\mathbf{D} \cdot \mathbf{CE} = 0$. Thus, $\mathbf{D}$ and $\mathbf{E}$ lie in either $\Sigma^+(\mathbf{C})$ or $\Sigma^-(\mathbf{C})$ [from the form of the proof, it is seen that if all orthogonal pairs in a plane are unsheared, then the plane is either $\Sigma^+(\mathbf{C})$ or $\Sigma^-(\mathbf{C})$].

Finally, assuming that the pair of line material elements along $\mathbf{M}$, $\mathbf{N}$ is unsheared in the deformation and that the stretches along $\mathbf{M}$ and $\mathbf{N}$ are equal, then both $\mathbf{M}$ and $\mathbf{N}$ lie in a plane of central circular section of the $\mathbf{C}$-ellipsoid (Boulanger and Hayes in press). Indeed, assume $\lambda_{(\mathbf{N})} = \lambda_{(\mathbf{M})} \equiv \lambda$ (with $\mathbf{N} \neq \pm\mathbf{M}$) and let $\mathbf{Y} = \alpha\mathbf{N} + \beta\mathbf{M}$ be any unit vector in the plane of $\mathbf{N}$ and $\mathbf{M}$. Then, using the condition (3.7) for unsheared pairs, we have

$$\begin{aligned}
\mathbf{Y} \cdot \mathbf{CY} &= \alpha^2\lambda_{(\mathbf{N})}^2 + \beta^2\lambda_{(\mathbf{M})}^2 + 2\alpha\beta\mathbf{N} \cdot \mathbf{CM} \\
&= \lambda^2(\alpha^2 + \beta^2 + 2\alpha\beta\mathbf{N} \cdot \mathbf{M}) = \lambda^2\mathbf{Y} \cdot \mathbf{Y} = \lambda^2.
\end{aligned} \tag{3.8}$$

Hence all radii to the $\mathbf{C}$-ellipsoid in the plane of $\mathbf{N}$ and $\mathbf{M}$ are of equal length λ^{-1}. Thus, this plane is a plane of central circular section of the $\mathbf{C}$-ellipsoid.

6.4 Unsheared pairs in a given plane

Here we consider material line elements lying in a definite given plane $\pi(\mathbf{S})$ with unit normal $\mathbf{S}$ at $\mathbf{X}$. We find those pairs of line elements which are not sheared. It was shown by Boulanger and Hayes (2000) that in general corresponding to any material line element in the plane $\pi(\mathbf{S})$, there is one and only one other "companion" material line element in the plane, such that the line element and its companion are unsheared in the deformation. However, in general, there are two line elements in $\pi(\mathbf{S})$ which are such that neither has a companion. The directions of these material line elements in $\pi(\mathbf{S})$ which do not have companion elements in $\pi(\mathbf{S})$ are called *limiting directions*. The limiting directions play a fundamental role in the analysis of strain. It is seen that if $\mathbf{S}$ is prescribed, then knowledge of $\mathbf{C}$ is sufficient for the determination of the angle between the limiting directions in $\pi(\mathbf{S})$.

6.4.1 Deformation of a material element

Consider the plane of material line elements $\pi(\mathbf{S})$ at $\mathbf{X}$: $\mathbf{S} \cdot d\mathbf{X} = 0$, where $\mathbf{S}$ is a definite given unit vector. This plane is deformed into the plane $\pi(\mathbf{s})$ (say) at $\mathbf{x}$: $\mathbf{s} \cdot d\mathbf{x} = 0$, where

$$\mathbf{s} = \mathbf{F}^{-T}\mathbf{S}, \quad s_i = \frac{\partial X_A}{\partial x_i} S_A. \tag{4.1}$$

Let $\mathbf{I}$ and $\mathbf{J}$ be material unit vectors along the principal axes of the elliptical section by the plane $\pi(\mathbf{S})$ of the C-ellipsoid at $\mathbf{X}$. Then

$$\mathbf{S} \cdot \mathbf{I} = \mathbf{S} \cdot \mathbf{J} = 0, \quad \mathbf{I} \cdot \mathbf{J} = 0, \quad \mathbf{I} \cdot \mathbf{CJ} = 0. \tag{4.2}$$

The last equation equation $(4.2)_4$ expresses the fact that $\mathbf{I}$ and $\mathbf{J}$ are conjugate with respect to the C-ellipsoid.

If $\mathbf{i}$ and $\mathbf{j}$ are material vectors into which $\mathbf{I}$ and $\mathbf{J}$ are deformed, so that

$$\mathbf{i} = \mathbf{FI}, \quad \mathbf{j} = \mathbf{FJ}, \tag{4.3}$$

then it follows from (4.1), (4.2), that

$$\mathbf{s} \cdot \mathbf{i} = \mathbf{s} \cdot \mathbf{j} = 0, \quad \mathbf{i} \cdot \mathbf{B}^{-1}\mathbf{j} = 0, \quad \mathbf{i} \cdot \mathbf{j} = 0. \tag{4.4}$$

Hence, $\mathbf{i}$ and $\mathbf{j}$ are along the principal axes of the elliptical section of the $\mathbf{B}^{-1}$-ellipsoid by the plane $\pi(\mathbf{s})$ at $\mathbf{x}$. We note that unit vectors along $\mathbf{i}, \mathbf{j}$ are $\hat{\mathbf{i}}, \hat{\mathbf{j}}$ given by

$$\hat{\mathbf{i}} = \mathbf{FI}/\lambda_{(\mathbf{I})}, \quad \hat{\mathbf{j}} = \mathbf{FJ}/\lambda_{(\mathbf{J})}, \tag{4.5}$$

where, using (2.15),

$$\lambda_{(\mathbf{I})}^2 = \mathbf{I} \cdot \mathbf{CI}, \quad \lambda_{(\mathbf{J})}^2 = \mathbf{J} \cdot \mathbf{CJ}. \tag{4.6}$$

Note that

$$\lambda_{(\mathbf{i})}^2 = \hat{\mathbf{i}} \cdot \mathbf{B}^{-1}\hat{\mathbf{i}} = 1/\lambda_{(\mathbf{I})}^2, \quad \lambda_{(\mathbf{j})}^2 = \hat{\mathbf{j}} \cdot \mathbf{B}^{-1}\hat{\mathbf{j}} = 1/\lambda_{(\mathbf{J})}^2. \tag{4.7}$$

Throughout we assume $\lambda_{(\mathbf{I})} \leq \lambda_{(\mathbf{J})}$, so that ϵ, defined by

$$\epsilon = \lambda_{(\mathbf{J})}/\lambda_{(\mathbf{I})} = \lambda_{(\mathbf{i})}/\lambda_{(\mathbf{j})}, \tag{4.8}$$

is such that

$$\epsilon \geq 1. \tag{4.9}$$

Here, $\mathbf{I}$ and $\mathbf{J}$ are, respectively, along the major and minor axes of the elliptical section of the C-ellipsoid by the plane $\pi(\mathbf{S})$. Also, $\mathbf{i}$ and $\mathbf{j}$ are, respectively, along the minor and major axes of the elliptical section of the $\mathbf{B}^{-1}$-ellipsoid by the plane $\pi(\mathbf{s})$.

Now let **N** be any material unit vector in the plane $\pi(\mathbf{S})$ at **X**:

$$\mathbf{N} = \cos \Phi \mathbf{I} + \sin \Phi \mathbf{J}, \tag{4.10}$$

where Φ is arbitrary. Following the deformation, this material element lies along the unit vector $\hat{\mathbf{n}}$ in the plane $\pi(\mathbf{s})$ at **x**:

$$\begin{aligned}
\hat{\mathbf{n}} &= \mathbf{FN}/\lambda_{(\mathbf{N})} = \{\cos \Phi \; \mathbf{FI} + \sin \Phi \; \mathbf{FJ}\}/\lambda_{(\mathbf{N})} \\
&= \{\lambda_{(\mathbf{I})} \cos \Phi \hat{\mathbf{i}} + \lambda_{(\mathbf{J})} \sin \Phi \hat{\mathbf{j}}\}/\lambda_{(\mathbf{N})},
\end{aligned} \tag{4.11}$$

where

$$\lambda_{(\mathbf{N})}^2 = |\mathbf{FN}|^2 = \mathbf{N} \cdot \mathbf{CN} = \lambda_{(\mathbf{I})}^2 \cos^2 \Phi + \lambda_{(\mathbf{J})}^2 \sin^2 \Phi, \tag{4.12}$$

on using (4.5). Thus

$$\hat{\mathbf{n}} = \cos \phi \hat{\mathbf{i}} + \sin \phi \hat{\mathbf{j}},$$

$$\cos \phi = \{\lambda_{(\mathbf{I})}/\lambda_{(\mathbf{N})}\} \cos \Phi, \quad \sin \phi = \{\lambda_{(\mathbf{J})}/\lambda_{(\mathbf{N})}\} \sin \Phi. \tag{4.13}$$

Because $\lambda_{(\mathbf{I})}, \lambda_{(\mathbf{J})}, \lambda_{(\mathbf{N})}$ are all positive quantities, we note that if **N** lies in a certain quadrant in the **I** − **J** plane (the plane $\pi(\mathbf{S})$), then $\hat{\mathbf{n}}$ lies in the same quadrant in the **i** − **j** plane (the plane $\pi(\mathbf{s})$). Also, it follows from (4.13) that the angle, Φ, that **N** makes with **I**, is deformed into the angle, ϕ, that $\hat{\mathbf{n}}$ makes with $\hat{\mathbf{i}}$, with

$$\tan \phi = \epsilon \tan \Phi. \tag{4.14}$$

In the case of plane deformation, Ramsay and Huber (1987) attribute the result (4.14) to Wettstein (1886). In this case, because $\epsilon \geq 1$ it follows that, for material elements in the first quadrant, $\phi \geq \Phi$, so that, as noted by Ramsay and Huber, material line elements move towards the principal direction of greatest stretch. In the geology literature the result (4.14) is known as *Wettstein's formula*.

Boulanger and Hayes (2000) pointed out that the formula (4.14) remains valid in general provided it is understood that Φ is the angle that the material element makes with the direction of **I** corresponding to the least stretch in the plane $\pi(\mathbf{S})$ and ϕ is the angle that the material element makes with the direction of **i** = **FI** in the plane $\pi(\mathbf{s})$. Of course, $\pi(\mathbf{s})$ is generally different from $\pi(\mathbf{S})$. Boulanger and Hayes called (4.14) the *generalized Wettstein's formula*.

6.4.2 Unsheared pairs. Limiting directions in the plane

Now introduce another arbitrary material unit vector $\mathbf{M}$ in the plane $\pi(\mathbf{S})$ at $\mathbf{X}$:

$$\mathbf{M} = \cos\Phi'\mathbf{I} + \sin\Phi'\mathbf{J}, \tag{4.15}$$

and the corresponding unit vector $\hat{\mathbf{m}}$ in the plane $\pi(\mathbf{s})$ at $\mathbf{x}$. Then

$$\hat{\mathbf{m}} = \cos\phi'\hat{\mathbf{i}} + \sin\phi'\hat{\mathbf{j}},$$
$$\cos\phi' = \{\lambda_{(\mathrm{I})}/\lambda_{(\mathrm{M})}\}\cos\Phi', \quad \sin\phi' = \{\lambda_{(\mathrm{J})}/\lambda_{(\mathrm{M})}\}\sin\Phi', \tag{4.16}$$
$$\tan\phi' = \epsilon\tan\Phi'.$$

Hence, from (4.13) and (4.16), it follows that

$$\lambda_{(\mathrm{M})}\lambda_{(\mathrm{N})}\sin(\phi'-\phi) = \lambda_{(\mathrm{I})}\lambda_{(\mathrm{J})}\sin(\Phi'-\Phi),$$
$$\tag{4.17}$$
$$\lambda_{(\mathrm{M})}\lambda_{(\mathrm{N})}\cos(\phi'-\phi) = \lambda_{(\mathrm{I})}^2\cos\Phi'\cos\Phi + \lambda_{(\mathrm{J})}^2\sin\Phi'\sin\Phi.$$

If the pair of line material elements $(\mathbf{M}, \mathbf{N})$ is not sheared, then $\Phi' - \Phi = \phi' - \phi$, or $\tan(\phi' - \phi) = \tan(\Phi' - \Phi)$, and hence, using equations (4.8), (4.14) and (4.16)$_4$, we have

$$(1 - \epsilon)(1 - \epsilon\tan\Phi\tan\Phi') = 0. \tag{4.18}$$

The first possibility, $\epsilon = 1$, means that $\lambda_{(\mathrm{I})} = \lambda_{(\mathrm{J})}$, so that the elliptical section is circular. Hence, in this case the plane $\pi(\mathbf{S})$ is a plane $\Sigma^{\pm}(\mathbf{C})$ of a central circular section of the $\mathbf{C}$-ellipsoid at $\mathbf{X}$ as was shown in Section 6.3. No pair of elements in this plane is sheared (Hayes 1988).

The second possibility gives

$$\epsilon\tan\Phi\tan\Phi' = 1, \quad \tan\phi\tan\phi' = \epsilon, \tag{4.19}$$

on using equations (4.14) and (4.16)$_4$. The first of these gives the product of the slopes (with respect to the major axis of the section of the $\mathbf{C}$-ellipsoid by $\pi(\mathbf{S})$) of those pairs of material elements in the plane $\pi(\mathbf{S})$ at $\mathbf{X}$ which are not sheared. Thus, given the slope of any material element, $\tan\Phi$, the slope, $\tan\Phi'$, of its only "companion" such that the pair is unsheared is given by equation (4.19)$_1$. Similarly, equation (4.19)$_2$ gives the product of the slopes (with respect to the major axis of the section of the $\mathbf{B}^{-1}$-ellipsoid by $\pi(\mathbf{s})$) of those pairs in the deformed state which have not been sheared in the deformation.

However, note that if

$$\tan\Phi = +\epsilon^{-1/2} \quad (\text{or } -\epsilon^{-1/2}), \tag{4.20}$$

then also

$$\tan \Phi' = +\epsilon^{-1/2} \quad (\textbf{or } -\epsilon^{-1/2}), \tag{4.21}$$

so that $\Phi' = \Phi$. Elements along these directions have no companions in the plane $\pi(\mathbf{S})$. Boulanger and Hayes (2000) called the corresponding directions of these elements *limiting directions*. The internal and external bisectors of the angles between the limiting directions are along $\mathbf{I}$ and $\mathbf{J}$, the principal axes of the elliptical section of the C-ellipsoid by $\pi(\mathbf{S})$. Because $\epsilon \neq 1$, the limiting directions are not at right angles. (If $\epsilon = 1$, then $\pi(\mathbf{S})$ coincides with $\Sigma^+(\mathbf{C})$ or $\Sigma^-(\mathbf{C})$.)

It may be concluded, therefore, that if $\pi(\mathbf{S})$ is not the plane of a central circular section of the C-ellipsoid at $\mathbf{X}$, then, apart from the pair of elements along the limiting directions, there is a companion element for each element such that the element and its companion suffer no shear.

Let $\hat{\Phi}$ and $-\hat{\Phi}$ be the angles that the pair of limiting directions make with the major ($\mathbf{I}$) axis of the elliptical section of the C-ellipsoid by $\pi(\mathbf{S})$ so that

$$\tan \hat{\Phi} = \epsilon^{-1/2}. \tag{4.22}$$

Then the limiting directions at $\mathbf{X}$ subtend the angle Θ_L (say) given by $\Theta_L = 2\hat{\Phi} = 2\tan^{-1}(\epsilon^{-1/2})$. Because $\epsilon \geq 1$ it follows that $\hat{\Phi} \leq \pi/4$. Also, if $\hat{\phi}$ and $-\hat{\phi}$ are the angles which the elements along the limiting directions at $\mathbf{X}$ make, when deformed, with the i-axis at $\mathbf{x}$, then, from equation (4.14), $\tan \hat{\phi} = \epsilon^{1/2}$, so that $\tan \hat{\Phi} \tan \hat{\phi} = 1$. Hence, the angle, $\hat{\Phi}$, which a limiting direction makes with the $\mathbf{I}$-axis at $\mathbf{X}$, is deformed into the angle $\hat{\phi} = \pi/2 - \hat{\Phi}$ which the deformed limiting direction makes with the i-axis at $\mathbf{x}$. The deformed limiting material elements at $\mathbf{x}$ subtend the angle θ_L (say) given by

$$\theta_L = 2\tan^{-1}(\epsilon^{1/2}) = \pi - 2\tan^{-1}(\epsilon^{-1/2}) = \pi - \Theta_L. \tag{4.23}$$

Thus, the elements along the pair of limiting directions are sheared by the amount γ_L (say), given by

$$\gamma_L = 2\tan^{-1}\epsilon^{1/2} - 2\tan^{-1}\epsilon^{-1/2} = \pi - 2\Theta_L. \tag{4.24}$$

Using the expression from equation (4.22) in equation (4.12), and using the definition (4.8) for ϵ, it follows immediately that the stretches along the limiting directions are both equal, to $\lambda_{(\mathbf{L})}$ (say) where $\lambda_{(\mathbf{L})} = (\lambda_{(\mathbf{I})}\lambda_{(\mathbf{J})})^{1/2}$.

It follows from $(4.17)_1$ that for any unsheared pair of material elements in $\pi(\mathbf{S})$, along the unit vectors $\mathbf{M}$ and $\mathbf{N}$ (say), the product of the stretches along $\mathbf{M}$ and $\mathbf{N}$ is equal to the product of the principal stretches in $\pi(\mathbf{S})$:

$$\lambda_{(\mathbf{M})}\lambda_{(\mathbf{N})} = \lambda_{(\mathbf{I})}\lambda_{(\mathbf{J})}. \tag{4.25}$$

Thus, for all unsheared pairs $\mathbf{M}$ and $\mathbf{N}$ in the plane $\pi(\mathbf{S})$, the product of the stretches $\lambda_{(\mathbf{M})}$ and $\lambda_{(\mathbf{N})}$ is a constant. The dual of the result (4.25) is

$$\lambda_{(\mathbf{m})}\lambda_{(\mathbf{n})} = \lambda_{(\mathbf{i})}\lambda_{(\mathbf{j})}, \tag{4.26}$$

where the pair of material elements which are along $\mathbf{m}$ and $\mathbf{n}$ following the deformation have not been sheared.

By using the identity

$$\begin{aligned}
\lambda_{(\mathbf{I})}\lambda_{(\mathbf{J})} &= (\mathbf{I} \cdot \mathbf{CI})^{1/2}(\mathbf{J} \cdot \mathbf{CJ})^{1/2} \\
&= [(\mathbf{I} \times \mathbf{J}) \cdot (\mathbf{CI} \times \mathbf{CJ})]^{1/2} \\
&= [\det \mathbf{C}(\mathbf{S} \cdot \mathbf{C}^{-1}\mathbf{S})]^{1/2},
\end{aligned} \tag{4.27}$$

equation (4.25) may also be written in the form

$$\lambda_{(\mathbf{M})}\lambda_{(\mathbf{N})} = [\det \mathbf{C}(\mathbf{S} \cdot \mathbf{C}^{-1}\mathbf{S})]^{1/2}. \tag{4.28}$$

For all unsheared pairs along $\mathbf{M}$ and $\mathbf{N}$, equation (4.25), or equivalently equation (4.28), is valid. Boulanger and Hayes (2000) have shown that the converse is not true: all pairs of material elements along $\mathbf{M}$ and $\mathbf{N}$ satisfying equation (4.25), or equivalently equation (4.28), are not unsheared. They showed that if $\mathbf{M}$ and $\mathbf{N}$ is a pair of material unit vector elements in $\pi(\mathbf{S})$ satisfying (4.25), or equivalently (4.28), then, **either** (*i*) $\pi(\mathbf{S})$ is a plane $\Sigma^{\pm}(\mathbf{C})$ of central circular section of the $\mathbf{C}$-ellipsoid, or (*ii*) the pair $(\mathbf{M}, \mathbf{N})$ is unsheared in the deformation, or (*iii*) the pair $(\mathbf{M}, \mathbf{N})$ is such that its angle is transformed into its supplement. We note that the limiting directions form a special pair of directions whose angle Θ_L is transformed into its supplement $\theta_L = \pi - \Theta_L$.

Because $\mathbf{S} = \mathbf{I} \times \mathbf{J}$, we have

$$\lambda_{(\mathbf{I})}^2 + \lambda_{(\mathbf{J})}^2 = \mathbf{I} \cdot \mathbf{CI} + \mathbf{J} \cdot \mathbf{CJ} = (\text{tr } \mathbf{C}) - \mathbf{S} \cdot \mathbf{CS}. \tag{4.29}$$

Hence, using (4.27)

$$\{\lambda_{(\mathbf{J})} \pm \lambda_{(\mathbf{I})}\}^2 = \text{tr } \mathbf{C} - \mathbf{S} \cdot \mathbf{CS} \pm 2[(\det \mathbf{C})(\mathbf{S} \cdot \mathbf{C}^{-1}\mathbf{S})]^{1/2}. \tag{4.30}$$

So the angle, Θ_L, between the limiting directions at $\mathbf{X}$, may be written

$$\Theta_L = 2\tan^{-1}(\epsilon^{-1/2}) = \cos^{-1}\left(\frac{\epsilon - 1}{\epsilon + 1}\right) = \cos^{-1}\left\{\frac{\lambda_{(\mathbf{J})} - \lambda_{(\mathbf{I})}}{\lambda_{(\mathbf{J})} + \lambda_{(\mathbf{I})}}\right\},$$
$$\tag{4.31}$$

$$\cos \Theta_L = \left\{\frac{\text{tr } \mathbf{C} - \mathbf{S} \cdot \mathbf{CS} - 2[\det \mathbf{C}(\mathbf{S} \cdot \mathbf{C}^{-1}\mathbf{S})]^{1/2}}{\text{tr } \mathbf{C} - \mathbf{S} \cdot \mathbf{CS} + 2[\det \mathbf{C}(\mathbf{S} \cdot \mathbf{C}^{-1}\mathbf{S})]^{1/2}}\right\}^{1/2}.$$

This expression involves $\mathbf{C}$ and $\mathbf{S}$ only. Hence, as pointed out in (Boulanger and Hayes 2000), the angle Θ_L between the pair of limiting directions in the plane

$\pi(\mathbf{S})$ at $\mathbf{X}$ is known once $\mathbf{C}$ is known. It does not involve the determination of the principal axes of $\mathbf{C}$ or the principal stretches. The limiting directions themselves are along the unit vectors $\mathbf{L}_+$ and $\mathbf{L}_-$ given by

$$\mathbf{L}_\pm = \frac{\lambda_{(J)}^{1/2}\mathbf{I} \pm \lambda_{(I)}^{1/2}\mathbf{J}}{(\lambda_{(I)} + \lambda_{(J)})^{1/2}}. \tag{4.32}$$

The dual of equation (4.31) is

$$\theta_L = 2\tan^{-1}(\epsilon^{1/2}) = \cos^{-1}\left(\frac{1-\epsilon}{1+\epsilon}\right) = \cos^{-1}\left\{\frac{\lambda_{(j)} - \lambda_{(i)}}{\lambda_{(j)} + \lambda_{(i)}}\right\}, \tag{4.33}$$

$$\cos\theta_L = \left\{\frac{\operatorname{tr}\mathbf{B}^{-1} - \hat{\mathbf{s}}\cdot\mathbf{B}^{-1}\hat{\mathbf{s}} - 2[\det\mathbf{B}^{-1}(\hat{\mathbf{s}}\cdot\mathbf{B}\hat{\mathbf{s}})]^{1/2}}{\operatorname{tr}\mathbf{B}^{-1} - \hat{\mathbf{s}}\cdot\mathbf{B}^{-1}\hat{\mathbf{s}} + 2[\det\mathbf{B}^{-1}(\hat{\mathbf{s}}\cdot\mathbf{B}\hat{\mathbf{s}})]^{1/2}}\right\}^{1/2}.$$

This expression involves $\mathbf{B}^{-1}$ and $\hat{\mathbf{s}}$ only. Hence the angle, θ_L, between the pair of elements in the plane $\pi(\mathbf{s})$ at $\mathbf{x}$ into which the pair of limiting directions is deformed, is known once $\mathbf{B}^{-1}$ is known. It has been seen (equation (4.23)) that $\theta_L = \pi - \Theta_L$. The elements, into which the limiting directions $\mathbf{L}_+$ and $\mathbf{L}_-$ are deformed, are along $\hat{\mathbf{l}}_+$ and $\hat{\mathbf{l}}_-$, given by

$$\hat{\mathbf{l}}_\pm = \mathbf{FL}_\pm/\lambda_{(\mathbf{L}_\pm)} = (\lambda_{(J)}^{1/2}\mathbf{FI} \pm \lambda_{(I)}^{1/2}\mathbf{FJ})/\{\lambda_{(I)}\lambda_{(J)}(\lambda_{(I)} + \lambda_{(J)})\}^{1/2}$$
$$= \frac{\lambda_{(j)}^{1/2}\hat{\mathbf{i}} \pm \lambda_{(i)}^{1/2}\hat{\mathbf{j}}}{(\lambda_{(i)} + \lambda_{(j)})^{1/2}}, \tag{4.34}$$

which is the dual of (4.32).

For any given plane $\pi(\mathbf{S})$ of material elements, which is not a plane of central circular section of the $\mathbf{C}$-ellipsoid, Boulanger and Hayes (2000) proved that it is those material elements along the limiting directions which suffer the maximum shear. That proof will be given later. By equation (4.24), the shear of the limiting directions is $\pi - 2\Theta_L$. This, then, is the amount of maximum shear in $\pi(\mathbf{S})$. Thus, by equation (4.31), the maximum shear in the plane $\pi(\mathbf{S})$ is determined once $\mathbf{C}$ at $\mathbf{X}$ is known.

Of course, for given $\mathbf{S}$, in order to determine the pair of limiting directions in $\pi(\mathbf{S})$, it is necessary to know $\mathbf{I}$ and $\mathbf{J}$, the principal axes of the elliptical section of the $\mathbf{C}$-ellipsoid by $\pi(\mathbf{S})$. Explicit expressions for $\mathbf{I}$ and $\mathbf{J}$ may be obtained (Boulanger and Hayes 1995) if the eigenvectors and eigenvalues of $\mathbf{C}$ are known.

6.5 Maximum shear in a given plane for given Θ

Let a pair of material line elements at $\mathbf{X}$ be along the unit vectors $\mathbf{M}$ and $\mathbf{N}$ which subtend the angle Θ $(0 < \Theta < \pi)$. The shear, $\gamma_{(\mathbf{M},\mathbf{N})}$, is given by

$$\gamma_{(\mathbf{M},\mathbf{N})} = \theta - \Theta, \tag{5.1}$$

where θ is the angle between the elements in the deformed state.

Now, alternatively,

$$\gamma_{(\mathbf{M},\mathbf{N})} = \phi' - \phi - (\Phi' - \Phi). \tag{5.2}$$

Thus, for given Θ, the shear, γ, will be a maximum when $\phi' - \phi$ is a maximum. Now, by equations (4.14) and (4.16)$_4$,

$$\tan\phi = \epsilon\tan\Phi, \quad \tan\phi' = \epsilon\tan\Phi'. \tag{5.3}$$

Boulanger and Hayes (2000) showed that the maximum shear for given Θ, denoted by $\gamma_M(\Theta)$, is given by

$$\gamma_M(\Theta) = 2\tan^{-1}(\epsilon\tan\frac{\Theta}{2}) - \Theta > 0. \tag{5.4}$$

It occurs when the bisector of the angle between $\mathbf{M}$ and $\mathbf{N}$ is along the I-axis - the major axis of the elliptical section of the C-ellipsoid which corresponds to the smaller of the stretches $\lambda_{(\mathbf{I})}$ and $\lambda_{(\mathbf{J})}$. After the deformation, $\hat{\mathbf{m}}$ and $\hat{\mathbf{n}}$ subtend an angle $\theta^* = \theta + \gamma_M(\Theta)$, greater than Θ, given by

$$\tan\frac{\theta^*}{2} = \epsilon\tan\frac{\Theta}{2}, \tag{5.5}$$

and the bisector of the angle between $\hat{\mathbf{m}}$ and $\hat{\mathbf{n}}$ is along the i-axis - the minor axis of the elliptical section of the $\mathbf{B}^{-1}$- ellipsoid by the plane $\pi(\mathbf{s})$.

Also, the minimum shear for given Θ, denoted by $\gamma_m(\Theta)$, is given by

$$\gamma_m(\Theta) = -2\tan^{-1}(\epsilon\cot\frac{\Theta}{2}) + \pi - \Theta < 0 \tag{5.6}$$

(Boulanger and Hayes 2000). It occurs when the bisector of the angle between $\mathbf{M}$ and $\mathbf{N}$ is along the J-axis - the minor axis of the elliptical section of the C-ellipsoid which corresponds to the larger of the stretches $\lambda_{(\mathbf{I})}$ and $\lambda_{(\mathbf{J})}$. After the deformation, $\hat{\mathbf{m}}$ and $\hat{\mathbf{n}}$ subtend an angle $\theta_* = \Theta + \gamma_m(\Theta)$, smaller than Θ, given by

$$\cot\frac{\theta_*}{2} = \epsilon\cot\frac{\Theta}{2}, \tag{5.7}$$

and the bisector of the angle between $\hat{\mathbf{m}}$ and $\hat{\mathbf{n}}$ is along the j-axis - the major axis of the elliptical section of the $\mathbf{B}^{-1}$-ellipsoid by the plane $\pi(\mathbf{s})$.

Physically, the minimum shear (representing a decrease of the angle) for a given angle Θ plus the maximum shear (representing an increase of the angle) for its supplement $(\pi - \Theta)$ should be zero:

$$\gamma_m(\Theta) = -\gamma_M(\pi - \Theta). \tag{5.8}$$

This is clear from equations (5.4) and (5.6).

Also, using these equations, Boulanger and Hayes (2000) showed that

$$\tan\frac{\gamma_M}{2}(\Theta) = \frac{(\epsilon - 1)\tan\frac{\Theta}{2}}{1 + \epsilon\tan^2\frac{\Theta}{2}}, \quad \sin\gamma_M(\Theta) = (\epsilon - 1)\sin\Theta\frac{\cos^2\frac{\Theta}{2} + \epsilon\sin^2\frac{\Theta}{2}}{\cos^2\frac{\Theta}{2} + \epsilon^2\sin^2\frac{\Theta}{2}},$$

$$\tag{5.9}$$

$$\tan\frac{\gamma_m}{2}(\Theta) = \frac{(1 - \epsilon)\cot\frac{\Theta}{2}}{1 + \epsilon\cot^2\frac{\Theta}{2}}, \quad \sin\gamma_m(\Theta) = (1 - \epsilon)\sin\Theta\frac{\sin^2\frac{\Theta}{2} + \epsilon\cos^2\frac{\Theta}{2}}{\sin^2\frac{\Theta}{2} + \epsilon^2\cos^2\frac{\Theta}{2}}.$$

In the special case of orthogonal shear, from (5.6) and (5.9), we have

$$\gamma_m(\frac{\pi}{2}) = -\gamma_M(\frac{\pi}{2}) = -2\tan^{-1}(\epsilon) + (\pi/2),$$

$$\sin\gamma_M(\frac{\pi}{2}) = \{\lambda_{(J)}^2 - \lambda_{(I)}^2\}/\{\lambda_{(J)}^2 + \lambda_{(I)}^2\}, \tag{5.10}$$

which is a plane version of Ericksen's formula, equation (1.1), for maximum orthogonal shear. In this case the pair $(\mathbf{M}, \mathbf{N})$ lie at $\pi/4$ to $\mathbf{I}$ and $\mathbf{J}$ in the plane $\pi(\mathbf{S})$.

6.6 Global extremal shear for given Θ

Having determined the extremal shears in a given plane for a definite angle Θ between the pair of material elements in $\pi(\mathbf{S})$, we turn now to evaluate the extremal shears for arbitrary $\pi(\mathbf{S})$ – the global extremal shears for given Θ.

From (5.4) the value of $\gamma_M(\Theta)$ in the plane $\pi(\mathbf{S})$ depends only upon $\epsilon = \lambda_{(J)}/\lambda_{(I)}$ and it is clear that $\gamma_M(\Theta)$ will have a global maximum when ϵ is as large as possible. This occurs when $\lambda_{(J)} = \lambda_3$, the largest principal stretch, so that $\mathbf{J}$ is along $\mathbf{E}_{(3)}$, the corresponding eigenvector of $\mathbf{C}$, and when $\lambda_{(I)} = \lambda_1$, the smallest principal stretch, so that $\mathbf{I}$ is along $\mathbf{E}_{(1)}$, the corresponding eigenvector of $\mathbf{C}$. Denoting the global maximum (minimum) value of γ, for given Θ, by $\gamma_G(\Theta)(\gamma_g(\Theta))$, Boulanger and Hayes (2000) showed that

$$\gamma_G(\Theta) = 2\tan^{-1}\left(\frac{\lambda_3}{\lambda_1}\tan\frac{\Theta}{2}\right) - \Theta. \tag{6.1}$$

The corresponding pair of material elements is in the principal $\mathbf{E}_{(1)} - \mathbf{E}_{(3)}$ plane of the $\mathbf{C}$-ellipsoid at $\mathbf{X}$, and has the $\mathbf{E}_{(1)}$-axis as internal bisector.

The global minimum shear $\gamma_g(\Theta)$, for given Θ, is

$$\gamma_g(\Theta) = -2\tan^{-1}\left(\frac{\lambda_3}{\lambda_1}\cot\frac{\Theta}{2}\right) + \pi - \Theta, \tag{6.2}$$

from equation (5.6). Again, the corresponding pair of material elements is in the principal $\mathbf{E}_{(1)} - \mathbf{E}_{(3)}$ plane of the **C**-ellipsoid at $\mathbf{X}$, but has the $\mathbf{E}_{(3)}$-axis as internal bisector.

When $\Theta = \pi/2$, then

$$\gamma_G(\frac{\pi}{2}) = 2\tan^{-1}\left(\frac{\lambda_3}{\lambda_1}\right) - \frac{\pi}{2}, \tag{6.3}$$

so that we recover Ericksen's formula (equation (1.1)):

$$\sin\gamma_G(\frac{\pi}{2}) = \frac{\lambda_3^2 - \lambda_1^2}{\lambda_3^2 + \lambda_1^2}, \tag{6.4}$$

and the corresponding pair of material elements is along $(\mathbf{E}_{(1)} \pm \mathbf{E}_{(3)})/\sqrt{2}$.

6.7 Maximum shear in a given plane. Global maximum

Here the angle Θ between the material elements in $\pi(\mathbf{S})$ may be arbitrary, subject only to $0 < \Theta < \pi$. The angle Θ is sought for which the shear is a maximum and the value of that shear.

Now the shear, γ, is given by

$$\gamma = \theta - \Theta = (\phi' - \Phi') - (\phi - \Phi). \tag{7.1}$$

Clearly γ will be a maximum when $(\phi' - \Phi')$ is a maximum and $(\phi - \Phi)$ is a minimum. Now, using equation (5.3), Boulanger and Hayes (2000) have shown that γ is maximum {or minimum} when $\tan\Phi = -\epsilon^{-1/2}$ {or $\epsilon^{-1/2}$} and $\tan\Phi' = \epsilon^{-1/2}$ {or $-\epsilon^{-1/2}$}, thus, when $\Phi = -\hat{\Phi}$ {or $+\hat{\Phi}$} and $\Phi = +\hat{\Phi}$ {or $-\hat{\Phi}$}, where $\hat{\Phi}$ is given by (4.22). Hence, γ is extremum when $\mathbf{M}$ and $\mathbf{N}$ are along the limiting directions in the plane $\pi(\mathbf{S})$.

Thus, if the maximum (minimum) values of γ in the plane are denoted by $\gamma^*(\gamma_*)$, then

$$\gamma^* = 2\tan^{-1}\{(\epsilon^{1/2} - \epsilon^{-1/2})/2\}, \quad \gamma_* = -\gamma^*, \tag{7.2}$$

and

$$\tan(\frac{\gamma^*}{2}) = \frac{\lambda_{(J)} - \lambda_{(I)}}{2(\lambda_{(J)}\lambda_{(I)})^{1/2}}, \quad \sin\gamma^* = \frac{4(\lambda_{(J)}\lambda_{(I)})^{1/2}(\lambda_{(J)} - \lambda_{(I)})}{(\lambda_{(J)} + \lambda_{(I)})^2}. \tag{7.3}$$

The maximum occurs when $\mathbf{M}$ and $\mathbf{N}$ are along the limiting directions and have their internal bisector along $\mathbf{I}$, the axis of the ellipse in $\pi(\mathbf{S})$ corresponding to the least stretch in that plane. Thus $\mathbf{M} = \mathbf{L}_+$ and $\mathbf{N} = \mathbf{L}_-$, where $\mathbf{L}_\pm$ are given by (4.32), suffer the maximum shear in the plane $\pi(\mathbf{S})$. Their angle before deformation is Θ_L given by (4.31) and becomes, after deformation, θ_L, given by (4.33). Because $\theta_L = \pi - \Theta_L$, the maximum shear (increase in angle) is

$$\gamma^* = \gamma_L = \pi - 2\Theta_L. \tag{7.4}$$

This is consistent with the result (5.4) for maximum shear for given angle, because it is easily seen that $\gamma_M(\Theta_L) = \gamma^*$.

The minimum occurs when $\mathbf{M}$ and $\mathbf{N}$ are along the limiting directions and have their internal bisector along $\mathbf{J}$, the axis of the ellipse in $\pi(\mathbf{S})$ corresponding to the greatest stretch in this plane. Thus $\mathbf{M} = \mathbf{L}_+$ and $\mathbf{N} = -\mathbf{L}_-$ suffer the minimum shear in the plane $\pi(\mathbf{S})$. Their angle before deformation is $\pi - \Theta_L$ and becomes, after deformation, $\pi - \theta_L$, so that the minimum shear (decrease in angle) is

$$\gamma_\star = -\gamma^*. \tag{7.5}$$

Global Maximum

The global maximum occurs when γ^* is as large as possible. Let the corresponding value of γ^* be γ_G^*. Clearly it occurs when $\epsilon = \lambda_3/\lambda_1$. Thus,

$$\tan\left(\frac{\gamma_G^*}{2}\right) = \frac{\lambda_3 - \lambda_1}{2(\lambda_1\lambda_3)^{1/2}}, \quad \sin\gamma_G^* = \frac{4(\lambda_1\lambda_3)^{1/2}(\lambda_3 - \lambda_1)}{(\lambda_1 + \lambda_3)^2}. \tag{7.6}$$

Let the corresponding material line elements at $\mathbf{X}$ lie along the unit vectors $\mathbf{L}_+^*$ and $\mathbf{L}_-^*$ (say). They lie in the limiting directions in the principal $\mathbf{E}_{(1)} - \mathbf{E}_{(3)}$ plane of the C-ellipsoid. Thus, recalling (4.32),

$$\mathbf{L}_\pm^* = \frac{\lambda_3^{1/2}\mathbf{E}_{(1)} \pm \lambda_1^{1/2}\mathbf{E}_{(3)}}{(\lambda_1 + \lambda_3)^{1/2}}. \tag{7.7}$$

They are symmetrically disposed about the $\mathbf{E}_{(1)}$-axis - corresponding to the least principal stretch, and subtend an angle Θ_G^*, where (recall (4.31))

$$\cos\Theta_G^* = \frac{\lambda_3 - \lambda_1}{\lambda_3 + \lambda_1}, \quad \tan\Theta_G^* = \frac{2(\lambda_1\lambda_3)^{1/2}}{\lambda_3 - \lambda_1}, \quad \tan\left(\frac{\Theta_G^*}{2}\right) = (\lambda_1/\lambda_3)^{1/2}. \tag{7.8}$$

From (7.6)$_1$ and (7.8)$_2$ it follows that

$$\gamma_G^* = \pi - 2\Theta_G^*, \tag{7.9}$$

so that the angle Θ_G^* in $\pi(\mathbf{S})$ is changed into its supplement, $\pi - \Theta_G^*$ at $\mathbf{x}$.

As a result of the deformation, the material line elements along the unit vectors $\mathbf{L}_+^*$ and $\mathbf{L}_-^*$ at $\mathbf{X}$ in $\pi(\mathbf{S})$ are deformed into material line elements along the unit vectors $\hat{\mathbf{l}}_+^*$ and $\hat{\mathbf{l}}_-^*$, where

$$\hat{\mathbf{l}}_\pm^* = \mathbf{F}\mathbf{L}_\pm^*/\lambda_{(\mathbf{L}_\pm^*)} = (\lambda_3^{1/2}\mathbf{F}\mathbf{E}_{(1)} \pm \lambda_1^{1/2}\mathbf{F}\mathbf{E}_{(3)})/\{\lambda_1\lambda_3(\lambda_1 + \lambda_3)\}^{1/2}$$

$$= \frac{\lambda_3^{-1/2}\hat{\mathbf{e}}_{(1)} \pm \lambda_1^{-1/2}\hat{\mathbf{e}}_{(3)}}{(\lambda_1^{-1} + \lambda_3^{-1})^{1/2}}. \tag{7.10}$$

These subtend the angle θ_G^*, where (recall(4.33))

$$\cos\theta_G^* = -\frac{\lambda_3 - \lambda_1}{\lambda_3 + \lambda_1} = \cos(\pi - \Theta_G^*) = \cos(\gamma_G^* + \Theta_G^*). \tag{7.11}$$

Examples

We present two illustrative examples.

1. Pure homogeneous deformation
   Let

$$x_\alpha = \lambda_\alpha X_\alpha \qquad (\alpha = 1, 2, 3), \tag{7.12}$$

where the constants λ_α are such that $\lambda_3 > \lambda_2 > \lambda_1$. In this case, $\mathbf{E}_{(\alpha)}$ are unit vectors along the coordinate axes and

$$\mathbf{F} = \text{diag}(\lambda_1, \lambda_2, \lambda_3). \tag{7.13}$$

Also, $\mathbf{e}_{(\alpha)} = \mathbf{F}\mathbf{E}_{(\alpha)} = \lambda_\alpha\mathbf{E}_{(\alpha)}$, so that $\hat{\mathbf{e}}_{(\alpha)} = \mathbf{E}_{(\alpha)}$. Thus, the material line elements that suffer the global maximum shear lie in the $\mathbf{E}_{(1)} - \mathbf{E}_{(3)}$ plane before and after deformation.

From equations (7.7) and (7.10),

$$\mathbf{L}_\pm^* = \frac{\lambda_3^{1/2}\mathbf{E}_{(1)} \pm \lambda_1^{1/2}\mathbf{E}_{(3)}}{(\lambda_1 + \lambda_3)^{1/2}}, \qquad \hat{\mathbf{l}}_\pm^* = \frac{\lambda_3^{-1/2}\mathbf{E}_{(1)} \pm \lambda_1^{-1/2}\mathbf{E}_{(3)}}{(\lambda_1^{-1} + \lambda_3^{-1})^{1/2}}. \tag{7.14}$$

so that $\tan(\Theta_G^*/2) = (\lambda_1/\lambda_3)^{1/2}$, with γ_G^* given by (7.6).

2. Simple Shear
   For the simple shear

$$x = X + KY, \quad y = Y, \quad z = Z, \tag{7.15}$$

where K is a constant, we have

$$\mathbf{F} = \begin{pmatrix} 1 & K & 0 \\ 0 & 1 & 0 \\ 0 & 0 & 1 \end{pmatrix}, \quad \mathbf{C} = \begin{pmatrix} 1 & K & 0 \\ K & 1+K^2 & 0 \\ 0 & 0 & 1 \end{pmatrix}. \tag{7.16}$$

The principal stretches λ_α are

$$\lambda_1 = \{\sqrt{K^2 + 4} - K\}/2, \quad \lambda_2 = 1, \quad \lambda_3 = \{\sqrt{K^2 + 4} + K\}/2, \tag{7.17}$$

with corresponding eigenvectors

$$\mathbf{E}_{(1)} = (\lambda_3 \mathbf{I}_X - \mathbf{I}_Y)/\sqrt{\lambda_3^2 + 1}, \quad \mathbf{E}_{(2)} = \mathbf{I}_Z, \quad \mathbf{E}_{(3)} = (\lambda_1 \mathbf{I}_X + \mathbf{I}_Y)/\sqrt{\lambda_1^2 + 1}, \tag{7.18}$$

where $\mathbf{I}_X$, $\mathbf{I}_Y$ and $\mathbf{I}_Z$ are unit vectors along the X, Y and Z-axes. The planes $\Sigma^\pm(\mathbf{C})$ are $Y = 0$ and $2X + KY = 0$, with unit normals

$$\mathcal{S}^+ = \mathbf{I}_Y, \quad \mathcal{S}^- = (2\mathbf{I}_X + K\mathbf{I}_Y)/\sqrt{4 + K^2}. \tag{7.19}$$

On using $(7.14)_1$, (7.17) and (7.18), we have the corresponding $\mathbf{L}_\pm^\star$ given by

$$\mathbf{L}_+^\star = \mathbf{I}_X, \quad \mathbf{L}_-^\star = (K\mathbf{I}_X - 2\mathbf{I}_Y)/\sqrt{K^2 + 4}. \tag{7.20}$$

Line elements along $\mathbf{L}_\pm^\star$ lie in $\Sigma^\pm(\mathbf{C})$. The angle, $\Theta_G^\star$, between $\mathbf{L}_+^\star$ and $\mathbf{L}_-^\star$, is given by

$$\cos\Theta_G^\star = K/\sqrt{K^2 + 4}, \quad \tan\Theta_G^\star = 2/K, \tag{7.21}$$

and consequently the global maximum shear, $\gamma_G^\star$, in the case of simple shear, is

$$\gamma_G^\star = \pi - 2\cos^{-1}\{K/\sqrt{K^2 + 4}\}, \tag{7.22}$$

so that

$$\sin\gamma_G^\star = 4K/(4 + K^2), \quad \tan(\gamma_G^\star/2) = K/2. \tag{7.23}$$

Note that the particle initially at $P(-K/2, 1, 0)$ is moved to $p(K/2, 1, 0)$, the angle POp being $\gamma_G^\star$.

The polar decomposition of the deformation gradient $\mathbf{F} = \mathbf{VR}$ given by $(7.16)_1$, is

$$\mathbf{F} = \begin{pmatrix} (2 + K^2)/\sqrt{4 + K^2} & K/\sqrt{4 + K^2} & 0 \\ K/\sqrt{4 + K^2} & 2/\sqrt{4 + K^2} & 0 \\ 0 & 0 & 1 \end{pmatrix} \begin{pmatrix} \sin\alpha & \cos\alpha & 0 \\ -\cos\alpha & \sin\alpha & 0 \\ 0 & 0 & 1 \end{pmatrix}, \tag{7.24}$$

where $\tan\alpha = 2/K$. Thus, from (7.21), the angle of the rotation, α, in the case of simple shear, is also the angle $\Theta_G^\star$ for which the shear is a maximum. As Boulanger and Hayes (2000) remarked, simple shear is exceptional in this respect. For the pure homogeneous deformation (7.13) it has been seen that $\tan(\Theta_G^\star/2) = (\lambda_1/\lambda_3)^{1/2}$. However, in this case $\mathbf{R} = 1$, so that $\alpha = 0$.

Boulanger and Hayes (2000) noted for simple shear:

$$\tan \alpha = \tan \Theta_G^* = 2/K, \quad \tan(\gamma_G^*/2) = K/2,$$
$$\tan \Theta_G^* \tan(\gamma_G^*/2) = 1, \quad (\gamma_G^*/2) + \Theta_G^* = \pi/2. \tag{7.25}$$

If $K \approx 0$, then $\Theta_G^* \approx 90°$, $\gamma_G^* \approx 0$. However, if $K = 2$, then $\Theta_G^* = 45°$, $\gamma_G^* = 90°$. If $K = 4$, which corresponds to principal stretches $\lambda_3 = 4.25$, $\lambda_2 = 1, \lambda_1 = \lambda_3^{-1} = 0.2353$, then $\Theta_G^* = 27°$, $\gamma_G^* = 126.9°$, so that the angle between the pair of material elements which suffer the maximum shear is $27°$ and the amount of that maximum shear is $126.9°$.

6.8 Pairs of unsheared material elements

It has been shown that, in general, in any plane $\pi(\mathbf{S})$ at $\mathbf{X}$, there is an infinity of pairs of material elements which are unsheared in the deformation. For any given element along $\mathbf{N}$ (say) its companion is along $\mathbf{M}$ (say). Here, for given $\mathbf{S}$, and for $\mathbf{N}$ lying in $\pi(\mathbf{S})$, the companion $\mathbf{M}$ is given explicitly in terms of $\mathbf{S}, \mathbf{N}$ and $\mathbf{C}$. This result is due to Joly (1905) who gave it in terms of quaternions.

The condition for the shear $\gamma(\mathbf{M}, \mathbf{N})$ to be zero is that (3.7) holds:

$$\mathbf{N} \cdot \mathbf{CM} = \mathbf{N} \cdot \mathbf{M}(\mathbf{N} \cdot \mathbf{CN})^{1/2}(\mathbf{M} \cdot \mathbf{CM})^{1/2}. \tag{8.1}$$

Let $\mathbf{S}$ be orthogonal to $\mathbf{M}$ and $\mathbf{N}$. Then

$$\mathbf{S} \cdot \mathbf{M} = \mathbf{S} \cdot \mathbf{N} = 0 , \quad \mathbf{S} = \mathbf{N} \times \mathbf{M}/\{1 - (\mathbf{N} \cdot \mathbf{M})^2\}^{1/2}. \tag{8.2}$$

It is assumed that $\mathbf{N}$ is given, and for each unit vector $\mathbf{S}$ orthogonal to $\mathbf{N}$, $\mathbf{M}$ is to be determined in the plane $\pi(\mathbf{S})$ with unit normal $\mathbf{S}$, so that (8.1) holds. Then, as shown in (Boulanger and Hayes 2000), Joly's formula is

$$m\mathbf{M} = \mathbf{S} \times \{\mathbf{CN} - (\det \mathbf{C})^{1/2}(\mathbf{S} \cdot \mathbf{C}^{-1}\mathbf{S})^{1/2}\mathbf{N}\}, \tag{8.3}$$

for some scalar m. It may also be written in a more useful form in terms of the orthogonal vectors, $\mathbf{N}$ and $\mathbf{S} \times \mathbf{N}$:

$$m\mathbf{M} = \{(\mathbf{S} \times \mathbf{CN}) \cdot \mathbf{N}\}\mathbf{N} + \{\mathbf{N} \cdot \mathbf{CN} - (\det \mathbf{C})^{1/2}(\mathbf{S} \cdot \mathbf{C}^{-1}\mathbf{S})^{1/2}\}\mathbf{S} \times \mathbf{N}. \tag{8.4}$$

For given $\mathbf{S}$, (8.3) or (8.4) gives the direction of the companion $\mathbf{M}$ of $\mathbf{N}$ in $\pi(\mathbf{S})$ such that the pair of material elements along $\mathbf{M}$ and $\mathbf{N}$ is not sheared. Note that the direction of $\mathbf{M}$ is determined in terms of $\mathbf{N}, \mathbf{S}$ and $\mathbf{C}$ – that is, specification of $\mathbf{S}$ and $\mathbf{N}$ such that $\mathbf{N} \cdot \mathbf{S} = 0$, and knowledge of $\mathbf{C}$, are sufficient for the determination of $\mathbf{M}$, the companion of $\mathbf{N}$ in $\pi(\mathbf{S})$. Two special cases may be noted.

Firstly, if **S** and **N** are orthogonal and such that

$$\mathbf{S} \times \mathbf{CN} = \{(\det \mathbf{C})\mathbf{S} \cdot \mathbf{C}^{-1}\mathbf{S}\}^{1/2}\mathbf{S} \times \mathbf{N}, \qquad (8.5)$$

then Joly's formula (8.3) yields **M** = **0**, which means that the direction of **M** is arbitrary in the plane $\pi(\mathbf{S})$. Boulanger and Hayes (2000) showed that in this case, **N** is in a plane of central circular section of the **C**-ellipsoid and **S** is orthogonal to this plane. The direction of the companion of **N** in $\pi(\mathbf{S})$ such that the pair of material elements along **M** and **N** is unsheared is then arbitrary. Thus, any pair of material directions **M** and **N** in either of the planes $\Sigma^{\pm}(\mathbf{C})$ suffer no shear in the deformation. This is the result of Hayes (1988).

Secondly, from equation (8.4), it is clear that **M** is along **N** provided

$$\mathbf{N} \cdot \mathbf{CN} = (\det \mathbf{C})^{1/2}(\mathbf{S} \cdot \mathbf{C}^{-1}\mathbf{S})^{1/2}. \qquad (8.6)$$

For given **S**, this gives the directions **N** in $\pi(\mathbf{S})$ which have no companion **M** (except for **N** itself) such that the pair of material elements along **M** and **N** suffer no shear. These are the limiting directions introduced in Section 6.4.2. Also, recall (Setion 6.7) that the pair of limiting directions in a plane $\pi(\mathbf{S})$ is the pair of material directions which suffer the maximum shear in that plane.

Finally, let **M**, **N** form an unsheared pair in the plane $\pi(\mathbf{S})$. Then it may be shown that the limiting directions **L** in $\pi(\mathbf{S})$ are given by

$$\lambda_{(M)}^{1/2}\mathbf{N} \pm \lambda_{(N)}^{1/2}\mathbf{M} = \frac{\lambda_{(I)}^{1/2}\cos\Phi \pm \lambda_{(J)}^{1/2}\sin\Phi}{(\lambda_{(I)}^2\cos^2\Phi + \lambda_{(J)}^2\sin^2\Phi)^{1/4}}(\lambda_{(J)}^{1/2}\mathbf{I} \pm \lambda_{(I)}^{1/2}\mathbf{J}). \qquad (8.7)$$

Thus, knowledge of **C** and of one unsheared pair in the plane $\pi(\mathbf{S})$ gives sufficient information for the determination of $\mathbf{L}_{\pm}$, the pair which suffers the maximum shear in the plane. Knowing the angle Θ_L subtended by $\mathbf{L}_{\pm}$ means that the maximum shear $\pi - 2\Theta_L$ is known.

Finally, we consider unsheared pairs (**N**, **M**) with **N** lying either along $\mathcal{N}^+$ or $\mathcal{N}^-$. Let $\mathbf{N} = \mathcal{N}^+$, the unit normal to a plane of central circular section of the $\mathbf{C}^{-1}$-ellipsoid. Boulanger and Hayes (2000) showed that the only vectors **M** such that the material line elements along **N**, **M** form an unsheared pair are all the vectors **M** lying in the plane $\pi(\mathcal{S}^-)$ of central circular section of the **C**-ellipsoid. Similarly, if $\mathbf{N} = \mathcal{N}^-$, the only vectors **M** such that the material line elements along **N**, **M** form an unsheared pair are all the vectors **M** lying in the plane $\pi(\mathcal{S}^+)$.

Indeed, for a given **N**, all the companions **M** forming with **N** an unsheared pair are given by Joly's formula (8.3) where the unit vector **S** must be varied in the plane orthogonal to **N** : $\mathbf{S} \cdot \mathbf{N} = 0$. With $\mathbf{N} = \mathcal{N}^+$, we have $\mathbf{S} \cdot \mathbf{C}^{-1}\mathbf{S} = \lambda_2^{-2}$, and hence $(\det \mathbf{C})^{1/2}(\mathbf{S} \cdot \mathbf{C}^{-1}\mathbf{S})^{1/2} = \lambda_1\lambda_3$. Then, using (2.21) and (2.22) it follows that

$$\mathbf{C}\mathcal{N}^+ - \lambda_1\lambda_3\mathcal{N}^+ = -\frac{\lambda_1\lambda_3(\lambda_3 - \lambda_1)^{1/2}}{\lambda_2(\lambda_3 + \lambda_1)^{1/2}}\mathcal{S}^-, \qquad (8.8)$$

so that here $\mathbf{C}\mathbf{N} - (\det \mathbf{C})^{1/2}(\mathbf{S} \cdot \mathbf{C}^{-1}\mathbf{S})^{1/2}\mathbf{N}$ is along $\mathcal{S}^-$, a normal to a plane of central circular of the C-ellipsoid. Hence, varying $\mathbf{S}$ in the plane orthogonal to $\mathbf{N} = \mathcal{N}^+$, Joly's formula (8.3) yields all the vectors $\mathbf{M}$ in the plane $\pi(\mathcal{S}^-)$ orthogonal to $\mathcal{S}^-$.

6.9 Unsheared triads

In the theory of finite strain, a classical result due to Cauchy is that at every particle $\mathbf{X}$ in the body before deformation there exists a triad of mutually orthogonal infinitesimal material line elements whose members are also mutually orthogonal after deformation, at $\mathbf{x}$, the position occupied by $\mathbf{X}$ as a result of this deformation. If the three principal stretches λ_1, λ_2, λ_3 at $\mathbf{X}$ are all unequal, as is assumed throughout this paper (they are ordered $\lambda_3 > \lambda_2 > \lambda_1$), then it may be shown that there is one and only one such triad of unsheared mutually orthogonal material line elements. In fact these material line elements lie along the principal axes of the right Cauchy-Green strain tensor $\mathbf{C}$ at $\mathbf{X}$, and, in the deformed state, they lie along the principal axes of the left Cauchy-Green strain tensor $\mathbf{B}$ at $\mathbf{x}$. In other words, the members of the triad lie along the principal axes of the C-*ellipsoid* at $\mathbf{X}$, before deformation, and lie along the principal axes of the $\mathbf{B}^{-1}$-*ellipsoid* at $\mathbf{x}$, after deformation. The related general result that Cauchy proved was that orthogonal material line elements at $\mathbf{X}$ are deformed into conjugate radii of the $\mathbf{B}^{-1}$-*ellipsoid* at $\mathbf{x}$.

Boulanger and Hayes (in press) introduced the concept of unsheared (oblique) triads of material line elements at a point $\mathbf{X}$ in the material. They sought triads of material line elements which are unsheared in the deformation, that is triads such that their three mutual angles are unchanged in the deformation. For such triads the lengths of the material line elements are generally changed, but the angle between any two elements is unaltered. They found that there is an infinity of unsheared triads. Moreover, they may be constructed from unsheared pairs in the sense that if such a pair of elements is given at $\mathbf{X}$, then, in general, a unique third material line element may be found such that the three material line elements form an unsheared triad. To be more specific, if the unit vectors $\mathbf{M}$ and $\mathbf{N}$ at $\mathbf{X}$ are along the arms of an unsheared pair of material line elements, then, in general, a unique unit vector $\mathbf{P}$ (by unique, we here mean unique up to a $\pm$ sign) may be found such that the two pairs of material line elements along $(\mathbf{N}, \mathbf{P})$ and $(\mathbf{P}, \mathbf{M})$ are each also unsheared. The triad is "genuine" in the sense that $\mathbf{M}, \mathbf{N}, \mathbf{P}$ are linearly independent. Because there is an infinity of unsheared pairs in any given plane, there is, in general, a corresponding infinity of genuine unsheared triads. Here the approach adopted by Boulanger and Hayes (in press) is briefly outlined.

It is assumed that an unsheared pair of material line elements is known, that is $\mathbf{M}, \mathbf{N}$ are known. Then, a formula for a third vector vector $\mathbf{P}$ is constructed such that the pairs of material line elements along $(\mathbf{N}, \mathbf{P})$ and $(\mathbf{P}, \mathbf{M})$ are also unsheared.

If $\mathbf{M}, \mathbf{N}$ and $\mathbf{P}$ form an unsheared triad at $\mathbf{X}$, then $\mathbf{M}, \mathbf{N}$ and $\mathbf{P}$ must satisfy

$$\mathbf{M} \cdot \mathbf{CN} = \lambda_{(\mathbf{N})}\lambda_{(\mathbf{M})}\mathbf{M} \cdot \mathbf{N},$$
$$\mathbf{N} \cdot \mathbf{CP} = \lambda_{(\mathbf{N})}\lambda_{(\mathbf{P})}\mathbf{N} \cdot \mathbf{P}, \tag{9.1}$$
$$\mathbf{P} \cdot \mathbf{CM} = \lambda_{(\mathbf{P})}\lambda_{(\mathbf{M})}\mathbf{P} \cdot \mathbf{M}.$$

It may be noted that the triad $\mathbf{M}, \mathbf{N}, \mathbf{P}$ of unit vectors along three material line elements at $\mathbf{X}$ is deformed into the triad $\mathbf{m}, \mathbf{n}, \mathbf{p}$ (say) at $\mathbf{x}$, where

$$\mathbf{m} = \mathbf{FM}, \quad \mathbf{n} = \mathbf{FN}, \quad \mathbf{p} = \mathbf{FP}. \tag{9.2}$$

The volume of the tetrahedron with edges $\mathbf{M}, \mathbf{N}, \mathbf{P}$ is $V_0/6$, where V_0 is given by

$$V_0 = \mathbf{M} \times \mathbf{N} \cdot \mathbf{P}, \tag{9.3}$$

and that of the tetrahedron with edges $\mathbf{m}, \mathbf{n}, \mathbf{p}$ is $V/6$, where V is given by

$$V = \mathbf{m} \times \mathbf{n} \cdot \mathbf{p} = \mathbf{FM} \times \mathbf{FN} \cdot \mathbf{FP} = (\det \mathbf{F})V_0. \tag{9.4}$$

Now, the length of $\mathbf{m}$ is $|\mathbf{m}| = |\mathbf{FM}| = \lambda_{(\mathbf{M})}$, that of $\mathbf{n}$ is $|\mathbf{n}| = \lambda_{(\mathbf{N})}$, and that of $\mathbf{p}$ is $|\mathbf{p}| = \lambda_{(\mathbf{P})}$. Because the angle subtended by $\mathbf{M}$ and $\mathbf{N}$ is equal to the angle subtended by $\mathbf{m}$ and $\mathbf{n}$, the pairs are unsheared, and, similarly, the angles subtended by $(\mathbf{N}, \mathbf{P})$ and $(\mathbf{P}, \mathbf{M})$ are equal to angles subtended by $(\mathbf{n}, \mathbf{p})$ and $(\mathbf{p}, \mathbf{m})$ respectively. It follows that

$$\frac{\mathbf{m}}{\lambda_{(\mathbf{M})}} \times \frac{\mathbf{n}}{\lambda_{(\mathbf{N})}} \cdot \frac{\mathbf{p}}{\lambda_{(\mathbf{P})}} = \mathbf{M} \times \mathbf{N} \cdot \mathbf{P}, \tag{9.5}$$

so that

$$V = \lambda_{(\mathbf{M})}\lambda_{(\mathbf{N})}\lambda_{(\mathbf{P})}V_0 = (\det \mathbf{F})V_0. \tag{9.6}$$

Thus, for any unsheared triad $\mathbf{M}, \mathbf{N}, \mathbf{P}$, with $V_0 \neq 0$,

$$\lambda_{(\mathbf{M})}\lambda_{(\mathbf{N})}\lambda_{(\mathbf{P})} = \det \mathbf{F} = \det \mathbf{C}^{1/2}. \tag{9.7}$$

Because it has been assumed $V_0 \neq 0$, this condition on the product of the stretches along the edges of any unsheared triad is valid only for triads consisting of three linearly independent vectors. Such triads will be said to be "genuine", whilst triads consisting of three linearly dependent vectors will be said to be "coplanar" (Boulanger and Hayes in press).

Using (9.7), the conditions $(9.1)_2$ and $(9.1)_3$ may be written, for any genuine triad,

$$\mathbf{P} \cdot (\mathbf{CN} - \lambda_{(\mathbf{P})}\lambda_{(\mathbf{N})}\mathbf{N}) = \mathbf{P} \cdot (\mathbf{CN} - \frac{\det \mathbf{C}^{1/2}}{\lambda_{(\mathbf{M})}}\mathbf{N}) = 0,$$

$$\mathbf{P} \cdot (\mathbf{CM} - \lambda_{(\mathbf{P})}\lambda_{(\mathbf{M})}\mathbf{M}) = \mathbf{P} \cdot (\mathbf{CM} - \frac{\det \mathbf{C}^{1/2}}{\lambda_{(\mathbf{N})}}\mathbf{M}) = 0,$$

(9.8)

Thus,

$$p\mathbf{P} = (\mathbf{CM} - \frac{\det \mathbf{C}^{1/2}}{\lambda_{(\mathbf{N})}}\mathbf{M}) \times (\mathbf{CN} - \frac{\det \mathbf{C}^{1/2}}{\lambda_{(\mathbf{M})}}\mathbf{N}),$$

(9.9)

where p is a scalar factor such that $\mathbf{P} \cdot \mathbf{P} = 1$.

Hence, the general procedure for obtaining unsheared triads $\mathbf{M}, \mathbf{N}, \mathbf{P}$ is as follows. First choose an unsheared pair $(\mathbf{M}, \mathbf{N})$. Then $\mathbf{P}$ is determined from equation (9.9).

Thus, in general, for any chosen unsheared pair $(\mathbf{M}, \mathbf{N})$, there is a unique unit vector $\mathbf{P}$ (by "unique", is meant unique up to a $\pm$ sign) such that $(\mathbf{M}, \mathbf{N}, \mathbf{P})$ is an unsheared triad. In other words, if an unsheared pair of material line elements is given at a point $\mathbf{X}$, then, in general, a unique third material line element at $\mathbf{X}$ may be found such that the three material line elements form an unsheared triad. However, special cases may occur. The reader is referred to (Boulanger and Hayes in press) for details.

Finally, as an example, we consider the case of simple shear, given by(7.15), and present an infinite set of unsheared triads of material line elements. The tensors $\mathbf{F}$ and $\mathbf{C}$ are given by (7.16). The unit vector $\mathbf{N} = (0,0,1)$ is along an invariant direction of the deformation. Let $\pi(\mathbf{S})$, with $\mathbf{S} = (\cos\phi, \sin\phi, 0)$ be any plane passing through $\mathbf{N}$. The companion $\mathbf{M}$ of $\mathbf{N}$ in $\pi(\mathbf{S})$ which forms with $\mathbf{N}$ an unsheared pair is given by Joly's formula (8.3), and, because here $\mathbf{N}$ is an eigenvector of $\mathbf{C}$, we have $\mathbf{M} = \mathbf{S} \times \mathbf{N}$. Then, assuming $\tan\phi \neq \kappa/2$, the vector $\mathbf{P}$ completing with $\mathbf{N}$ and $\mathbf{M}$ an unsheared triad is given by (9.9), so that we obtain the infinity of unsheared triads (ϕ is arbitrary)

$$\mathbf{N} = (0,0,1), \quad \mathbf{M} = (\sin\phi, -\cos\phi, 0), \quad \mathbf{P} = (\kappa\cos\phi - \sin\phi, -\cos\phi, 0)/\lambda_{(\mathbf{M})},$$

(9.10)

with

$$\lambda_{(\mathbf{N})}^2 = 1, \quad \lambda_{(\mathbf{M})}^2 = 1 + \kappa\cos\phi(\kappa\cos\phi - 2\sin\phi), \quad \lambda_{(\mathbf{P})}^2 = 1/\lambda_{(\mathbf{M})}^2.$$

(9.11)

Further details and other examples may be found in (Boulanger and Hayes in press).

6.10 Shear of planar elements

In describing shear at a particle $\mathbf{X}$, we have examined the change, as a result of the deformation, in the angle between two infinitesimal material line elements at $\mathbf{X}$. Following (Boulanger and Hayes 2000), we consider two infinitesimal material planar elements at $\mathbf{X}$ and determine the change, as a result of the deformation, in the angle between this pair of planar elements. This change in angle is called the *planar shear* of the pair of material planar elements.

A preliminary consideration of planar shear is presented here. Further details may be found in (Boulanger and Hayes 2000). It is seen that there is a correspondence between the results for planar shear and the results obtained for the shear of line elements. The determination of the shear of a pair of material planar elements may be obtained from the results for a shear of material line elements by replacing $\mathbf{C}$ by $\mathbf{C}^{-1}$, $\mathbf{B}^{-1}$ by $\mathbf{B}$, ...

Consider the infinitesimal material planar element $d\mathbf{A}$ at $\mathbf{X}$. Let it be spanned by two infinitesimal material line elements $d\mathbf{X}^{(1)}$ and $d\mathbf{X}^{(2)}$ so that $d\mathbf{A} = d\mathbf{X}^{(1)} \times d\mathbf{X}^{(2)}$. Let $\mathbf{P}$ be the unit normal to the planar element, and hence along $d\mathbf{A}$, so that $\mathbf{P} \cdot d\mathbf{X}^{(1)} = \mathbf{P} \cdot d\mathbf{X}^{(2)} = 0$. Following the deformation, we have $P_A F_{Ai}^{-1} dx_i^{(1)} = P_A F_{Ai}^{-1} dx_i^{(2)} = 0$, where $dx^{(\alpha)} = \mathbf{F} \, d\mathbf{X}^{(\alpha)}$, $(\alpha = 1, 2)$, are the line elements into which $d\mathbf{X}^{(\alpha)}$ are deformed. Let $d\mathbf{a}$ be the planar element at $\mathbf{x}$ into which $d\mathbf{A}$ has been deformed. It has normal $\mathbf{p}$ (say), along $d\mathbf{a}$, given by

$$\mathbf{p} = \mathbf{F}^{-T}\mathbf{P}, \qquad p_i = P_A F_{Ai}^{-1}. \tag{10.1}$$

The unit vector $\hat{\mathbf{p}}$ along $\mathbf{p}$ is given by

$$\hat{\mathbf{p}} = \mathbf{F}^{-T}\mathbf{P}/\mu_{(\mathbf{P})}, \tag{10.2}$$

where, analogously to $\lambda_{(\mathbf{N})}$ defined by (2.15), $\mu_{(\mathbf{P})}$ is defined by

$$\mu_{(\mathbf{P})} = |\mathbf{F}^{-T}\mathbf{P}| = (\mathbf{P} \cdot \mathbf{C}^{-1}\mathbf{P})^{1/2} = (P_A C_{AB}^{-1} P_B)^{1/2}. \tag{10.3}$$

Similarly, that material planar element at $\mathbf{x}$ with unit normal $\hat{\mathbf{p}}$ following the deformation, had unit normal $\mathbf{P} = \mathbf{F}^T \hat{\mathbf{p}}/\mu_{(\mathbf{p})}$ prior to the deformation, where (see (2.16))

$$\mu_{(\mathbf{p})} = |\mathbf{F}^T \hat{\mathbf{p}}| = (\hat{\mathbf{p}} \cdot \mathbf{B}\hat{\mathbf{p}})^{1/2} = (\hat{p}_i B_{ij} \hat{p}_j)^{1/2}. \tag{10.4}$$

We note that (see (2.17))

$$\mu_{(\mathbf{P})} = 1/\mu_{(\mathbf{p})}. \tag{10.5}$$

We note that $L/\mu_{(\mathbf{P})}$ is the length of the radius along $\mathbf{P}$ to the $\mathbf{C}^{-1}$-ellipsoid {see (2.6)}. Also, if $\mathbf{P}$ is a unit vector normal to a planar element of infinitesimal area A, before deformation, at $\mathbf{X}$, then, after deformation, the area of this element is $A\mu_{(\mathbf{P})}\det\mathbf{F}$. Similarly, if $\hat{\mathbf{p}}$ is a unit vector normal to a planar element of infinitesimal area a, after deformation, at $\mathbf{x}$, then, before deformation, the area of this element was $a\mu_{(\mathbf{p})}\det\mathbf{F}^{-1}$. Boulanger and Hayes (2000) called $\mu_{(\mathbf{P})}\det\mathbf{F}$ the *areal stretch* of the planar element whose normal is $\mathbf{P}$.

Let $\mathbf{P}$ and $\mathbf{Q}$ be unit normals to two planar elements at $\mathbf{X}$, and Ψ the angle between them, $(0 < \Psi < \pi)$. As a result of the deformation, the unit normals are deformed into $\mathbf{p}$ and $\mathbf{q}$ and the angle becomes $\psi = \Psi + \alpha_{(\mathbf{P},\mathbf{Q})}$. In (Boulanger and Hayes 2000), $\alpha_{(\mathbf{P},\mathbf{Q})}$ is called the *planar shear* of the material planar elements with normals $\mathbf{P}$ and $\mathbf{Q}$ at $\mathbf{X}$. We have (see equation (2.19))

$$\cos\psi = \cos(\Psi + \alpha_{(\mathbf{P},\mathbf{Q})}) = \hat{\mathbf{p}} \cdot \hat{\mathbf{q}} = \mathbf{P} \cdot \mathbf{C}^{-1}\mathbf{Q}/\{\mu_{(\mathbf{P})}\mu_{(\mathbf{Q})}\}, \tag{10.6}$$

and also (see equation (2.20))

$$\cos\Psi = \mathbf{P} \cdot \mathbf{Q} = \hat{\mathbf{p}} \cdot \mathbf{B}\hat{\mathbf{q}}/\{\mu_{(\mathbf{p})}\mu_{(\mathbf{q})}\}. \tag{10.7}$$

Here we consider material planar elements intersecting along a given unit vector $\mathbf{T}$. The normals to these planar elements thus lie in the plane $\pi(\mathbf{T})$, with normal $\mathbf{T}$, at $\mathbf{X}$. The vector $\mathbf{T}$, being the intersection of material planar elements, is thus a material vector. After deformation, $\mathbf{T}$ becomes $\mathbf{t}$, given by

$$\mathbf{t} = \mathbf{FT}, \qquad t_i = \frac{\partial x_i}{\partial X_A}T_A, \tag{10.8}$$

and planar elements intersecting along $\mathbf{T}$ before deformation intersect along $\mathbf{t}$, at $\mathbf{x}$, after deformation. Thus, normals to planar elements lying in $\pi(\mathbf{T})$ before deformation, lie in $\pi(\mathbf{t})$, the plane with normal $\mathbf{t}$, at $\mathbf{x}$, after deformation.

Let $\mathbf{G}$ and $\mathbf{H}$ be unit vectors, respectively along the major and minor axes of the elliptical section by the plane $\pi(\mathbf{T})$ of the $\mathbf{C}^{-1}$-ellipsoid at $\mathbf{X}$. Here, $\mathbf{G}$ and $\mathbf{H}$ are unit normals to planar material elements at $\mathbf{X}$. In the deformed state, let the corresponding normals be along $\mathbf{g}$ and $\mathbf{h}$, respectively. These are along the principal axes of the elliptical section of the $\mathbf{B}$-ellipsoid by the plane $\pi(\mathbf{t})$ (Boulanger and Hayes 2000).

Now, let $\mathbf{P}$ be any unit normal in the plane $\pi(\mathbf{T})$ at $\mathbf{X}$:

$$\mathbf{P} = \cos\Omega\mathbf{G} + \sin\Omega\mathbf{H}, \tag{10.9}$$

where Ω is arbitrary. The planar element with unit normal $\mathbf{P}$, before deformation, is deformed into the planar element with unit normal $\hat{\mathbf{p}}$ in $\pi(\mathbf{t})$:

$$\begin{aligned}
\hat{\mathbf{p}} &= \mathbf{F}^{-1^T}\mathbf{P}/\mu_{(\mathbf{P})} = \{\cos\Omega\ \mathbf{F}^{-T}\mathbf{G} + \sin\Omega\ \mathbf{F}^{-T}\mathbf{H}\}/\mu_{(\mathbf{P})} \\
&= \{\mu_{(\mathbf{G})}\cos\Omega\hat{\mathbf{g}} + \mu_{(\mathbf{H})}\sin\Omega\hat{\mathbf{h}}\}/\mu_{(\mathbf{P})},
\end{aligned} \tag{10.10}$$

where

$$\mu^2_{(\mathbf{P})} = |\mathbf{F}^{-T}\mathbf{P}|^2 = \mathbf{P} \cdot \mathbf{C}^{-1}\mathbf{P} = \mu^2_{(\mathbf{G})}\cos^2\Omega + \mu^2_{(\mathbf{H})}\sin^2\Omega. \tag{10.11}$$

Thus, from (10.10),

$$\hat{\mathbf{p}} = \cos\omega\,\hat{\mathbf{g}} + \sin\omega\,\hat{\mathbf{h}},$$
$$\cos\omega = \{\mu_{(\mathbf{G})}/\mu_{(\mathbf{P})}\}\cos\Omega, \quad \sin\omega = \{\mu_{(\mathbf{H})}/\mu_{(\mathbf{P})}\}\sin\Omega. \tag{10.12}$$

Hence, the angle Ω, that the normal $\mathbf{P}$ makes with $\mathbf{G}$, becomes ω, the angle that the normal $\hat{\mathbf{p}}$ makes with $\hat{\mathbf{g}}$, with

$$\tan\omega = \kappa\tan\Omega, \tag{10.13}$$

where

$$\kappa = \mu_{(\mathbf{H})}/\mu_{(\mathbf{G})} = \mu_{(\mathbf{g})}/\mu_{(\mathbf{h})} \geq 1. \tag{10.14}$$

Because $\kappa \geq 1$, it follows that $\omega \geq \Omega$.

Boulanger and Hayes (2000) called equation (10.13) the *generalized Wettstein planar formula* because it is the analogue for planar shear of the *generalized Wettstein's formula* (4.14) for the shear of material line elements.

In the case of plane deformation, it follows that, because $\omega \geq \Omega$, the normals to material planar elements will tend to be aligned after deformation towards the direction of least areal stretch and thus towards the direction of greatest stretch in the plane.

Let $\mathbf{Q}$ be another arbitrary unit normal in $\pi(\mathbf{T})$, so that

$$\mathbf{Q} = \cos\Omega'\mathbf{G} + \sin\Omega'\mathbf{H}. \tag{10.15}$$

The corresponding unit normal in $\pi(\mathbf{t})$ at $\mathbf{x}$ is $\hat{\mathbf{q}}$, given by

$$\hat{\mathbf{q}} = \cos\omega'\,\hat{\mathbf{g}} + \sin\omega'\,\hat{\mathbf{h}},$$
$$\cos\omega' = \{\mu_{(\mathbf{G})}/\mu_{(\mathbf{Q})}\}\cos\Omega', \quad \sin\omega' = \{\mu_{(\mathbf{H})}/\mu_{(\mathbf{Q})}\}\sin\Omega', \tag{10.16}$$
$$\tan\omega' = \kappa\tan\Omega',$$

so that

$$\mu_{(\mathbf{P})}\mu_{(\mathbf{Q})}\sin(\omega' - \omega) = \mu_{(\mathbf{G})}\mu_{(\mathbf{H})}\sin(\Omega' - \Omega),$$

$$\mu_{(\mathbf{P})}\mu_{(\mathbf{Q})}\cos(\omega' - \omega) = \{\mu_{(\mathbf{G})}^2\cos\Omega'\cos\Omega + \mu_{(\mathbf{H})}^2\sin\Omega'\sin\Omega\}. \tag{10.17}$$

If the angle between the planar elements does not change in the deformation, then $\omega' - \omega = \Omega' - \Omega$, and hence (see (4.18) and (4.25))

$$\mu_{(\mathbf{P})}\mu_{(\mathbf{Q})} = \mu_{(\mathbf{G})}\mu_{(\mathbf{H})}, \qquad (\kappa - 1)(\kappa\tan\Omega\tan\Omega' - 1) = 0. \tag{10.18}$$

The first possibility, $\kappa = 1$, means that $\mu_{(\mathbf{G})} = \mu_{(\mathbf{H})}$, and hence $\pi(\mathbf{T})$ is a plane $\Sigma^{\pm}(\mathbf{C}^{-1})$ of a central circular section of the $\mathbf{C}^{-1}$-ellipsoid. No pair of planar elements with normals in this plane is sheared (Hayes 1988).

The second possibility gives

$$\kappa \tan\Omega \tan\Omega' = 1, \qquad \tan\omega \tan\omega' = \kappa, \qquad (10.19)$$

Equation $(10.19)_1$, gives the product of the slopes (with respect to the major axis of the section of the $\mathbf{C}^{-1}$-ellipsoid by $\pi(\mathbf{T})$) of those pairs of normals to material planar elements at $\mathbf{X}$ which are not sheared in the deformation. Equation $(10.19)_2$ gives the product of the slopes (with respect to the major axis of the section of the $\mathbf{B}$-ellipsoid by $\pi(\mathbf{t})$) of those pairs of normals to material planar elements at $\mathbf{x}$ which have not been sheared in the deformation. Equations (10.19) are the counterparts, for normals to planar elements, of equations (4.19) for tangents to line elements.

These results point to the directions of analysis of planar shear. Results for the planar shear of material planar elements run parallel to results for the shear of material line elements, by replacing $\mathbf{C}$ by $\mathbf{C}^{-1}$, $\mathbf{B}^{-1}$ by $\mathbf{B}$, λ by μ, γ by α,... Limiting directions have corresponding *limiting normals*; there is an analogue of Joly's formula for planar shear; there are pairs of unsheared planar elements and triads of unsheared planar elements... Details may be found in (Boulanger and Hayes 2000, Boulanger and Hayes in press).

References

Boulanger, Ph. and Hayes, M. 1993 *Bivectors and Waves in Mechanics and Optics.* Chapman and Hall: London.

Boulanger, Ph. and Hayes, M. 1995 The common conjugate directions of plane sections of concentric ellipsoids. *Z. angew. Math. Phys.* **46**, 356-371.

Boulanger, Ph. and Hayes, M. 2000 On finite shear. *Arch. Rat. Mech. Anal.* **151**, 125-185.

Boulanger, Ph. and Hayes, M. in press Unsheared triads and extended polar decompositions of the deformation gradient. *Int. J. Nonlinear Mech.*

Cauchy, A.-L. 1841 Mémoire sur les dilatations, les condensations et les rotations produits par un changement de forme dans un système de points matériels. *Ex. d'An. Phys. Math.* **2** (1841) = Oeuvres (2) **12**, 343-377.

Hayes, M. 1988 On strain and straining. *Arch. Rat. Mech. Anal.* **100**, 265-273.

Joly, Ch. J. 1905 *A Manual of Quaternions.* MacMillan and Co.: New-York.

Love, A. E. H. 1927 *A Treatise on the Mathematical Theory of Elasticity* (4th ed.). Cambridge University Press: Cambridge.

Ramsay, J. G. and Huber, M. I. 1987 *The Techniques of Modern Structural Geology. Vol. I.* Academic Press: New-York.

Tait, P. G. 1859 *An Elementary Treatise on Quaternions.* Clarendon Press: Oxford.

Thomas, T. Y. 1961 *Plastic Flow and Fracture in Solids.* Academic Press: New-York.

Thomson (Lord Kelvin), W. and Tait, P. G. 1879 *Treatise on Natural Philosophy.* Cambridge University Press: Cambridge.

Truesdell, C. and Toupin, R. 1960 *The Classical Field Theories*, Handbuch der Physik III/1. Springer Verlag: Berlin.

7

Elastic membranes

D.M. Haughton

Department of Mathematics
University of Glasgow, Glasgow G12 8QW, U.K.
Email: d.haughton@maths.gla.ac.uk

In this chapter we give a simple account of the theory of isotropic nonlinear elastic membranes. Firstly we look at both two-dimensional and three-dimensional theories and highlight some of the differences. A number of examples are then used to illustrate the application of various aspects of the theory. These include basic finite deformations, bifurcation problems, wrinkling, cavitation and existence problems.

7.1 Introduction

The aim of this chapter is to give a simple basic account of the theory of isotropic hyperelastic membranes and to illustrate the application of the theory through a number of examples. We do not aim to supply an exhaustive list of all relevant references, but, conversely, we give only a few selected references which should nevertheless provide a suitable starting point for a literature search.

The basic equations of motion can be formulated in two distinct ways; either by starting from the three-dimensional theory as outlined in Chapter 1 of this volume and then making assumptions and approximations appropriate to a very thin sheet; or from first principles by forming a theory of two-dimensional sheets. The former approach leads to what might be called the three-dimensional theory and can be found in Green and Adkins (1970), for example. A clear derivation of the two-dimensional theory can be found in the paper of Steigmann (1990). Since there are two different theories attempting to model the same physical entities it is natural to compare and contrast these two theories. Naghdi and Tang (1977) successfully showed that the two theories may be regarded as being equivalent, subject to certain compatibility conditions. Unfortunately, it is quite common in the literature to see the two theories being regarded as equivalent without the additional information contained in the three-dimensional theory being taken into account nor the correct comparisons being made. Here we derive both versions of the theory using the notation of Haughton and Ogden (1978) and then we compare the two theories from a constitutive point of view.

The three-dimensional membrane theory uses ordinary three-dimensional elastic constitutive equations through the usual strain-energy function. The two-dimensional theory uses a different, purely two-dimensional, strain energy. We think that if the two theories are to be equivalent for a specific problem then a thin sheet of three-dimensional material should give the same response to given loads as an equivalent two-dimensional sheet. We use this as a definition of equivalent two- and three-dimensional strain-energy functions. To show how differences in the two theories can influence the interpretation of results we look at the cavitation problem for a stretched disc from the point of view of both theories; see Haughton (1986, 1990) and Steigmann (1992). We can then explain the paradox that Steigmann (1992) discovered.

Next we look at the incremental equations as applied to membrane theory. We then consider the problem of inflating an infinitely long cylindrical membrane and show how the basic equilibrium equations can be solved to give both the trivial (cylindrical) shape and also the post-bifurcation (bulged) shape. The actual bifurcation point is supplied by the incremental equations.

Another important topic that applies to membrane theory but which has no counterpart in three-dimensional elasticity is the phenomena of wrinkling. Here we give an account of the theory derived by Pipkin (1986) and show how it can be applied to the radial deformation of a circular disc, as discussed in Haughton and McKay (1995).

An important problem within non-linear elasticity generally is the derivation of exact solutions. In the final part of this chapter we look at a method of generating exact solutions which is applicable only to membranes. We also look at a more general method which maximises the information about a known solution.

7.2 The general theory

We start by giving an account of the two different membrane theories. Firstly membrane theory derived from the three-dimensional theory of elasticity to be found in Chapter 1 and secondly the direct two-dimensional theory.

7.2.1 The three-dimensional theory

Here we just outline the main features of the theory. In particular we do not attempt to give a rigorous error analysis. For a more detailed account, including an error analysis, see Haughton and Ogden (1978). We shall consider unconstrained materials since the extension to the incompressible case is quite straightforward.

Consider a thin shell of isotropic hyperelastic material. A material point of the shell in the reference configuration has position vector

$$\mathbf{X} = \mathbf{A}(\xi^1, \xi^2) + \xi^3 \mathbf{A}_3(\xi^1, \xi^2), \qquad (2.1)$$

where (ξ^1, ξ^2, ξ^3) are curvilinear coordinates of the point and the vector $\mathbf{A}(\xi^1, \xi^2)$ is the position vector of a point on the "middle surface" of the shell. The unit vector $\mathbf{A}_3(\xi^1, \xi^2)$ is the positive normal to this middle surface and ξ^3 is the parameter measured in this direction. The shell is then defined by

$$-H/2 \le \xi^3 \le H/2, \qquad (2.2)$$

where $H(\xi^1, \xi^2)$ is the thickness of the shell in the reference configuration, together with appropriate restrictions on ξ^1 and ξ^2. The in-plane linear dimensions of the shell are assumed to be very much larger than the thickness. If R is the minimum of the principal radii of curvature of the middle surface of the shell in the reference configuration and ϵ is defined to be the ratio of the maximum value of H (over the whole surface) divided by R then we assume that

$$\epsilon = \frac{H_{\max}}{R} \ll 1. \qquad (2.3)$$

Henceforth Greek indices take the values $1, 2$ and we adopt the usual summation convention for repeated indices. From (2.1) we obtain the basis vectors

$$\mathbf{E}_\mu = \frac{\partial \mathbf{X}}{\partial \xi^\mu} = \mathbf{A}_\mu + \xi^3 \mathbf{A}_{3,\mu}, \qquad \mu = 1, 2, \qquad (2.4)$$

$$\mathbf{E}_3 = \mathbf{E}^3 = \mathbf{A}_3, \qquad (2.5)$$

where $\mathbf{A}_\mu = \mathbf{A}_{,\mu} = \partial \mathbf{A}/\partial \xi^\mu$.

To first order in ϵ we have reciprocal basis vectors

$$\mathbf{E}^\mu = \mathbf{A}^\mu - \xi^3 (\mathbf{A}^\mu . \mathbf{A}_{3,\nu}) \mathbf{A}^\nu, \qquad (2.6)$$

where $\mathbf{A}^\mu$ is the vector reciprocal to $\mathbf{A}_\mu$, i.e.

$$\mathbf{A}^\nu . \mathbf{A}_\mu = \delta^\nu_\mu. \qquad (2.7)$$

Suppose that the shell now undergoes a deformation into the current configuration. The material particles of the shell are now described by curvilinear coordinates $(\zeta^1, \zeta^2, \zeta^3)$ and the position vector of a particle is written

$$\mathbf{x} = \mathbf{a}(\zeta^1, \zeta^2) + \zeta^3 \mathbf{a}_3(\zeta^1, \zeta^2). \qquad (2.8)$$

We assume that the undeformed middle surface $\xi^3 = 0$ is mapped into the surface $\zeta^3 = 0$. This will not in general be the middle surface of the deformed shell since the deformation cannot be assumed to be symmetrical about $\xi^3 = 0$. The vector $\mathbf{a}(\zeta^1, \zeta^2)$ is then the position vector of a point on the surface into which the "middle surface" of the shell in the reference configuration deforms. The unit vector $\mathbf{a}_3(\zeta^1, \zeta^2)$ is the positive normal to this surface and ζ^3 is the parameter measured in this direction. From (2.8) we can write

$$\mathbf{e}_\mu = \frac{\partial \mathbf{x}}{\partial \zeta^\mu} = \mathbf{a}_\mu + \zeta^3 \mathbf{a}_{3,\mu}, \qquad \mu = 1, 2, \tag{2.9}$$

with

$$\mathbf{e}_3 = \mathbf{e}^3 = \mathbf{a}_3, \tag{2.10}$$

and

$$\mathbf{e}^\mu = \mathbf{a}^\mu - \zeta^3 (\mathbf{a}^\mu . \mathbf{a}_{3,\nu}) \mathbf{a}^\nu. \tag{2.11}$$

Taking $\zeta^i = \zeta^i(\xi^1, \xi^2, \xi^3)$, $(i = 1, 2, 3)$ we can expand about $\xi^3 = 0, \pm H$ and, in particular, we note that the deformed thickness of the shell h can be written

$$h = H\zeta^3{}_{,3}(\xi^1, \xi^2, 0), \tag{2.12}$$

to lowest order in ϵ, having used $\zeta^3(\xi^1, \xi^2, 0) = 0$. Also,

$$\zeta^3(\xi^1, \xi^2, \pm H/2) = \pm h/2. \tag{2.13}$$

Hence, to within the order of approximations being made, we can in fact regard (ζ^1, ζ^2) as parameters describing the middle surface of the deformed shell.

The deformation gradient, defined in Chapter 1, with (2.1) and (2.8) can now be written

$$\mathbf{F} = \frac{\partial \mathbf{x}}{\partial \mathbf{X}} = \frac{\partial}{\partial \xi^i}(\mathbf{a} + \zeta^3 \mathbf{a}_3) \otimes \mathbf{E}^i$$

$$= \frac{\partial \mathbf{a}}{\partial \mathbf{A}} + \zeta^3{}_{,3}(\xi^1, \xi^2, 0)\mathbf{a}_3 \otimes \mathbf{A}^3 + \frac{\partial \zeta^\mu}{\partial \xi^3}\mathbf{a}_\mu \otimes \mathbf{A}^3 + \frac{\partial \zeta^3}{\partial \xi^\mu}\mathbf{a}_3 \otimes \mathbf{A}^\mu, \tag{2.14}$$

where in the last equation only the leading order terms are kept.

To further the analysis we now make the assumption that there is no shear through the thickness of the shell at least on the middle surface. That is, we assume

$$\frac{\partial \zeta^\mu}{\partial \xi^3}(\xi^1, \xi^2, 0) = 0. \tag{2.15}$$

Also, from a Taylor series expansion of ζ^3 about the middle surface ($\xi^3 = 0$), we have

$$\frac{\partial \zeta^3}{\partial \xi^\mu} = \xi^3 \frac{\partial \zeta^3_{,3}}{\partial \xi^\mu}(\xi^1, \xi^2, 0), \qquad (2.16)$$

to lowest order. We assume that this term is of $O(\epsilon)$ at most, which requires that $\dfrac{\partial(h/H)}{\partial \xi^\mu}$ is of $O(1)$ or smaller. In this case, to lowest order

$$\mathbf{F} = \frac{\partial \mathbf{a}}{\partial \mathbf{A}} + \zeta^3_{,3}(\xi^1, \xi^2, 0)\mathbf{a}_3 \otimes \mathbf{A}^3. \qquad (2.17)$$

We deduce from (2.17) that $\mathbf{a}_3$ is a principal Eulerian direction and the corresponding principal stretch $\lambda_3 = \zeta^3_{,3}(\xi^1, \xi^2, 0)$.

It is often convenient to work with averaged quantities in membrane theory and we define

$$\overline{\mathbf{F}} = \frac{1}{H}\int_{-H/2}^{H/2} \mathbf{F}\,d\xi^3 = \frac{1}{h}\int_{-h/2}^{h/2}\mathbf{F}\,d\zeta^3. \qquad (2.18)$$

Generally, the averaged value of a variable is the same as the value of the variable evaluated on the middle surface of the membrane, correct to an error of order ϵ^2. See Haughton and Ogden (1978) for more details.

Provided that the material is homogeneous in the direction normal to the middle surface in the undeformed configuration we have,

$$\overline{W(\mathbf{F})} = W(\overline{\mathbf{F}}). \qquad (2.19)$$

From Chapter 1 the equilibrium equations for a three-dimensional body in the absence of body forces can be written in the form

$$\mathrm{Div}\,\mathbf{S} = \mathbf{0}, \qquad (2.20)$$

where $\mathbf{S}$ is the nominal stress tensor and Div denotes differentiation in the reference configuration. Here we shall refer $\mathbf{S}$ to reference coordinates so that the component form of $\mathbf{S}$ is

$$\mathbf{S} = S^{ij}\mathbf{E}_i \otimes \mathbf{E}_j. \qquad (2.21)$$

In component form (2.20) is then

$$S^{\mu i}_{,\mu} + S^{3i}_{,3} + S^{\mu i}\mathbf{E}^\nu.\mathbf{E}_{\mu,\nu} + S^{3i}\mathbf{E}^\nu.\mathbf{E}_{3,\nu} + S^{\mu\nu}\mathbf{E}^i.\mathbf{E}_{\nu,\mu}$$

$$+ S^{3\nu}\mathbf{E}^i.\mathbf{E}_{\nu,3} + S^{\mu 3}\mathbf{E}^i.\mathbf{E}_{3,\mu} = 0, \quad i = 1, 2, 3, \qquad (2.22)$$

where we have used

$$\mathbf{E}_{3,3} = 0, \qquad \mathbf{E}^3.\mathbf{E}_{\mu,3} = \mathbf{E}^3.\mathbf{E}_{3,\mu} = 0, \qquad (2.23)$$

which follow from (2.5), (2.1) and (2.4). Averaging (2.22) and using (2.4), (2.5) and (2.6) gives

$$\overline{S}^{\mu i}{}_{,\mu} + \overline{S}^{\mu i}\mathbf{A}^\nu.\mathbf{A}_{\mu,\nu} + \overline{S}^{3i}\mathbf{A}^\nu.\mathbf{A}_{3,\nu} + \overline{S}^{\mu\nu}\mathbf{A}^i.\mathbf{A}_{\nu,\mu} + \overline{S}^{3\nu}\mathbf{A}^i.\mathbf{A}_{3,\nu}$$

$$+\overline{S}^{\mu 3}\mathbf{A}^i.\mathbf{A}_{3,\mu} + [S^{3i}]_{-H/2}^{H/2}/H = 0, \qquad i = 1, 2, 3. \tag{2.24}$$

We now show how boundary conditions on the surfaces $\xi^3 = \pm H/2$ are incorporated. For simplicity we shall assume that we have a hydrostatic pressure loading

$$\mathbf{S}^T\mathbf{N} = -JP\mathbf{F}^{-T}\mathbf{N}, \tag{2.25}$$

where P is a hydrostatic pressure applied to the surface $\xi^3 = -H/2$ with zero pressure applied to the outer surface. On the surfaces $\xi^3 = \pm H/2$ the tangent vectors on the inner and outer major surfaces are

$$\mathbf{Y}_{,\mu} = \mathbf{E}_\mu \pm H_{,\mu}\,\mathbf{E}_3/2, \qquad \xi^3 = \pm H/2. \tag{2.26}$$

The unit normals to these surfaces are then

$$\mathbf{N} = \frac{\mathbf{Y}_{,1}\times\mathbf{Y}_{,2}}{|\mathbf{Y}_{,1}\times\mathbf{Y}_{,2}|}. \tag{2.27}$$

Using (2.26) this can be written

$$\mathbf{N} = \frac{\mathbf{E}^3 \mp H_{,\mu}\mathbf{E}^\mu/2}{(1 + H_{,\kappa}\,H_{,\nu}\,\mathbf{E}^\kappa.\mathbf{E}^\nu/4)^{1/2}}, \qquad \xi^3 = \pm H/2. \tag{2.28}$$

to lowest order. If we suppose that $H_{,\mu}$ is of order H then we can write the outward unit normal as

$$\mathbf{N} = \pm\mathbf{A}^3 - \frac{1}{2}H_{,\mu}\mathbf{A}^\mu, \qquad \xi^3 = \pm H/2. \tag{2.29}$$

It then follows that the pressure P, the components $\overline{S}^{3i}$ and the term $[S^{3i}]_{-H/2}^{H/2}$ must all be of order ϵ. See Haughton and Ogden (1978) for details. To lowest order the equilibrium equations (2.24) become

$$\overline{S}^{\mu i}{}_{,\mu} + \overline{S}^{\mu i}\mathbf{A}^\nu.\mathbf{A}_{\mu,\nu} + \overline{S}^{\mu\nu}\mathbf{A}^i.\mathbf{A}_{\nu,\mu} + \overline{S}^{\mu 3}\mathbf{A}^i.\mathbf{A}_{3,\mu}$$

$$+[S^{3i}]_{-H/2}^{H/2}/H = 0, \quad i = 1, 2, 3, \tag{2.30}$$

where

$$[S^{3i}]_{-H/2}^{H/2} = PJ\mathbf{A}^i.\mathbf{F}^{-T}\mathbf{A}_3 + H_{,\mu}\overline{S}^{\mu i}, \tag{2.31}$$

to lowest order.

It is often convenient to arrange for the basis vectors $\mathbf{a}_i$ to be orthonormal. In this case

$$\mathbf{a}_\mu = \frac{\partial \mathbf{x}}{\partial \zeta^\mu} \bigg/ \left| \frac{\partial \mathbf{x}}{\partial \zeta^\mu} \right|, \tag{2.32}$$

and we interpret $()_{,\mu}$ as $\left| \dfrac{\partial \mathbf{x}}{\partial \zeta^\mu} \right|^{-1} \dfrac{\partial ()}{\partial \zeta^\mu}$. If we use this orthonormal basis as the reference configuration then the equilibrium equations (2.30) with (2.31) for an isotropic material become

$$\overline{\sigma}_{\mu\kappa,\mu} + \overline{\sigma}_{\mu\kappa} \mathbf{a}_\nu . \mathbf{a}_{\mu,\nu} + \overline{\sigma}_{\mu\nu} \mathbf{a}_\kappa . \mathbf{a}_{\mu,\nu} + h_{,\mu} \overline{\sigma}_{\mu\kappa}/h = 0, \qquad \kappa = 1,2, \tag{2.33}$$

$$\overline{\sigma}_{\mu\nu} \mathbf{a}_3 . \mathbf{a}_{\mu,\nu} + \frac{P}{h} = 0, \tag{2.34}$$

where $\overline{\sigma}$ is the averaged Cauchy stress tensor, and we deduce that $\overline{\sigma}_{\mu 3} = 0$, $\mu = 1, 2$, from the form of $\overline{\mathbf{F}}$ in (2.17). From the assumptions made it follows that $\overline{\sigma}_{33}$ is of $O(\epsilon)$ at most and so it does not appear in the equilibrium equations (2.33) or (2.34). Effectively this can be written

$$\overline{\sigma}_{33} = 0, \tag{2.35}$$

and is often called the *membrane assumption*. We note that since $\overline{\sigma}_{33}$ is a principal stress the three dimensional theory for isotropic membranes is a plane stress theory.

Equations (2.33), (2.34) and (2.35) are then the equilibrium equations for three dimensional, isotropic, elastic membranes with hydrostatic loading. The constitutive theory for $\overline{\sigma}$ follows that given in Chapter 1 of this volume (with (2.19)) and allows us to distinguish between compressible and incompressible membranes. Further details are given in the examples which follow below.

7.2.2 The two-dimensional theory

Here we follow Steigmann (1990) but using the notation introduced above. Quantities that apply only to the two-dimensional theory are indicated by a hat. We note that the elastic sheet is unconstrained, i.e. the concepts of compressibility/incompressibility do not apply here. First we suppose that there exists a strain-energy function $\hat{W}(\hat{\mathbf{F}}, \mathbf{A})$ per unit reference area of the two-dimensional elastic sheet, where

$$\hat{\mathbf{F}} = \frac{\partial \mathbf{a}}{\partial \mathbf{A}}, \tag{2.36}$$

is the two-dimensional deformation gradient and $\mathbf{a}$, $\mathbf{A}$ are the position vectors of a material point as given in (2.1) and (2.8). We note that the sheet may be inhomogeneous but henceforth we will omit any explicit reference to position **A** in $\hat{W}$. If we consider the case of a closed homogeneous membrane subjected to an inflating pressure $\hat{P}$ then the energy can be written

$$E(\mathbf{x}) = \iint\limits_{S_0} \hat{W}(\hat{\mathbf{F}})dS - \int_{V_0}^{V} \hat{P}(v)dv, \qquad (2.37)$$

where S_0 is the surface of the sheet in the reference configuration and V_0 can be taken as the original volume enclosed by the sheet, which is arbitrary, and

$$V(\mathbf{x}) = \frac{1}{3}\iint\limits_{s} \mathbf{x}.\mathbf{n}\,ds. \qquad (2.38)$$

Here s is the surface of the sheet in the current configuration and **n** is the current unit outward normal to the sheet. In order to take the first variation of the energy we note that the two-dimensional form of Nanson's equation (Chapter 1) gives

$$\mathbf{n} = \hat{\mathbf{F}}^{-T}\mathbf{N}, \qquad (2.39)$$

where **N** is the unit normal in the reference configuration and

$$\hat{J} = \det\hat{\mathbf{F}} = \frac{ds}{dS}. \qquad (2.40)$$

Writing (2.38) in terms of the reference configuration, the first variation of the energy (2.37) is then

$$\delta E = \iint\limits_{S_0} \left\{ \mathrm{tr}\{\frac{\partial \hat{W}}{\partial \hat{\mathbf{F}}}\delta\hat{\mathbf{F}}\} - \frac{\hat{P}}{3}(\delta\hat{J}\mathbf{x}.\mathbf{n} + \hat{J}\delta\mathbf{x}.\mathbf{n} + \hat{J}\mathbf{x}.\delta\mathbf{n}) \right\} dS. \qquad (2.41)$$

Using (2.39) and integrating by parts reduces this to

$$\delta E = \iint\limits_{S_0} \left\{ \mathrm{tr}\{\frac{\partial \hat{W}}{\partial \hat{\mathbf{F}}}\delta\hat{\mathbf{F}}\} - \hat{P}\hat{J}\mathbf{n}.\delta\mathbf{x} \right\} dS. \qquad (2.42)$$

Writing $\hat{\mathbf{S}} = \partial\hat{W}/\partial\hat{\mathbf{F}}$ and integrating by parts again, this can be written

$$\delta E = -\iint\limits_{S_0} (\mathrm{Div}\,\hat{\mathbf{S}} + \hat{P}\hat{J}\mathbf{n}).\delta\mathbf{x}\,dS. \qquad (2.43)$$

The Euler-Lagrange equations are then

$$\mathrm{Div}\,\hat{\mathbf{S}} + \hat{P}\hat{J}\mathbf{a_3} = \mathbf{0}. \qquad (2.44)$$

Referring $\hat{\mathbf{S}}$ to the reference basis vectors, the component form of (2.44) is

$$\hat{S}^{\mu\kappa}{}_{,\mu} + \hat{S}^{\mu\kappa}\mathbf{A}^{\nu}.\mathbf{A}_{\mu,\nu} + \hat{S}^{\mu\nu}\mathbf{A}^{\kappa}.\mathbf{A}_{\nu,\mu} + \hat{P}\hat{J}\mathbf{A}^{\kappa}.\mathbf{a_3} = 0, \qquad \kappa = 1, 2, \quad (2.45)$$

$$\hat{S}^{\mu\nu}\mathbf{A}^3.\mathbf{A}_{\nu,\mu} + \hat{P}\hat{J}\mathbf{A}^3.\mathbf{a}_3 = 0. \qquad (2.46)$$

If we now choose the current configuration with orthonormal basis vectors (2.32) as our reference configuration the equilibrium equations (2.45) and (2.46) become

$$\hat{\sigma}_{\mu\kappa,\mu} + \hat{\sigma}_{\mu\kappa}\mathbf{a}_\nu.\mathbf{a}_{\mu,\nu} + \hat{\sigma}_{\mu\nu}\mathbf{a}_\kappa.\mathbf{a}_{\nu,\mu} = 0, \qquad \kappa = 1,2, \qquad (2.47)$$

$$\hat{\sigma}_{\mu\nu}\mathbf{a}_3.\mathbf{a}_{\nu,\mu} + \hat{P} = 0. \qquad (2.48)$$

We note that there are many circumstances when equations (2.47) and (2.48) can be extracted from the equilibrium equations for full three-dimensional elasticity by simply setting $\sigma_{i3} = 0$, $i = 1,2,3$ and $\lambda_3 = 1$ (see the final section of this chapter for an explicit example). This means that the two-dimensional theory, when looked at from a three-dimensional point of view, can be regarded as a plane stress and/or a plane strain theory. This will inevitably mean that there are solutions available to the two-dimensional theory that will have no counterpart in three-dimensional membrane theory.

7.2.3 Equivalent two and three dimensional materials

Although the two sets of equilibrium equations (2.33), (2.34) and (2.47), (2.48) are superficially the same, there are differences. In the three-dimensional case we have the additional equation (2.35) and the constitutive equations for the Cauchy stresses will contain $\lambda_3 = h/H$, the out-of-plane stretch. Naghdi and Tang (1977) have shown that if λ_3 is eliminated from the three-dimensional equations by using the membrane assumption (2.35) (or for incompressible three-dimensional membranes, by using the incompressibility condition $\lambda_1\lambda_2\lambda_3 = 1$ and (2.35) to eliminate λ_3 and the hydrostatic pressure p) then the equilibrium equations for the two theories are identical *provided* that we identify $\hat{P}$ with P/h and $\hat{\sigma}$ with $h\overline{\sigma}$. The latter identity has implications for the strain energies used in the two different theories and we shall look in more detail at this aspect.

We note that an inhomogeneous two-dimensional sheet may be equivalent to an inhomogeneous three-dimensional membrane or a homogeneous membrane with varying thickness, or a combination of the two, but we shall restrict attention to the comparison of homogeneous two-dimensional sheets with homogeneous three-dimensional membranes with uniform thickness. The two-dimensional strain-energy function, when regarded as a function of the two in-plane principal stretches λ_1 and λ_2, takes the form

$$\hat{W} = \hat{W}(\lambda_1, \lambda_2). \qquad (2.49)$$

The principal Cauchy stresses are then

$$\hat{\sigma}_{11} = \frac{\hat{W}_1}{\lambda_2}, \qquad \hat{\sigma}_{22} = \frac{\hat{W}_2}{\lambda_1}, \tag{2.50}$$

where $\hat{W}_\mu = \partial \hat{W}/\partial \lambda_\mu$. In the three-dimensional case the strain-energy

$$\overline{W} = \overline{W}(\lambda_1, \lambda_2, \lambda_3), \tag{2.51}$$

where the principal stretches are now averaged values (or equivalently, the values of the stretches on the middle surface). In terms of the strain-energy function $\overline{W}(\lambda_1, \lambda_2, \lambda_3)$ the averaged principal Cauchy stresses for an unconstrained material can be written

$$\overline{\sigma}_{ii} = \frac{\lambda_i}{J} \frac{\partial \overline{W}}{\partial \lambda_i}, \quad i = 1, 2, 3. \tag{2.52}$$

The stretch normal to the middle surface, λ_3, is given implicitly in terms of the two in-plane stretches by the membrane assumption (2.35). Using (2.52), this can be written

$$\frac{\partial \overline{W}}{\partial \lambda_3}(\lambda_1, \lambda_2, \lambda_3) = 0. \tag{2.53}$$

The principal Cauchy stresses can then be written

$$h\overline{\sigma}_{11} = H \frac{W_1^*}{\lambda_2}, \qquad h\overline{\sigma}_{22} = H \frac{W_2^*}{\lambda_1}, \tag{2.54}$$

where

$$W^*(\lambda_1, \lambda_2) = \overline{W}(\lambda_1, \lambda_2, \lambda_3(\lambda_1, \lambda_2)). \tag{2.55}$$

Hence, for a three-dimensional membrane with constant undeformed thickness, we deduce that the strain-energy functions $\overline{W}(\lambda_1, \lambda_2, \lambda_3)$ and $\hat{W}(\lambda_1, \lambda_2)$ are equivalent, in that they will lead to the same equilibrium equations. However, it should be emphasised that even in this case the extra information contained in the three-dimensional theory may lead to different conclusions. For example, the membrane assumption (2.35) may predict physically unacceptable behaviour through the thickness of the sheet. The cavitation problem considered below provides an example of this.

If we use the above as a definition of equivalent two- and three-dimensional strain energies it is very easy to calculate the two-dimensional equivalent of any three-dimensional strain-energy but very much more difficult to do the opposite. As an example; the harmonic material defined by John (1960) has the strain-energy

$$W(\lambda_1, \lambda_2, \lambda_3) = 2\mu(f(\lambda_1 + \lambda_2 + \lambda_3) - \lambda_1 \lambda_2 \lambda_3), \tag{2.56}$$

where $\mu > 0$ and the function f is arbitrary apart from the conditions $f(3) = 1$, $f'(3) = 1$ and $f''(3) > 2/3$ which ensure that the undeformed configuration is energy and stress free and that the bulk modulus is positive. To calculate the two-dimensional energy equivalent to (2.56) we first apply the membrane assumption (2.35) ($\sigma_{33} = 0$), which gives

$$f'(\lambda_1 + \lambda_2 + \lambda_3) - \lambda_1\lambda_2 = 0, \tag{2.57}$$

and we can formally solve this for λ_3 as

$$\lambda_3 = g(\lambda_1\lambda_2) - \lambda_1 - \lambda_2, \tag{2.58}$$

where g is the function inverse to f', provided that f is suitably well behaved. Substituting for λ_3 in (2.56) gives

$$\hat{W}(\lambda_1, \lambda_2) = 2\mu(F(\lambda_1\lambda_2) + \lambda_1\lambda_2(\lambda_1 + \lambda_2)), \tag{2.59}$$

where $F(x) = f(g(x)) - xg(x)$. In the literature, Li and Steigmann (1993), for example, the two-dimensional strain energy

$$W(\lambda_1, \lambda_2) = 2\mu(h(\lambda_1 + \lambda_2) - \lambda_1\lambda_2), \tag{2.60}$$

is often called the two-dimensional harmonic material. It is obvious why this name is used but a comparison of (2.60) and (2.59) shows that a sheet of this two-dimensional material will *not* behave in the same way as a thin sheet of harmonic material (2.56). In this sense the name is not well chosen.

We have seen that the three-dimensional harmonic material (2.56) is equivalent to the two-dimensional material (2.59) through a simple calculation. Since (2.60) is used in the literature it is natural to ask which three-dimensional materials are equivalent to this. A systematic derivation does not seem to be possible. As an example of one family of materials which reduce to (2.60) we have

$$W(\lambda_1, \lambda_2, \lambda_3) = 2\mu(f_1(\lambda_1 + \lambda_2 + \lambda_3) - \lambda_1\lambda_2 - \lambda_1\lambda_3 - \lambda_2\lambda_3), \tag{2.61}$$

with $f_1(3) = 3$, $f_1'(3) = 2$ and $f_1''(3) > 2/3$. This is not, of course, a three-dimension harmonic material (2.56). We return to the problem of finding three-dimensional strain-energy functions that are equivalent to given two-dimensional strain energies in the last section of this chapter.

We now look in more detail at the different interpretations of the results predicted by the different membrane theories through a non-trivial example.

7.2.4 An example: cavitation

Here we compare and contrast the results given in Haughton (1986, 1990) and
Steigmann (1992) for the existence or non-existence of cavitation in stretched
elastic sheets.

We start by considering the unconstrained three-dimensional membrane theory. We consider a uniformly thick sheet

$$0 \le R \le B, \qquad 0 \le \Theta \le 2\pi, \tag{2.62}$$

with $-H/2 \le Z \le H/2$, where (R, Θ, Z) are cylindrical coordinates. This
sheet is radially stretched into the disc

$$0 \le a \le r \le b, \qquad 0 \le \theta \le 2\pi \tag{2.63}$$

with deformed thickness $-h/2 \le z \le h/2$, where (r, θ, z) are again cylindrical
coordinates. The deformation is given by

$$r = r(R), \qquad \theta = \Theta, \tag{2.64}$$

with $h = h(R)$. Using cylindrical coordinates (r, θ, z) for $(\zeta_1, \zeta_2, \zeta_3)$ in (2.8) the
position vector of a point on the middle surface can be written

$$\mathbf{a} = r\mathbf{e}_r, \tag{2.65}$$

and (2.32) gives

$$\mathbf{a}_1 = \frac{\partial}{\partial r}(r\mathbf{e}_r) = \mathbf{e}_r, \quad \mathbf{a}_2 = \frac{\partial}{r\partial\theta}(r\mathbf{e}_r) = \mathbf{e}_\theta, \tag{2.66}$$

with similar results for the undeformed configuration. The deformation gradient (2.14) becomes

$$\begin{aligned}
\mathbf{F} &= \frac{\partial\mathbf{a}}{\partial R} \otimes \mathbf{E}_R + \frac{1}{R}\frac{\partial\mathbf{a}}{\partial\Theta} \otimes \mathbf{E}_\Theta + \lambda_3\mathbf{e}_z \otimes \mathbf{E}_Z, \\
&= \frac{dr}{dR}\mathbf{e}_r \otimes \mathbf{E}_R + \frac{r}{R}\mathbf{e}_\theta \otimes \mathbf{E}_\Theta + \lambda_3\mathbf{e}_z \otimes \mathbf{E}_Z.
\end{aligned} \tag{2.67}$$

Cylindrical coordinates are then the principal directions and the radial and
azimuthal principal stretches λ_1 and λ_2 respectively are then given by

$$\lambda_1 = \frac{dr}{dR}, \quad \lambda_2 = \frac{r}{R}. \tag{2.68}$$

The stretch through the thickness of the membrane is $\lambda_3 = h/H$. The equilibrium equations (2.33) and (2.34) reduce to the single equation

$$\frac{d\bar{\sigma}_{11}}{dr} + \frac{\bar{\sigma}_{11} - \bar{\sigma}_{22}}{r} = 0. \tag{2.69}$$

Using (2.52), the equilibrium equation (2.69) can be written

$$Rr''\overline{W}_{11} + \overline{W}_{12}(\lambda_1 - \lambda_2) + \overline{W}_1 - \overline{W}_2 = 0, \qquad (2.70)$$

where subscripts denote partial derivatives with respect to λ_i and we have used the membrane assumption (2.35), which, from (2.52), is equivalent to $\overline{W}_3 = 0$.

We shall assume simple displacement boundary conditions at the outer surface, i.e.

$$r(B) = b. \qquad (2.71)$$

On the inner surface $R = 0$ we shall either require

$$r(0) = a = 0 \qquad (2.72)$$

for the trivial solution, or that the stress vector $\mathbf{t} = \boldsymbol{\sigma}\mathbf{n}$ is zero on the newly created inner surface when $a > 0$. Since $\mathbf{n} = -\mathbf{e}_r$ on any newly created cavity we require that

$$\bar{\sigma}_{11} = 0, \quad a > 0, \quad R = 0 \qquad (2.73)$$

to ensure a traction-free cavity in the non-trivial cavitated solution.

We shall assume that the disc is composed of (compressible) Varga material with strain-energy function

$$\overline{W}(\lambda_1, \lambda_2, \lambda_3) = 2\mu(\lambda_1 + \lambda_2 + \lambda_3 - g(J)), \qquad (2.74)$$

where $\mu > 0$ is the constant ground state shear modulus and the function g is arbitrary apart from

$$g(1) = 3, \quad g'(1) = 1, \quad g''(1) < -2/3, \qquad (2.75)$$

where a prime denotes differentiation of a function with respect to its argument. These constraints are necessary and sufficient to ensure that the undeformed configuration is energy and stress free and that the bulk modulus is positive, respectively. The equilibrium equation (2.69) then becomes

$$g'(J)\{g'(J) + 2Jg''(J)\}[rr'' + \lambda_1(\lambda_1 - \lambda_2)] = 0, \qquad (2.76)$$

having used

$$\lambda_3 = Jg'(J), \qquad (2.77)$$

which follows from the membrane assumption. The first factor of (2.76) does not lead to a solution since $g'(J) > 0$ by (2.77) and the assumption of positive principal stretches. Setting the second factor to be zero gives $J = $ constant. From (2.77) we see that $\lambda_3 = $ constant and hence $\lambda_1\lambda_2 = $ constant. Now using (2.68) we have

$$r^2 - b^2 = J(R^2 - B^2), \qquad (2.78)$$

where the constant of integration can be identified as the dilatation $J = \lambda_1\lambda_2\lambda_3 = $ constant. The trivial solution is then given by

$$r^2 - b^2 = \frac{b^2(R^2 - B^2)}{B^2}. \tag{2.79}$$

For nontrivial solutions satisfying $r(B) = b$ and $r(0) = a > 0$ we have

$$r^2 - b^2 = \frac{(b^2 - a^2)(R^2 - B^2)}{B^2}. \tag{2.80}$$

In this case we now have the boundary condition (2.73) to determine the radius of the cavity (if one exists). If, in (2.76), we consider $g'(J) + 2Jg''(J) \equiv 0$ the resulting $g(J)$ are not consistent with (5.2). The third factor of (2.76) again gives (2.78).

Hence all solutions are given by (2.80) and the deformed thickness is uniform throughout the sheet. (It does, however change with the deformation b). Satisfying the non-trivial boundary condition (2.73) is now the problem. From (2.52) and (2.74), (2.73) can be written

$$\bar{\sigma}_{11} = 2\mu((\lambda_2\lambda_3)^{-1} - g'(J)) = 0 \quad \text{as} \quad R \to 0. \tag{2.81}$$

From (2.68) we see that $\lambda_2^{-1} \to 0$ as $R \to 0$ but $\lambda_3(R) = $ constant and $J(R) = $ constant so $g'(J) = $ constant. We do not have the option of setting $g'(J) = 0$ since this is prescribed by the requirement of positive principal stretches. Hence cavitating solutions for the class of materials (2.74) do not exist.

We now look at the same problem using the direct two-dimensional theory. In terms of the two-dimensional strain-energy function $\hat{W}(\lambda_1, \lambda_2)$ the equilibrium equation (2.69) can be written

$$Rr''\hat{W}_{11} + \hat{W}_{12}(\lambda_1 - \lambda_2) + \hat{W}_1 - \hat{W}_2 = 0, \tag{2.82}$$

where the principal stretches are again given by (2.68). The boundary conditions are again (2.72) and (2.73) or

$$\frac{\hat{W}_1}{\lambda_2} = 0, \qquad R = 0. \tag{2.83}$$

Following Steigmann (1992) we consider the class of materials

$$\hat{W}(\lambda_1, \lambda_2) = 2\mu(\lambda_1 + \lambda_2 + f(\lambda_1\lambda_2)), \tag{2.84}$$

where $\mu > 0$ is a positive constant of dimensions force per unit length, and the constitutive function f is arbitrary apart from the conditions

$$f(1) = -2, \qquad f'(1) = -1, \tag{2.85}$$

which are necessary and sufficient to give zero energy and stress in the undeformed configuration. The equilibrium equation (2.82) becomes

$$\lambda_2[rr'' + \lambda_1(\lambda_1 - \lambda_2)]f''(\lambda_1\lambda_2) = 0. \tag{2.86}$$

If $f'' \equiv 0$ then the equilibrium equation would be satisfied for an arbitrary deformation. Since this is not reasonable we exclude this case. Hence the solution to (2.86) is again (2.78), where the constant of integration can now be identified as the two-dimensional dilatation $\hat{J} = \lambda_1\lambda_2 = $ constant. The trivial solution is then given by (2.79).

For nontrivial solutions satisfying $r(0) = a > 0$ we again have (2.80), the radius of the cavity being determined by the boundary condition (2.73). From (2.84) we require

$$f'(\lambda_1\lambda_2) + \lambda_2^{-1} = 0, \quad R = 0. \tag{2.87}$$

Using (2.80) with (2.68) and taking the limit $R \to 0$ the surface of the cavity if traction free if and only if

$$f'(\hat{J}) = 0. \tag{2.88}$$

In the undeformed configuration $\hat{J} = 1$; then, following the trivial solution $\hat{J} = b^2/B^2 > 1$. If, for some critical value of $b^2/B^2 = \hat{J}_0$, say, the function f reaches a turning point then the cavitating solution becomes possible. The solution with a cavity will then progress according to the equation

$$\hat{J} = \frac{(b^2 - a^2)}{B^2} = \hat{J}_0 = b_{\text{crit}}^2/B^2, \tag{2.89}$$

where b_{crit} is the value of b at the turning point of f. That is

$$a^2 = b^2 - b_{\text{crit}}^2. \tag{2.90}$$

Hence the existence of cavitating solutions depends on the existence of a turning point of the constitutive function f and there appears to be no reason why a constitutive function f with a turning point could not be chosen.

Steigmann (1992) states that it is paradoxical that solutions should exist in the two-dimensional theory but not in the three-dimensional theory. However, this difference is easy to explain, it is just that the two- and three-dimensional materials that are being compared are not equivalent, as the following will show. In the above we are comparing results for the two material classes (2.74) and (2.84). As can be seen from equations (2.81) and (2.87) with (2.68) the difference in the two different results hinges on the fact that we can set $f(\lambda_1\lambda_2) = 0$ in (2.87), but the physical requirement $\lambda_3 > 0$ prevents us from considering $g'(J) = 0$ in (2.81). Further, if we start with (2.74), set $\overline{W}_3 = 0$, solve for λ_3 and substitute back in we can write (2.74) as

$$W^*(\lambda_1, \lambda_2) = 2\mu(\lambda_1 + \lambda_2 + F(\lambda_1\lambda_2)), \tag{2.91}$$

where

$$F(x) = x^{-1}(g')^{-1}(x^{-1}) - g((g')^{-1}(x^{-1})), \tag{2.92}$$

and $(g')^{-1}$ is used to denote the inverse function to g'. Comparing (2.91) with (2.84) we can identify the function f with F. Differentiating (2.92) gives the result

$$F'(x) = -x^{-2}(g')^{-1}(x^{-1}).$$

For a true comparison of the materials (2.74) and (2.84) we should restrict attention to those functions f in (2.84) which are derivable from some function g in (2.74). In this case we see from (2.77) that the function $g' : \mathbb{R}^+ \to \mathbb{R}^+$ and so $(g')^{-1} \neq 0$ purely on physical grounds looking at λ_3. Hence we should restrict attention to those functions f in (2.83) that are monotonic. In this case the two theories then give the same result, namely that cavitating solutions do not exist. Conversely, we might ask if there are any three-dimensional strain energies that reduce to (2.83) and allow $f'(x) = 0$. In this case the two theories would now agree that cavitation is possible for that class of materials.

7.3 Incremental equations

Many problems such as the investigation of infinitesimal wave propagation on finitely deformed membranes or the stability and bifurcation behaviour of deformed membranes requires the use of incremental equations, that is the equations of small deformations superposed on large deformations. The full three-dimensional theory of incremental deformations is included in Chapter 1 and in Appendix A of Chapter 10 but a special treatment is required for membranes. Here we develop the incremental equations based on the membrane equilibrium equations (2.20) with (2.21) which were derived from three-dimensional theory. The inclusion of the inertia terms for dynamic problems is straightforward. See Haughton and Ogden (1978, 1980) for more details. Since there are important differences we shall treat the incompressible and compressible cases separately. We suppose that the position vector $\mathbf{x}$ of a point of the membrane in the finitely deformed configuration is perturbed by an amount $\dot{\mathbf{x}}$ and similarly for all other variables. Taking the increment of the equilibrium equations (2.20) gives

$$\mathrm{Div}\,\dot{\mathbf{S}} = \mathbf{0}. \tag{3.1}$$

If we choose the finitely deformed configuration as our reference configuration, (3.1) becomes div $\Sigma = 0$, where the incremental stress tensor Σ is defined by equation (5.4) in Chapter 1. From equation (5.6) in Chapter 1, we have

$$\Sigma = \mathcal{A}_0^1 \Gamma, \qquad \Sigma = \mathcal{A}_0^1 \Gamma + p\Gamma - \dot{p}I, \tag{3.2}$$

for compressible and incompressible materials respectively, where p is the hydrostatic pressure, Γ is the displacement gradient defined by equation (5.2) in Chapter 1, $\mathcal{A}_0^1$ is the fourth-order tensor of instantaneous moduli and I is the identity tensor. The components of $\mathcal{A}_0^1$ can be found in Chadwick and Ogden (1971) for both compressible and incompressible materials. See also Ogden (1997). With respect to the orthonormal basis (2.32) we have

$$\left.\begin{array}{c} J\mathcal{A}_{0ijij}^1 = \lambda_i^2 \dfrac{\lambda_i W_i - \lambda_j W_j}{\lambda_i^2 - \lambda_j^2}\,, \qquad \lambda_i \neq \lambda_j\,, \\[2mm] J\mathcal{A}_{0iijj}^1 = J\mathcal{A}_{0jjii}^1 = \lambda_i \lambda_j \dfrac{\partial^2 W}{\partial \lambda_i \partial \lambda_j}\,, \\[2mm] J\mathcal{A}_{0ijij}^1 - J\mathcal{A}_{0ijji}^1 = J\mathcal{A}_{0ijij}^1 - J\mathcal{A}_{0jiij}^1 = \lambda_i W_i\,, \quad i \neq j\,, \end{array}\right\} \tag{3.3}$$

which are valid for both compressible and incompressible materials (we simply set $J = 1$ for incompressible materials).

If we use the orthonormal basis defined by (2.32), then $\Sigma = \Sigma_{ij} \mathbf{a}_i \otimes \mathbf{a}_j$, and div $\Sigma = 0$ yields

$$\Sigma_{ji,j} + \Sigma_{ji}\mathbf{a}_k.\mathbf{a}_{j,k} + \Sigma_{kj}\mathbf{a}_i.\mathbf{a}_{j,k} = 0, \qquad i = 1,2,3, \tag{3.4}$$

where $(),_j = \dfrac{1}{|\mathbf{e}_j|} \dfrac{\partial()}{\partial \zeta^j}$, no sum.

From (2.18) it is easily shown that $\overline{\mathbf{F}} = \dot{\mathbf{F}}$ and similarly we can show that $\overline{\dot{\mathbf{S}}} = \dot{\overline{\mathbf{S}}}$ and $(\overline{\dot{\mathbf{S}}})_{,\mu} = \dot{\overline{\mathbf{S}}}_{,\mu}$ to within an error of order ϵ^2. Similar results hold for the components of these tensors on any basis. Further, we can show that

$$\overline{\Sigma} = \overline{\mathcal{A}}_0^1 \overline{\Gamma}, \qquad \text{or} \qquad \overline{\Sigma} = \overline{\mathcal{A}}_0^1 \overline{\Gamma} + \overline{p}\,\overline{\Gamma} - \dot{\overline{p}}I, \tag{3.5}$$

where $\overline{\mathcal{A}}_0^1$ is the average of $\mathcal{A}_0^1$. From (2.17) we have

$$\overline{\Gamma} = \overline{\dot{\mathbf{F}}\mathbf{F}^{-1}} = \frac{\partial \dot{\mathbf{a}}}{\partial \mathbf{a}} + \overline{\dot{\lambda}}_3 \mathbf{a}_3 \otimes \mathbf{a}_3 + \dot{\mathbf{a}}_3 \otimes \mathbf{a}_3. \tag{3.6}$$

We also have

$$\overline{\Gamma}_{3\mu} = \mathbf{a}_3.\dot{\mathbf{a}}_\mu = -\mathbf{a}_\mu.\dot{\mathbf{a}}_3 = -\overline{\Gamma}_{\mu 3}, \tag{3.7}$$

which gives $\overline{\Gamma}_{\mu 3}$, and

$$\overline{\Gamma}_{33} = \dot{\overline{\lambda}}_3. \tag{3.8}$$

For incompressible materials we have

$$\text{tr}(\overline{\Gamma}) = 0, \tag{3.9}$$

and so

$$\overline{\Gamma}_{33} = -\overline{\Gamma}_{11} - \overline{\Gamma}_{22}, \tag{3.10}$$

which determines $\overline{\Gamma}_{33}$. From (2.31) we obtain

$$[\Sigma_{3i}]_{-h/2}^{h/2} = \{\text{tr}(\overline{\Gamma}) + \dot{P}\}\delta_{i3} - P\overline{\Gamma}_{3i} + h_{,\mu}\,\overline{\Sigma}_{\mu i}. \tag{3.11}$$

Taking the average of (3.4) then gives

$$\overline{\Sigma}_{\mu i,\mu} + \overline{\Sigma}_{\mu i}\mathbf{a}_{\nu}.\mathbf{a}_{\mu,\nu} + \overline{\Sigma}_{\mu\nu}\mathbf{a}_i.\mathbf{a}_{\nu,\mu} + \overline{\Sigma}_{\mu 3}\mathbf{a}_i.\mathbf{a}_{3,\mu}$$
$$+ \{[\text{tr}(\overline{\Gamma}) + \dot{P}]\delta_{i3} - P\overline{\Gamma}_{3i} + h_{,\mu}\overline{\Sigma}_{\mu i}\}/h = 0, \quad i = 1, 2, 3, \tag{3.12}$$

with the obvious reductions for an incompressible material. We note that the symmetries in the fourth-order moduli combined with (3.8) and the membrane assumption ($\overline{\sigma}_{33} = 0$) leads to

$$\overline{\Sigma}_{3i} = 0, \quad i = 1, 2, 3, \qquad \overline{\Sigma}_{\mu 3} = \overline{\sigma}_{\mu\mu}\overline{\Gamma}_{3\mu}, \quad \mu = 1, 2. \tag{3.13}$$

7.3.1 Inflating a cylindrical membrane

To demonstrate the use of the incremental equations to determine bifurcation points we consider the problem of inflating an infinitely long cylinder composed of incompressible material. As the inflating pressure is increased the pressure may reach a bifurcation point and a local maximum point (depending on the constitutive equation of the material). This problem also provides a further example of how to apply the basic equilibrium equations to find both the trivial solution (a larger right circular cylinder) and the post bifurcation shape (a cylinder with an axi-symmetric bulge).

In the reference configuration we use cylindrical coordinates (R, Θ, Z) to describe a point on the middle surface of the membrane, where $0 \le \Theta \le 2\pi$, $-\infty < Z < \infty$ and the uniform thickness is taken to be H. We inflate the cylinder with a pressure P and adjust end loads to ensure that the length of the cylinder remains unchanged. (When the deformed shape is non-cylindrical we take the length of sections well away from the origin to remain unchanged.) The deformed shape of the cylinder is again described by cylindrical coordinates (r, θ, z), where $0 \le \theta \le 2\pi$, $-\infty < z < \infty$ and the non-uniform deformed thickness is taken to be h. This choice of coordinate system is adequate for the problem under consideration but the resulting differential equations are not suitable to describe (numerically) sections of the tube with parallel sides. (Here we shall integrate the equations analytically.) Kydoniefs and Spencer (1969) introduced a more complicated coordinate system for axisymmetric membrane problems but their approach suffers from the same problem when the sides are parallel.

We shall use the three-dimensional membrane theory described above but we will omit overbars for the averaged values.

The deformed membrane is assumed to be axisymmetric and we write

$$r = r(Z), \quad \theta = \Theta, \quad z = z(Z). \tag{3.14}$$

The position vectors of material points on the middle surface before and after deformation are

$$\mathbf{A}(\Theta, Z) = R\mathbf{E}_R + Z\mathbf{E}_Z, \quad \mathbf{a}(\theta, z) = r(Z)\mathbf{e}_r + z(Z)\mathbf{e}_z. \tag{3.15}$$

Using subscripts $(1, 2, 3)$ for the (θ, z, r) directions respectively we define the orthonormal basis vectors $\mathbf{A}_i$ and $\mathbf{a}_i$ to be

$$\mathbf{A}_1 = \mathbf{E}_\Theta, \quad \mathbf{A}_2 = \mathbf{E}_Z, \quad \mathbf{A}_3 = \mathbf{E}_R, \tag{3.16}$$

and

$$\mathbf{a}_1 = \mathbf{e}_\theta, \quad \mathbf{a}_2 = \frac{r'\mathbf{e}_r + z'\mathbf{e}_z}{\sqrt{(r'^2 + z'^2)}}, \quad \mathbf{a}_3 = \frac{z'\mathbf{e}_r - r'\mathbf{e}_z}{\sqrt{(r'^2 + z'^2)}}, \tag{3.17}$$

where the prime denotes differentiation with respect to Z. The deformation gradient $\mathbf{F}$, given by (2.17), can therefore be calculated with respect to the base vectors defined in (3.16) and (3.17) as

$$\mathbf{F} = \begin{bmatrix} r/R & 0 & 0 \\ 0 & (r'^2 + z'^2)^{\frac{1}{2}} & 0 \\ 0 & 0 & \lambda_3 \end{bmatrix}. \tag{3.18}$$

The principal stretches are then

$$\lambda_1 = \frac{r}{R}, \quad \lambda_2 = (r'^2 + z'^2)^{\frac{1}{2}}, \quad \lambda_3 = \frac{h}{H}. \tag{3.19}$$

From (3.17) the non-zero components of $\mathbf{a}_i.\mathbf{a}_{j,k}$ are

$$\mathbf{a}_2.\mathbf{a}_{1,1} = -\mathbf{a}_1.\mathbf{a}_{2,1} = \frac{-r'}{r\lambda_2}, \quad \mathbf{a}_3.\mathbf{a}_{1,1} = \frac{-z'}{r\lambda_2}, \quad \mathbf{a}_3.\mathbf{a}_{2,2} = \frac{r''z' - r'z''}{\lambda_2^3}. \tag{3.20}$$

Substituting into (2.33) and (2.34) the equilibrium equations are

$$\frac{1}{h}\frac{d(h\sigma_{22})}{dZ} + \frac{r'}{r}(\sigma_{22} - \sigma_{11}) = 0, \tag{3.21}$$

$$\frac{\sigma_{22}(r''z' - r'z'')}{(r'^2 + z'^2)^{\frac{3}{2}}} - \frac{z'\sigma_{11}}{r(r'^2 + z'^2)^{\frac{1}{2}}} + \frac{P}{h} = 0, \tag{3.22}$$

where the third equation is trivially satisfied. The principal Cauchy stresses σ_{ii} for an incompressible material can be written

$$\sigma_{ii} = \lambda_i W_i - p, \quad i = 1, 2, 3. \tag{3.23}$$

By using the incompressibility constraint $\lambda_1 \lambda_2 \lambda_3 = 1$ and the membrane assumption $\sigma_{33} = 0$ we can introduce a function $\hat{W}(\lambda_1, \lambda_2) = W(\lambda_1, \lambda_2, (\lambda_1 \lambda_2)^{-1})$ and write

$$\sigma_{\mu\mu} = \lambda_\mu \hat{W}_\mu, \quad \mu = 1, 2. \tag{3.24}$$

Substituting the above into the equilibrium equations (3.21) and (3.22) we have

$$\frac{d\hat{W}_2}{dZ} - \frac{r'\hat{W}_1}{R\lambda_2} = 0, \tag{3.25}$$

and

$$\frac{(r''z' - r'z'')\hat{W}_2}{\lambda_2^2} - \frac{z'\lambda_1 \hat{W}_1}{r\lambda_2} + \frac{P\lambda_1 \lambda_2}{H} = 0. \tag{3.26}$$

Pipkin (1968) pointed out that the equilibrium equations for an axisymmetric deformation of an axisymmetric membrane can always be integrated at least once. With $(3.19)_2$, (3.25) can be seen to be a perfect differential which is integrated to give

$$\hat{W} - \lambda_2 \hat{W}_2 = C_1, \tag{3.27}$$

where C_1 is the constant of integration. It turns out that, for this problem, we may also integrate the second equation. Substituting $\hat{W}_1$ from (3.25) into (3.26) we have

$$\frac{(r''z' - r'z'')\hat{W}_2}{\lambda_2^2} - \frac{z'}{r'} \frac{d\hat{W}_1}{dZ} + \frac{P\lambda_1 \lambda_2}{H} = 0, \tag{3.28}$$

which integrates to

$$\frac{\hat{W}_2 z'}{\lambda_2} - \frac{P\lambda_1^2 R}{2H} = C_2, \tag{3.29}$$

where C_2 is a second constant of integration.

To complete the formulation of the problem we need to specify appropriate boundary conditions. First we consider the case where the inflated cylinder is a right circular cylinder. We then have $r = \lambda R$, where λ is constant, and $z = \lambda_z Z$, where λ_z is the constant axial stretch. For convenience we shall take $\lambda_z = 1$. The equilibrium equations (3.25) and (3.26) reduce to the single equation

$$\frac{PR}{H} = \frac{\hat{W}_1}{\lambda_1},\qquad(3.30)$$

which determines the inflated radius (λ_2) for a given strain energy and pressure. We denote this inflated radius by r_∞ and we use this to formulate the boundary conditions when the deformed shape of the cylinder is non-trivial. For all deformations we suppose that there is symmetry about the origin so that, without loss of generality,

$$r'(0) = 0, \quad z(0) = 0,\qquad(3.31)$$

together with

$$r \to r_\infty, \quad z' \to 1 \quad \text{as} \quad Z \to \pm\infty.\qquad(3.32)$$

Due to the assumed symmetry we need only consider $Z \geq 0$.

The solution procedure for the non-trivial state is now quite straightforward. First the variables are non-dimensionalised with respect to the undeformed radius R and shear modulus, as appropriate. Now choose the pressure P and apply the conditions at infinity (3.32) to determine C_1 and C_2, recalling (3.19). Now we guess a value of $r(0)$ and, using (3.19), (3.27) and (3.29), we can find the corresponding value of $r'(0)$. Our estimated value of $r(0)$ is corrected to make $r'(0) = 0$. Once we have the correct value of $r(0)$ we can use this with $z(0) = 0$ as initial conditions to numerically integrate (3.27) and (3.29) to obtain $r(Z)$ and $z(Z)$. To give an explicit calculation we have used the incompressible Varga strain-energy function

$$W(\lambda_1, \lambda_2, \lambda_3) = 2\mu(\lambda_1 + \lambda_2 + \lambda_3 - 3).\qquad(3.33)$$

This gives

$$\hat{W}(\lambda_1, \lambda_2) = 2\mu(\lambda_1 + \lambda_2 + (\lambda_1\lambda_2)^{-1} - 3).\qquad(3.34)$$

Using the above, a plot of the deformed shape of the cylinder is given in Figure 1. We plot the radius $r(Z)$ against the deformed axial coordinate $z(Z)$ for cylinders inflated to give $r_\infty = 1.25$, 1.35 and 1.45. To give a clearer picture of the post bifurcation behaviour we plot in Figure 2 the difference of the maximum and minimum radii $(r(0) - r_\infty)$ against the radius at infinity r_∞. Surprisingly, we see that this material gives a sub-critical post bifurcation curve. A simple calculation shows that $\lambda_1 \to \infty$ when $r(0) - r_\infty \to 2/r_\infty$ and so the post-bifurcation curve does in fact terminate on the left (at $r_\infty = r* \simeq 1.22$). Thus there is a range of values $r^* \leq r_\infty \leq r_c$, where r_c is the critical value of r_∞, for which two solutions exist.

D.M. Haughton

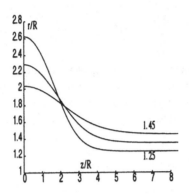

Fig. 1. Plot of the post bifurcation shape r against z for $r_\infty = 1.25$, 1.35, 1.45.

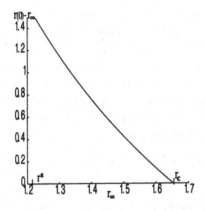

Fig. 2. Plot of the post bifurcation curve $|r_{max} - r_{min}|$ against r_∞.

The strain-energy function (3.33) was chosen for its mathematical simplicity and so we should not expect the results predicted to correspond with what might be observed for an actual cylindrical membrane. Nevertheless, it is interesting to speculate what the results shown in Figures 1 and 2 might mean. Since we do observe bulges in inflated cylindrical membranes Figure 2 suggests that for the Varga material (3.33) the trivial (cylindrical) solution is unstable for $r* < r_\infty < r_c$, and so, as the cylinder is inflated we reach a value of

$r_\infty = r*$ when a large bulge would suddenly form. As we continue inflating r_∞ gets larger and the size of the bulge diminishes (see Figure 1) until we arrive at $r_\infty = r_c$ when we regain the cylindrical shape. Any further inflation retains the cylindrical shape.

To determine the bifurcation point $(r_\infty = r_c)$ explicitly we now turn to the incremental equations. The bifurcation problem for inflated cylinders has been considered in some detail by several authors. See Haughton and Ogden (1979) and the references therein.

In this case we are concerned with the cylindrical shape of the membrane and so we take cylindrical coordinates. The principal stretches are now given by

$$\lambda_1 = \frac{r}{R}, \quad \lambda_2 = 1, \quad \lambda_3 = \frac{h}{H}, \tag{3.35}$$

where we recall that the $(1, 2, 3)$ directions correspond to (θ, z, r) respectively. The incremental displacement can then be written

$$\mathbf{u} = v\mathbf{a}_1 + w\mathbf{a}_2 + u\mathbf{a}_3, \tag{3.36}$$

where

$$\mathbf{a}_1 = \mathbf{e}_\theta, \quad \mathbf{a}_2 = \mathbf{e}_z, \quad \mathbf{a}_3 = \mathbf{e}_r. \tag{3.37}$$

The incremental displacements u, v and w are, in general, functions of θ and z. However, in this example we are looking for an axisymmetric bulging mode only. This suggests that we restrict attention to the case where u, v and w are functions of z only. Further, the displacement $v(z)$ will only add a torsional displacement which does not seem to be appropriate to a bulging mode so we will take $v = 0$.

Using (3.7) and (3.10) the displacement gradient referred to the basis (3.37) is

$$\overline{\Gamma} = \begin{bmatrix} \dfrac{u}{r} & 0 & 0 \\ 0 & w_z & -u_z \\ 0 & u_z & -\left(\dfrac{u}{r} + w_z\right) \end{bmatrix}. \tag{3.38}$$

The inflating pressure P is given by (3.30) and so

$$\frac{\dot{P}r^2}{h} = u\left(\frac{\hat{W}_{11}}{\lambda_1} - \frac{\hat{W}_1}{\lambda_1^2}\right), \tag{3.39}$$

and this equation governs turning points of the pressure. For the Varga material (3.33) with $\lambda_2 = 1$ (since we are considering the perfectly cylindrical configuration of the shell) we see that the pressure reaches a local maximum when $\lambda_1 = 3$. If we refer to Figure 2 we see that the bifurcation point occurs well before this value. Using (3.36) the only non-zero components of $\mathbf{a}_i.\mathbf{a}_{j,k}$ are

$$\mathbf{a}_1.\mathbf{a}_{3,1} = -\mathbf{a}_3.\mathbf{a}_{1,1} = \frac{1}{r}. \tag{3.40}$$

Substituting (3.38), (3.39) and (3.40) into the incremental equations (3.12) with $(3.5)_2$ and using (3.3) together with $\lambda_2 = 1$ we get

$$\xi_1 u_z + r\xi_2 w_{zz} = 0, \tag{3.41}$$

$$\xi_3 u - r^2\xi_4 u_{zz} + r\xi_1 w_z = 0, \tag{3.42}$$

having eliminated $\dot{p}$ by using $\dot{\bar{s}}_{033} = 0$, see (3.7), where

$$\begin{aligned} \xi_1 &= \lambda_1(\hat{W}_{12} - \hat{W}_1), & \xi_2 &= \hat{W}_{22}, \\ \xi_3 &= \lambda_1^2\hat{W}_{11} + \lambda_1\hat{W}_1, & \xi_4 &= \hat{W}_2. \end{aligned} \tag{3.43}$$

Since the membrane will be in tension we note that $\xi_4 > 0$. We can integrate these equations and write the solution as

$$\begin{aligned} u &= C_1\cos\beta z + C_2\sin\beta z + C_3, \\ w &= \frac{\xi_1}{r\beta\xi_2}\{-C_1\sin\beta z + C_2\cos\beta z\} - \frac{\xi_3 C_3 z}{r\xi_1} + C_4, \end{aligned} \tag{3.44}$$

where C_i, $i = 1,..,4$ are constants and

$$\beta^2 = \frac{\xi_1^2 - \xi_2\xi_3}{r^2\xi_2\xi_4}. \tag{3.45}$$

We have assumed that $\xi_1 \neq 0$ and $\xi_2 \neq 0$ (the cases $\xi_1 = 0$, $\xi_2 = 0$ can be treated separately but do not lead to anything substantial). We would now like to apply boundary conditions on u and w and the criterion for the existence of non-trivial solutions would then be the bifurcation criterion. Unfortunately, taking an infinitely long cylinder directly creates problems. Since the length of the cylinder has not yet been used in the (cylindrical) inflation or incremental equations we use an indirect route, taking a cylinder of length ℓ, finding the bifurcation criterion in this case and letting $\ell \to \infty$. Taking $0 \leq z \leq \ell$ we impose the fixed ends conditions

$$u = w = 0, \quad z = 0, \ell. \tag{3.46}$$

In this case there are non-trivial solutions to (3.44) provided that

$$\sin y\left[\frac{\xi_1}{r\xi_2}\sin y - \frac{\xi_3 y}{r\xi_1}\cos y\right] = 0, \tag{3.47}$$

where

$$y = \frac{\beta \ell}{2}. \tag{3.48}$$

The first factor has solution

$$\beta = \frac{2n\pi}{\ell}, \tag{3.49}$$

and in the limit as $\ell \to \infty$ our bifurcation criterion is

$$\xi_1^2 - \xi_2 \xi_3 = 0. \tag{3.50}$$

Using (3.33) and (3.43) this becomes

$$\left(\frac{2}{\lambda_1} - \lambda_1 \right)^2 - \frac{2}{\lambda_1} \left(\frac{3}{\lambda_1} - \lambda_1 \right) = 0, \tag{3.51}$$

or

$$\lambda_1 = (1 + \sqrt{3})^{\frac{1}{2}} \simeq 1.6529. \tag{3.52}$$

This value is then the intercept on the axis in Figure 1. We note that the second factor in (3.47) also has $y = 0$ as a solution which again leads to (3.49) and (3.52).

7.4 Wrinkling theory

The membrane theory given above is a tension theory. If either of the two principal in-plane stresses were to become negative we would expect the results predicted by the membrane theory to be incorrect. This is because membrane theory requires a membrane to have no bending stiffness and so a compressive in-plane load should result in the buckling of the shell. In practice this buckling will take on the appearance of a wrinkled region. There have been several attempts to model the compressive behaviour of membranes, but it was Pipkin (1986) who formulated a workable theory.

Tension field theory assumes that no energy is required to fold the membrane and therefore the total strain-energy of the membrane can be minimised when in such a state. Minimum-energy (equilibrium) states can involve a continuous distribution of infinitesimal wrinkles and the stress distribution associated with this wrinkled region is known as the tension field. Tension field theory does not provide details about the local distribution of wrinkles but does give more realistic global results than ordinary membrane theory.

Here we shall restrict attention to the basic method and show how to use the theory. A wrinkled region forms when one of the two in-plane principal stresses become negative. Ordinary membrane theory is no longer applicable and we assume that this compressive stress is in fact zero throughout the wrinkled region. Pipkin (1986) showed how the strain-energy function of the material can be replaced by a relaxed strain-energy function which satisfies all the required convexity conditions and ensures that the in-plane principal stresses are non-negative. Using this, together with the membrane assumption (2.35), we have only one in-plane stress non-zero. When this occurs the membrane is defined as being in a state of simple tension. The idea is to modify the constitutive equation of the material so that the membrane remains in a state of simple tension throughout the wrinkled region. This is achieved as follows.

Firstly, if we have a three-dimensional membrane, which may be compressible or incompressible, with averaged strain-energy function $\overline{W}(\lambda_1, \lambda_2, \lambda_3)$, the membrane assumption (2.35) allows us to define a new strain energy $W(\lambda_1, \lambda_2)$ obtained from $\overline{W}$ by replacing λ_3 as a function of λ_1 and λ_2. We can then write

$$\sigma_{11} = \frac{\lambda_1 W_1(\lambda_1, \lambda_2)}{J}, \qquad \sigma_{22} = \frac{\lambda_2 W_2(\lambda_1, \lambda_2)}{J}. \tag{4.1}$$

Here W may be interpreted as either a two-dimensional energy or either an incompressible or compressible three-dimensional energy provided that we take $J = \lambda_1 \lambda_2$, $J = 1$ or $J = \lambda_1 \lambda_2 \lambda_3$ respectively. For simple tension in the 1-direction we then require $\sigma_{22} = 0$. From (4.1) this can be written

$$W_2(\lambda_1, \lambda_2) = 0, \tag{4.2}$$

assuming positive principal stretches. Equation (4.2) gives the "natural width" of the membrane. That is, λ_2 as a function of the prescribed stretch λ_1. To emphasise this we will follow Pipkin (1986) and write $\lambda_2 = n(\lambda_1)$.

We define the relaxed strain energy

$$w(\lambda_1) = W(\lambda_1, n(\lambda_1)). \tag{4.3}$$

For the whole range of possible strains we then have

$$W = \begin{cases} W(\lambda_1, \lambda_2) & \text{if } \lambda_1 \geq n(\lambda_2) \text{ and } \lambda_2 \geq n(\lambda_1), \\ w(\lambda_1) & \text{if } \lambda_1 \geq 1 \text{ and } \lambda_2 \leq n(\lambda_1), \\ w(\lambda_2) & \text{if } \lambda_1 \leq n(\lambda_2) \text{ and } \lambda_2 \geq 1, \\ 0 & \text{if } \lambda_1 \leq 1 \text{ and } \lambda_2 \leq 1. \end{cases} \tag{4.4}$$

The transition from one region to another is continuous although not necessarily smooth.

7.4.1 Wrinkling of an annular disc

Here we consider the problem of an isotropic elastic membrane annulus of uniform thickness which is subjected to a displacement on the inner boundary while the outer boundary is held fixed. Related problems were originally considered by Rivlin and Thomas (1951) and have been studied by many other authors since, but the problems have always been formulated to avoid the possibility of wrinkling. Haughton and Mckay (1995) give further references and studied such problems specifically from the wrinkling point of view. This is one of the problems considered in that paper.

We shall consider a uniform circular membrane which occupies the region

$$0 < A \leq R \leq B, \quad 0 \leq \Theta \leq 2\pi \qquad (4.5)$$

in the undeformed configuration of the body. If we regard the membrane from the three-dimensional theory we shall assume that the disc has uniform thickness $-H/2 \leq Z \leq H/2$, where (R, Θ, Z) are cylindrical coordinates. The membrane is then deformed symmetrically to occupy the region

$$0 \leq a \leq r \leq b, \quad 0 \leq \theta \leq 2\pi, \quad \frac{-h}{2} \leq z \leq \frac{h}{2}, \qquad (4.6)$$

where (r, θ, z) are cylindrical coordinates with $r = r(R)$, $\theta = \Theta$ and $h = h(R)$. We shall take the outer boundary to be fixed so that

$$b = B. \qquad (4.7)$$

The deformation is then identical to that for the cavitation problem considered above (with different boundary conditions). This allows us to read off many results from those already given. The principal stretches are given by (2.68) and the equilibrium equation is again (2.69) which can also be written as (2.70).

Here we shall consider the (incompressible) Varga material (3.33). In this case

$$W(\lambda_1, \lambda_2) = 2\mu(\lambda_1 + \lambda_2 + (\lambda_1 \lambda_2)^{-1} - 3). \qquad (4.8)$$

Substituting (4.8) into (2.70) the equilibrium equation becomes

$$Rrr'' + Rr'^2 - rr' = 0, \qquad (4.9)$$

which has the solution (2.78)

$$r^2 - B^2 = C(R^2 - B^2), \qquad (4.10)$$

where C is a constant. Taking $r(A) = a$ as the second displacement boundary condition we have

$$C = \frac{B^2 - a^2}{B^2 - A^2}.$$ (4.11)

From (4.1) and (4.8) the principal Cauchy stresses are

$$\sigma_{11} = 2\mu\lambda_1 \left(1 - \frac{1}{\lambda_1{}^2\lambda_2}\right) = 2\mu\left(\frac{CR}{r} - \frac{1}{C}\right),$$ (4.12)

$$\sigma_{22} = 2\mu\lambda_2 \left(1 - \frac{1}{\lambda_1\lambda_2{}^2}\right) = 2\mu\left(\frac{r}{R} - \frac{1}{C}\right).$$ (4.13)

The physics of the deformation suggest that σ_{11} will always be positive and so the existence of a wrinkled region, if any, will be governed by the solutions to the equations $\sigma_{22} = 0$.

We define the critical radius, R^* (or r^*), to be the radius at which the wrinkled and non-wrinkled regions meet and hence where $\sigma_{22}(R^*) = 0$. Using (4.13), (4.10) and (4.11) we can write

$$R^{*2} = \frac{C^2 B^2}{C^2 + C + 1} = \frac{B^2(B^2 - a^2)^2}{(3B^4 - 3B^2a^2 - 3B^2A^2 + a^2A^2 + a^4 + A^4)}.$$ (4.14)

In particular, if we take the limiting value $a = A$ then $R^* = B/\sqrt{3}$. This means that the region $A \leq R < B/\sqrt{3}$ will immediately wrinkle as the inner boundary is moved inwards. Hence there will always be some wrinkling if $B/A > \sqrt{3}$. If $B/A < \sqrt{3}$ there may still be some wrinkling if a is made small enough.

For the incompressible Varga material (3.33) and (4.8) $n(\lambda_1) = 1/\sqrt{\lambda_1}$ and the relaxed strain-energy function is

$$w(\lambda_1) = 2\mu\left(\lambda_1 + \frac{2}{\sqrt{\lambda_1}} - 3\right).$$ (4.15)

The equilibrium equation (2.69) with $\sigma_{22} \equiv 0$ becomes

$$\frac{d}{dr}(h\sigma_{11}) + \frac{h}{r}\sigma_{11} = 0.$$ (4.16)

Hence we obtain

$$rh\sigma_{11} = \text{constant}.$$ (4.17)

From (4.1) and (4.15) we have

$$\sigma_{11} = \lambda_1 w_1,$$ (4.18)

and so the solution for the wrinkled region can be written

$$r(R) = \int \left(\frac{R}{R - \alpha_1}\right)^{\frac{2}{3}} dR + \alpha_2,$$ (4.19)

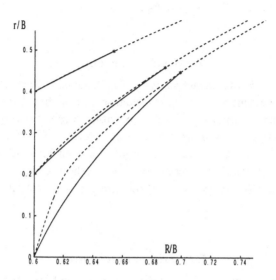

Fig. 3. Plot of the non-dimensionalised deformed radius r/B against the undeformed radius R/B for an annulus with $A/B = 0.6$ and $a/B = 0.0(0.2)0.4$. Both the wrinkled (——) and membrane solutions (- - - -) are shown.

where α_1, α_2 are constants. This integral can be evaluated analytically (see, e.g., Haughton and Mckay 1995), but is not very informative. The boundary conditions to determine the constants α_1 and α_2 are

$$r(R^*) = r^*, \qquad r(A) = a, \qquad (4.20)$$

where R^* is found from (4.14) and r^* is given by $r(R^*)$ using (4.10) and (4.11).

For the purpose of graph plotting we non-dimensionalise the variables with respect to the undeformed outer radius B and the shear modulus μ as appropriate. In Figure 3 we consider a disc with undeformed radii such that $A/B = 0.6$. The disc undergoes three separate deformations so that the inner radius of the disc is pulled in to give a deformed inner radius of (0, 0.2 or 0.4). For each deformation we plot the different solutions $r(R)$ against the undeformed radius R described by two separate theories. Firstly, the dashed curves are obtained from ordinary membrane theory, which gives (4.10) with (4.11) and ignores the fact that one of the principal stresses becomes negative. Secondly, the wrinkled solution (solid line), which is given by (4.19). In each case there is just the one solution (normal membrane theory) to the right of (*) since the membrane is then in tension. Clearly, the effect of the wrinkling theory becomes more pronounced as the amount of wrinkling increases.

7.5 Exact solutions

The importance of finding exact solutions to problems in finite elasticity is discussed elsewhere in this volume. In this final section we firstly look at a method of obtaining exact solutions which is applicable to membrane problems only. This method enables well-known plane strain solutions from three-dimensional elasticity to be interpreted as membrane solutions but for a different material. Secondly, we illustrate an interesting *inverse* method which can optimise the value of new exact solutions for full three-dimensional elasticity or membrane problems.

7.5.1 Using plane strain

As we have pointed out above, two-dimensional membrane theory can be regarded as full three-dimensional elasticity when the behaviour in one of the principal directions is ignored. This means that there is a source of solutions (from full three-dimensional elasticity) readily available to the two-dimensional membrane theory.

If we briefly consider the full three-dimensional (not membrane) theory for plane strain problems in isotropic materials with $\lambda_3 = 1$ then, by referring the full equilibrium equations (2.22) to the orthonormal basic (2.32) we can write div $\sigma = 0$ as

$$\sigma_{\mu\kappa,\mu} + \sigma_{\mu\kappa} a_\nu . a_{\mu,\nu} + \sigma_{\mu\nu} a_\kappa . a_{\nu,\mu} = 0, \qquad \kappa = 1, 2, \qquad (5.1)$$

$$\sigma_{33,3} + \sigma_{33} a_\nu . a_{3,\nu} + \sigma_{\mu\nu} a_3 . a_{\nu,\mu} = 0. \qquad (5.2)$$

Comparing with the two-dimensional membrane equations (2.47), (2.48) and recalling that σ_{33} has no role in the (two-dimensional) membrane case, we can see that there are fully three-dimensional plane strain solutions that can be interpreted as two-dimensional membrane solutions. We can take this analysis one stage further and (try to) compute the three-dimensional membrane strain energy equivalent to the two-dimensional energy. We then have the possibility of using a fully three-dimensional plane strain deformation as the solution to a plane stress (membrane) problem for a different material. To illustrate this we consider the following trivial example.

A homogeneous rectangular block is subjected to a plane strain along the coordinate axes. We suppose that the block has a compressible neo-Hookean strain-energy function

$$W(\lambda_1, \lambda_2, \lambda_3) = \mu(\lambda_1^2 + \lambda_2^2 + \lambda_3^2 + J^{-2} - 4)/2. \tag{5.3}$$

This material is highly compressible with a bulk modulus of $8\mu/3$. The equilibrium equations are then trivially satisfied and the current loads on the sides of the block are $\mathbf{t} = \boldsymbol{\sigma}\mathbf{n}$, where $\mathbf{n}$ is the current unit normal, that is

$$t_1 = \sigma_{11} = \frac{\mu}{\lambda_1\lambda_2}\left(\lambda_1^2 - \frac{1}{\lambda_1^2\lambda_2^2}\right), \quad t_2 = \sigma_{22} = \frac{\mu}{\lambda_1\lambda_2}\left(\lambda_2^2 - \frac{1}{\lambda_1^2\lambda_2^2}\right), \tag{5.4}$$

and

$$t_3 = \sigma_{33} = \frac{\mu}{\lambda_1\lambda_2}\left(1 - \frac{1}{\lambda_1^2\lambda_2^2}\right). \tag{5.5}$$

Now consider the case of a two-dimensional plane sheet with strain energy

$$\hat{W}(\lambda_1, \lambda_2) = \hat{\mu}(\lambda_1^2 + \lambda_2^2 + \frac{1}{(\lambda_1\lambda_2)^2} - 3)/2, \tag{5.6}$$

where μ now has different units, which we simply write down and assume that it represents some physical elastic sheet. For a plane sheet subjected to in-plane edge loads the equilibrium equations (2.47) and (2.48) are satisfied and the edge loads are given by (2.50). These can be written

$$\hat{t}_1 = \hat{\sigma}_{11} = \frac{\hat{\mu}}{\lambda_1\lambda_2}\left(\lambda_1^2 - \frac{1}{\lambda_1^2\lambda_2^2}\right), \quad \hat{t}_2 = \hat{\sigma}_{22} = \frac{\hat{\mu}}{\lambda_1\lambda_2}\left(\lambda_2^2 - \frac{1}{\lambda_1^2\lambda_2^2}\right), \tag{5.7}$$

measured per unit length. Hence the three-dimensional plane strain elasticity solution (5.4) can also be interpreted as a two-dimensional membrane solution. However, the question still remains as to how to interpret the strain energy (5.6).

Suppose that we would now like to interpret the two-dimensional membrane solution (5.7) in the context of three-dimensional membrane theory. We must calculate the three-dimensional membrane strain energy equivalent to (5.6). Clearly, one solution is to take the incompressible neo-Hookean material

$$\overline{W}(\lambda_1, \lambda_2, \lambda_3) = \mu(\lambda_1^2 + \lambda_2^2 + \lambda_3^2 - 3)/2, \tag{5.8}$$

since the substitution $\lambda_3 = 1/\lambda_1\lambda_2$ in (5.8) recovers (5.6). In this case

$$\bar{t}_1 = \bar{\sigma}_{11} = \lambda_1\overline{W}_1 - \lambda_3\overline{W}_3 = \mu(\lambda_1^2 - \lambda_3^2) = \mu\left(\lambda_1^2 - \frac{1}{\lambda_1^2\lambda_2^2}\right),$$

$$\bar{t}_2 = \bar{\sigma}_{22} = \lambda_2\overline{W}_2 - \lambda_3\overline{W}_3 = \mu(\lambda_2^2 - \lambda_3^2) = \mu\left(\lambda_2^2 - \frac{1}{\lambda_1^2\lambda_2^2}\right), \tag{5.9}$$

where $\bar{t}_1$ and $\bar{t}_2$ are the averaged tractions per unit area, having used the membrane assumption (2.35). To compare with the two-dimensional values we must calculate $\lambda_3 \bar{\sigma}_{11}$ and $\lambda_3 \bar{\sigma}_{22}$, see the paragraph following (2.48), which then gives (5.7). Hence we have used a fully three-dimensional plane strain solution for a compressible neo-Hookean material to generate a three-dimensional membrane (plane stress) solution for an incompressible neo-Hookean material. It is unlikely that (5.8) will be the only three-dimensional strain energy that reduces to (5.6) on use of the membrane assumption (2.35) but we do not pursue this here. This approach to generate solutions can be used in more complicated cases; see Haughton (1998), for example.

7.5.2 *An inverse method*

Often the choice of a specific strain-energy function allows the integration of the equilibrium equations (full three-dimensional theory or two or three-dimensional membrane theory) and hence an exact solution. The method we shall illustrate below does not lead to further solutions but it (potentially) maximises information about which materials have the given solution. The process involves taking a solution and inserting it into the general equilibrium equations. The equilibrium equations then become equations for the class of strain-energy functions having the known solution; see Murphy (1996).

As an example of the general process, suppose that we have the disc problem of Section 7.2.3 or 7.4.1 above where we are not concerned with the possibility of cavitation or wrinkling. We have seen that both the incompressible and compressible Varga strain-energy functions (3.33) and (2.74) lead to the trivial solution

$$r = C_1 R^2 + C_2 , \qquad (5.10)$$

where the constants C_1 and C_2 are to be determined by the boundary conditions. We now pose the inverse problem and ask what is the maximal class of strain-energy functions $W(\lambda_1, \lambda_2)$ which give rise to the solution (5.10) for the radial deformation of a disc?

We note that (5.10) with (2.68) gives

$$Rr'' = \lambda_1(\lambda_2 - \lambda_1)/\lambda_2 . \qquad (5.11)$$

Substituting this into the equilibrium equation (2.70) gives

$$\frac{\lambda_1}{\lambda_2}(\lambda_2 - \lambda_1)W_{11}(\lambda_1, \lambda_2) + (\lambda_1 - \lambda_2)W_{12}(\lambda_1, \lambda_2)$$

$$+ W_1(\lambda_1, \lambda_2) - W_2(\lambda_1, \lambda_2) = 0 , \qquad (5.12)$$

where we interpret the strain-energy function $W = W(\lambda_1, \lambda_2)$ as a two-dimensional material. If we now change notation in the above (interchange λ_1 and λ_2) then the equilibrium equation can also be written

$$\frac{\lambda_2}{\lambda_1}(\lambda_1 - \lambda_2)W_{11}(\lambda_2, \lambda_1) + (\lambda_2 - \lambda_1)W_{12}(\lambda_2, \lambda_1)$$

$$+ W_1(\lambda_2, \lambda_1) - W_2(\lambda_2, \lambda_1) = 0. \tag{5.13}$$

If we now use the symmetries required to hold for W, i.e.

$$W_1(\lambda_2, \lambda_1) = W_2(\lambda_1, \lambda_2),$$

$$W_{12}(\lambda_2, \lambda_1) = W_{12}(\lambda_1, \lambda_2), \quad W_{11}(\lambda_2, \lambda_1) = W_{22}(\lambda_1, \lambda_2), \tag{5.14}$$

then the above becomes

$$\frac{\lambda_2}{\lambda_1}(\lambda_1 - \lambda_2)W_{22}(\lambda_1, \lambda_2) + (\lambda_2 - \lambda_1)W_{12}(\lambda_1, \lambda_2)$$

$$- W_1(\lambda_1, \lambda_2) + W_2(\lambda_1, \lambda_2) = 0. \tag{5.15}$$

Subtracting (5.15) from (5.13) gives a necessary condition for the equilibrium equation to be satisfied, namely

$$\lambda_1^2 W_{11} = \lambda_2^2 W_{22}, \tag{5.16}$$

which must hold when

$$\lambda_1 = C_1/\lambda_2, \quad \lambda_2 = \sqrt{C_1 + C_2/R^2}. \tag{5.17}$$

It is physically meaningless for a solution to hold at isolated points so we require that (5.16) holds in general. The solution to (5.16) is

$$W(\lambda_1, \lambda_2) = g_1(\lambda_1 \lambda_2) + \{\lambda_1 \lambda_2\}^{1/2} g_2(\lambda_2/\lambda_1), . \tag{5.18}$$

where g_1 and g_2 are arbitrary functions. Since we have only a necessary condition for equilibrium we substitute (5.18) into the equilibrium equation (5.12) to get

$$g_2''(x) + \frac{x - 3}{2x(x - 1)}g_2'(x) = 0, \tag{5.19}$$

where $x = \lambda_2/\lambda_1$. The function g_1 remains arbitrary. Integrating the above equation and imposing the symmetry constraints on W leads to a final solution

$$W(\lambda_1, \lambda_2) = g_1(\lambda_1 \lambda_2) + C_1(\lambda_1 + \lambda_2), \tag{5.20}$$

where g_1 is an arbitrary function and C_1 is an arbitrary constant.

For the two-dimensional membrane theory this may be a satisfactory conclusion to the problem (subject to imposing conditions on g_1 and C_1 to ensure that the undeformed configuration is stress and energy free and that the shear modulus is positive). However, for the three-dimensional membrane theory it remains to determine the class of strain-energy functions $\overline{W}(\lambda_1, \lambda_2, \lambda_3)$ which reduce to (5.20) when we solve $\overline{W}_3 = 0$ for λ_3 and evaluate $\hat{W}(\lambda_1, \lambda_2) = \overline{W}(\lambda_1, \lambda_2, \lambda_3(\lambda_1, \lambda_2))$. The evaluation of $\lambda_3(R)$ may also be an important consideration.

We would like to find all three-dimensional strain energies which reduce to (5.20) but this is not practical. We confine our attention to finding a reasonably general class of suitable strain-energy functions that reduce to (5.20). Starting with a general $\overline{W}(\lambda_1, \lambda_2, \lambda_3)$ is too open ended so we ask if it is possible to choose a function g and constants C and D so that

$$\overline{W}(\lambda_1, \lambda_2, \lambda_3) = g(J) + C(\lambda_1 + \lambda_2 + \lambda_3) + D(\lambda_1\lambda_2 + \lambda_1\lambda_3 + \lambda_2\lambda_3) , \quad (5.21)$$

reduces to (5.20). (This is motivated by considering the strain-energy function as a function of the invariants of $\mathbf{V}$, where $\mathbf{V}^2 = \mathbf{F}\mathbf{F}^T$.) In this case we require $\overline{W}_3 = 0$ which, from (5.21), gives

$$g'(J) = \frac{-C - D(\lambda_1 + \lambda_2)}{\lambda_1\lambda_2} . \quad (5.22)$$

Using $\hat{W}(\lambda_1, \lambda_2) = \overline{W}(\lambda_1, \lambda_2, \lambda_3(\lambda_1, \lambda_2))$ with (5.21) and (5.20) gives

$$C(\lambda_1 + \lambda_2 + \lambda_3(\lambda_1, \lambda_2)) + D(\lambda_1\lambda_2 + (\lambda_1 + \lambda_2)\lambda_3(\lambda_1, \lambda_2))$$
$$+ g(\lambda_1\lambda_2\lambda_3(\lambda_1, \lambda_2)) = g_1(\lambda_1\lambda_2) + C_1(\lambda_1 + \lambda_2) . \quad (5.23)$$

Differentiating (5.23) with respect to λ_1 and using (5.22) leads to

$$\lambda_3(\lambda_1, \lambda_2) = - \left[\frac{g_1(\lambda_1\lambda_2)\lambda_1\lambda_2 + \lambda_1(C_1 - C)}{C + D\lambda_2} \right] . \quad (5.24)$$

Since λ_3 must be symmetric in λ_1 and λ_2 it follows that $D = 0$ and $C = C_1$. Hence (5.21) reduces to the compressible Varga material (2.74).

Clearly there is scope for using alternatives to (5.21) as our starting point. However, the problem of finding general three-dimensional strain energies from a given two-dimensional one remains an open problem.

Finally, if we interpret (5.20) as an incompressible material then one obvious three-dimensional strain-energy function is the incompressible Varga material (3.33) and (3.34). Again, finding other incompressible strain-energy functions which reduce to (5.20), or proving that there are no others, is not going to be easy.

References

Chadwick, P. and Ogden, R. W. 1971 On the definition of elastic moduli, *Arch. Rat. Mech. Anal.* **44**, 41–53.

Green, A. E. and Adkins, J. E. 1970 *Large Elastic Deformations*. Second Edition. Clarendon Press Oxford.

Haughton, D. M. 1986 On non-existence of cavitation in incompressible elastic membranes. *Q. J. Mech. Appl. Math.* **39**, 289–296.

Haughton, D. M. 1990 Cavitation in compressible elastic membranes. *Int. J. Engng. Sci.* **28**, 163–168.

Haughton, D. M. 1998 Exact solutions for elastic membrane disks. *Math. Mech. Solids* **4**, 393–410.

Haughton, D. M. and McKay, B. A. 1995 Wrinkling of annular discs subjected to radial displacements. *Int. J. Engng. Sci.* **33**, 335–350.

Haughton, D. M. and Ogden, R. W. 1978 On the incremental equations in non–linear elasticity–I Membrane theory, *J. Mech. Phys. Solids* **26**,93–110.

Haughton, D. M. and Ogden, R. W. 1979 Bifurcation of inflated circular cylinders of elastic material under axial loading–I Membrane theory for thin walled tubes, *J. Mech. Phys. Solids* **27**, 179-212.

Haughton, D. M. and Ogden, R. W. 1980 Bifurcation of rotating circular cylindrical elastic membranes. *Math. Proc. Camb. Phil. Soc.* **87**, 357-376.

Li, X. and Steigmann, D. J. 1993 Finite plane twist of an annular membrane. *Q.J. Mech. Appl. Maths.* **46**, 601–626.

John, F. 1960 Plane strain problems for a perfectly elastic material of harmonic type. *Comm. Pure Appl. Math.* **13**, 239-296.

Kydoniefs, A. D. and Spencer, A. J. M. 1969 Finite axisymmetric deformations of an initially cylindrical elastic membrane. *Q. J. Mech. appl. Math.* **22**, 87–95.

Murphy, J. G. 1996 A family of solutions describing plane strain cylindrical inflation in finite compressible elasticity. *J. Elasticity* **45**, 1–11.

Naghdi, P. M. and Tang, P. Y. 1977 Large deformation possible in every isotropic elastic membrane. *Phil. Trans. R. Soc. Lond.* A **287**, 145–187.

Ogden, R. W. 1997 *Non-Linear Elastic Deformations*. New York: Dover Publications.

Pipkin, A. C. 1968 Integration of an equation in membranes theory. *ZAMP* **19**, 818–819.

Pipkin, A. C. 1986 The relaxed energy density for isotropic elastic membranes, *IMA J. appl. Math.* **36**, 85–99.

Rivlin, R. S. and Thomas, A. G. 1951 Large elastic deformations of isotropic materials *VII*. Starin distribution around a hole in a sheet. *Phil. Trans. Roy. Soc. Lond.* A **243**, 289–298.

Steigmann, D. J. 1990 Tension field theory. *Proc. Roy. Soc. Lond.* A **429**, 141–173.

Steigmann, D. J. 1992 Cavitation in elastic membranes–an example. *J. Elasticity* **28**, 277–287.

8

Elements of the theory of elastic surfaces

David J. Steigmann

Department of Mechanical Engineering
University of California
Berkeley, CA. 94720, U.S.A.
Email: steigman@newton.me.berkeley.edu

I present a development of the modern theories of elastic shells, regarded as mathematical surfaces endowed with kinematical and constitutive structures deemed sufficient to represent many of the features of the response of thin shell-like bodies. The emphasis is on Cosserat theory, specialized to obtain a model of the Kirchhoff-Love type through the introduction of appropriate constraints. Noll's concept of material symmetry, adapted to surface theory by Cohen and Murdoch, is used to derive new constitutive equations for elastic surfaces having hemitropic, isotropic and unimodular symmetries. The last of these furnishes a model for fluid films with local bending resistance, which may be used to describe the response of certain fluid microstructures and biological cell membranes.

8.1 Introduction

I use the nonlinear Kirchhoff-Love theory of shells to describe the mechanics of a number of phenomena including elastic surface-substrate interactions and the equilibria of fluid-film microstructures. The Kirchhoff-Love shell may be interpreted as a one-director Cosserat surface (Naghdi 1972) with the director field constrained to coincide with the local orientation field.

The phenomenology of surfactant fluid-film microstructures interspersed in bulk fluids poses significant challenges to continuum theory. By using simple models of elastic surfaces, chemical physicists have been partially successful in describing the qualitative features of the large variety of equilibrium structures observed (Kellay *et al.* 1994, Gelbart *et al.* 1994). The basic constituent of such a surface is a polar molecule composed of hydrophilic head groups attached to hydrophobic tail groups. At low concentrations in a bulk fluid such as water, the surfactant molecules, or *amphiphiles*, migrate to free surfaces and arrange themselves as monomolecular films with the tail groups shielded from the bulk phase. This process is associated with a dramatic reduction of the apparent surface tension, and continues with increasing amphiphile concentration until the *critical micelle concentration* is reached at which the molecules begin to form microstructures interspersed in the bulk (Gelbart *et al.* 1994). These can assume a variety of forms due to the range of mechanisms available for shielding the tail groups. The basic component of these fluid microstructures is the monolayer. At fixed temperature and at amphiphile concentrations exceeding the critical value, the monolayers typically form rod-like cylinders or spherical micelles. The latter structure is also associated with surfactant-stabilized microemulsions of immiscible fluids such as water and oil, in which small droplets of one fluid are suspended in the bulk phase of the other. The interfaces separating the fluids consist of amphiphilic monolayers with the tail groups directed away from the water phase. Emulsification inversion, in which the interior and exterior phases are exchanged, is thought to be facilitated by the emergence of an intermediate *bicontinuous* microstructure conceived by Scriven (1976, 1977). This consists of compact oriented films of high genus having a spongy or porous structure.

Bilayer surfaces composed of oppositely oriented monolayers furnish another mechanism for shielding tail groups. These occur in a variety of forms including planar lamellae, spherical vesicles, and variants of the bicontinuous topology (Gelbart *et al.* 1994). Bilayers of biological surfactants, called phospholipids, are also known to be of fundamental importance to the structure and function of cell walls (Evans and Skalak 1980, Collings 1990).

Typical length scales for these microstructures are so small that local interactions due to changes in the alignment of the amphiphilic molecules have a pronounced influence on surface morphology. Cosserat surface theory has been the preferred continuum model since surfactant systems do not exist in bulk. Local effects associated with alignment are represented by assigning elastic resistance to the configurations of a director field representing molecular orientation. Variants of this model have been developed by Helfrich (1973), Ericksen (1979), Jenkins (1977), and Krishnaswamy (1996). The general theory accounts for surface strain, director extension, and director tilt, the latter being associated with misalignment of the director and the surface normal. However, there is a preference in the physical chemistry and cell biology literatures for a simpler model based on surface geometry alone (Helfrich 1973, Jenkins 1977). This is justified by the belief that local Van der Waal's and electrostatic forces tend to act in such a way as to suppress misalignment and maintain roughly fixed tail lengths, particularly in bilayers (Helfrich 1973, Charvolin and Sadoc 1994). The Kirchhoff-Love constraints are intended to represent the suppression of director tilt and extension or contraction in accordance with the generally accepted phenomenology.

In the setting of elastic solid surface-substrate interactions, Gurtin and Murdoch (1975) presented a theory for the mechanics of surface-stressed solids based on the idea of a two-dimensional membrane bonded to the surface of a bulk substrate material. Their work generalizes the classical notion of surface tension in solids and allows for the systematic theoretical description of general states of residual surface stress. In principle, this theory may be used to describe the mechanical behavior of a bulk substrate material coated with a thin film of a different substance. It is also sufficiently general to accommodate inelastic response and dynamical effects. The Gurtin-Murdoch theory was motivated in part by empirical observations pointing to the presence of compressive surface stress in certain types of crystals. It was later used by Andreussi and Gurtin (1977) to model wrinkling at the free surface of a solid as a bifurcation of the undeformed configuration induced by compressive residual surface stress. Steigmann and Ogden (1997) have reconsidered the latter problem in the context of a plane-strain theory for elastic solids with surface energies that incorporate local elastic resistance to flexure in addition to the strain resistance included in the Gurtin-Murdoch theory. The findings reveal a marked departure from the earlier predictions, including the presence of dispersion due to an intrinsic length scale associated with the constitutive relations for the surface. A variational argument was also given demonstrating the impossibility of compressive surface stress in energy-minimizing configurations of the entire film-substrate combination. Results like these are to be expected in view of the fact that flexural resistance singularly perturbs the membrane model equations and regularizes the associated variational problem. From the physical viewpoint it may therefore be argued that a surface model that does not account for flexural resistance cannot be used to simulate local surface features engendered by the response of solids to compressive surface stress of any magnitude.

Conventional treatments of film/substrate combinations are usually based on shell theories derived from the three-dimensional elastic properties of the material of which the film is composed. In this regard I note the substantial progress achieved in the justification of such theories through the use of formal asymptotic expansions (see, for example, Fox *et al.* 1993), or, alternatively, through the use of Γ-convergence theory (De Giorgi and Dal Maso 1983) to characterize the limit of a sequence of variational problems for a thin body with thickness tending to zero (Le Dret and Raoult 1995). Here, however, I have in mind applications to systems in which the film may not occur naturally in bulk, as in the condensation, cooling, and solidification of films from the vapor phase of a particular substance. In such circumstances the properties of the solid film may differ markedly from the bulk properties of the same substance. The suitability of a theory derived from three-dimensional considerations would then be open to question.

I adapt the extension of Noll's (1958) theory of material symmetry to elastic surfaces, as presented by Murdoch and Cohen (1979), to obtain new canonical constitutive equations for films having *hemitropic, isotropic* and *unimodular* symmetries. My view is that the Murdoch-Cohen theory is the most logically appealing among the various alternative symmetry theories for elastic surfaces that have been advanced, and that it furnishes a particularly satisfactory framework for the description of the response of film/substrate systems. To my knowledge this theory has not been used in the subsequent literature on developments in shell theory. Finally, I discuss necessary conditions for a deformation to be energy-minimizing and use them to obtain restrictions on the equilibrium values of certain response functions.

This chapter is a synopsis of some of my recent work on the subject, much of which has been carried out with my colleague Ray Ogden and associates. Applications of the theory to the solution of simple finite-deformation problems are presented by Steigmann and Ogden (1999) and Ogden *et al.* (1997).

8.2 The relationship between the Cosserat and Kirchhoff-Love theories of elastic shells

I endeavour here to clarify the relationship of the Kirchhoff-Love theory of elastic shells to the more general Cosserat theory of deformable surfaces with a single director. The latter has as its kinematical basis two vector fields defined on the surface, one that defines the particle position, and the other, the director, that is intended to account for finite-thickness effects.

Specifically, I obtain the Kirchhoff-Love theory by imposing constraints on the director field and deriving the general forms of the associated response functions through a careful application of the rigorous Lagrange multiplier rule. Although this rule is standard, the constraints considered are of an unusual type and the development thus includes more detail than one usually finds in the literature on constrained elasticity. I consider the theory in its local form, rather than in the more traditional variational setting initiated by Kirchhoff (1850) and advanced to its modern standard by Antman (1995) and associates.

In Naghdi's treatment of this subject (Naghdi 1972, Sections 10 and 15) the Kirchhoff-Love theory is not derived from the Cosserat theory but instead considered separately on the basis of distinct balance and invariance postulates. This contrasts with my view that the Cosserat theory should reduce to the Kirchhoff-Love theory upon the introduction of appropriate constraints. Nevertheless the resulting model is substantially equivalent to Naghdi's and thus the differences between the two approaches are primarily conceptual, the main distinction being that the logical framework of the present development conforms to that of finite elasticity theory.

8.3 Cosserat theory

I recall the balance laws for the purely mechanical Cosserat theory of elastic shells with a single director. An exhaustive account is given in Naghdi's treatise (Naghdi 1972).

Let $r(\theta^\alpha, t)$ be the position in 3-space at time t of a particle with convected coordinates θ^α. Here and elsewhere Greek indices take values in $\{1,2\}$, Latin indices take values in $\{1,2,3\}$ and standard tensor notation is used. The position function and the director field $d(\theta^\alpha, t)$ constitute the basic kinematical variables of the theory. Let ω be an open simply-connected region of the surface in a given configuration with piecewise smooth boundary $\partial\omega$. The global balance postulates consist of the conservations of mass

$$\frac{d}{dt} \int_\omega \rho \mathrm{d}a = 0, \tag{3.1}$$

momentum

$$\frac{d}{dt} \int_\omega \rho(v + \alpha w) \mathrm{d}a = \int_\omega \rho f \mathrm{d}a + \int_{\partial\omega} N \mathrm{d}s, \tag{3.2}$$

director momentum

$$\frac{d}{dt} \int_\omega \rho(\alpha v + \beta w) \mathrm{d}a = \int_\omega (\rho l - k) \mathrm{d}a + \int_{\partial\omega} M \mathrm{d}s, \tag{3.3}$$

and moment of momentum

$$\frac{d}{dt} \int_\omega \rho A \mathrm{d}a = \int_\omega \rho(r \times f + d \times l) \mathrm{d}a + \int_{\partial\omega} (r \times N + d \times M) \mathrm{d}s, \tag{3.4}$$

where

$$A = r \times (v + \alpha w) + d \times (\alpha v + \beta w). \tag{3.5}$$

The notation and terminology are similar to those used by Naghdi. Thus, ρ is the mass per unit area, $v = \dot{r}$ and $w = \dot{d}$, respectively, are the particle and director velocities, where the superposed dot is used to denote the time derivative at fixed values of the θ^α, f and l, respectively, are the distributed force and distributed director force, k is the intrinsic director force, N and M, respectively, are the traction and the director traction, and α, β are prescribed functions of θ^α.

As remarked by Naghdi the foregoing postulates may be motivated by appealing to conventional shell theory wherein the position function in a thin three-dimensional body is assumed to be an affine function of a third convected coordinate associated with thickness. However, the mathematical status of the latter theory *vis-à-vis* three-dimensional elasticity is uncertain and thus (3.1)–(3.4) are regarded here as being fundamental.

As in conventional continuum mechanics the traction and director traction are assumed to depend on the curve $\partial \omega$ through its local orientation. This leads by the usual argument, outlined in Naghdi (1972) and Gelbart *et al.* (1994), and developed in detail in Scriven (1976, 1977), to the existence of stress vectors N^α and M^α such that

$$N = N^\alpha \nu_\alpha, \qquad M = M^\alpha \nu_\alpha, \tag{3.6}$$

where $\nu_\alpha = \varepsilon_{\alpha\beta} d\theta^\beta / ds$ are the covariant components of the unit normal to the curve, $\theta^\alpha(s)$ is the arclength parametrization of the curve, and $\varepsilon_{\alpha\beta}$ are the covariant components of the permutation tensor density on the closure of ω.

Assuming sufficient smoothness, and recalling that ω is simply connected, the global balances (3.2), (3.3) together with (3.6) furnish the local equations (Naghdi 1972)

$$N^\alpha_{;\alpha} + \rho \bar{f} = 0, \qquad M^\alpha_{;\alpha} - k + \rho \bar{l} = 0, \tag{3.7}$$

where

$$\bar{f} = f - (\dot{v} + \alpha \dot{w}), \qquad \bar{l} = l - (\alpha \dot{v} + \beta \dot{w}), \tag{3.8}$$

and the semicolon is used to denote covariant differentiation on ω. With these results the local form of (3.4) may be written (Naghdi 1972):

$$a_\alpha \times N^\alpha + g_\alpha \times M^\alpha + d \times k = 0, \tag{3.9}$$

where

$$a_\alpha = r_{,\alpha}, \qquad g_\alpha = d_{,\alpha} \tag{3.10}$$

and $(\cdot)_{,\alpha} = \partial(\cdot)/\partial\theta^\alpha$. The local form of the conservation of mass is

$$(\rho J)^\cdot = 0, \qquad J = (a/A)^{1/2}, \tag{3.11}$$

where $a \, (> 0)$ is the determinant of the symmetric, positive-definite matrix

$$C_{\alpha\beta} = a_\alpha \cdot a_\beta \tag{3.12}$$

and A is the value of a in a fixed reference configuration.

Following Naghdi I impose the condition

$$a_1 \times a_2 \cdot d > 0 \tag{3.13}$$

which implies that the a_α are not collinear and the director is nowhere tangent to the surface ω. This is the surface analog of the local impenetrability inequality of conventional continuum mechanics. I further assume that under arbitrary rigid body motions the transformation rules

$$a_\alpha \rightarrow Qa_\alpha, \quad d \rightarrow Qd, \quad g_\alpha \rightarrow Qg_\alpha \qquad (3.14)$$

apply with Q a rotation.

A mechanical power identity may be constructed by scalar multiplying the first and second equations in (3.7) with v and w, respectively, adding the resulting equations, and then integrating over the simply-connected region ω. This yields

$$\frac{dK}{dt} + S = P, \qquad (3.15)$$

where

$$K = \frac{1}{2} \int_\omega \rho(v \cdot v + 2\alpha v \cdot w + \beta w \cdot w) da \qquad (3.16)$$

is the kinetic energy of the material occupying the closure of ω,

$$P = \int_\omega \rho(f \cdot v + l \cdot w) da + \int_{\partial \omega} (N \cdot v + M \cdot w) ds \qquad (3.17)$$

is the power of the forces and director forces acting on the material, and

$$S = \int_\omega (N^\alpha \cdot \dot{a}_\alpha + M^\alpha \cdot \dot{g}_\alpha + k \cdot \dot{d}) da \qquad (3.18)$$

is the stress power.

Elasticity is defined by the existence of a strain energy Ψ per unit mass which, together with N^α, M^α and k, depends on a_α, g_α and d and is such that $S = dU/dt$ for any subregion ω of the surface, where

$$U = \int_\omega \rho \Psi da. \qquad (3.19)$$

This yields

$$\frac{dH}{dt} = P, \qquad (3.20)$$

where $H = K + U$ is the mechanical energy. The same assumption, combined with conservation of mass, also implies that

$$\dot{W} = JN^\alpha \cdot \dot{a}_\alpha + JM^\alpha \cdot \dot{g}_\alpha + Jk \cdot \dot{d} \qquad (3.21)$$

pointwise, where $W = J\rho\Psi$ is the strain energy per unit area of the fixed surface used in the definition of J. Here W is regarded as a function of the coefficient matrices in the expressions

$$a_\alpha = F_{i\alpha} e_i, \quad g_\alpha = G_{i\alpha} e_i, \quad d = d_i e_i, \tag{3.22}$$

where $\{e_i\}$ is a set of fixed orthonormal vectors in 3-space. The values of the first pair of variables depend on the choice of coordinates θ^α but the strain energy is required to be an absolute scalar field with values that are unaffected by this choice. Thus the *form* of the function W must depend on coordinates.

On substituting (3.22) into (3.21), and assuming that $\dot{a}_\alpha$, g_α and $\dot{d}$ may be assigned independently and arbitrarily, I obtain the constitutive equations for unconstrained elastic Cosserat shells (Naghdi 1972), namely

$$JN^\alpha = \frac{\partial W}{\partial F_{i\alpha}} e_i = \frac{\partial W}{\partial a_\alpha}, \quad JM^\alpha = \frac{\partial W}{\partial G_{i\alpha}} e_i = \frac{\partial W}{\partial g_\alpha},$$

$$Jk = \frac{\partial W}{\partial d_i} e_i = \frac{\partial W}{\partial d}, \tag{3.23}$$

wherein the second equality in each expression is to be regarded as a definition.

8.4 Constraints

In the Kirchhoff-Love theory the strain energy is determined by the local deformation through the values of a_α and $n_{,\alpha}$, where $n = a_1 \times a_2 / \|a_1 \times a_2\|$ is the local orientation field of the surface evaluated at the particle in question. It is also assumed that the gradient of n satisfies the Weingarten equation $n_{,\alpha} = -b_{\beta\alpha} a^\beta$, where $b_{\alpha\beta} = n \cdot a_{\alpha,\beta} = b_{\beta\alpha}$ are the symmetric coefficients of the second fundamental form of the surface. Here, the a^α are the unique vectors on the tangent plane such that $a^\alpha \cdot a_\beta = \delta^\alpha_\beta$ and $n \cdot a^1 \times a^2 > 0$ where δ^α_β is the Kronecker delta.

A model with these features, regarded as being distinct from the Cosserat theory, is described in Naghdi's treatise (Naghdi 1972, Sections 10 and 15) where it is referred to as the *restricted director theory*. In particular, Naghdi's treatment involves dynamical variables formally similar to those appearing in (3.1)–(3.4) above, but with certain components set to zero *a priori*. In addition, the director momentum balance (3.3) is dropped from the list of postulates. However, in view of the fact that the constitutive assumptions of the Kirchhoff-Love theory are subsumed under those of the Cosserat theory, it seems more natural to consider the former theory within the framework of the latter rather than as a distinct model. This is the viewpoint adopted by Antman (1995, Chapter 14, Sections 10 and 13), who incorporated the constraint $d = n$ in a global virtual work statement but also omitted the director momentum balance.

In this work I recover the Kirchhoff-Love theory by imposing the constraints

$$d = n \quad \text{and} \quad g_\alpha = n_{,\alpha} = -b_{\beta\alpha}a^\beta \tag{4.1}$$

at the constitutive level. Constraints on d *and* its derivatives g_α are needed since the constitutive equations of the Cosserat theory involve the independent values of d and $d_{,\alpha}$ at a material point. Moreover the $b_{\alpha\beta}$ are symmetric but otherwise arbitrary. This requires that $n \cdot a_{\alpha,\beta} = n \cdot a_{\beta,\alpha}$ but $a_{\alpha,\beta}$ and $a_{\beta,\alpha}$ need not be equal. Thus there is no need to restrict attention to holonomic a_α in the development of the constitutive theory. This allows for the possibility, not considered here, of a non-Riemannian material manifold (Wang and Cross 1977).

To obtain constraints involving only the variables that appear in the Cosserat theory I note that in view of (3.13) equation $(4.1)_1$ is equivalent to the conditions $\phi = 0$ and $\phi_\alpha = 0$, where

$$\phi = d \cdot d - 1, \qquad \phi_\alpha = d \cdot a_\alpha. \tag{4.2}$$

If these are satisfied then $(4.1)_2$ is equivalent to the conditions $\psi = 0$ and $\psi_\alpha = 0$, where

$$\psi = e^{\alpha\beta}a_\beta \cdot g_\alpha, \qquad \psi_\alpha = d \cdot g_\alpha \tag{4.3}$$

and $e^{\alpha\beta}$ is the unit alternator ($e^{12} = +1$). These constraints are invariant under the transformations (3.14).

Naghdi (1981) discussed the application of $(4.2)_2$ and $(4.3)_1$ to a model for a special Cosserat shell with a deformable director d constrained to be aligned with n. An alternative development of the Kirchhoff-Love theory, obtained by descent from three-dimensional theory, may be found in Libai and Simmonds (1998).

I use the standard Lagrange multiplier rule to derive representation formulas for the various dynamical variables in the presence of the constraints. To apply this rule it is necessary to verify (Steigmann 2000) that the gradients of (4.2) and (4.3) are linearly independent vectors in a vector space E with elements of the type $x = (a_\alpha, g_\alpha, d)$. In view of (3.22) I regard E as $\mathbb{R}^{15}$ endowed with the usual vector space structure and the Euclidean inner product. It is also necessary to show that the constraint equations are solvable in terms of variables that locally parametrize a constraint manifold $\mathcal{M} \subset E$ obtained by restricting x such that all the constraints are satisfied simultaneously. Verification of the latter requirement is trivial. The general solution of the constraint equations (4.2) and (4.3) is $d = a_1 \times a_2 / \|a_1 \times a_2\|$ and $g_\alpha = K_{\beta\alpha}a^\beta$, where

$$2K_{\beta\alpha} = g_\alpha \cdot a_\beta + g_\beta \cdot a_\alpha. \tag{4.4}$$

Thus $\mathcal{M}$ is a nine-dimensional manifold parametrized by the variables $F_{i\alpha}$, $K_{\beta\alpha}$ with typical point $y = (a_\alpha, K_{\beta\alpha}a^\beta, n)$. In view of the remarks following (4.1) I identify $-K_{\beta\alpha}$ with the coefficients of the second fundamental form of ω at the particle in question.

Let $x(t)$ be a curve on $\mathcal{M}$ obtained by specifying $F_{i\alpha}(t)$ and $K_{\beta\alpha}(t)$ for t in an open interval. By differentiating (4.2) and (4.3) with respect to t it is verified that $\nabla\phi(x(t)) \cdot \dot{x}(t)$, etc., vanish for all $\dot{x}(t) \in T(x(t))$, where $T(y)$ is the unique nine-dimensional vector subspace of E tangent to $\mathcal{M}$ at y. Thus, E is the direct sum of the orthogonal subspaces $T(y)$ and $N(y) = \text{Span}\{\nabla\phi(y), \nabla\psi(y), \nabla\phi_\alpha(y), \nabla\psi_\alpha(y)\}$.

In (3.21) the strain energy is regarded as a function defined on $\mathcal{M}$ which I denote by $W(y)$. To obtain $\dot{W}$ I introduce a smooth extension $W'(x)$ of W to E such that $W' = W$ when $x \in \mathcal{M}$. Then for t in an open interval and $x(t) \in \mathcal{M}$ the chain rule yields $\dot{W} = \dot{W}' = \nabla W'(x(t)) \cdot \dot{x}(t)$ with $\dot{x}(t) \in T(x(t))$ and $\nabla W'(x) = (\partial W'/\partial a_\alpha, \partial W'/\partial g_\alpha, \partial W'/\partial d)$. Here the entries are defined as in (3.23) so that $\nabla W'(x) \in E$. Note that, given W, there is a degree of arbitrariness in the extension W'. To see this let $\bar{W}$ be another extension, let t belong to an open interval containing zero, and consider an arbitrary curve $x(t) \in \mathcal{M}$ with $x(0) = x_0$. Both extensions are equal to W when restricted to $\mathcal{M}$ and the difference of their time derivatives vanishes on $x(t)$. It follows that $\nabla\bar{W}(x_0) - \nabla W'(x_0) \in N(x_0)$. This result is exploited in the next section.

The remaining response functions in (3.21) are also defined on $\mathcal{M}$. By decomposing them as in (3.22) it is easily verified that $S(y) \in E$, where $S(y) = J(N^\alpha, M^\alpha, k)$ is the response vector.

Equation (3.21) may now be written as $B(x_0) \cdot \dot{x} = 0$ for $x_0 \in \mathcal{M}$ and $\dot{x}$ an arbitrary element of $T(x_0)$, where $B(x_0) = S(x_0) - \nabla\bar{W}(x_0)$. Since S and $\nabla\bar{W}$ belong to E it follows that $B(x_0) \in N(x_0)$. This is the Lagrange multiplier rule in the present context. Of course, the same conclusion applies with $\bar{W}$ replaced by W'. I use this result to obtain

$$N^\alpha - \bar{N}^\alpha = q^\alpha n + q\varepsilon^{\beta\alpha}n_{,\beta},$$
$$M^\alpha - \bar{M}^\alpha = q\varepsilon^{\alpha\beta}a_\beta + r^\alpha n,$$
$$k - \bar{k} = q^\alpha a_\alpha + rn + r^\alpha n_{,\alpha}, \qquad (4.5)$$

where, in view of the physical context, the Lagrange multipliers (q, r), (q^α, r^α) respectively are the values at θ^α and t of arbitrary scalar fields and contravariant vector fields on the closure of ω, $\varepsilon^{\alpha\beta} = a^{-1/2}e^{\alpha\beta}$ is the associated permutation tensor density, and

$$J\bar{N}^\alpha = \frac{\partial\bar{W}}{\partial a_\alpha}, \quad J\bar{M}^\alpha = \frac{\partial\bar{W}}{\partial g_\alpha}, \quad J\bar{k} = \frac{\partial\bar{W}}{\partial d}, \qquad (4.6)$$

where the derivatives are evaluated on $\mathcal{M}$.

8.5 Invariance and the reduced constitutive equations

Since the constraint manifold is the domain of the constitutive functions the
latter are determined by the variables that parametrize $\mathcal{M}$, namely a_α and
$K_{\alpha\beta} = -b_{\alpha\beta}$. In particular, the strain energy W at a particle is the value of a
function $F(a_\alpha, -b_{\alpha\beta})$.

I assume W to be unaltered by superposed rigid motions. In view of $(3.14)_1$
I thus require that $F(a_\alpha, -b_{\alpha\beta}) = F(Qa_\alpha, -b_{\alpha\beta})$ for all rotations Q, in which
the $b_{\alpha\beta}$, being invariant (Naghdi 1972), play a passive role. From (3.12) and
Cauchy's theorem on hemitropic functions (e.g. Antman 1995) there follows
the necessary and sufficient condition

$$W = \bar{F}(C_{\alpha\beta}, -b_{\alpha\beta}) \tag{5.1}$$

for some function $\bar{F}$ whose domain is the set consisting of all ordered pairs of
symmetric matrices, the first members of which are positive definite. To ensure
that W is a scalar field, the form of the function $\bar{F}$ must reflect the choice of
surface coordinates θ^α. Usually this condition is met by requiring the energy
to be a scalar-valued function of the tensors induced by the arguments of $\bar{F}$ on
the tangent plane of a reference surface at the particle in question.

An extension $\bar{W}(a_\alpha, g_\alpha, d)$ of W to E may be obtained by using (3.12) and
(4.4) to define extensions of $C_{\alpha\beta}$ and $K_{\alpha\beta}$ for $x \in E$. I then set

$$\bar{W} = \bar{F}(C_{\alpha\beta}, K_{\alpha\beta}). \tag{5.2}$$

Equations (3.12), (3.14) and (4.4) imply that $\bar{W}$ thus defined is invariant under
superposed rigid motions. Since the restrictions to $\mathcal{M}$ of all extensions are
given by (5.1) it follows that the former are necessarily invariant under such
motions. This requirement is satisfied by adopting as an additional assumption
the invariance of all extensions of W for all $x \in E$.

For $x_0 \in \mathcal{M}$, let $S(x_0) \in E$ be the response vector obtained by using
the extension (5.2). From the analysis of the previous section I conclude that
another invariant extension generates a response vector that differs from $S(x_0)$
by an element of $N(x_0)$. This difference may be absorbed into the Lagrange
multipliers, and thus no generality is lost by using the extension $\bar{W}$ defined by
(5.2). I demonstrate this explicitly in Steigmann (2000).

Finally, I use (4.4) and (4.6) together with (5.2) to derive

$$\bar{N}^\alpha = \bar{N}^{\beta\alpha}a_\beta, \quad \bar{M}^\alpha = \bar{M}^{\beta\alpha}a_\beta \quad \text{and} \quad \bar{k} = 0, \tag{5.3}$$

where

$$\bar{N}^{\beta\alpha} = \bar{\sigma}^{\beta\alpha} - b^\beta_\lambda \bar{M}^{\lambda\alpha}, \quad J\bar{\sigma}^{\beta\alpha} = \partial\bar{F}/\partial C_{\alpha\beta} + \partial\bar{F}/\partial C_{\beta\alpha}$$
$$J\bar{M}^{\beta\alpha} = \tfrac{1}{2}(\partial\bar{F}/\partial K_{\alpha\beta} + \partial\bar{F}/\partial K_{\beta\alpha}). \tag{5.4}$$

These are evaluated at $K_{\alpha\beta} = -b_{\alpha\beta}$ and $(4.1)_2$ has been used in the form $g_\alpha = -b^\beta_\alpha a_\beta$.

8.6 Equations of motion and the Kirchhoff edge conditions

The equations of motion for a Kirchhoff-Love shell have a well-known divergence structure that may be recovered by first eliminating certain combinations of the Lagrange multipliers from the director-momentum equation $(3.7)_2$. To this end I use $(4.5)_2$ together with the Gauss equation $a_{\beta;\alpha} = b_{\beta\alpha}n$ (Naghdi 1972), the identity $\varepsilon^{\alpha\beta}_{;\alpha} = 0$ and the skew symmetry of $\varepsilon^{\alpha\beta}$ to derive

$$M^\alpha_{;\alpha} - k = \bar{M}^\alpha_{;\alpha} - S^\alpha a_\alpha - pn, \tag{6.1}$$

where S^α and p are defined by

$$S^\alpha = q^\alpha + \varepsilon^{\alpha\beta}q_{,\beta} \quad \text{and} \quad p = r - r^\alpha_{;\alpha}. \tag{6.2}$$

Using $(4.6)_2$, $(3.7)_2$ may then be solved to obtain

$$S^\alpha = \bar{M}^{\alpha\beta}_{;\beta} + \rho\bar{l}^\alpha \quad \text{and} \quad p = \bar{M}^{\alpha\beta}b_{\alpha\beta} + \rho\bar{l}, \tag{6.3}$$

where $\bar{l}^\alpha = \bar{l} \cdot a^\alpha$ and $\bar{l} = \bar{l} \cdot n$ respectively are the tangential and normal components of $\bar{l}$. With the constraints in force $(3.8)_2$ becomes $l - \bar{l} = \alpha\dot{v} + \beta\dot{n}$, which is determined by the position function $r(\theta^\alpha, t)$.

Following Naghdi (1981, Section 6) I use $(4.5)_1$, $(6.2)_1$ and the symmetry of $n_{;\beta\alpha}$ to obtain

$$N^\alpha_{;\alpha} = \bar{N}^\alpha_{;\alpha} + q^\alpha_{;\alpha}n + S^\alpha n_{,\alpha}. \tag{6.4}$$

The symmetry of $q_{;\beta\alpha}$ further implies that $q^\alpha_{;\alpha} = S^\alpha_{;\alpha}$. Consequently,

$$N^\alpha_{;\alpha} = T^\alpha_{;\alpha}, \quad \text{where} \quad T^\alpha = \bar{N}^\alpha + S^\alpha n. \tag{6.5}$$

It is of interest to note, from $(4.5)_1$ and $(6.2)_1$, that

$$N^\alpha - T^\alpha = \varepsilon^{\beta\alpha}(qn)_{,\beta}, \tag{6.6}$$

with divergence $\varepsilon^{\beta\alpha}(qn)_{;\beta\alpha}$ that vanishes identically, in accordance with $(6.5)_1$. Thus, the divergence form of the final equations, obtained by combining the foregoing with $(3.7)_1$, is

$$T^\alpha_{;\alpha} + \rho\bar{f} = 0. \tag{6.7}$$

If the constraints are operative then $\bar{f}$ is expressible in terms of $r(\theta^\alpha, t)$ and the assigned field f. Equations (5.3), (5.4), (6.3)$_1$, (6.5)$_2$ and (6.7) then furnish a formally determinate system of three equations for the three components of the position function, granted suitable boundary and initial conditions. The sole function of (6.3)$_2$ is to evaluate the combination of Lagrange multipliers defined by (6.2)$_2$ after the fact. These multipliers do not appear elsewhere in the balance laws. Therefore (6.2)$_2$ and (6.7) are decoupled. The former equation, which may now be regarded as irrelevant to the problem of determining $r(\theta^\alpha, t)$, has no counterpart in alternative treatments of the Kirchhoff-Love theory.

It is straightforward to use Stokes' theorem with (6.7) to generate a global balance law identical in form to the linear-momentum balance (3.2) in which $N = N^\alpha \nu_\alpha$ is replaced by $T^\alpha \nu_\alpha$. The two balance laws are equivalent since the difference between N^α and T^α is divergence free.

Naghdi's form of the equations (Naghdi 1972, Section 10) may be recovered by setting the inertia coefficients α and β to zero and taking $\bar{l} = l$ to be a tangential vector field. Indeed it is apparent from (6.3), (6.5) and (6.7) that in general only the tangential components of $\bar{l}$ contribute to the final equations for $r(\theta^\alpha, t)$. However, in the present framework it is not necessary to suppress the normal component of $\bar{l}$ as the latter merely affects the values of p. Further, it is not possible to eliminate this component if the inertia coefficients are present. Mathematical arguments for the retention of these coefficients are given in Antman (1995).

In Section 15 of Naghdi (1972) and in Section 6 of Naghdi (1981) Naghdi also stipulates that the skew part of the matrix of tangential components of M^α is to be set to zero to obtain a determinate system of equations. This is tantamount to requiring that $q = 0$ (cf. (4.5), (5.3) and (5.4)). However, the gradient of q appears only in the combination (6.3)$_1$, and q itself appears only in a divergence-free term that does not affect the local or global forms of the linear momentum balance. We will show that it has no direct effect on the Kirchhoff edge tractions or couples either. Thus it is not necessary to eliminate q, although one may do so without loss of generality insofar as the problem of calculating the mechanical response of the shell is concerned. Variables with this property are called *null-Lagrangians* in variational treatments of mechanics.

To obtain the Kirchhoff conditions I consider the part of the mechanical power P in (3.15) arising from boundary terms. According to (3.17) this is P_b, where

$$P_b = \int_{\partial\omega} (N \cdot v + M \cdot \dot{n}) \mathrm{d}s, \qquad (6.8)$$

in which the constraint $w = \dot{n}$ $(d = n)$ has been imposed. I now use

$$\dot{n} = -(n \cdot v_{,\alpha})a^{\alpha} \tag{6.9}$$

(e.g. Naghdi 1972) together with (3.6), (4.5)$_2$, (5.3)$_2$ and (6.6) to write

$$M \cdot \dot{n} = -(\bar{M}^{\beta\alpha} + q\varepsilon^{\alpha\beta})\nu_{\alpha}n \cdot v_{,\beta} \quad \text{and} \quad N \cdot v = [T^{\alpha} + \varepsilon^{\beta\alpha}(qn)_{,\beta}]\nu_{\alpha} \cdot v, \tag{6.10}$$

and combine these to obtain

$$N \cdot v + M \cdot \dot{n} = T^{\alpha}\nu_{\alpha} \cdot v - \bar{M}^{\beta\alpha}\nu_{\alpha}n \cdot v_{,\beta} + \varepsilon^{\beta\alpha}(qn \cdot v)_{,\beta}\nu_{\alpha}. \tag{6.11}$$

The vector field contracted with ν_{α} in the last term has zero divergence and thus makes no contribution to (6.8). In the second term I substitute

$$v_{,\beta} = \nu_{\beta}v_{\nu} + \tau_{\beta}v', \tag{6.12}$$

where τ_{β} are the covariant components of the unit tangent $\tau = n \times \nu$ to $\partial\omega$ and v', v_{ν} respectively are the independent arclength and normal derivatives of v on $\partial\omega$ (e.g. Naghdi 1972). The part of the resulting expression involving τ_{β} is

$$\bar{M}^{\beta\alpha}\nu_{\alpha}\tau_{\beta}n \cdot v' = (\bar{M}^{\beta\alpha}\nu_{\alpha}\tau_{\beta}n \cdot v)' - (\bar{M}^{\beta\alpha}\nu_{\alpha}\tau_{\beta}n)' \cdot v, \tag{6.13}$$

in which the first term on the right is regarded as a distributional derivative if $\partial\omega$ is piecewise smooth with a finite number of corners where ν and τ are discontinuous. It follows that

$$P_b = \int_{\partial\omega} (t \cdot v - Mn \cdot v_{\nu})\mathrm{d}s + \sum f_i \cdot v_i, \tag{6.14}$$

with

$$t = T^{\alpha}\nu_{\alpha} + (\bar{M}^{\beta\alpha}\nu_{\alpha}\tau_{\beta}n)', \quad M = \bar{M}^{\beta\alpha}\nu_{\beta}\nu_{\alpha}, \tag{6.15}$$

and

$$f = -\bar{M}^{\beta\alpha}[\nu_{\alpha}\tau_{\beta}]n, \tag{6.16}$$

where f_i is the point force at the ith corner of $\partial\omega$ required to support the motion, v_i is the particle velocity at the corner, the sum ranges over all the corners, and the notation $[\cdot]$ is used to denote the forward jump of the enclosed quantity as a corner is traversed in the sense of increasing s.

A further reduction may by achieved by using (6.9) and (6.12) to derive

$$\dot{n} = -(n \cdot v_{\nu})\nu - (n \cdot v')\tau. \tag{6.17}$$

Therefore $-Mn \cdot v_{\nu}$ may be replaced by $M\nu \cdot \dot{n}$ in (6.14). Then M is power-conjugate to $\nu \cdot \dot{n}$ and hence equal to the bending couple along the edge. Independent justification of this interpretation is provided below.

Finally, t is conjugate to v and thus provides the Kirchhoff edge traction. On substituting $(5.3)_1$ and $(6.5)_2$ into $(6.15)_1$ I find

$$n \cdot t = S^\alpha \nu_\alpha + (\bar{M}^{\beta\alpha} \nu_\alpha \tau_\beta)', \tag{6.18}$$

and conclude that the S^α contribute to the transverse shear tractions across material curves. The multiplier q does not appear explicitly and therefore does not directly affect the mechanical response of any subregion of the shell, as previously noted. Thus S^α may be identified with q^α (cf. $(6.3)_1$) without loss of generality. Using $(6.15)_1$ and (6.16) I may also write the global balance law associated with (6.7) in the form (3.2) with

$$\int_{\partial\omega} N \mathrm{d}s = \int_{\partial\omega} t \mathrm{d}s + \sum f_i. \tag{6.19}$$

8.7 Moment of momentum

In finite elasticity theory it is well known that the moment-of-momentum equation is identically satisfied if the response functions are derived from a strain energy that is insensitive to superposed rigid motions. The corresponding result for unconstrained Cosserat surfaces follows easily by evaluating (3.21) on the motion described by (3.14), with Q an arbitrary time-dependent rotation. The resulting equality is equivalent to the moment-of-momentum equation (3.9) by virtue of the arbitrariness of the axial vector of $\dot{Q}Q^T$.

For the Kirchhoff-Love theory (3.9) remains valid with the response functions replaced by the constitutive functions $\bar{N}^\alpha$, $\bar{M}^\alpha$ and $\bar{k}$. The same result follows directly by substituting (4.5) into (7.1), and with (5.3) this ultimately yields

$$\varepsilon_{\alpha\beta} \bar{N}^{\alpha\beta} = \varepsilon_{\alpha\beta} b_\lambda^\beta \bar{M}^{\alpha\lambda}. \tag{7.1}$$

In Naghdi's treatment (Naghdi 1972, Section 15) the counterpart of this equation is used to determine the skew part of the matrix corresponding to our $\bar{N}^{\alpha\beta}$. Here, I use $(5.4)_1$ and the symmetry of $\bar{\sigma}^{\alpha\beta}$ to obtain $\varepsilon_{\alpha\beta} \bar{N}^{\alpha\beta} = -\varepsilon_{\alpha\beta} b_\lambda^\alpha \bar{M}^{\lambda\beta}$, which is identically equal to the right-hand side of (7.1) by virtue of the symmetry of $\bar{M}^{\lambda\beta}$.

The global form of the moment-of-momentum identity for Kirchhoff-Love shells is given by (3.4) and (3.5) in which the boundary integral is

$$I = \int_{\partial\omega} (r \times N^\alpha \nu_\alpha + n \times M^\alpha \nu_\alpha) \mathrm{d}s. \tag{7.2}$$

I combine $(4.5)_2$ with (6.6) to reduce this to

$$I = \int_{\partial\omega} (r \times T^\alpha \nu_\alpha + n \times \bar{M}^\alpha \nu_\alpha) \mathrm{d}s, \tag{7.3}$$

wherein I have used $r \times (qn)_{,\beta} = (r \times qn)_{,\beta} + qn \times a_\beta$ to write the term in (6.6) involving the multiplier q as the sum of a divergence-free vector field and another term that cancels its counterpart in $(4.5)_2$. I now invoke $(6.15)_1$ together with $r' = \tau$ and $\tau \times n = \nu$ to write

$$r \times T^\alpha \nu_\alpha = r \times t + (\bar{M}^{\beta\alpha}\nu_\alpha \tau_\beta)\nu - (\bar{M}^{\beta\alpha}\nu_\alpha\tau_\beta r \times n)'. \tag{7.4}$$

Similarly, $(5.3)_2$ may be used with the identity $a_\beta = \nu_\beta\nu + \tau_\beta\tau$ and $(6.15)_3$ to obtain

$$n \times \bar{M}^\alpha \nu_\alpha = c - (\bar{M}^{\beta\alpha}\nu_\alpha\tau_\beta)\nu, \tag{7.5}$$

where

$$c = M\tau. \tag{7.6}$$

Thus, for piecewise smooth $\partial\omega$, I takes the simple form

$$I = \int_{\partial\omega} (r \times t + c)\mathrm{d}s + \sum r_i \times f_i, \tag{7.7}$$

which implies that c is a bending couple along $\partial\omega$ in accordance with the interpretation of M given in the previous section.

8.8 Summary of the Kirchhoff-Love theory

In view of the large amount of detail contained in the foregoing considerations, I pause here to recapitulate the *equilibrium* theory for elastic surfaces under the Kirchhoff-Love constraints.

Local equilibrium of forces may be expressed concisely as

$$T^\alpha_{;\alpha} + \rho f = 0, \tag{8.1}$$

where ρ is the mass of the film measured per unit area of ω, f is the distributed force per unit mass, T^α are stress vectors that contribute to the tractions transmitted across material curves, and the semi-colon is used to denote the surface covariant derivative using the metric of the coordinates induced by $r(\theta^\alpha)$. The stress vectors are given by

$$T^\alpha = N^\alpha + S^\alpha n, \tag{8.2}$$

where N^α are constitutively determined tangential vector fields, S^α is a contravariant vector field of Lagrange multipliers, and

$$n = \tfrac{1}{2}\varepsilon^{\alpha\beta}a_\alpha \times a_\beta \tag{8.3}$$

is the local orientation of ω. For the sake of clarity, the overbar notation for constitutive functions is suppressed in this section and in what follows. Further,

$a_\alpha = r_{,\alpha}$ are the tangent vectors induced by the coordinates, commas denoting partial derivatives, $\varepsilon^{\alpha\beta} = a^{-1/2}e^{\alpha\beta}$ is the permutation tensor density, $e^{\alpha\beta}$ ($= e_{\alpha\beta}$) is the unit alternator ($e^{12} = +1$), and $a = \det(C_{\alpha\beta})$, where $C_{\alpha\beta} = a_\alpha \cdot a_\beta$ is the induced metric, non-negative definite in general and assumed here to be positive definite. It is well known that the surface divergence in (8.1) may be written $T^\alpha_{;\alpha} = a^{-1/2}(a^{1/2}T^\alpha)_{,\alpha}$, allowing one to avoid Christoffel symbols.

The constitutively determinate term in (8.2) is expressible in the form

$$N^\alpha = N^{\beta\alpha}a_\beta, \tag{8.4}$$

with

$$N^{\beta\alpha} = \sigma^{\beta\alpha} - b^\beta_\mu M^{\mu\alpha}, \tag{8.5}$$

where

$$\sigma^{\beta\alpha} = \rho\left(\frac{\partial\Psi}{\partial C_{\alpha\beta}} + \frac{\partial\Psi}{\partial C_{\beta\alpha}}\right), \qquad M^{\beta\alpha} = -\frac{\rho}{2}\left(\frac{\partial\Psi}{\partial b_{\alpha\beta}} + \frac{\partial\Psi}{\partial b_{\beta\alpha}}\right), \tag{8.6}$$

and $b_{\alpha\beta} = n \cdot r_{,\alpha\beta}$ are the symmetric coefficients of the second fundamental form on ω. In the absence of distributed couples the normal components of (8.2) are given by

$$S^\alpha = M^{\alpha\beta}_{;\beta}. \tag{8.7}$$

The mixed components b^α_β in (8.5) are related to $b_{\alpha\beta}$ through $b_{\alpha\beta} = C_{\alpha\lambda}b^\lambda_\beta$. Further, the coordinate-dependent function $\Psi(C_{\alpha\beta}, b_{\alpha\beta})$ is the Galilean-invariant energy per unit mass of the film. I temporarily suppress the dependence of Ψ on the particle x. The *form* of this function is such that its *values* are independent of the coordinate system as the energy is required to be an absolute scalar field. One way to accommodate this requirement and the notion of fixed configuration embodied in the Noll-Murdoch-Cohen theory of material symmetry is to assume that

$$\Psi(C_{\alpha\beta}, b_{\alpha\beta}) = \hat\Psi(C, \kappa), \tag{8.8}$$

where $\hat\Psi$ is a coordinate-independent function,

$$C = C_{\alpha\beta}A^\alpha \otimes A^\beta \tag{8.9}$$

is the invertible symmetric surface strain tensor, and

$$\kappa = b_{\alpha\beta}A^\alpha \otimes A^\beta \tag{8.10}$$

is the symmetric relative curvature tensor. This is the viewpoint adopted in most works on elastic surfaces. In these definitions A^α are dual vectors to the induced tangent vectors $A_\alpha = x_{,\alpha}$ at the particle x, where $x(\theta^\alpha)$ is the local parametrization of a reference surface Ω. Given the parametrization, these tensors, together with $\hat{\Psi}$, are functions of the matrices $C_{\alpha\beta}$ and $b_{\alpha\beta}$ as suggested by the notation of (8.8).

Murdoch and Cohen introduced a primitive notion of material surface that includes (8.8) as a special case. In their work the local constitutive response is defined by Galilean-invariant functions of

$$F = a_\alpha \otimes A^\alpha \tag{8.11}$$

and its gradient. This maps the tangent space of Ω to that of ω at x. For hyperelastic surfaces, their constitutive equations follow from (8.8) if dependence on strain gradient is suppressed.

The local mass conservation law is

$$\rho_0 = J\rho, \quad \text{where} \quad J = (a/A)^{1/2}; \tag{8.12}$$

A and ρ_0 are the values of a and ρ respectively on a fixed reference surface Ω.

Many writers study the response of *fluid* films subject to the two-dimensional incompressibility constraint $J = 1$. This implies that deformations preserve surface area, and may be added to the list of constraints already imposed in the foregoing to obtain the local Kirchhoff-Love response functions from those of the Cosserat theory. The same procedure yields equations identical to those obtained by using the formal Lagrange multiplier rule

$$\Psi = \bar{\Psi}(a_{\alpha\beta}, b_{\alpha\beta}) - \gamma/\rho, \tag{8.13}$$

in (8.6), where $\bar{\Psi}$ is a constitutive function and $\gamma(\theta^\alpha)$ is a constitutively-indeterminate scalar field.

The foregoing equations are well known in principle, but rarely stated in forms that illuminate the underlying physics. To aid in the interpretation of the various terms it is helpful to relate them to the tractions and moments transmitted across material curves. To this end let $\theta^\alpha(s)$ be an arclength parametrization of such a curve on ω and let τ be the unit tangent in the direction of increasing s. Then, $\nu = \tau \times n$ is the rightward unit normal as the curve is traversed in the same direction. This has components $\nu_\alpha = \varepsilon_{\alpha\beta}\tau^\beta$, where $\tau^\alpha = d\theta^\alpha/ds$ are the components of τ and $\varepsilon_{\alpha\beta} = a^{1/2}e_{\alpha\beta}$ are the covariant components of the permutation tensor density. The traction transmitted by the material on the right to the material on the left is then given by

$$t = T^\alpha \nu_\alpha + (M^{\beta\alpha}\nu_\alpha\tau_\beta n)', \tag{8.14}$$

where $\tau_\alpha = C_{\alpha\beta}\tau^\beta$ and the prime denotes the derivative with respect to s. This furnishes the force per unit arclength. One then uses (8.2) to interpret $S^\alpha\nu_\alpha$ as a transverse shear traction across the curve. The moment per unit length is

$$m = r \times t + M\tau, \quad \text{where} \quad M = M^{\beta\alpha}\nu_\beta\nu_\alpha \tag{8.15}$$

is the bending couple.

Global forms of the equations for a simply-connected region $r \subset \omega$ are obtained by using Stokes' theorem in the form

$$\int_r T^\alpha_{;\alpha} da = \int_{\partial r} T^\alpha \nu_\alpha ds \qquad (8.16)$$

together with (8.1) and (8.14). The resulting force balance is

$$\int_r \rho f da + \int_{\partial r} t ds + \sum f_i = 0, \qquad (8.17)$$

where

$$f = -M^{\beta\alpha}[\nu_\alpha \tau_\beta] n \qquad (8.18)$$

is the force acting at a vertex of ∂r if the latter is piecewise smooth with a finite number of points where τ and ν are discontinuous and the sum ranges over all the vertices. Further, a straightforward but involved calculation yields the global identity

$$\int_r \rho r \times f da + \int_{\partial r} m ds + \sum r_i \times f_i = 0, \qquad (8.19)$$

which may be regarded as the specialization to equilibrium of the moment-of-momentum balance.

8.9 Material symmetry

Material symmetry theory for surfaces is not settled. This appears to be due to the difference between coordinate *form invariance*, as advocated by Rivlin (Barenblatt and Joseph 1997) and Naghdi (1972), and Noll's invariance of response under distinguished compositions of maps. For plate and shell theories, a number of alternative proposals have been advanced, some incorporating elements of Noll's approach (Murdoch and Cohen 1979, Ericksen 1972, Carroll and Naghdi 1972). Among them, I find that of Murdoch and Cohen (1979) to be the most satisfactory extension of Noll's concept. This is based on the idea that local configurations of the body are to be regarded as the restrictions to surfaces of diffeomorphisms of 3-space. Symmetries are associated with local maps among *fixed* surfaces that leave the energy invariant in a given diffeomorphism. I present a brief summary of the Murdoch-Cohen theory here, with adaptations tailored to the narrower class of material surfaces considered.

I remark that the present concept of symmetry is not equivalent to that adopted by Naghdi (1972); his theory presumes the *form invariance* of the response functions of the surface under certain coordinate transformations. This requires that the strain energy, written as a function of tensor components referred to a particular coordinate system, be the same function of the components obtained by a specific transformation of coordinates that reflects the underlying notion of symmetry. It appears that the two theories yield equivalent results for models of simple materials and surfaces (membranes) but not for models that incorporate curvature or strain-gradient effects. My preference for Noll's framework derives from its accessibility to empirical test and its coordinate invariance. However, coordinate parametrizations are freely used in its description.

Preliminary to this, I examine certain local properties of maps between two fixed surfaces Ω and Ω^*, with local parametrizations $x(\theta^\alpha)$ and $x^*(\theta^\alpha)$ respectively, occupied by the same material body. Thus, let $\phi(X)$ be a C^2 orientation-preserving diffeomorphism of 3-space to itself defined on an open neighborhood of a material point x with coordinates θ^α. Let $N^* \subset \Omega^*$ be the intersection of this neighborhood with Ω^*, and suppose $x = \phi(x^*)$ for $x^* \in N^*$. Then, $N = \phi(N^*) \subset \Omega$ is the intersection of the same set of material points with Ω.

If A_α^* and A_α are the tangent vectors induced by θ^α on Ω^* and Ω at x, then

$$A_\alpha = (\nabla\phi)A_\alpha^* \quad \text{and} \quad A_{\alpha,\beta} = (\nabla\nabla\phi)[A_\alpha^* \otimes A_\beta^*] + (\nabla\phi)A_{\alpha,\beta}^*, \tag{9.1}$$

where $\nabla\phi$ and $\nabla\nabla\phi$ are the first and second gradients of $\phi(X)$ evaluated at $x \in N^*$. The operation in the second expression is defined, using Cartesian notation, by

$$(\nabla\nabla\phi)[u \otimes v] = (\partial^2\phi_i/\partial X_A \partial X_B)u_A v_B e_i, \tag{9.2}$$

with $\{e_i\}$ an orthonormal basis for 3-space.

In the remainder of this section I assume the tangent spaces to the various surfaces occupied by the body to coincide at the particle x. Galilean invariance implies that this entails no loss of generality in the characterization of constitutive response (Murdoch and Cohen 1979). With this adjustment $(9.1)_1$ is then equivalent to

$$A_\alpha = HA_\alpha^* = H_{.\alpha}^\lambda A_\lambda^*, \tag{9.3}$$

where

$$H = A_\alpha \otimes A^{*\alpha} = H_{.\beta}^\alpha A_\alpha^* \otimes A^{*\beta}, \qquad H_{.\beta}^\alpha = A^{*\alpha} \cdot (\nabla\phi)A_\beta^*, \tag{9.4}$$

and the $A^{*\alpha}$ are dual to A_α^*. The properties of ϕ ensure that H is an invertible linear transformation from the tangent space to itself. Accordingly, there is a tensor R of the same type such that $R^T = H^{-1}$, and it is straightforward to show that

$$A^\alpha = RA^{*\alpha}. \tag{9.5}$$

Let N be the orientation of Ω at x. Then,

$$\mu_{\alpha\beta}N = A_\alpha \times A_\beta, \tag{9.6}$$

where $\mu_{\alpha\beta} = A^{1/2}e_{\alpha\beta}$ is the associated permutation tensor density. I combine this with a similar formula for the orientation N^* of Ω^* and use (9.3) with $\mu^*_{\lambda\gamma}H^\lambda_{.\alpha}H^\gamma_{.\beta} = (\det H)\mu^*_{\alpha\beta}$ to derive

$$\mu_{\alpha\beta}N = (\det H)\mu^*_{\alpha\beta}N^*. \tag{9.7}$$

This result and $A/A^* = (\det H)^2$, which follows from (9.3), yield $\det H = \pm(A/A^*)^{1/2}$ according as $N = \pm N^*$. In particular, $\det R$ $(= 1/\det H)$ is positive if and only if Ω and Ω^* have the same orientation; otherwise it is negative.

Let B be the curvature tensor of Ω at x. Then $B = B_{\alpha\beta}A^\alpha \otimes A^\beta$, where $B_{\alpha\beta} = N \cdot A_{\alpha,\beta}$. Let B^*, defined similarly, be the curvature tensor of Ω^* at the same particle. The relationship between the two curvatures may be inferred from $(9.1)_{1,2}$, (9.5) and the fact that $\nabla\phi$ maps the tangent space to itself:

$$R^{-1}BR^{-T} = (N \cdot \nabla\nabla\phi[A_\alpha \otimes A_\beta])A^\alpha \otimes A^\beta + (N \cdot (\nabla\phi)N^*)B^*, \tag{9.8}$$

wherein the first term on the right has the same value regardless of which set of tangent bases is used.

Consider a configuration ω of the film parametrized locally by $r(\theta^\alpha)$, and let C, κ and C^*, κ^* be the strains and curvatures of ω relative to Ω and Ω^* respectively. Since the first and second fundamental forms on ω are determined by its parametric representation, it follows from (8.9), (8.10) and (9.5) that

$$C = RC^*R^T \quad \text{and} \quad \kappa = R\kappa^*R^T. \tag{9.9}$$

The energy per unit mass is presumed to be a property of the body in a given state. As such its values at x are not dependent on the reference surface used to compute them. In the notation of (8.8),

$$\hat{\Psi}^*(C^*, \kappa^*) = \hat{\Psi}(C, \kappa) = \hat{\Psi}(RC^*R^T, R\kappa^*R^T), \tag{9.10}$$

where $\hat{\Psi}$ and $\hat{\Psi}^*$ are constitutive functions defined on Ω and Ω^*.

Consider another diffeomorphism $\xi(X)$ of the same kind as ϕ but with the property that $r = \xi(x)$ for $x \in N$. This induces at x a strain

$$C = ((\nabla\xi)A_\alpha \cdot (\nabla\xi)A_\beta)A^\alpha \otimes A^\beta \tag{9.11}$$

relative to Ω and a relative curvature obtained with the aid of the formula

$$\kappa = F^T b F, \tag{9.12}$$

which follows from (8.10) and (8.11). With the tangent spaces aligned at x, F plays the same role in the local map from Ω to ω as that played by H in the map from Ω^* to Ω. Accordingly, (9.8) and (9.12) give

$$\kappa = (n \cdot \nabla\nabla\xi[A_\alpha \otimes A_\beta])A^\alpha \otimes A^\beta + (n \cdot (\nabla\xi)N)B, \tag{9.13}$$

where n is the orientation of ω at x. For definiteness, and without loss of generality, I choose the orientations of the reference and distorted surfaces to coincide at x. Thus, $n = N$ in (9.13).

To characterize the relationship between N and N^* due to symmetry, it is necessary to determine the strain $\bar{C}$ and curvature $\bar{\kappa}$ relative to Ω^* induced at x by $r^* = \xi(x^*)$ for $x^* \in N^*$. These are given by the obvious modifications to (9.11) and (9.13) in which $n^* = N^*$. The values of $\nabla\xi$ and $\nabla\nabla\xi$ are the same in both sets of formulas, and it follows easily that the strains are also equal, but $\bar{\kappa}$ and κ differ in a manner that depends on the relative orientations of N^* and N:

$$\bar{C} = C, \quad \bar{\kappa} = \begin{cases} \kappa + (N \cdot (\nabla\xi)N)(B^* - B); & \det R > 0, \\ -\kappa + (N \cdot (\nabla\xi)N)(B^* + B); & \det R < 0, \end{cases} \tag{9.14}$$

wherein B and B^* are connected by the map $\phi(X)$ through (9.8).

Following Noll (1958), Murdoch and Cohen (1979) regard N and N^* as being related by symmetry if they respond identically to the *same* $\xi(X)$. Then, $\hat{\Psi}(C, \kappa) = \hat{\Psi}^*(\bar{C}, \bar{\kappa})$, which, when combined with (9.10), yields

$$\hat{\Psi}(C, \kappa) = \hat{\Psi}(R\bar{C}R^T, R\bar{\kappa}R^T). \tag{9.15}$$

Murdoch and Cohen have shown that the pairs $(R, B^* \pm B)$ satisfying (9.15), with the sign chosen in accordance with (9.14), are elements of a group, which I denote by $\mathcal{G}$. Thus, arguments used in conventional elasticity may be used here to restrict $\mathcal{G}$ to the unimodular group (det $R = \pm 1$).

Remark 1. Equation (9.15) has a counterpart among the formulas obtained by Murdoch and Cohen (1979, equations (5.26) and (5.27)). It is apparent that the dependence of the function $\hat{\Psi}$ on its second argument generates a non-standard symmetry condition involving the geometry of the reference surface. This has no analogue in Noll's theory of simple materials, nor in the concept of coordinate form invariance, but is nevertheless readily understood. To illustrate this, let det $R > 0$ and consider a local configuration having the geometry of a circular cylinder. A map by the two-dimensional rotation

$R = -1$ ($\det R = 1$) may be regarded as the projection onto the tangent plane of a three-dimensional rotation through the angle π about the normal to the surface. This map preserves generators and latitudinal curves on the cylinder, and thus yields $B^* = B$. If the response of the surface to *strain* is unaffected by such a transformation, then intuitively one would expect the original and the mapped neighborhoods to respond identically to a given deformation. However, a map R corresponding to an arbitrary rotation about the normal yields a neighborhood that may not be congruent to the original, and there is no reason to expect that the two neighborhoods would now respond identically to the same deformation. Thus, the presence of $B - B^*$ in (6.17) is plausible on physical grounds. In this example $\{\pm 1, 0\} \subset \mathcal{G}$; the associated specialization of (9.15) is an identity and thus yields no restrictions on the form of the function $\hat{\Psi}$. I refer to such surfaces as having *trivial* symmetry.

Remark 2. In an addendum to their work, Murdoch and Cohen (1981) propose the use of a limited class of maps $\phi(X)$ in the definition of the symmetry group. This is motivated by the observation that the arguments C and κ of the surface strain-energy function involve only the restriction $\phi|_\Omega$ of ϕ to the surface Ω. If this restriction is chosen to be the identity transformation on Ω, then the induced deformation tensors of the neighborhoods $N \subset \Omega$ and $N^* \subset \Omega^*$ are both equal to 1, the unit tensor on the common tangent planes of Ω and Ω^* at the particle in question. Further, the component $(\nabla\phi)N \cdot N$ is unaffected by the specification of $\phi|_\Omega$. It then follows from (9.10) that $\hat{\Psi}(1, -B) = \hat{\Psi}^*(1, \pm[(\nabla\phi)N \cdot N - 1]B - [(\nabla\phi)N \cdot N]B^*)$ whenever N and N^* are symmetry related, with the choice of sign depending on the orientation of N^* relative to N. Since infinitely many values of $(\nabla\phi)N \cdot N$ may be associated with $\phi|_\Omega$, it follows that $\hat{\Psi}^*(1, \bar{\kappa})$ takes the same value at infinitely many $\bar{\kappa}$. This is clearly undesirable. Exceptionally, if $B^* = \pm B$, as appropriate, or if $\phi(X)$ is chosen such that $(\nabla\phi)N \cdot N = 1$ on Ω, then the symmetry condition yields $\hat{\Psi}(1, -B) = \hat{\Psi}^*(1, -B^*)$ and the foregoing conclusion is avoided. If neither alternative applies then it is likely that the symmetry set is trivial.

In Murdoch and Cohen (1979) a constraint equivalent to $(\nabla\phi)N \cdot N = 1$ is adopted to allow for the existence of non-trivial symmetry sets without restrictions on the embedding geometry of N^* relative to that of N. In their framework, such a constraint may be imposed without loss of generality since the function ϕ merely plays the role of an orientation-preserving diffeomorphism of the enveloping 3-space, and it is only the function $\phi|_\Omega$ that is operative in the associated constitutive theory. Murdoch and Cohen impose the constraint for arbitrary deformations of the surface. However, in the setting of surface-substrate interactions the response of the film-substrate system is affected by the properties of $\phi(X)$ in a three-dimensional neighborhood of a point on Ω. Thus it is more natural to regard the map $\phi(X) = X$ as the appropriate identity transformation in the present theory. This choice automatically satisfies the Murdoch-Cohen constraint for the trivial deformation. However, the imposition of the constraint in an arbitrary deformation of the surface is unnatural to the extent that it imposes undue restrictions on the substrate deformation. It is then likely that the symmetry set is trivial for arbitrary film-substrate deformations unless $B^* = \pm B$.

Remark 3. For those surfaces having $\mathcal{G} \supset \{R, 0\}$ with $\det R = 1$, (9.15) reduces to the functional equation $\hat{\Psi}(C, \kappa) = \hat{\Psi}(RCR^T, R\kappa R^T)$, which is amenable to analysis by available representation theory. In particular, it is then equivalent to the coordinate form invariance of $\hat{\Psi}$. Planes are obvious examples of this, since all neighborhoods of a point on a plane have zero curvature. In the course of seeking other examples I am led to re-examine the nature of the maps between the neighborhoods N^* and N discussed previously. In particular, I note that H in (9.4) is the value of the gradient of the map at a particular point x. Consider a three-dimensional neighborhood of the point whose intersection with the surface is N^*, and let the neighborhood undergo an affine three-dimensional rotation about the normal, with N^* embedded, the projection of which onto the tangent plane is H at the particle in question. Then, N is the image of a rigid map of N^*, and it follows that the curvatures B and B^* of the two neighborhoods are related by $B = RB^*R^T$, where $R = H^{-T}$ (cf. (9.8)). To ensure that $B = B^*$ (such that $\mathcal{G} \supset \{R, 0\}$ with $\det R = 1$) I then require that $RB = BR$. If this should hold for *arbitrary* rotations R then it is necessary and sufficient that B be a scalar multiple of the surfacial unit tensor 1. In this case the local embedding geometry is that of a plane or a sphere. Of course, it is possible to consider other kinds of symmetry in the same framework. However, if non-affine transformations are admitted then symmetry considerations are further complicated by the fact that $B \neq RB^*R^T$.

Remark 4. It is to be expected that the strain energy will be affected by the curvature of the reference surface for reasons other than those implied in (9.14) and (9.15), and that this will generally be manifested by an explicit coordinate dependence in the function $\hat{\Psi}$. An illustrative example is furnished by a coiled cylinder. The tightness of the coil, as measured by the reference curvature tensor, varies from one generator to the next; thus, the local response to a prescribed deformation is expected to vary as the orthogonal trajectories are traversed, even if the corresponding *membrane* energy, obtained by suppressing the curvature dependence of $\hat{\Psi}$, is uniform. The presence of this type of non-uniformity, reflected in the θ^α-dependence of $\hat{\Psi}(C, \kappa; x)$, is thus natural in surfaces with flexural resistance.

I note that it is common practice in shell theory to suppose that the strain energy depends parametrically on reference curvature (see, for example, Simmonds 1985). The foregoing example furnishes motivation for such an assumption. Typically this is introduced via a *curvature strain* $\Delta = \kappa + B$ (Naghdi 1972, Section 13; Simmonds 1985) that vanishes in the reference configuration, where $\kappa = -B$. In the context of the present theory Δ may be introduced as a constitutive variable through an explicit dependence of the function $\hat{\Psi}$ on coordinates of the form $\hat{\Psi}(C, \kappa + B(\theta^\alpha))$. This modification does not alter the foregoing conclusions regarding symmetry. Thus, examination of the symmetry restrictions given in the works cited reveals that they are generally incompatible with (9.14) and (9.15). Exceptionally, if the constraint $(\nabla\phi)N \cdot N = 1$ is imposed, and if $B = RB^*R^T$, then the argument leading to (9.15) may be used to derive the restriction $\hat{\Psi}(C, \Delta) = \hat{\Psi}(RCR^T, \pm R\Delta R^T)$, with the sign chosen in accordance with that of $\det R$. The present theory then reduces to the alternatives cited when $\det R > 0$.

I remark that the concept of symmetry advanced by Carroll and Naghdi (1972) reflects the view that reference curvature should enter into the local constitutive response in a fundamental way. However, they furnished no primitive notion of symmetry upon which to base a deductive line of reasoning leading to a characterization of the nature of this dependence. The Murdoch-Cohen generalization of Noll's theory yields a precise and readily understood restriction on the strain energy that reveals the role of reference curvature explicitly.

(a) Hemitropy relative to planes and spheres

I specialize the general theory to *hemitropic* films. These are defined to be surfaces for which a reference configuration can be found such that $\mathcal{G} = \{R, 0\}$ with R proper orthogonal ($\det R = +1$) but otherwise arbitrary at every point. I take Ω to be such a configuration. In the light of the third of the foregoing remarks, planes and spherical sectors furnish examples for which the characterization of hemitropy does not involve the non-standard terms in (9.15) associated with the curvature of the reference surface. Thus, I seek to characterize those functions $\hat{\Psi}$ with the property

$$\hat{\Psi}(C, \kappa) = \hat{\Psi}(RCR^T, R\kappa R^T); \qquad \det R = 1, \quad RR^T = 1. \qquad (9.16)$$

According to standard representation theory, this is satisfied for all such R if and only if $\hat{\Psi}$ is expressible as a function U of the elements of the functional basis (Zheng 1993, Table 2)

$$I = \{I_1, I_2, ..., I_6\}, \tag{9.17}$$

where

$$
\begin{aligned}
I_1 &= \text{tr}\,C = C_{\alpha\beta}A^{\alpha\beta}, \quad I_2 = \det C = a/A, \quad I_3 = \text{tr}\,\kappa = \kappa_{\alpha\beta}A^{\alpha\beta}, \\
I_4 &= \det \kappa = \tfrac{1}{2}\mu^{\alpha\beta}\mu^{\lambda\gamma}\kappa_{\alpha\lambda}\kappa_{\beta\gamma}, \quad I_5 = \text{tr}(C\kappa) = C_{\alpha\beta}\kappa^{\alpha\beta} = C^{\alpha\beta}\kappa_{\alpha\beta}, \\
I_6 &= \text{tr}(C\kappa\mu) = C_{\alpha\beta}D^{\alpha\beta} = \kappa_{\alpha\beta}E^{\alpha\beta}. \tag{9.18}
\end{aligned}
$$

and $A^{\alpha\beta}$ is the dual metric induced by the coordinates on Ω. Here, I use the definitions

$$
\begin{aligned}
C^{\alpha\beta} &= G^{\alpha\gamma}G^{\beta\delta}C_{\gamma\delta}, \quad \kappa^{\alpha\beta} = G^{\alpha\gamma}G^{\beta\delta}\kappa_{\gamma\delta}, \\
D^{\alpha\beta} &= \kappa^{\alpha\sigma}\mu^{\gamma\beta}G_{\gamma\sigma}, \quad E^{\alpha\beta} = \mu^{\alpha\gamma}C^{\sigma\beta}G_{\sigma\gamma}, \tag{9.19}
\end{aligned}
$$

where $\mu = \mu^{\alpha\beta}A_\alpha \otimes A_\beta$ is the permutation tensor density on Ω induced by the components $\mu^{\alpha\beta} = A^{-1/2}e^{\alpha\beta}$.

The constitutive relations for the surface follow from

$$
\frac{\partial I_1}{\partial C_{\alpha\beta}} = A^{\alpha\beta}, \quad \frac{\partial I_2}{\partial C_{\alpha\beta}} = A^{-1}\frac{\partial a}{\partial C_{\alpha\beta}} = \tilde{C}^{\alpha\beta}, \quad \frac{\partial I_3}{\partial \kappa_{\alpha\beta}} = A^{\alpha\beta},
$$

$$
\frac{\partial I_4}{\partial \kappa_{\alpha\beta}} = \mu^{\alpha\gamma}\mu^{\beta\lambda}\kappa_{\gamma\lambda} = \tilde{\kappa}^{\alpha\beta}, \quad \frac{\partial I_5}{\partial C_{\alpha\beta}} = \kappa^{\alpha\beta}, \quad \frac{\partial I_5}{\partial \kappa_{\alpha\beta}} = C^{\alpha\beta},
$$

$$
\frac{\partial I_6}{\partial C_{\alpha\beta}} = D^{\alpha\beta}, \quad \frac{\partial I_6}{\partial \kappa_{\alpha\beta}} = E^{\alpha\beta}, \tag{9.20}
$$

wherein the notation $(\tilde{\cdot})$ is used to denote the adjugate of a surface tensor. An alternative formula for the latter may be obtained from the Cayley-Hamilton theorem. Thus, for any such tensor T, the adjugate is given by

$$\tilde{T} = (\text{tr}\,T)\mathbf{1} - T, \tag{9.21}$$

which is valid whether or not the determinant of T vanishes. Equations (8.6) then yield the constitutive equations

$$
\frac{1}{2}\rho^{-1}\sigma^{\alpha\beta} = \frac{\partial U}{\partial I_1}A^{\alpha\beta} + \frac{\partial U}{\partial I_2}\tilde{C}^{\alpha\beta} + \frac{\partial U}{\partial I_5}\kappa^{\alpha\beta} + \frac{1}{2}\frac{\partial U}{\partial I_6}(D^{\alpha\beta} + D^{\beta\alpha}),
$$

$$
-\rho^{-1}M^{\alpha\beta} = \frac{\partial U}{\partial I_3}A^{\alpha\beta} + \frac{\partial U}{\partial I_4}\tilde{\kappa}^{\alpha\beta} + \frac{\partial U}{\partial I_5}C^{\alpha\beta} + \frac{1}{2}\frac{\partial U}{\partial I_6}(E^{\alpha\beta} + E^{\beta\alpha}). \tag{9.22}
$$

The dependence of the energy on the invariants I_5 and I_6 allows for the modelling of thin three-dimensional bodies composed of lamella having different properties. These generate coupling of the local (two-dimensional) response to strain and flexure. Related issues are discussed in Libai and Simmonds (1998).

(b) Isotropy relative to planes

I define a surface to be isotropic relative to a plane reference surface Ω if the symmetry set $\mathcal{G} = \{R, 0\}$ for all two-dimensional orthogonal R. According to (9.15) it then follows that

$$\hat{\Psi}(C, \kappa) = \hat{\Psi}(RCR^T, \pm R\kappa R^T); \quad \det R = \pm 1, \quad RR^T = 1. \tag{9.23}$$

This does not conform to the standard definition of isotropy, according to which the first branch of (9.14) applies with $\det R = \pm 1$. The conventional definition is incompatible with the theory of symmetry adopted in this work unless $\hat{\Psi}$ happens to be insensitive to curvature. The latter specialization furnishes a model for ideal membranes.

I further assume the strain energy to be an even function of curvature since this implies that the couples $M^{\alpha\beta}$ vanish at $\kappa = 0$ provided that $\hat{\Psi}$ is differentiable, and this in turn requires that plane deformations be supported without bending moments. This restriction is consistent with the behavior of conventional isotropic plates, regarded as thin, uniform, three-dimensional prismatic bodies.

The functional basis for isotropic response differs from that for hemitropy. To construct it I note, with reference to (9.18), that for arbitrary orthogonal R the scalars I_1 to I_5 are invariant under replacement of C and κ by RCR^T and $R\kappa R^T$ respectively, whereas $I_6 \to (\det R)I_6$. Among these, only I_3 and I_5 are altered (to the extent of a change of sign) if κ is replaced by $-R\kappa R^T$, while I_6 is unaffected by this replacement provided that $\det R = -1$. The functional basis for isotropic response under the additional restriction that $\hat{\Psi}$ be an even function of curvature thus consists of the independent invariants

$$I_1, I_2, I_3^2, I_4, I_5^2, I_6^2, I_7 \equiv I_3 I_5. \tag{9.24}$$

The response functions are now given by

$$\frac{1}{2\rho}\sigma^{\alpha\beta} = \frac{\partial U}{\partial I_1}A^{\alpha\beta} + \frac{\partial U}{\partial I_2}\tilde{C}^{\alpha\beta} + (2I_5\frac{\partial U}{\partial I_5^2} + I_3\frac{\partial U}{\partial I_7})\kappa^{\alpha\beta} + I_6\frac{\partial U}{\partial I_6^2}(D^{\alpha\beta} + D^{\beta\alpha}),$$

$$-\frac{1}{\rho}M^{\alpha\beta} = (2I_3\frac{\partial U}{\partial I_3^2} + I_5\frac{\partial U}{\partial I_7})A^{\alpha\beta} + \frac{\partial U}{\partial I_4}\tilde{\kappa}^{\alpha\beta} + (2I_5\frac{\partial U}{\partial I_5^2} + I_3\frac{\partial U}{\partial I_7})C^{\alpha\beta}$$

$$+ I_6\frac{\partial U}{\partial I_6^2}(E^{\alpha\beta} + E^{\beta\alpha}). \tag{9.25}$$

(c) Fluid films

The structure of amphiphilic fluid bilayers and the high degree of in-plane mobility observed in equilibrium states suggest a definition of fluidity analogous to that of Noll (1958) for conventional bulk fluids. Thus, suppose the embedding geometry of $N^* \subset \Omega^*$ is that of a plane, so that $B^* = 0$ at x. For such N^*, I define fluidity by the requirement that (9.15) be satisfied for all affine $\phi(X)$ $(\nabla\nabla\phi \equiv 0)$ with the properties that $\nabla\phi$ is proper unimodular $(\det \nabla\phi = +1)$ and maps the subspaces $\mathcal{T}$ and $\text{span}\{N\}$ to themselves, where $\mathcal{T}$ is the common tangent space at x. The induced surface tensor R fulfills the requirement $|\det R| = 1$, and (9.8) implies that N is related to N^* by symmetry only if $B = 0$. Accordingly, $(9.14)_2$ simplifies to $\bar{\kappa} = \pm\kappa$ and (9.15) becomes

$$\hat{\Psi}(C, \kappa) = \hat{\Psi}(RCR^T, \pm R\kappa R^T); \quad \det R = \pm 1. \tag{9.26}$$

This definition of fluidity is meant to reflect the small-scale three-dimensional structure of bilayers in configurations in which the interfaces between the bilayer and the bulk fluid are parallel planes, as depicted, for example, in Figure 4.2 of Charvolin and Sadoc (1994). Its implications for response in arbitrary configurations are of primary interest here.

To obtain the canonical form of the energy function, I note that surface rotations $(R^{-1} = R^T, \det R = +1)$ are admitted by the definition. The appropriate specialization of (9.26) is satisfied for all such R if and only if $\hat{\Psi}$ is expressible as a function of the elements of the hemitropic function basis (9.20). Invariance under arbitrary proper-unimodular transformations implies that the energy is expressible as a function of all independent invariants formed from the list (9.18) that are also proper-unimodular invariants. That this class of functions is the most general one to fulfill the stated requirement follows from the fact that invariance under the larger set of transformations implies invariance under the smaller set. Thus, while any such function is necessarily expressible in terms of the elements of the set I, it cannot be an arbitrary function of these elements.

The obvious candidates for inclusion are $\det C$ and $\det \kappa$. Another invariant having the required property is

$$\sigma = (\text{tr } C)(\text{tr } \kappa) - \text{tr}(C\kappa). \tag{9.27}$$

To prove this I use the Cayley-Hamilton theorem (9.21). Thus,

$$\sigma = \text{tr}(\tilde{C}\kappa) = \text{tr}(C\tilde{\kappa}). \tag{9.28}$$

Since C is presumed invertible, the first alternative gives $\sigma = (\det C) \text{tr}(C^{-1}\kappa)$, in which the second factor is invariant under the replacements $C \to RCR^T$ and $\kappa \to R\kappa R^T$ for all invertible surface tensors R. The result then follows by the unimodular invariance of $\det C$. I have not succeeded in generating additional independent proper-unimodular invariants from the set I and thus conjecture that the three discussed comprise the maximal set.

It has already been noted that $\det C = J^2$. Granted the truth of the conjecture, it is thus necessary that

$$\hat{\Psi}(C, \kappa) = G(J(C), \sigma(C, \kappa), \kappa(\kappa); x);$$

$$J(C) \doteq (\det C)^{1/2}, \quad \sigma(C, \kappa) \doteq \operatorname{tr}(\tilde{C}\kappa), \quad \kappa(\kappa) \doteq \det \kappa, \qquad (9.29)$$

where G is scalar valued and parametric dependence on the particle is indicated explicitly. Conversely, if (9.29) holds, then, since κ and σ are even and odd functions of κ respectively, (9.26) is satisfied for all unimodular R ($\det R = \pm 1$) provided that G is an even function of σ. Each of these generates a proper-unimodular $\nabla\phi$ through $\nabla\phi = R^{-T} \pm N \otimes N$, with the sign chosen as appropriate. This in turn is the general form of $\nabla\phi$ compatible with the definition of fluidity and so the bilayer is fluid if and only if (9.29) holds with σ replaced by $|\sigma|$.

A monolayer film may be viewed as half a bilayer. For these an appropriate definition of fluidity is obtained from that for bilayers by restricting ϕ so as to preserve the local orientation of the surface. In this case the second branch of (9.26) is not applicable and the necessary and sufficient condition for fluidity is again given by (9.29), but without the requirement that G be an even function of σ.

To determine the response functions relative to arbitrarily curved local reference configurations, I re-write (9.29) in the form

$$\hat{\Psi}_\lambda(C_\lambda, \kappa_\lambda) = G_\lambda(J_\lambda, \sigma_\lambda, \kappa_\lambda; x);$$

$$J_\lambda = J(C_\lambda), \quad \sigma_\lambda = \sigma(C_\lambda, \kappa_\lambda), \quad \kappa_\lambda = \kappa(\kappa_\lambda), \qquad (9.30)$$

where the subscript λ is used to identify the reference configuration. With ω fixed, the transformation from λ to another local reference configuration, μ say, yields the composition formula $J_\lambda = J_\mu D(x)$, where $J_\mu = J(C_\mu)$ and $D(x)$ is the positive square root of the determinant of the strain of μ relative to λ at x. Next, I observe that κ is related to κ in the same way that the Gaussian curvature K is related to b. Thus,

$$\kappa = \tfrac{1}{2}\mu^{\alpha\beta}\mu^{\lambda\mu}b_{\alpha\lambda}b_{\beta\mu} = J^2 K, \qquad (9.31)$$

and so $\kappa_\lambda = \kappa_\mu D^2$ where $\kappa_\mu = \kappa(\kappa_\mu)$. The function σ may likewise be expressed in terms of the mean curvature H by using (9.12) and $C^{-1} = a^{\alpha\beta}A_\alpha \otimes A_\beta$ in (9.28), yielding

$$\sigma = 2J^2 H. \qquad (9.32)$$

Hence, $\sigma_\lambda = \sigma_\mu D^2$ where $\sigma_\mu = \sigma(C_\mu, \kappa_\mu)$.

If $\hat{\Psi}_\mu$ is the response function with μ as reference, then

$$\hat{\Psi}_\mu(C_\mu, \kappa_\mu) = \hat{\Psi}_\lambda(C_\lambda, \kappa_\lambda), \tag{9.33}$$

and (9.30) furnishes

$$\hat{\Psi}_\mu = G_\mu(J_\mu, \sigma_\mu, \kappa_\mu; x), \tag{9.34}$$

where

$$G_\mu(J_\mu, \sigma_\mu, \kappa_\mu; x) = G_\lambda(J_\mu D(x), \sigma_\mu D^2(x), \kappa_\mu D^2(x); x). \tag{9.35}$$

This holds without restrictions on the embedding geometry of μ or its tangent space. Equation (9.29) therefore implies that the response relative to any local reference configuration is sensitive to the strain and relative curvature through the associated values of J, σ and κ. In particular, the canonical form of the energy function, derived on the basis of a particular class of reference placements, is insensitive to the reference placement. This result is in accord with the intuitive idea of fluidity. Explicit forms of the response functions $\sigma^{\alpha\beta}$ and $M^{\alpha\beta}$ are given in Steigmann (1999).

8.10 Energy minimizers

Most treatments (Antman 1995, Simmonds 1985) of the Kirchhoff-Love equations are based on stationary- or minimum-energy considerations in the spirit of Kirchhoff's original work (Kirchhoff 1850). I record here an energy functional which is rendered stationary by films coexisting in equilibrium with bulk materials. This furnishes a variational framework for diverse phenomena ranging from elastic surface-substrate interactions to surfactant fluid-film microstructures interspersed in a bulk fluid phase. The energy is used in the discussion of necessary conditions for minimizing states. It is sufficient for this purpose to confine attention to local compactly-supported variations defined on a closed connected region R consisting of the bulk materials and a number of films. These films consist either of compact (closed) surfaces contained entirely within the interior of R, or of surfaces that intersect ∂R along certain curves. In the latter case I avoid the complicated and irrelevant boundary terms associated with the film traction and moment by considering variations that vanish together with their gradients on ∂R.

The appropriate energy functional is

$$E[y, r] = \sum_i \int_{v_i} \varrho \Phi dv + \sum_j \int_{\omega_j} \rho \Psi da, \tag{10.1}$$

where v_i and ω_j respectively are the volumes and surfaces occupied by the bulk fluids and films in R, Φ is the energy of a bulk material phase per unit mass, ϱ is the associated mass density, r is the position field on a film, and y is the position field in the bulk. I assume the bulk materials to be simple solids or fluids in the sense that Φ is determined by y through its gradient with respect to a suitable reference placement.

That equilibria render this energy stationary under the stated conditions may be verified by evaluating the Gâteaux differential of $E[y_\epsilon, r_\epsilon]$ with respect to a parameter ϵ at the value $\epsilon = 0$ (say) associated with the equilibrium state. Here, y_ϵ and r_ϵ are the parametrized positions of particles of the bulk materials and the films respectively. I refer to Steigmann and Ogden (1999) for details of the lengthy formal argument.

I obtain the quasiconvexity condition and related algebraic inequalities associated with necessary conditions for energy minimizers. Although the minimum energy test is inconclusive with respect to the dynamical stability of equilibria, it nevertheless furnishes a formal necessary condition for asymptotic stability if the associated dynamics are strictly dissipative (Ericksen 1966, Koiter 1976). Granted this it is then also necessary that stable equilibria furnish non-negative values of the second variation of the energy. Again, let superposed dots denote derivatives with respect to a parameter ϵ that labels configurations, evaluated at the equilibrium state $\epsilon = 0$. The second variation of the energy functional (10.1) may then be written

$$\ddot{E} = \sum_i \int_{V_i} \ddot{U} \, dV + \sum_j \int_{\Omega_j} \ddot{W} \, dA, \tag{10.2}$$

where V_i and Ω_j are fixed reference configurations of the bulk fluids and the films, and $U = \varrho_0 \Phi$ and $W = \rho_0 \Psi$ are the bulk and film energies per unit reference volume and area respectively. The first of these is a function of the bulk material deformation function $y(X; \epsilon)$; the second, a function of the metric and curvature induced by the parametrization $r(\theta^\alpha; \epsilon)$.

The second variation at an *equilibrium* state is a homogeneous quadratic functional of the *first-order* derivatives $\dot{y}$ and $\dot{r}$. Using this state as reference, I write

$$\ddot{E} = \sum_i \int_{v_i} A(\mathrm{grad}\, u) \, dv + \sum_j \int_{\omega_j} B(\dot{a}_{\alpha\beta}, \dot{b}_{\alpha\beta}) \, da, \tag{10.3}$$

where $u(y) = \dot{y}$,

$$\dot{a}_{\alpha\beta} = a_\alpha \cdot v_{,\beta} + a_\beta \cdot v_{,\alpha}, \qquad \dot{b}_{\alpha\beta} = n \cdot v_{;\alpha\beta} \tag{10.4}$$

where $v(\theta^\alpha) = \dot{r}$, and the covariant derivative is based on the metric induced by $r(\theta^\alpha)$ at $\epsilon = 0$ (Kirchhoff 1850). Further, $A(\cdot)$ and $B(\cdot, \cdot)$ are homogeneous quadratic functions involving the second derivatives of U and W with respect to their arguments.

In many applications the bulk material to which the surface films are attached may be regarded as practically incompressible. To accommodate this possibility I consider variations of the form

$$u(y) = \operatorname{curl} w(y); \quad w(y) = \delta^3 \hat{w}(z(y)), \quad \text{and} \quad v(\theta^\alpha) = \delta^2 \hat{v}(\eta(\theta^\alpha)), \quad (10.5)$$

where

$$z(y) = \delta^{-1}(y - r_0) \quad \text{and} \quad \eta(\theta^\alpha) = z(r(\theta^\alpha)). \quad (10.6)$$

Here, δ is a positive number and the functions $\hat{w}(\cdot)$, $\hat{v}(\cdot)$ have compact support in a three-dimensional region D containing a point r_0 on one of the films, ω_0 say. I assume δ to be small enough that the intersection of D with any other film is empty.

Since $u(y)$ is solenoidal, it automatically satisfies the variational form of the incompressibility constraint in the bulk material. Further, it is natural to assume that there be no flux of bulk material across the film; hence,

$$u \cdot n \mid_{\omega_0} = v \cdot n \doteq v(\theta^\alpha), \quad (10.7)$$

where n is the local orientation field on ω_0. If the film is incompressible in the sense that local surface area is preserved, then the surface divergence of v vanishes, and the representation $v = v^\alpha a_\alpha + vn$ yields

$$v^\alpha_{;\alpha} = 2Hv, \quad (10.8)$$

where H is the mean curvature of ω_0.

Local normal coordinates (Naghdi 1972) may be used with $(7.4)_1$ to reduce (10.7) to the form

$$\varepsilon^{\alpha\beta} w_{\beta;\alpha} = v, \quad (10.9)$$

where $w_\beta(\theta^\alpha) = a_\beta \cdot w \mid_{\omega_0}$ and v vanishes on the curve $c = \partial D \cap \omega_0$. With v prescribed, the existence of a covariant vector field satisfying this equation may be proved by writing $\varepsilon^{\alpha\beta} w_\beta = a^{\alpha\beta} \phi_{,\beta}$, which has a unique solution w_α. Then, for a given parametrization of ω_0, (10.9) reduces to a second-order linear elliptic equation for ϕ. For sufficiently smooth Dirichlet data on c the existence of a unique solution follows from Theorem 21(I) of Miranda (1970). Unfortunately, such a scheme does not yield the existence of v^α satisfying (10.8) for incompressible films. For example, setting $v^\alpha = a^{\alpha\beta} \phi_{,\beta}$ I again obtain a linear elliptic equation for ϕ, but the additional requirement that v^α vanish on c entails the simultaneous specification of homogeneous Dirichlet and Neumann data for ϕ. Exceptionally, if ω_0 is a minimal surface ($H \equiv 0$) in a neighborhood of r_0, then

v^α may be any divergence-free vector field that vanishes on c. Alternatively, one may specify a vector field v^α which vanishes together with its divergence on c and use (10.8) to calculate v at points where $H \neq 0$, but in the absence of detailed information about the surface ω_0 it may not be feasible to ensure that v then possesses the properties required to generate the Legendre-Hadamard condition from the quasiconvexity inequality. This is due to the fact that it is the component of v normal to the tangent plane at r_0, rather than v, that is relevant, as shown below.

Let $\nabla(\cdot)$ and $\nabla \times (\cdot)$ denote the gradient and curl with respect to z. Then,

$$u(y) = \delta^2 \nabla \times \hat{w}(z), \quad \text{grad}\, u(y) = \delta \nabla(\nabla \times \hat{w}),$$

$$v_{,\alpha} = \delta(\nabla \hat{v})a_\alpha, \quad v_{,\alpha\beta} = (\nabla\nabla\hat{v})[a_\alpha \otimes a_\beta] + \delta(\nabla\hat{v})a_{\alpha,\beta}. \tag{10.10}$$

Now, let u^α be smooth extensions of the coordinates θ^α onto the plane tangent to ω_0 at r_0. I take these to be *affine* coordinates such that $\partial\eta/\partial u^\alpha = \mathring{a}_\alpha$, the superposed circle identifying the values of functions at r_0. Thus, $\partial^2\eta/\partial u^\alpha \partial u^\beta = 0$, and it follows that

$$(\nabla\nabla\hat{v})[\mathring{a}_\alpha \otimes \mathring{a}_\beta] = \hat{v}_{,\alpha\beta}, \tag{10.11}$$

where, here and henceforth, commas denote derivatives with respect to u^α.

I change variables in accordance with $(10.6)_1$ and use (10.4), (10.10) and (10.11). Holding D fixed, I divide the second-variation inequality by δ^2 and pass to the limit to obtain the quasiconvexity condition for the film

$$B^{\alpha\beta\lambda\mu} \int_{\omega^*} w_{,\alpha\beta} w_{,\lambda\mu} da \geq 0, \tag{10.12}$$

where $w = \mathring{n}\cdot\hat{v}$, ω^* is the intersection of D with the tangent plane of ω_0 at r_0, and

$$B^{\alpha\beta\lambda\mu} = \rho\frac{\partial^2\Psi}{\partial b_{\alpha\beta}\partial b_{\lambda\mu}} \tag{10.13}$$

is evaluated at r_0 in the configuration ω_0. The symmetries inherent in this tensor ensure that (10.12) is equivalent to the inequality obtained by replacing the integrand with $w_{,\alpha\beta}\bar{w}_{,\lambda\mu}$, where w is now complex-valued and $\bar{w}$ is its conjugate. The resulting inequality is in standard form for generating the relevant Legendre-Hadamard condition (e.g. Giaquinta and Hildebrandt 1996, pp. 229–231). For w, I choose

$$w(u^\alpha) = a\phi(u^\beta) \exp[i\tau(\varsigma_\alpha u^\alpha)], \tag{10.14}$$

where a, $\tau(> 0)$ and ς_α are real constants and ϕ is a real C^∞ function supported in ω^*. Then,

$$w_{,\alpha\beta}\bar{w}_{,\lambda\mu} = a^2\phi^2\varsigma_\alpha\varsigma_\beta\varsigma_\lambda\varsigma_\mu + O(\tau^{-1}), \tag{10.15}$$

and passing to the limit $\tau \to \infty$ in (10.12) yields the necessary condition

$$B^{\alpha\beta\lambda\mu}\varsigma_\alpha\varsigma_\beta\varsigma_\lambda\varsigma_\mu \geq 0 \tag{10.16}$$

for all ς_α. I conjecture that this condition is also necessary in the presence of the two-dimensional incompressibility constraint.

For hemitropic surfaces, (10.16) is equivalent to

$$\sum_{i,j=3}^{6} U_{ij}I_i'I_j' \geq 0, \tag{10.17}$$

where the $U_{ij} \equiv \partial^2 U/\partial I_i \partial I_j$ are evaluated at $\theta = 0$, and

$$I_3' = |v|^2, \quad I_4' = v \cdot \tilde{\kappa}v, \quad I_5' = v \cdot Cv, \quad I_6' = \text{tr}(C(v \otimes v)\mu). \tag{10.18}$$

I do not present a detailed analysis of this inequality here, except to note that *sufficient* conditions are furnished by the non-negativity of the principal minors of the matrix U_{ij}. The latter conditions are not necessary because it is not possible to specify the I_i' independently

For fluid films I use (8.12), (9.31) and (9.32) to define

$$F(\rho, H, K) = G(J, \sigma, \kappa). \tag{10.19}$$

The appropriate specialization of (10.16) may be shown to be (Steigmann 1999)

$$\tfrac{1}{4}F_{HH} + 2xF_{HK} + x^2 F_{KK} \geq 0, \quad \text{where} \quad x = \tilde{b}^{\alpha\beta}\varsigma_\alpha\varsigma_\beta. \tag{10.20}$$

In this expression the first-order derivative of F with respect to H does not appear because H is a linear function of the matrix $b_{\alpha\beta}$; the first-order derivative with respect to K does not appear because the second derivatives of K with respect to this matrix involve the permutation tensor density in such a way as to make no contribution to (10.16).

For incompressible films, several forms of the function $\bar{F}(H, K)$ compatible with (10.20) have been proposed. Among them, Helfrich's function (Helfrich 1973)

$$\rho\bar{F} = \alpha(H - H_0)^2 + \beta K, \quad \text{with} \quad \alpha > 0, \tag{10.21}$$

has been the most widely applied and studied. The *spontaneous curvature* H_0 is a parameter introduced to render the energy a non-even function of the curvature tensor. The resulting formulation is thought to furnish an appropriate model for monolayer films such as those associated with oil-in-water or water-in-oil emulsions (Naghdi 1972). Bilayer response is recovered by setting $H_0 = 0$. Some writers model morphological phase transformations by allowing H_0 to depend on amphiphile concentration, which in turn is the principal factor influencing local film chemistry (Safran 1994).

Existence theory for the local equilibrium equations based on Helfrich's function and related functions has been discussed by Nitsche (1993). However, from the viewpoint of variational theory, Helfrich's model is deficient in the sense that the energy of a given film is generally not bounded below. This is easily seen with reference to compact orientable films by applying the Gauss-Bonnet formula

$$\int_\omega K \mathrm{d}a = 4\pi(1 - g),\qquad\qquad(10.22)$$

where g is the genus (Hopf 1980). For $\beta > 0$, the second term in (10.21) contributes a term to the total film energy that decreases without bound as g increases. The same term contributes only a fixed constant to the energy in the presence of the topological constraints imposed by some writers (Helfrich 1973, Jenkins 1977).

A simple alternative model for bilayers, as yet unexplored, may be based on the assumption that $\bar{F}$ is a function of K depending parametrically on temperature and amphiphile concentration. At fixed concentration and temperature, the graph of such a function might exhibit local minima at $K > 0$, $K = 0$, and $K < 0$. These correspond to points of convexity of the energy in accordance with (10.20). The first alternative promotes the formation of spherical vesicles interspersed in the bulk fluid; the second, developable surfaces, including the cylindrical and lamellar phases; and the third, the bicontinuous phases associated with compact orientable surfaces having large genus (Scriven 1976, 1977). In the latter case the minimizing value of K cannot be achieved at all points of the film as there are no surfaces in 3-space with constant negative Gaussian curvature (Stoker 1969). Nevertheless it is appropriate to conjecture that configurations with high genus are promoted by a sufficiently deep and wide energy well spanning an interval of the domain of $\bar{F}$ in which $K < 0$. The structure of the energy wells might depend on concentration and temperature in such a way as to favor some of these structures over others in accordance with the observed phase behavior of the particular system at hand.

References

Antman, S.S. 1995 *Nonlinear Problems of Elasticity.* Springer-Verlag: New York.
Andreussi, F. and Gurtin, M.E. 1977 On the wrinkling of a free surface. *J. Appl. Phys.* **48**, 3798-3799.
Carroll, M.M. and Naghdi, P.M. 1972 The influence of reference geometry on the response of elastic shells. *Arch. Rat. Mech. Anal.* **48**, 302-318.
Charvolin, J. and Sadoc, J.-F. 1994 Geometrical foundation of mesomorphic polymorphism. In Kellay *et al.* (1994) listed below, pp. 218-249.
Collings, P.J. 1990 *Liquid Crystals: Nature's Delicate Phase of Matter.* Princeton University Press: Princeton, N.J..

Le Dret, H. and Raoult, A. 1995 The nonlinear membrane model as a variational limit of nonlinear three-dimensional elasticity. *J. Math. Pures Appl.* **74**, 549-578.

Ericksen, J.L. 1966 A thermo-kinetic view of elastic stability theory, *Int. J. Solids Structures.* **2**, 573-580.

Ericksen, J.L. 1972 Symmetry transformations for thin elastic shells, *Arch. Rat. Mech. Anal.* **47**, 1-14.

Ericksen, J.L. 1979 Theory of Cosserat surfaces and its applications to shells, interfaces and cell membranes. In *Proc. Int. Symp. on Recent Developments in the Theory and Application of Generalized and Oriented Media* (P.G. Glockner, M. Epstein and D.J. Malcolm, eds.), pp. 27-39, Calgary.

Evans, E.A. and Skalak, R. 1980 *Mechanics and thermodynamics of biomembranes.* CRC Press: Boca Raton, FL..

Fox, D.D., Raoult, A. and Simo, J.C. 1993 A justification of nonlinear properly invariant plate theories. *Arch. Rat. Mech. Anal.* **124**, 157-199.

Gelbart, W.M., Ben-Shaul, A. and Roux, D. (editors) 1994 *Micelles, Membranes, Microemulsions, and Monolayers.* Springer Series on Partially Ordered Systems. Springer-Verlag: New York.

Giaquinta, M. and Hildebrandt, S. 1996 *Calculus of Variations I*, Springer-Verlag: Berlin.

De Giorgi, E. and Dal Maso, G. 1983 Γ-convergence and the calculus of variations. In *Mathematical Theories of Optimization.* (J.P. Cecconi, T. Zolezzi, eds.) Lecture Notes in Mathematics no. 979. Berlin: Springer.

Gurtin, M.E. and Murdoch, A.I. 1975 A continuum theory of elastic material surfaces. *Arch. Rat. Mech. Anal.* **57**, 291-323.

Helfrich, W. 1973 *Elastic properties of lipid bilayers: theory and possible experiments.* *Z. Naturforsch.* 28c, 693-703.

Hopf, H. 1980 Differential geometry in the large, Lecture Notes in Mathematics, Vol. 1000 (Lectures delivered at New York University, 1946, and Stanford University, 1956), Springer-Verlag: New York.

Jenkins, J.T. 1977 The equations of mechanical equilibrium of a model membrane, *SIAM J. Applied Math.* **32**, 755-764.

Jenkins, J.T. 1977 Static equilibrium configurations of a model red blood cell. *J. Math. Biology* **4**, 149-169.

Kellay, H., Binks, B.P., Hendrikx, Y., Lee, L.T. and Meunier, J. 1994 Properties of surfactant monolayers in relation to microemulsion phase behaviour. *Advances in Colloid and Interface Science* **49**, 85-112.

Kirchhoff, G. 1850 Über das gleichgewicht und die bewegung einer elastischen scheibe. *J. Reine u. Angew. Math.* **40**, 51-88.

Koiter, W.T. 1976 A basic open problem in the theory of elastic stability. In *Proc. IUTAM/IMU Symp. on Applications of Methods of Functional Analysis to Problems in Mechanics* (Germain, P. and Nayroles, B., eds.), Springer-Verlag: Berlin, pp. 366-373.

Krishnaswamy, S. 1996 A Cosserat-type model for the red blood cell wall, *Int. J. Engng. Sci.* **34**, 873-899.

Libai, A. and Simmonds, J.G. 1998 *The Nonlinear Theory of Elastic Shells (2nd ed.)*, Cambridge University Press: Cambridge.

Miranda, C. 1970 *Partial Differential Equations of Elliptic Type*, Springer-Verlag: Berlin.

Murdoch, A.I. and Cohen, C. 1979 Symmetry considerations for material surfaces. *Arch. Rat. Mech. Anal.* **72**, 61-89.

Murdoch, A.I. and Cohen, H. 1981 Symmetry considerations for material surfaces: Addendum. *Arch. Rat. Mech. Anal.* **76**, 393-400.

Naghdi, P.M. 1972 Theory of Shells and Plates. In *Handbuch der Physik*, Vol. VI/2 (C. Truesdell, ed.), Springer-Verlag: Berlin, pp. 425-640.

Naghdi, P.M. 1981 Finite deformation of elastic rods and shells. In *Proc. IUTAM Symposium on Finite Elasticity* (Lehigh University, August 10-15, 1980) (D.E. Carlson and R.T. Shield, eds.), Martinus Nijhoff: The Hague, pp. 47-103.

Nitsche, J.C.C. 1993 Periodic surfaces that are extremal for energy functionals containing curvature functions. In *Statistical Thermodynamics and Differential Geometry of Microstructured Materials* (Davis, H.T. and Nitsche, J.C.C., eds.), Inst. for Math. and its Applications, Vol. 51, Springer-Verlag: New York, pp. 69-98.

Noll, W. 1958 A mathematical theory of the mechanical behavior of continuous media, *Arch. Rat. Mech. Anal.* **2**, 197-226.

Ogden, R.W., Steigmann, D.J. and Haughton, D.M. 1997 The effect of elastic surface coating on the finite deformation and bifurcation of a pressurized circular annulus. *J. Elasticity* **47**, 121-145.

Barenblatt, G.I. and Joseph, D.D (eds.) 1997 *Collected Papers of R.S. Rivlin*, Vols. I, II. Springer-Verlag: New York.

Safran, S.A. 1994 Fluctuating interfaces and the structure of microemulsions. In Kellay *et al.* (1994) listed above, pp. 427-484.

Scriven, L.E. 1976 Equilibrium bicontinuous structure. *Nature* **263**, 123-125.

Scriven, L.E. 1977 Equilibrium bicontinuous structures. In *Micellization, Solubization, and Microemulsions* (K.L. Mittal, ed.), vol. 2, Plenum Press: New York, pp. 877-893.

Simmonds, J.G. 1985 The strain energy density of rubber-like shells. *Int. J. Solids Structures* **21**, 67-77.

Steigmann, D.J. 1999 Fluid films with curvature elasticity, *Arch. Rat. Mech. Anal.* **150**, 127-152.

Steigmann, D.J. 2000 On the relationship between the Cosserat and Kirchhoff-Love theories of elastic shells, *Mathematics and Mechanics of Solids.* To appear.

Steigmann, D.J. and Ogden, R.W. 1997 Plane deformations of elastic solids with intrinsic boundary elasticity. *Proc. R. Soc. Lond.* **A453**, 853-877.

Steigmann, D.J. and Ogden, R.W. 1999 Elastic surface-substrate interactions, *Proc. R. Soc. Lond.* **A455**, 437-74.

Stoker, J.J. 1969 *Differential Geometry.* Wiley-Interscience: New York.

Wang, C.-C and Cross, J.J. 1977 On the field equations of motion for a smooth, materially uniform, elastic shell. *Arch. Rat. Mech. Anal.* **65**, 57-72.

Zheng, Q.S. 1993 Two-dimensional tensor function representation for all kinds of material symmetry, *Proc. R. Soc. Lond.* **A443**, 127-138.

9

Singularity theory and nonlinear bifurcation analysis

Yi-chao Chen

Department of Mechanical Engineering
University of Houston
Houston, Texas 77204, U.S.A.
Email: Chen@uh.edu

In this chapter we provide an introductory exposition of singularity theory and its application to nonlinear bifurcation analysis in elasticity. Basic concepts and methods are discussed with simple mathematics. Several examples of bifurcation analysis in nonlinear elasticity are presented in order to demonstrate the solution procedures.

9.1 Introduction

Singularity theory is a useful mathematical tool for studying bifurcation solutions. By reducing a singular function to a simple normal form, the properties of multiple solutions of a bifurcation equation can be determined from a finite number of derivatives of the singular function. Some basic ideas of singularity theory were first conjectured by R. Thom, and were then formally developed and rigorously justified by J. Mather (1968, 1969a, b). The subject was extended further by V. I. Arnold (1976, 1981). In two volumes of monographs, M. Golubitsky and D. G. Schaeffer (1985), and M. Golubitsky, I. Stewart and D. G. Schaeffer (1988) systematized the development of singularity theory, and combined it with group theory in treating bifurcation problems with symmetry. Their work establishes singularity theory as a comprehensive mathematical theory for nonlinear bifurcation analysis.

The purpose of this chapter is to give a brief exposition of singularity theory for researchers in elasticity. The emphasis is on providing a working knowledge of the theory to the reader with minimal mathematical prerequisites. It can also serve as a handy reference source of basic techniques and useful formulae in bifurcation analysis. Throughout this chapter, important solution techniques are demonstrated through examples and case studies. In most cases, theorems are stated without proof. No attempt has been made to provide a complete

Y.-C. Chen

bibliographical survey of the field, nor an accurate account of the contributions made by various researchers.

To illustrate some topics dealt with in singularity theory, we begin with a classical example of nonlinear bifurcation in elasticity – the problem of the *elastica*. Consider the deformation of a slender elastic rod, subjected to a pair of compressive forces at the ends. By the Bernoulli-Euler beam theory, the bending moment is proportional to the curvature. This leads to the governing equation

$$EIu''(s) + \lambda \sin u(s) = 0, \quad 0 < s < l, \tag{1.1}$$

where u is the angle between the undeformed rod and the tangent of the deformed rod, s the material coordinate, E the elastic modulus, I the moment of inertia, λ the compressive applied force, and l the length of the rod. The rod is hinged at its two ends, with the boundary conditions

$$u'(0) = u'(l) = 0. \tag{1.2}$$

The solution of this boundary value problem can be represented by using elliptic integrals (see, e.g., Timoshenko and Gere 1961, Section 2.7). In an elementary analysis, one considers the linearized equation of (1.1), namely

$$EIu''(s) + \lambda u(s) = 0. \tag{1.3}$$

The linearized boundary value problem (1.3) and (1.2) has a non-trivial solution

$$u(s) = C \cos \frac{n\pi s}{l} \tag{1.4}$$

if and only if

$$\lambda = \frac{n^2 \pi^2 EI}{l^2}, \tag{1.5}$$

where n is an integer. The smallest non-zero value $P_{cr} = \pi^2 EI/l^2$ is usually called the *critical load* at which the rod is thought to start to buckle.

As the solution (1.4) and (1.5) of the linearized equation provides certain important information about the exact solution of the nonlinear problem, there are questions which cannot be answered by the linearized analysis. Some of the issues are discussed below, demonstrated with simple examples.

(i) It is well-known that the existence of non-trivial solutions of the linearized equation is a necessary condition for bifurcation. It is, however, not sufficient. Consider the algebraic equation

$$x^3 + \lambda^2 x = 0 \tag{1.6}$$

with a real state variable x and a real bifurcation parameter λ. Equation

(1.6) has the trivial solution $x = 0$ for all values of λ. The linearized equation about the trivial solution is

$$\lambda^2 x = 0,$$

which, at $\lambda = 0$, has non-trivial solutions $x = C$ for arbitrary C. The original nonlinear equation, however, does not have a bifurcation solution branch at $\lambda = 0$.

(ii) When a bifurcation solution branch does exist, little can be said about its qualitative behavior on the basis of the solution of the linearized equation. Qualitative behavior includes, for example, the number of the solution branches, and how they evolve as the bifurcation parameter varies. (Do they disappear or continue as the parameter decreases or increases from the bifurcation point?) Consider, for instance, the equation

$$x^5 - 3\lambda x^3 + 2\lambda^2 x = 0. \tag{1.7}$$

It again has the trivial solution $x = 0$ for all values of λ. The linearized equation about the trivial solution is

$$2\lambda^2 x = 0,$$

which, at $\lambda = 0$, has one non-trivial solution branch $x = C, C$ being an arbitrary constant. No information can be deduced from this linearized equation in regard to how many solution branches the nonlinear equation (1.7) actually has, and to the behavior of these solution branches in a neighborhood of the bifurcation point. By solving (1.7) directly, one finds that it has only the trivial solution when $\lambda < 0$, and has four non-trivial solution branches when $\lambda > 0$.

(iii) A mathematical equation often represents an idealization of a real physical system, which may have imperfections that are not accounted for by the equation. As a result, the behavior of the real system may be different, sometimes drastically, from that predicted by the solution. In reality, there are likely to be infinitely many ways in which imperfections could be present in a physical system. For the example of the compressed rod, possible imperfections include eccentric loads, imperfect supports, imperfect geometry, non-uniform elastic modulus, etc. Nevertheless, mathematically it is possible to describe the qualitative behavior of the imperfect system by introducing only a finite number of parameters. This is the subject of universal unfolding theory. As an example, a universal unfolding of (1.6) is given by

$$x^3 + \lambda^2 x + \alpha x = 0. \tag{1.8}$$

Y.-C. Chen

The solution diagrams of equation (1.8) are shown in Figure 1 for various values of α. These diagrams give a complete description of the qualitative behavior of *all* physical systems that are obtained by introducing small perturbations to the idealized system which is described by equation (1.6).

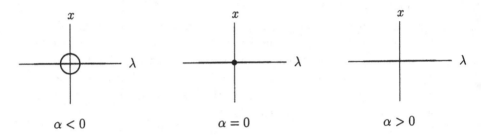

Figure 1. Bifurcation diagrams of equation (1.8).

Singularity theory is developed to address systematically the issues raised above. It provides efficient tools for studying nonlinear bifurcation problems. For example, by using the Liapunov-Schmidt reduction it can be shown that the solution set of equation (1.1), which is defined on an infinite-dimensional space, is equivalent to that of an algebraic equation with one state variable. By solving a recognition problem, it can be shown further that this algebraic equation is equivalent to a polynomial equation that has a pitchfork bifurcation. This implies that in a neighborhood of $(u(s), \lambda) = (0, n^2\pi^2 EI/l^2)$, the boundary value problem (1.1) and (1.2) has only the trivial solution when $\lambda < n^2\pi^2 EI/l^2$, and three solutions when $\lambda > n^2\pi^2 EI/l^2$ corresponding to the unbuckled state and two buckled states. Moreover, through an analysis of universal unfolding, it can be shown that two additional parameters are all that is needed to describe the qualitative behavior of the solutions in the presence of *all* possible imperfections.

Another topic dealt with in singularity theory is the bifurcation of systems with symmetry. The governing equations of a system with symmetry are equivariant under certain group actions. Singularity theory, when combined with group theory, provides a powerful tool in gaining insight into the persistence of or change of the symmetry that a solution branch possesses at the bifurcation point.

Singularity theory also concerns the stability of bifurcation solution branches. Often used is the so-called linear stability criterion. Under such a criterion, the reduced algebraic equation is treated as the governing equation for the equilibrium states of a dynamical system, and the stability of the solutions is determined by examining the behavior of this system under small dynamic disturbances.

In the next section, we introduce the basic terminologies and formulation of bifurcation problems. The derivation of the linearized equation is discussed. The procedure of the Liapunov-Schmidt reduction is then presented. Section 9.3 is devoted to the solution method of the recognition problem, which constitutes the main technical development of singularity theory. The notions of equivalence and normal form are introduced. The normal form of pitchfork bifurcation is then derived explicitly through elementary analysis, and related concepts are discussed in a general context. The solutions of some recognition problems are listed. Also given is a brief discussion of universal unfolding theory. Three case studies are presented in the remaining sections. In Section 9.4, bifurcation analysis of pure homogeneous deformations with Z_2 symmetry is presented. A similar problem with D_3 symmetry is studied in Section 9.5. This difference in symmetry, however, results in distinct representations of the singular function and bifurcation diagrams from those in Section 9.4. In the concluding Section 9.6, the solution of an infinite-dimensional bifurcation problem, that of the inflation of a spherical membrane, is discussed.

9.2 Bifurcation equation and Liapunov-Schmidt reduction

In this section, we formulate bifurcation problems and discuss a necessary condition for the existence of bifurcation solution branches by examining the linearized equation. A brief exposition is given of the Liapunov-Schmidt reduction scheme, which reduces the bifurcation equation to a finite-dimensional or lower dimensional algebraic equation.

9.2.1 Bifurcation equation. Linearized equation

Let $\mathcal{X}$ and $\mathcal{Y}$ be Banach spaces (i.e. normed linear spaces), $\mathcal{U}$ an open subset of $\mathcal{X}$, and Λ an open subset of $\mathbf{R}^n$. Consider a smooth mapping $f : \mathcal{U} \times \Lambda \to \mathcal{Y}$. We assume that the equation

$$f(u, \lambda) = 0 \tag{2.1}$$

determines the state of a physical system with n parameters. For example, (2.1) can be the equilibrium equation for an elastic body. In this connection, u can be a function that describes the deformation of the body, f a differential operator, and λ a set of parameters that specify, for example, the loads, the geometry and the material properties of the body. The variable u is called the *state variable*, and λ the *bifurcation parameter*.

As an example, let us re-formulate the elastica problem discussed in Section 9.1. Let

$$\mathcal{U} \subset \mathcal{X} \equiv \{u \in C^2([0, l]; \mathbf{R}) : u'(0) = u'(l) = 0\}, \tag{2.2}$$

and let $f : \mathcal{U} \times \mathbf{R} \to C^0([0, l]; \mathbf{R})$ be given by

$$f(u(s), \lambda) = EIu''(s) + \lambda \sin u(s). \tag{2.3}$$

Then the boundary value problem (1.1) and (1.2) can be expressed by (2.1) with $u \in \mathcal{U}$. In this example, the elastic modulus E and the moment of inertia I are taken to be constant. Alternatively, they can be treated as additional bifurcation parameters, although here their effects are essentially inseparable from that of λ.

Suppose that $(u_0, \lambda_0) \in \mathcal{U} \times \Lambda$ satisfies (2.1). If the number of solutions of (2.1) in an arbitrarily small neighborhood of (u_0, λ_0) changes as λ varies, the pair (u_0, λ_0) is called a *bifurcation point*. The solutions in this neighborhood are referred to as *bifurcation solution branches*, and a graphical representation of the bifurcation solution branches is called a *bifurcation diagram*.

The mapping $f(u, \lambda)$ is assumed to be smooth in the sense that it has Fréchet derivatives of any order. The first-order Fréchet derivative $D_u f(u_0, \lambda_0)$ of $f(u, \lambda)$ with respect to u at (u_0, λ_0) is a linear operator from $\mathcal{X}$ to $\mathcal{Y}$ such that

$$f(u, \lambda_0) = f(u_0, \lambda_0) + D_u f(u_0, \lambda_0)(u - u_0) + o(\|u - u_0\|) \quad \text{as } \|u - u_0\| \to 0.$$

Higher-order Fréchet derivatives of $f(u, \lambda)$ are defined similarly. By the implicit function theorem, a necessary condition for (u_0, λ_0) to be a bifurcation point is that $D_u f(u_0, \lambda_0)$ be not invertible.

For the example of f given by (2.3), the pair $(0, \lambda_0)$ is a solution of (2.1) for each $\lambda_0 \in \mathbf{R}$. The Fréchet derivative of f at $(0, \lambda_0)$ is given by

$$D_u f(0, \lambda_0)u(s) = EIu''(s) + \lambda_0 u(s).$$

The operator $D_u f(0, \lambda_0)$ is not invertible if and only if the boundary value problem (1.3) and (1.2) has a non-trivial solution. This occurs if and only if

$$\lambda_0 = \frac{n^2 \pi^2 EI}{l^2},$$

as observed earlier.

9.2.2 Liapunov-Schmidt reduction

For many bifurcation problems in elasticity, the space $\mathcal{X}$ of the state variable is high dimensional or even infinite dimensional, as is the case with (2.2). This is one of the sources of difficulty in solving bifurcation problems. There is, however, a standard procedure, called the *Liapunov-Schmidt reduction*, that may effectively reduce an infinite-dimensional bifurcation problem to one with finite dimensions, or reduce a high-dimensional problem to one with lower dimensions.

The basic idea of the Liapunov-Schmidt reduction is to decompose (2.1) into two equivalent equations. One has finite dimensions, or lower dimensions if (2.1) is already finite dimensional. The other equation can be solved by using the implicit function theorem. Substitution of the solution of the second equation into the first equation results in a reduced equation that is equivalent to (2.1).

The Liapunov-Schmidt reduction is applicable when the Fréchet derivative of f at the bifurcation point is a Fredholm operator. Precisely, a bounded linear operator $L : \mathcal{X} \to \mathcal{Y}$ is *Fredholm* if the kernel of L, defined by ker $L \equiv \{u \in \mathcal{X} : L(u) = 0\}$, is finite-dimensional, and if the range of L, defined by range $L \equiv \{y \in \mathcal{Y} : L(u) = y \text{ for some } u \in \mathcal{X}\}$, is closed in $\mathcal{Y}$ with a finite-dimensional complement. The *index* $i(L)$ of a Fredholm operator is given by

$$i(L) = \dim \ker L - \operatorname{codim} \operatorname{range} L,$$

where codim denotes the dimension of the complement. Most operators encountered in elasticity are Fredholm of index 0, which we shall assume in this chapter. A consequence of this assumption is that the spaces $\mathcal{X}$ and $\mathcal{Y}$ have the orthogonal decompositions

$$\mathcal{X} = \ker L \oplus \mathcal{K}, \qquad \mathcal{Y} = \operatorname{range}\mathcal{L} \oplus \mathcal{R}, \tag{2.4}$$

with $\dim \mathcal{R} = \dim \ker L$.

We now consider equation (2.1). Suppose that $(0, \lambda_0) \in \mathcal{X} \times \Lambda$ is a solution of (2.1), so that

$$f(0, \lambda_0) = 0, \tag{2.5}$$

and that $L \equiv D_u f(0, \lambda_0)$ is a Fredholm operator of index 0 with $\dim \ker L > 0$. Let $P : \mathcal{Y} \to \operatorname{range} L$ be the orthogonal projection associated with the decomposition (2.4). We can write equation (2.1) as an equivalent pair of equations

$$Pf(v + w, \lambda) = 0, \tag{2.6}$$

$$(I - P)f(v + w, \lambda) = 0, \tag{2.7}$$

where I is the identity operator, and we have replaced u by $v + w$ with $v \in \ker L$

and $w \in \mathcal{K}$. Define $F : \ker L \times \mathcal{K} \times \Lambda \to \operatorname{range} L$ by

$$F(v, w, \lambda) \equiv Pf(v + w, \lambda).$$

It is observed that the linear operator $D_w F(0, 0, \lambda_0) : \mathcal{K} \to \operatorname{range} L$ is the restriction of L on $\mathcal{K}$. This linear operator is invertible. By the implicit function theorem, one can solve equation (2.6) locally for w. That is, there is a smooth function W defined in a neighborhood $\mathcal{N}$ of $(0, \lambda_0)$ in $\ker L \times \Lambda$, such that

$$Pf(v + W(v, \lambda), \lambda) = 0 \quad \forall (v, \lambda) \in \mathcal{N}. \tag{2.8}$$

Roughly speaking, by projecting the state variable and the value of f on $\mathcal{K}$ and range L, respectively, equation (2.6) effectively factors out the non-invertible part of f. If follows from (2.5) and (2.8) that

$$W(0, \lambda_0) = 0. \tag{2.9}$$

Now define $g : \mathcal{N} \to \mathcal{R}$ by

$$g(v, \lambda) = (I - P)f(v + W(v, \lambda), \lambda).$$

By the construction of W, equation

$$g(v, \lambda) = 0 \tag{2.10}$$

is then equivalent to (2.6) and (2.7), and therefore equivalent to equation (2.1). Equation (2.10) is called the *reduced bifurcation equation*. Its solutions are in one-to-one correspondence with those of (2.1). It is noted that the state variable v of the reduced equation is in the finite-dimensional space $\ker L$.

In loose terms, the fact that equation (2.1) admits multiple solutions near a bifurcation point means that it is impossible to express all components of the state variable uniquely in terms of the bifurcation parameter. The advantage offered by (2.6) is that a unique solution of this equation is ensured by the implicit function theorem. This amounts to solving (2.1) for as many components of the state variable as possible in terms of the bifurcation parameter and the remaining components of the state variable. Substituting the resulting components back into (2.1) yields (2.10).

It is noted that equation (2.6), just like (2.1), may be a nonlinear equation and cannot be solved explicitly. As a result, the explicit form of g in (2.10) cannot be obtained in general. This information, however, is not at all necessary when one studies the qualitative behavior of the solution of (2.10) by using singularity theory. As shall be discussed in Section 9.3, what one needs are no more than the values of a few derivatives of g at the bifurcation point. These derivatives can be found by applying the chain rule and the implicit function theorem to (2.6) and (2.7).

To this end, let $\{e_i\}$ be an orthogonal basis of ker L, where $i \in \{1,..,n\}, n$ being the dimension of ker L. An element v of ker L can be written as

$$v = x_i e_i.$$

Here the usual summation convention for repeated indices is assumed. We can rewrite equation (2.8) as

$$Pf(x_i e_i + W(x_i, \lambda), \lambda) = 0. \tag{2.11}$$

Here we have used W for different functions. Differentiating (2.11) with respect to x_i and evaluating the resulting equation at $(v, \lambda) = (0, \lambda_0)$, we find that

$$PL(e_i + W_i) = 0, \quad \text{where } W_i \equiv \frac{\partial W}{\partial x_i}(0, \lambda_0). \tag{2.12}$$

Similar notation will be used in the sequel. Equation (2.12) implies that $L(e_i + W_i) = 0$. Hence, $(e_i + W_i) \in$ ker L. This further implies, since $W_i \in \mathcal{K}$, that

$$W_i = 0. \tag{2.13}$$

Next, differentiating (2.11) with respect to λ leads to

$$P(LW_\lambda + f_\lambda) = 0, \tag{2.14}$$

where f_λ is the Fréchet derivative of f with respect to λ, again evaluated at $(v, \lambda) = (0, \lambda_0)$. Equation (2.14) implies that

$$W_\lambda = -L^{-1} P f_\lambda, \tag{2.15}$$

where L^{-1} is the inverse of the restriction of L on $\mathcal{K}$. A similar calculation by differentiating (2.11) with respect to x_i and x_j gives

$$W_{ij} = -L^{-1} P f_{uu} e_i e_j. \tag{2.16}$$

Let $\{e_m^*\}$ be an orthogonal basis of $\mathcal{R}$ and define

$$g_m(x_i, \lambda) \equiv (e_m^*, f(x_i e_i + W(x_i, \lambda), \lambda)), \quad m = 1,..,n, \tag{2.17}$$

where $(\cdot, \cdot)$ is the inner product on $\mathcal{Y}$. Equation (2.10) then can be written as

$$g_m(x_i, \lambda) = 0, \quad m = 1,..,n.$$

It is obvious from (2.5), (2.9) and (2.17) that

$$g_m(0, \lambda_0) = 0, \tag{2.18}$$

which simply restates that $(0, \lambda_0)$ is a solution of (2.1). We now compute the derivatives of $g_m(x_i, \lambda)$. Differentiating (2.17) with respect to x_i and using the fact that e_m^* is orthogonal to range L, we find that

$$g_{m,i} = (e_m^*, L(e_i + W_i)) = 0. \tag{2.19}$$

Here and henceforth, unless otherwise stated, the derivatives of g_m are evaluated at $(v, \lambda) = (0, \lambda_0)$, and we use the notation

$$g_{m,i} \equiv \frac{\partial g_m}{\partial x_i}, \ g_{m,\lambda} \equiv \frac{\partial g_m}{\partial \lambda}, \ \text{etc.} \tag{2.20}$$

Similar calculations, with the help of (2.13) and (2.15), lead to

$$g_{m,\lambda} = (e_m^*, LW_\lambda + f_\lambda) = (e_m^*, f_\lambda), \tag{2.21}$$

$$g_{m,ij} = (e_m^*, f_{uu}(e_i + W_i)(e_j + W_j) + LW_{ij}) = (e_m^*, f_{uu}e_ie_j), \tag{2.22}$$

$$\begin{aligned} g_{m,i\lambda} &= (e_m^*, f_{uu}(e_i + W_i)W_\lambda + f_{u\lambda}(e_i + W_i) + LW_{i\lambda}) \\ &= (e_m^*, -f_{uu}e_i(L^{-1}Pf_\lambda) + f_{u\lambda}e_i), \end{aligned} \tag{2.23}$$

$$\begin{aligned} g_{m,ijk} &= (e_m^*, f_{uuu}(e_i + W_i)(e_j + W_j)(e_k + W_k) \\ &\quad + f_{uu}[(e_i + W_i)W_{jk} + (e_j + W_j)W_{ik} + (e_k + W_k)W_{ij}] + LW_{ijk}) \\ &= (e_m^*, f_{uuu}e_ie_je_k + f_{uu}(e_iW_{jk} + e_jW_{ik} + e_kW_{ij})), \end{aligned} \tag{2.24}$$

where W_{ij} is given by (2.16).

As an example, let us examine equation (2.1) with $\mathcal{X}$ being given by (2.2), $\mathcal{Y} = C^0([0,l]; \mathbf{R}), \Lambda = \mathbf{R}$, and f given by (2.3). The Fréchet derivative L of f with respect to u at $(u(s), \lambda) = (0, \lambda_0)$ is given by

$$Lu(s) = EIu''(s) + \lambda_0 u(s).$$

By solving the two-point boundary value problem

$$Lu = 0, \quad u \in \mathcal{X},$$

one finds that

$$\dim \ker L = \begin{cases} 1 & \text{if } \lambda_0 = n^2\pi^2 EI/l^2 \\ 0 & \text{otherwise.} \end{cases}$$

We shall consider the case where $\lambda_0 = n^2\pi^2 EI/l^2$. The subspace $\ker L$ is given by

$$\ker L = \{u \in \mathcal{X} : u(s) = C\cos\frac{n\pi s}{l}, C \in \mathbf{R}\}.$$

We shall employ the standard inner product

$$(u, v) = \int_0^l u(s)v(s)ds.$$

The orthogonal complement $\mathcal{K}$ of ker L in $\mathcal{X}$ is then given by

$$\mathcal{K} = \{w \in \mathcal{X} : \int_0^l w(s) \cos \frac{n\pi s}{l} ds = 0\}.$$

Moreover, an element $y(s)$ in the orthogonal complement $\mathcal{R}$ of range L in $\mathcal{Y}$ satisfies

$$\begin{aligned}(y, Lu) &= \int_0^l y(s)[EIu''(s) + \lambda_0 u(s)]ds \\ &= -[EIy'(s)u(s)]_0^l + \int_0^l [EIy''(s) + \lambda_0 y(s)]u(s)ds \\ &= 0 \quad \forall u \in \mathcal{X}\end{aligned}$$

Hence, $y(s)$ must satisfy

$$EIy''(s) + \lambda_0 y(s) = 0, \quad y'(0) = y'(l) = 0.$$

This result also follows readily from the fact that the linear operator L is self-adjoint. Therefore, for this particular example, we have

$$\mathcal{R} = \ker L, \quad \text{range } L = \mathcal{K}.$$

Furthermore, the orthogonal projection of $\mathcal{Y}$ onto range L is given by

$$Py(s) = y(s) - [\frac{2}{l}\int_0^l y(t) \cos \frac{n\pi t}{l} dt] \cos \frac{n\pi s}{l}.$$

The subspaces $\ker L$ and $\mathcal{R}$ are one dimensional and spanned by the function

$$e = e^* = \cos \frac{n\pi s}{l}.$$

Hence, we shall use $g(x, \lambda)$ in place of $g_m(x_i, \lambda)$ in (2.17), etc. It follows from (2.3) that

$$f_\lambda = 0, \ f_{uu} = 0, \ f_{u\lambda} = 1, \ f_{uuu} = -\lambda_0. \tag{2.25}$$

Substituting $(2.25)_{1,2}$ into (2.15) and (2.16) gives

$$W_\lambda = 0, \ W_{xx} = 0.$$

By using the above results and (2.18), (2.19), (2.21)–(2.24), we find that

$$g = g_x = g_\lambda = g_{xx} = 0, \ g_{x\lambda} = (e^*, e) = \frac{l}{2}, \ g_{xxx} = (e^*, -\lambda_0 e^3) = -\frac{3l\lambda_0}{8}. \tag{2.26}$$

Other derivatives of $g(x, \lambda)$ can be calculated similarly. However, we shall see in the next section that the derivatives in (2.26) are sufficient to determine the qualitative behavior of the bifurcation solution branches in a neighborhood of the bifurcation point.

9.3 The recognition problem

The material presented in this section constitutes an essential part of singularity theory. A recognition problem for a given algebraic equation is to find a polynomial equation, as simple as possible, of which the solution is in one-to-one correspondence with that of the given equation in a neighborhood of the bifurcation point. This polynomial, called the *normal form* of the given function, can be determined solely by the values of a finite number of derivatives of the given function at the bifurcation point.

9.3.1 Equivalence and normal form

In this section, we consider the reduced bifurcation equation of a single state variable and a single bifurcation variable,

$$g(x, \lambda) = 0. \tag{3.1}$$

For convenience, we move the anticipated bifurcation point $(0, \lambda_0)$ to the origin by a translation of λ. Let $\mathcal{N}$ be a small neighborhood of the origin in $\mathbf{R} \times \mathbf{R}$. The function $g : \mathcal{N} \to \mathbf{R}$ is assumed to be of class C^∞ with

$$g(0,0) = 0. \tag{3.2}$$

Henceforth, all the derivatives of g are evaluated at the origin. The analysis presented here is local. All conclusions are valid in $\mathcal{N}$.

A necessary condition for the existence of bifurcation of (3.1) at the origin is that

$$g_x(0,0) = 0. \tag{3.3}$$

A function g that satisfies (3.2) and (3.3) is said to be singular at the origin. In this case, one cannot apply the implicit function theorem to solve (3.1) for x in terms of λ. The primary goal here is to find an equation, as simple as possible, whose solution has the same qualitative behavior as that of (3.1), and whose bifurcation diagram can easily be obtained by elementary computation. As commented earlier, the information that we have, and need, concerning g is no more than the values of its derivatives at the origin.

We first make precise the meaning of the phrase that the solutions of two equations have the same qualitative behavior. Two smooth functions g, h : $\mathcal{N} \to \mathbf{R}$ are said to be *strongly equivalent*† if there exist smooth functions $X, S : \mathcal{N} \to \mathbf{R}$, such that

$$X(0,0) = 0, \; X_x(x, \lambda) > 0, \; S(x, \lambda) > 0, \tag{3.4}$$

and that

$$g(x, \lambda) = S(x, \lambda)h(X(x, \lambda), \lambda) \quad \forall (x, \lambda) \in \mathcal{N}. \tag{3.5}$$

A direct consequence of this definition is that there is a one-to-one correspondence between the solutions of (3.1) and

$$h(x, \lambda) = 0; \tag{3.6}$$

in particular, the number of solutions is preserved when (3.1) is replaced by (3.6). To see this, suppose that, for a fixed λ, equation (3.1) has exactly n solutions $x_1 < x_2 < ... < x_n$, i.e.

$$g(x, \lambda) = 0 \text{ iff } x = x_i, i = 1, .., n.$$

This and (3.5) imply that

$$h(X, \lambda) = 0 \text{ iff } X = X(x_i, \lambda), i = 1, .., n.$$

It is obvious from $(3.4)_2$ that $X(x_1, \lambda) < X(x_2, \lambda) < ... < X(x_n, \lambda)$.

As an example, we examine a function $g(x, \lambda)$ that satisfies (2.26). It will be shown in the next section that there are functions $X(x, \lambda)$ and $S(x, \lambda)$, satisfying (3.4), such that

$$g(x, \lambda) = S(x, \lambda)[\lambda X(x, \lambda) - X^3(x, \lambda)] \tag{3.7}$$

in a neighborhood of $(x, \lambda) = (0, 0)$. Hence, function $g(x, \lambda)$ is equivalent to

$$h(x, \lambda) = \lambda x - x^3. \tag{3.8}$$

The solution of (3.1), and therefore the solution of the elastica problem, is in one-to-one correspondence with the solution of

$$\lambda x - x^3 = 0. \tag{3.9}$$

The bifurcation diagram of (3.9) corresponds to a pitchfork bifurcation as sketched in Figure 2. The function $h(x, \lambda)$ is called a *normal form* of $g(x, \lambda)$. Much of singularity theory is devoted to determining the simplest normal form for a function of which a certain number of derivatives are given at the bifurcation point. Some comments are in order here. Firstly, for an equation

† In the definition of general equivalence, the transformation of the bifurcation parameter λ is also allowed. See Golubitsky and Schaeffer (1985).

Y.-C. Chen

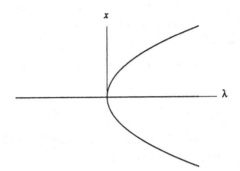

Fig. 2. Pitchfork bifurcation $\lambda x - x^3 = 0$.

as simple as (3.7), one can, by using the solution of cubic equations, actually find $X(x, \lambda)$ and $S(x, \lambda)$ for a given $g(x, \lambda)$ that satisfies (2.26). However, this elementary approach, aside from other matters such as the degree of sophistication, is not practical for more complicated problems. Secondly, it may appear plausible that the function $h(x, \lambda)$ could be obtained by inspecting (2.26) and retaining the non-zero leading terms of the Taylor expansion of $g(x, \lambda)$. A further examination of this idea quickly reveals that the situation is not as simple as it may appear at first glance. For example, it is not intuitive at all why $h(x, \lambda)$ need not contain the term λ^2 while $g_{\lambda\lambda}$ may not be zero. Moreover, other third order derivatives of g, i.e., $g_{xx\lambda}, g_{x\lambda\lambda}$ and $g_{\lambda\lambda\lambda}$, are not calculated and listed in (2.26). Why these derivatives can be safely omitted is entirely beyond what one can deduce from the usual Taylor expansion argument.

These are exactly the issues that singularity theory addresses, and resolves in a complete and elegant manner. Specifically, the conditions for the equivalence of two functions are established through a thorough examination of the tangent spaces of the two functions. It can be shown that function $h(x, \lambda)$ in (3.8) is in fact the simplest normal form of function $g(x, \lambda)$ that satisfies (2.26). Additional terms such as λ^2 are redundant. It can be shown further that all functions that are equivalent to $h(x, \lambda)$ of form (3.8) are characterized by the derivatives listed in (2.26). No additional derivatives are necessary.

9.3.2 Recognition problem for pitchfork bifurcation

The development of singularity theory draws heavily upon algebra theory. Some basic concepts would be best expounded with the theories of groups and rings. It is impractical to go into these theories in an introductory exposition. Here, we shall illustrate some pertinent ideas by solving the recognition problem for the pitchfork bifurcation using elementary mathematics. In particular, we shall show that a function $g(x, \lambda)$ is strongly equivalent to $h(x, \lambda)$ given by (3.8) if and only if

$$g = 0, \ g_x = 0, \ g_\lambda = 0, \ g_{xx} = 0, \ g_{x\lambda} > 0, \ g_{xxx} < 0. \tag{3.10}$$

Equations $(3.10)_{1-4}$ are called *defining conditions* for the normal form $\lambda x - x^3$, while inequalities $(3.10)_{5,6}$ are called *nondegeneracy conditions*.

The proof of necessity is elementary. Suppose that $g(x, \lambda)$ is strongly equivalent to $h(x, \lambda)$. By (3.5), there exist $X(x, \lambda)$ and $S(x, \lambda)$ satisfying (3.4), such that

$$g(x, \lambda) = S(x, \lambda)[\lambda X(x, \lambda) - X^3(x, \lambda)]. \tag{3.11}$$

Taking the derivatives of (3.11) successively and evaluating them at the origin with the help of (3.4) leads to (3.10).

We now turn to sufficiency. Let a smooth function $g(x, \lambda)$ satisfying (3.10) be given. By (3.10) and Taylor's theorem, $g(x, \lambda)$ can be written as

$$g(x, \lambda) = c_1 \lambda x + c_2 \lambda^2 + a_1(x, \lambda) x^3 + a_2(x, \lambda) \lambda x^2 + a_3(x, \lambda) \lambda^2 x + a_4(x, \lambda) \lambda^3, \tag{3.12}$$

where c_1 and c_2 are constants, and $a_1(x, \lambda), \dots, a_4(x, \lambda)$ are smooth functions with

$$c_1 > 0, \ a_1(x, \lambda) < 0. \tag{3.13}$$

Here, we re-emphasize that the analysis is local, and all statements, such as $(3.13)_2$, are valid in the small neighborhood $\mathcal{N}$. We can rearrange the terms in (3.12) to give

$$g(x, \lambda) = a_5(x, \lambda) \lambda x + a_6(x, \lambda) \lambda^2 + a_1(x, \lambda) x^3, \tag{3.14}$$

where

$$a_5(x, \lambda) \equiv c_1 + a_2(x, \lambda) x + a_3(x, \lambda) \lambda > 0, \ a_6(x, \lambda) \equiv c_2 + a_4(x, \lambda) \lambda.$$

Assume that the right-hand side of (3.14) is strongly equivalent to

$$f(x, \lambda) \equiv a_5(x, \lambda) \lambda x + a_1(x, \lambda) x^3. \tag{3.15}$$

Then by choosing

$$X(x, \lambda) = \sqrt{-\frac{a_1(x, \lambda)}{a_5(x, \lambda)}} x, \ S(x, \lambda) = \sqrt{-\frac{a_5^3(x, \lambda)}{a_1(x, \lambda)}} \tag{3.16}$$

one has

$$f(x, \lambda) = S(x, \lambda)[\lambda X(x, \lambda) - X^3(x, \lambda)].$$

That is, $f(x,\lambda)$, and therefore $g(x,\lambda)$, is strongly equivalent to $h(x,\lambda)$. It is readily checked that functions $X(x,\lambda)$ and $S(x,\lambda)$ defined in (3.16) satisfy (3.4).

It remains to show that functions $g(x,\lambda)$ and $f(x,\lambda)$ in (3.14) and (3.15) are strongly equivalent. These two functions differ by a term of λ^2. The desired conclusion follows from a theorem (see, e.g., Theorem 2.2 in Golubitsky and Schaeffer 1985) which states that given two smooth functions g and p, if

$$T(g+p) = T(g),$$

then $g+p$ is strongly equivalent to g. Here $T(g)$ is the restricted tangent space† of g, which consists of all smooth functions of the form

$$a(x,\lambda)g(x,\lambda) + [b(x,\lambda)x + c(x,\lambda)\lambda]g_x(x,\lambda), \tag{3.17}$$

where $a(x,\lambda), b(x,\lambda)$ and $c(x,\lambda)$ are smooth functions. The theorem can be proved (Golubitsky and Schaeffer 1985, II.11), by construction of appropriate S and X in the equivalence transformation and solving certain ordinary differential equations. With the theorem, it now suffices to show that

$$T(g) = T(f).$$

We shall give the proof of $T(f) \subset T(g)$. The proof of the reverse containment is similar. By the definition of the restricted tangent space, a function in $T(g)$ has the form of (3.17). Using (3.14) and (3.15), we find that

$$\begin{aligned} ag + (bx+c\lambda)g_x &= a(f + a_6\lambda^2) + (bx+c\lambda)(f_x + a_{6,x}\lambda^2) \\ &= \tilde{a}f + (\tilde{b}x + \tilde{c}\lambda)f_x \end{aligned} \tag{3.18}$$

where $\tilde{a}, \tilde{b}$ and $\tilde{c}$ are smooth functions given by

$$\tilde{a} \equiv a - \frac{[aa_6 + (bx+c\lambda)a_{6,x}](3a_1 + a_{1,x}x)^2 x}{[2a_1 a_5 + (a_5 a_{1,x} - a_1 a_{5,x})x](a_5 + a_{5,x}x)},$$

$$\tilde{b} \equiv b + \frac{a_1[aa_6 + (bx+c\lambda)a_{6,x}](3a_1 + a_{1,x}x)x}{[2a_1 a_5 + (a_5 a_{1,x} - a_1 a_{5,x})x](a_5 + a_{5,x}x)},$$

$$\tilde{c} \equiv c + \frac{aa_6 + (bx+c\lambda)a_{6,x}}{a_5 + a_{5,x}x}.$$

The last expression in (3.18) is a function in $T(f)$.

Although the above example of the pitchfork bifurcation is one of the simplest, it does demonstrate some preliminary yet essential ideas used for solving a recognition problem. A full exposition of these ideas requires sophisticated

† The definition of general tangent space pertains to the general equivalence.

mathematical treatment. Here we offer only a brief discussion of these ideas in elementary language.

The Taylor series expansion of a smooth function g consists of a collection of monomials, which can be divided into three classes: low-, intermediate-, and higher-order terms. We shall discuss in sequence the treatment of these terms in solving a recognition problem.

The low-order terms are those monomials $x^k \lambda^l$ such that $\partial^{k+l} g / \partial x^k \partial \lambda^l = 0$ in the defining conditions. In the case of (3.10), they are $1, x, \lambda$ and x^2 terms. These terms do not appear in the Taylor expansion, as evident in (3.12), and hence need no treatment in the equivalence transformation. The exclusion of the low-order terms amounts to identifying the smallest intrinsic ideal containing the given function g. In the terminology of algebra, an *intrinsic ideal* is a linear space which is closed under multiplication by smooth functions, and which is invariant under strong equivalence transformation.

The higher-order terms are those monomials that can be transformed away through the strong equivalence transformation. In the case of (3.10), they are $\lambda^2, \lambda x^2, \lambda^2 x, \lambda^3, x^4, \ldots$. These terms can be determined by using the theorem stated above and by examining the tangent spaces of a normal form and its perturbations, as we did for the λ^2 term above. Sometimes, these terms can be more simply identified by the operation of absorbing them in the Taylor expansion through proper redefinition of the coefficients, as we did for the $\lambda x^2, \lambda^2 x$ and λ^3 terms when getting from (3.12) to (3.14). The higher-order terms are endowed with certain algebraic structures. In particular, they form an intrinsic ideal which can be generated by a finite number of functions.

It is worth noting that it is in general incorrect to identify the higher-order terms as those monomials $x^k \lambda^l$ such that $\partial^{k+l} g / \partial x^k \partial \lambda^l$ do not appear in the defining conditions and the nondegeneracy conditions. Otherwise, the solution of a recognition problem would be all too simple. While this happens to be the case for the pitchfork bifurcation problem, as well as for some other recognition problems, there are examples for which it is not true. For instance, the defining conditions and the nondegeneracy conditions for the normal form $x^2 - \lambda^2$ are

$$g = g_x = g_\lambda = 0, \quad g_{xx} > 0, \quad g_{xx} g_{\lambda\lambda} - g_{x\lambda}^2 < 0. \qquad (3.19)$$

Although $g_{x\lambda}$ appears in (3.19), the $x\lambda$ term is of higher order, and can be absorbed by x^2 and λ^2 terms using an equivalence transformation.

The monomials that are neither low-order nor higher-order are intermediate-order terms. These are the terms that survive in the normal form. After reducing the given function to finite intermediate-order terms, one only needs to transform their coefficients, which are smooth functions, into constants, usually 1 or -1, to reach the final expression of the normal form. This often can be done by elementary calculations. The essence of such calculations is associated with the representation of a certain Lie group of strong equivalence transformations.

9.3.3 Solutions of some recognition problems

In Table 1 below, we list the normal form, the defining conditions and the nondegeneracy conditions for several recognition problems that are often encountered in bifurcation analysis. This serves as perhaps the most important practical data that one would need when working on bifurcation problems in elasticity. Detailed derivation of these solutions can be found in Golubitsky and Schaeffer (1985, IV. 2).

Table 1. Solution of the Recognition Problem for Several Singular Functions

Nomenclature	Normal form	Defining conditions	Nondegeneracy conditions
Limit point	$\epsilon x^2 + \delta\lambda$	$g = g_x = 0$	$\epsilon = \mathrm{sgn}(g_{xx})$, $\delta = \mathrm{sgn}(g_\lambda)$
Simple bifurcation	$\epsilon(x^2 - \lambda^2)$	$g = g_x = g_\lambda = 0$	$\epsilon = \mathrm{sgn}(g_{xx})$, $g_{xx}g_{\lambda\lambda} - g_{x\lambda}^2 < 0$
Isola center	$\epsilon(x^2 + \lambda^2)$	$g = g_x = g_\lambda = 0$	$\epsilon = \mathrm{sgn}(g_{xx})$, $g_{xx}g_{\lambda\lambda} - g_{x\lambda}^2 > 0$
Hysteresis	$\epsilon x^3 + \delta\lambda$	$g = g_x = g_{xx} = 0$	$\epsilon = \mathrm{sgn}(g_{xxx})$, $\delta = \mathrm{sgn}(g_\lambda)$
Asymmetric cusp	$\epsilon x^2 + \delta\lambda^3$	$g = g_x = g_\lambda = 0$, $g_{xx}g_{\lambda\lambda} - g_{x\lambda}^2 = 0$	$\epsilon = \mathrm{sgn}(g_{xx})$, $\delta = \mathrm{sgn}(g_{vvv})$
Pitchfork	$\epsilon x^3 + \delta\lambda x$	$g = g_x = g_\lambda = 0$, $g_{xx} = 0$	$\epsilon = \mathrm{sgn}(g_{xxx})$, $\delta = \mathrm{sgn}(g_{x\lambda})$
Quartic fold	$\epsilon x^4 + \delta\lambda$	$g = g_x = g_{xx} = 0$, $g_{xxx} = 0$	$\epsilon = \mathrm{sgn}(g_{xxxx})$, $\delta = \mathrm{sgn}(g_\lambda)$
——	$\epsilon x^2 + \delta\lambda^4$	$g = g_x = g_\lambda = 0$, $g_{xx}g_{\lambda\lambda} - g_{x\lambda}^2 = 0$, $g_{vvv} = 0$	$\epsilon = \mathrm{sgn}(g_{xx})$, $\delta = \mathrm{sgn}(g_{vvvv} - 3g_{vvx}^2/g_{xx})$
Winged cusp	$\epsilon x^3 + \delta\lambda^2$	$g = g_x = g_\lambda = 0$, $g_{xx} = g_{x\lambda} = 0$	$\epsilon = \mathrm{sgn}(g_{xxx})$, $\delta = \mathrm{sgn}(g_{\lambda\lambda})$
——	$\epsilon x^4 + \delta\lambda x$	$g = g_x = g_\lambda = 0$, $g_{xx} = g_{xxx} = 0$	$\epsilon = \mathrm{sgn}(g_{xxxx})$, $\delta = \mathrm{sgn}(g_{x\lambda})$
——	$\epsilon x^5 + \delta\lambda$	$g = g_x = g_{xx} = 0$, $g_{xxx} = g_{xxxx} = 0$	$\epsilon = \mathrm{sgn}(g_{xxxxx})$, $\delta = \mathrm{sgn}(g_\lambda)$

Here, the function g and its derivatives are again evaluated at the origin, the anticipated bifurcation point. The equation $\epsilon = \mathrm{sgn}(f)$ means that $f \neq 0$ and

$$\epsilon = \begin{cases} 1 & \text{if } f > 0 \\ -1 & \text{if } f < 0. \end{cases}$$

In the defining conditions for the asymmetric cusp singularity, the equation $g_{xx}g_{\lambda\lambda} - g_{x\lambda}^2 = 0$ means that the Hessian matrix of g, given by

$$\begin{pmatrix} g_{xx} & g_{x\lambda} \\ g_{x\lambda} & g_{\lambda\lambda} \end{pmatrix},$$

has a zero eigenvalue with eigenvector

$$v = \begin{pmatrix} a \\ 1 \end{pmatrix}, \qquad a \equiv -\frac{g_{x\lambda}}{g_{xx}}.$$

A subscript v of g denotes the derivative of g in the direction of v. One may take

$$\frac{\partial}{\partial v} = a\frac{\partial}{\partial x} + \frac{\partial}{\partial \lambda},$$

and hence have

$$g_{vvv} = a^3 g_{xxx} + 3a^2 g_{xx\lambda} + 3a g_{x\lambda\lambda} + g_{\lambda\lambda\lambda}.$$

The expressions g_{vvvv} and g_{vvx} for the normal form $\epsilon x^2 + \delta\lambda^4$ are defined similarly.

9.3.4 Universal unfolding

The theory of universal unfoldings deals with imperfect bifurcation in which the bifurcation equation is subjected to small perturbations. While perturbations can be introduced to the bifurcation equation in infinitely many ways, the qualitative behavior of all possible perturbed bifurcation diagrams can be captured by introducing a finite number of parameters in the normal form of the unperturbed bifurcation equation. A comprehensive treatment of universal unfolding theory can be found in Golubitsky and Schaeffer (1985). Here we only state the definitions of unfolding and universal unfolding, and compile the universal unfoldings of several normal forms.

Let $g(x, \lambda)$ be a smooth function defined in a neighborhood $\mathcal{N}$ of the origin in $\mathbf{R} \times \mathbf{R}$, and let $G(x, \lambda, \alpha)$ be a smooth function defined in a neighborhood of the origin in $\mathbf{R} \times \mathbf{R} \times \mathbf{R}^k$, k being a positive integer. The function G is said to be an *unfolding* of g if

$$G(x, \lambda, 0) = g(x, \lambda) \quad \forall(x, \lambda) \in \mathcal{N}.$$

For example, the function

$$G(x, \lambda, \alpha) = \lambda x - x^3 + \alpha_1 + \alpha_2 x + \alpha_3 x^2 \tag{3.20}$$

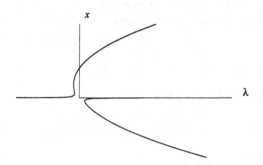

Fig. 3. Bifurcation diagram of an unfolding of the pitchfork.

is an unfolding of the function

$$g(x, \lambda) = \lambda x - x^3,$$

which is the normal form of the pitchfork bifurcation. We note that the bifurcation diagrams of these two functions can be drastically different even for small values of α. This is evident by comparing Figure 2 with Figure 3, which is the bifurcation diagram for $G(x, \lambda, \alpha) = 0$ with $\alpha_1 > 0, \alpha_2 = 0, \alpha_3 = 4\alpha_1^{1/3}$. The unperturbed equation has continuous bifurcation solutions, while the perturbed equation has discontinuous bifurcation solutions.

An unfolding G of g is *universal* if any other unfolding of g is equivalent to G and if G has the minimum number of parameters α_i. Roughly speaking, a universal unfolding $G(x, \lambda, \alpha)$ of $g(x, \lambda)$ is a function with the minimum dimension of parameter α, such that any small perturbation of g is equivalent to G for some small α. The unfolding $G(x, \lambda, \alpha)$ given by (3.20) is not universal because the term $\alpha_2 x$ is redundant. On the other hand, the function

$$G(x, \lambda, \alpha) = \lambda x - x^3 + \alpha_1 + \alpha_2 x^2$$

is a universal unfolding of the pitchfork. In general, the form of a universal unfolding is not unique. For example, another universal unfolding of the pitchfork is given by

$$G(x, \lambda, \alpha) = \lambda x - x^3 + \alpha_1 + \alpha_2 \lambda.$$

In Table 2, we list universal unfoldings of the singular functions whose normal form were listed in Table 1. The derivation of these universal unfoldings can be found in Golubitsky and Schaeffer (1985, IV. 3).

Thus far, we have discussed the recognition problem of bifurcation equations with one state variable. Bifurcation equations with multiple variables and bifurcation equations with symmetry have been treated in singularity theory. We shall not discuss the theoretical development pertaining to these issues. Instead, in the subsequent sections we shall present the analyses of a few bifurcation problems in elasticity with these features. This is not intended to be a complete exposition of the theory. We shall only introduce necessary definitions, discuss the related solution techniques, and give relevant references.

Table 2. Universal Unfoldings for Several Singular Functions

Nomenclature	Normal Form	Universal Unfolding
Limit point	$\epsilon x^2 + \delta\lambda$	$\epsilon x^2 + \delta\lambda$
Simple bifurcation	$\epsilon(x^2 - \lambda^2)$	$\epsilon(x^2 - \lambda^2 + \alpha)$
Isola center	$\epsilon(x^2 + \lambda^2)$	$\epsilon(x^2 + \lambda^2 + \alpha)$
Hysteresis	$\epsilon x^3 + \delta\lambda$	$\epsilon x^3 + \delta\lambda + \alpha x$
Asymmetric cusp	$\epsilon x^2 + \delta\lambda^3$	$\epsilon x^2 + \delta\lambda^3 + \alpha_1 + \alpha_2\lambda$
Pitchfork	$\epsilon x^3 + \delta\lambda x$	$\epsilon x^3 + \delta\lambda x + \alpha_1 + \alpha_2 x^2$
Quartic fold	$\epsilon x^4 + \delta\lambda$	$\epsilon x^4 + \delta\lambda + \alpha_1 x + \alpha_2 x^2$
——	$\epsilon x^2 + \delta\lambda^4$	$\epsilon x^2 + \delta\lambda^4 + \alpha_1 + \alpha_2\lambda + \alpha_3\lambda^2$
Winged cusp	$\epsilon x^3 + \delta\lambda^2$	$\epsilon x^3 + \delta\lambda^2 + \alpha_1 + \alpha_2 x + \alpha_3\lambda x$
——	$\epsilon x^4 + \delta\lambda x$	$\epsilon x^4 + \delta\lambda x + \alpha_1 + \alpha_2\lambda + \alpha_3 x^2$
——	$\epsilon x^5 + \delta\lambda$	$\epsilon x^5 + \delta\lambda + \alpha_1 x + \alpha_2 x^2 + \alpha_3 x^3$

9.4 Bifurcation of pure homogeneous deformations with Z_2 symmetry

In this section, we present a bifurcation analysis for pure homogeneous deformations of a homogeneous, isotropic, incompressible elastic body under dead load tractions with Z_2 symmetry. Physically, this is the situation of an elastic sheet being stretched on its four edges by two pairs of perpendicular, uniformly distributed forces of equal magnitude. This is perhaps the simplest bifurcation problem with two state variables and a symmetry structure. In particular, we are interested in symmetry breaking bifurcation solutions. The existence of asymmetric equilibrium deformations under symmetric loads is suggested by the experiment of Treloar (1948), and has been studied by a number of authors, including Ogden (1985), Kearsley (1986), Chen (1991), and MacSithigh and Chen (1992a, b).

The equations of equilibrium are (3.6) in Chapter 1 and repeated here as

$$\hat{W}_1(\lambda_1, \lambda_2) = t, \ \hat{W}_2(\lambda_1, \lambda_2) = t, \tag{4.1}$$

where λ_1 and λ_2 are two principal stretches of a pure homogeneous deformation, $\hat{W}(\lambda_1, \lambda_2)$ is the reduced strain-energy function for isotropic, incompressible materials, as defined in (3.4) of Chapter 1, and the subscript i denotes the partial derivative with respect to λ_i. Also, for equi-biaxial stretch, we have set the two principal Biot stresses to t. The strain-energy function $\hat{W}$ is assumed to be C^∞.

By the isotropy, the strain-energy function has the symmetry

$$\hat{W}(\lambda_1, \lambda_2) = \hat{W}(\lambda_2, \lambda_1). \tag{4.2}$$

This symmetry, as well as the symmetry of the load, is passed to the bifurcation equations (4.1), which are hence invariant under the action of the interchange permutation group. Now we explore such a symmetry.

Let $\lambda_1 = \lambda_2 = \lambda$ be a symmetric equilibrium solution of (4.1) for $t = T$. We make the change of variables

$$x = \frac{\lambda_1 + \lambda_2}{2} - \lambda, \ y = \frac{\lambda_1 - \lambda_2}{2}, \ \tau = t - T.$$

It is observed that y is a measure of the departure of a homogeneous deformation from the symmetric deformation. The equations of equilibrium (4.1) can be written in an equivalent form

$$\hat{W}_1(\lambda + x + y, \lambda + x - y) + \hat{W}_2(\lambda + x + y, \lambda + x - y) - 2(T + \tau) = 0, \quad (4.3)$$

$$\hat{W}_1(\lambda + x + y, \lambda + x - y) - \hat{W}_2(\lambda + x + y, \lambda + x - y) = 0. \tag{4.4}$$

It follows from (4.2) that the left-hand side of (4.3) is even in y. By a theorem due to Whitney (1943), there exists a smooth function p, such that

$$p(x, y^2, \tau) = \hat{W}_1(\lambda + x + y, \lambda + x - y) + \hat{W}_2(\lambda + x + y, \lambda + x - y) - 2(T + \tau). \tag{4.5}$$

Similarly, the left-hand side of (4.4) is an odd function of y, and there exists a smooth function q, such that

$$yq(x, y^2) = \hat{W}_1(\lambda + x + y, \lambda + x - y) - \hat{W}_2(\lambda + x + y, \lambda + x - y). \tag{4.6}$$

Now the equations of equilibrium (4.3) and (4.4) can be written as

$$g(x, y, \tau) = 0, \tag{4.7}$$

where $g : \mathcal{N} \to \mathbf{R}^2$ is given by

$$g(x, y, \tau) \equiv (p(x, y^2, \tau), yq(x, y^2)), \tag{4.8}$$

$\mathcal{N}$ being a neighborhood of $\mathbf{R}^2 \times \mathbf{R}$.

The symmetry property of the function g can be described by the Z_2 symmetry group that has the following linear representation on $\mathbf{R}^2$:

$$Z_2 = \left\{ \begin{pmatrix} 1 & 0 \\ 0 & 1 \end{pmatrix}, \begin{pmatrix} 1 & 0 \\ 0 & -1 \end{pmatrix} \right\}.$$

In general terms, a mapping $f : \mathbf{R}^2 \to \mathbf{R}^2$ is said to be Z_2-*equivariant* if

$$f(\gamma z) = \gamma f(z) \quad \forall \gamma \in Z_2, \ z \in \mathbf{R}^2.$$

By our choice of λ and T, the origin is a solution of (4.7). Henceforth, we omit the arguments of p, q and g and their derivatives when they are evaluated at $x = y = \tau = 0$. It follows from the discussion in Section 9.2.1 that a necessary condition for the origin to be a bifurcation point is that the Fréchet derivative of g with respect to the state variable (x, y) be not invertible, i.e.,

$$p_1 q = 0.$$

Here and henceforth, a subscript i of p or q denotes the partial derivative with respect to the ith argument.

We now introduce the definition of strong equivalence of Z_2-equivariant functions. Two smooth Z_2-equivariant functions $g, h : \mathcal{N} \to \mathbf{R}^2$ are said to be *strongly Z_2-equivalent* if there exist smooth functions $S : \mathcal{N} \to \mathbf{M}^{2 \times 2}$ and $Z : \mathcal{N} \to \mathbf{R}^2$, such that

$$Z(0,0) = 0, \ \det S > 0, \ \mathrm{tr}\, S > 0, \ \det \nabla Z > 0, \ \mathrm{tr}\, \nabla Z > 0, \tag{4.9}$$

$$S(\gamma z, \tau)\gamma = \gamma S(z, \tau), \ Z(\gamma z, \tau) = \gamma Z(z, \tau) \quad \forall \gamma \in Z_2, \ (z, \tau) \in \mathcal{N}, \tag{4.10}$$

and that

$$g(z, \tau) = S(z, \tau) h(Z(z, \tau), \tau). \tag{4.11}$$

An immediate consequence of the above definition is that strong Z_2-equivalence preserves the solution set and its symmetry near the origin.

Recognition problems for strong Z_2-equivalence can be solved by using the techniques discussed in Section 9.3. Here we list, in Table 3, the normal form, defining conditions and nondegeneracy conditions of several recognition problems. Detailed derivation of the last three solutions can be found in Golubitsky, Stewart and Shaeffer (1988, XIX. 2). Below we give a direct verification of the second solution, which corresponds to a pitchfork bifurcation in two state variables. The verification of the first solution is similar.

We first assume that the right-hand side of (4.8) is strongly Z_2-equivalent to $(\epsilon_1 x, \epsilon_2 y^3 + \epsilon_3 \tau y)$. Then there exist transformation functions S and Z that satisfy (4.9) and (4.10). By (4.10) and Whitney's theorem (1943), S and Z must have the forms

$$S = \begin{pmatrix} S_{11}(x,y^2,\tau) & yS_{12}(x,y^2,\tau) \\ yS_{21}(x,y^2,\tau) & S_{22}(x,y^2,\tau) \end{pmatrix}, \quad Z = \begin{pmatrix} Z_1(x,y^2,\tau) \\ yZ_2(x,y^2,\tau) \end{pmatrix},$$

where S_{ij} and Z_i are smooth functions. By (4.9), these functions satisfy

$$Z_1(0,0,0) = 0, \quad S_{11} > 0, \quad S_{22} > 0, \quad Z_{1,1} > 0, \quad Z_2 > 0. \tag{4.12}$$

Equation (4.11) now reads

$$\begin{pmatrix} p \\ yq \end{pmatrix} = \begin{pmatrix} S_{11} & yS_{12} \\ yS_{21} & S_{22} \end{pmatrix} \begin{pmatrix} \epsilon_1 Z_1 \\ \epsilon_2 y^3 Z_2^3 + \epsilon_3 \tau y Z_2 \end{pmatrix}. \tag{4.13}$$

Solving (4.13) for p and q, taking their derivatives and evaluating at the origin with the aid of (4.12), we find the listed defining conditions and nondegeneracy conditions.

Table 3. Solution of the Recognition Problem for Several Z_2-equivariant Singular Functions

Normal Form	Defining Conditions	Nondegeneracy Conditions
$(\epsilon_1 x^2 + \epsilon_2 \tau, \epsilon_3 y)$	$p = p_1 = 0$	$\epsilon_1 = \mathrm{sgn}(p_{11})$, $\epsilon_2 = \mathrm{sgn}(p_3)$, $\epsilon_3 = \mathrm{sgn}(q)$
$(\epsilon_1 x, \epsilon_2 y^3 + \epsilon_3 \tau y)$	$p = q = 0$	$\epsilon_1 = \mathrm{sgn}(p_1)$, $\epsilon_2 = \mathrm{sgn}(q_2 - p_2 q_1/p_1)$, $\epsilon_3 = \mathrm{sgn}(-p_3 q_1/p_1)$
$(\epsilon_1 x^2 + \epsilon_2 y^2 + \epsilon_3 \tau, \epsilon_4 xy)$	$p = q = p_1 = 0$	$\epsilon_1 = \mathrm{sgn}(p_{11})$, $\epsilon_2 = \mathrm{sgn}(p_2)$, $\epsilon_3 = \mathrm{sgn}(p_3)$, $\epsilon_4 = \mathrm{sgn}(q_1)$
$(\epsilon_1 x^3 + \epsilon_2 y^2 + \epsilon_3 \tau, \epsilon_4 xy)$	$p = q = p_1 = p_{11} = 0$	$\epsilon_1 = \mathrm{sgn}(p_{111})$, $\epsilon_2 = \mathrm{sgn}(p_2)$, $\epsilon_3 = \mathrm{sgn}(p_3)$, $\epsilon_4 = \mathrm{sgn}(q_1)$
$(\epsilon_1 x^2 + \epsilon_2 y^4 + \epsilon_3 \tau, \epsilon_4 xy)$	$p = q = p_1 = p_2 = 0$	$\epsilon_1 = \mathrm{sgn}(p_{11})$, $\epsilon_2 = \mathrm{sgn}(p_{11}q_2^2 - 2p_{12}q_1 q_2 + p_{22}q_1^2)$, $\epsilon_3 = \mathrm{sgn}(p_3)$, $\epsilon_4 = \mathrm{sgn}(q_1)$

Next, let the functions $p(x,y^2,\tau)$ and $q(x,y^2)$ be given that satisfy the given defining conditions and the nondegeneracy conditions. Then they can be written, by Taylor's theorem, as

$$p = p_1 x + p_2 y^2 + p_3 \tau, \quad q = q_1 x + q_2 y^2,$$

where p_i and q_i are smooth functions of x, y^2 and τ, the notation being so chosen that the values of these functions at the origin are identical to those appearing in the nondegeneracy conditions. Now we choose

$$S \equiv \begin{pmatrix} 1 & 0 \\ y\frac{q_1}{p_1} & \frac{|p_3 q_1|^{3/2}}{|p_1||p_1 q_2 - p_2 q_1|^{1/2}} \end{pmatrix}, \quad Z \equiv \begin{pmatrix} \epsilon_1 p \\ y\frac{|p_1 q_2 - p_2 q_1|^{1/2}}{|p_3 q_1|^{1/2}} \end{pmatrix},$$

which satisfy conditions (4.9) and (4.10). Furthermore, a straightforward calculation shows that

$$\begin{pmatrix} p \\ yq \end{pmatrix} = S \begin{pmatrix} \epsilon_1(\epsilon_1 p) \\ \epsilon_2(y\frac{|p_1 q_2 - p_2 q_1|^{1/2}}{|p_3 q_1|^{1/2}})^3 + \epsilon_3 \tau y \frac{|p_1 q_2 - p_2 q_1|^{1/2}}{|p_3 q_1|^{1/2}} \end{pmatrix}.$$

This establishes the desired equivalence.

As a particular example, we consider the strain-energy function of the Mooney-Rivlin form

$$\hat{W}(\lambda_1, \lambda_2) = C_1(\lambda_1^2 + \lambda_2^2 + \frac{1}{\lambda_1^2 \lambda_2^2} - 3) + C_2(\lambda_1^2 \lambda_2^2 + \frac{1}{\lambda_1^2} + \frac{1}{\lambda_2^2} - 3), \quad (4.14)$$

where C_1 and C_2 are nonnegative constants and are not both zero. When $C_2 = 0$, the strain-energy function is said to be of neo-Hookean form. A symmetric solution of (4.1) is given by

$$\lambda_1 = \lambda_2 = \lambda, \ t = T \equiv \frac{2(C_1 + C_2\lambda^2)(\lambda^6 - 1)}{\lambda^5}. \quad (4.15)$$

It is obvious that T is monotone increasing in λ.

Substitution of (4.14) and (4.15) into (4.5) and (4.6) leads to functions $p(x, y^2, \tau)$ and $q(x, y^2, \tau)$. Evaluating these functions and their derivatives at the origin, we find that

$$p = 0, \ q = 4C_1(1 + \frac{1}{\lambda^6}) + 4C_2(-\lambda^2 + \frac{3}{\lambda^4}),$$

$$p_1 = 4C_1(1 + \frac{5}{\lambda^6}) + 12C_2(\lambda^2 + \frac{1}{\lambda^4}), \ p_2 = -12\frac{C_1}{\lambda^7} - 4C_2(\lambda + \frac{6}{\lambda^5}), \ p_3 = -2.$$

$$q_1 = -24\frac{C_1}{\lambda^7} - 8C_2(\lambda + \frac{6}{\lambda^5}), \ q_2 = 12\frac{C_1}{\lambda^8} + 4C_2(1 + \frac{10}{\lambda^6}).$$

It is observed that $p_1 > 0$. Therefore, of the normal forms in Table 3, only the second one is possible. When $C_2 = 0, q$ is always positive. This means that no bifurcation exists for neo-Hookean materials. It is further observed that q is monotone decreasing in λ, and that when $C_2 > 0, q$ becomes negative for sufficiently large λ. Therefore, there is a unique λ for which $q = 0$. This corresponds to a pitchfork bifurcation point. The coefficients in the normal form are readily calculated. It is found that $\epsilon_1 = 1, \epsilon_3 = -1$, and $\epsilon_2 = 1$ when

330 Y.-C. Chen

$C_1/C_2 > 0.08731$ and $\epsilon_2 = -1$ when $C_1/C_2 < 0.08731$. In the former case, equations (4.3) and (4.4) are strongly equivalent to

$$x = 0, \; y^3 - \tau y = 0.$$

That is, when t is less than T, there is only the symmetric solution, and as t increases from T, there exists one symmetric solution branch and two asymmetric solution branches. It has been shown by Chen (1991) with use of an energy stability criterion that the symmetric solution branch ceases to be stable at the bifurcation point, while the two asymmetric solution branches are stable.

9.5 Bifurcation of pure homogeneous deformations with D_3 symmetry

The problem to be discussed in this section is similar, in spirit, to that in the previous section. Again, we consider the bifurcation of pure homogeneous deformations of a homogeneous, isotropic, incompressible elastic body under dead load tractions. The difference is that the loads considered in this section have a D_3 symmetry. As a result, the characteristics of bifurcation for the present problem are quite different from those for the previous problem with Z_2 symmetry. Among other things, the present problem demonstrates the utility of universal unfolding, which reveals the existence of a secondary bifurcation from the solution branch with Z_2 symmetry to solution branches with no symmetry, while the primary bifurcation is from the D_3 symmetric solutions to the Z_2 symmetric solutions. Physically, this analysis describes pure homogeneous deformations of an elastic cube being stretched on its six faces by three pairs of perpendicular, uniformly distributed forces of equal magnitude. This problem was first studied by Rivlin (1948, 1974). The analysis presented here was given by Ball and Schaeffer (1983).

The equations of equilibrium are (2.96) in Chapter 1 and repeated here as

$$t = W_i(\lambda_1, \lambda_2, \lambda_3) - p\lambda_i^{-1} \quad i = 1, 2, 3, \tag{5.1}$$

where t is the magnitude of the principal Biot stresses, p the hydrostatic pressure required by the incompressibility constraint, and λ_1, λ_2 and λ_3 are again the principal stretches, which satisfy the incompressibility constraint

$$\lambda_1 \lambda_2 \lambda_3 = 1. \tag{5.2}$$

We eliminate p from (5.1) to obtain

$$\lambda_1(W_1 - t) - \lambda_2(W_2 - t) = 0, \quad \lambda_1(W_1 - t) - \lambda_3(W_3 - t) = 0, \tag{5.3}$$

By the isotropy, the strain-energy function is symmetric, i.e.

$$W(\lambda_1, \lambda_2, \lambda_3) = W(\lambda_2, \lambda_1, \lambda_3) = W(\lambda_1, \lambda_3, \lambda_2). \tag{5.4}$$

Endowed with this material symmetry and the symmetry of loads, the equations of equilibrium (5.1) are equivariant under the action of the interchange permutation and cyclic permutation group D_3. We now explore such a symmetry. First, it follows from (5.3) and (5.4) that $(\lambda_1, \lambda_2, \lambda_3) = (1, 1, 1)$ is an equilibrium solution for any t, reflecting the inability of an incompressible material to deform under hydrostatic pressure. Let T be the value of load at which bifurcation is anticipated. We make the change of variables

$$\lambda_1 = e^{2x}, \quad \lambda_2 = e^{-x+\sqrt{3}y}, \quad \lambda_3 = e^{-x-\sqrt{3}y}, \quad t = T + \tau. \tag{5.5}$$

It is observed that λ_1, λ_2 and λ_3 in (5.5) satisfy (5.2) for any values of x and y. The equations of equilibrium (5.3) can be rewritten in an equivalent form

$$g(z, \tau) = 0,$$

where $g : \mathbb{C} \times \mathbb{R} \to \mathbb{C}$ is a smooth function given by

$$g(x + iy, \tau) \equiv 2\lambda_1(W_1 - t) - \lambda_2(W_2 - t) - \lambda_3(W_3 - t)$$
$$+i\sqrt{3}\,[\lambda_2(W_2 - t) - \lambda_3(W_3 - t)], \tag{5.6}$$

where $\lambda_1, \lambda_2, \lambda_3$ and t are related to x, y and τ through (5.5). We note that the statement that g is a smooth function means only that the real and imaginary parts of g are C^∞ functions of x and y. Indeed, as a complex function, g is not in general analytic. The action of the D_3 group on a complex number z is given by

$$\{z, \bar{z}, e^{2\pi i/3}z, e^{2\pi i/3}\bar{z}, e^{-2\pi i/3}z, e^{-2\pi i/3}\bar{z}\},$$

where a superimposed bar denotes complex conjugate. It is a simple exercise to verify that g defined in (5.6) is D_3-equivariant, i.e.,

$$g(\gamma z, \tau) = \gamma g(z, \tau) \quad \forall \gamma \in D_3, z \in \mathbb{C}.$$

It is obvious from (5.6) and (5.5) that

$$g(0, \tau) = 0, \tag{5.7}$$

confirming the earlier observation that the identity deformation is an equilibrium solution for each value of the symmetric applied load. It has been shown by Golubitsky and Schaeffer (1982) that a D_3-equivariant function g satisfying (5.7) has the representation

$$g(z, \tau) = a(|z|^2, \operatorname{Re} z^3, \tau)z + b(|z|^2, \operatorname{Re} z^3, \tau)\bar{z}^2, \tag{5.8}$$

where $a, b : \mathbf{R}^2 \times \mathbf{R} \to \mathbf{R}$ are smooth functions.

The definition of strong equivalence of D_3-equivariant functions is the same as that of Z_2-equivariant functions stated in (4.9)–(4.11) with proper replacement of the symmetry group. In Table 4, we list two normal forms of g that are of interest to the particular elasticity problem at hand.

As an example, we once again consider the strain-energy function of Mooney-Rivlin form, now written as

$$W(\lambda_1, \lambda_2, \lambda_3) = C_1(\lambda_1^2 + \lambda_2^2 + \lambda_3^2 - 3) + C_2(\lambda_1^{-2} + \lambda_2^{-2} + \lambda_3^{-2} - 3). \quad (5.9)$$

Table 4. Solution of the Recognition Problem for Two D_3-equivariant
Singular Functions

Normal Form	Defining Conditions	Nondegeneracy Conditions
$\epsilon_1 \tau z + \epsilon_2 \bar{z}^2$	$a = 0$	$\epsilon_1 = \mathrm{sgn}(a_3),\ \epsilon_2 = \mathrm{sgn}(b)$
$(\epsilon_1 |z|^2 + \epsilon_2 \tau)z$	$a = b = 0$	$\epsilon_1 = \mathrm{sgn}(a_1),\ \epsilon_2 = \mathrm{sgn}(a_3),$
$+(\epsilon_3 |z|^2 + \Delta \mathrm{Re}\, z^3)\bar{z}^2$		$\epsilon_3 = \mathrm{sgn}(\frac{a_3 b_1 - a_1 b_3}{a_3}),$
		$\Delta = \frac{(a_1 b_2 - a_2 b_1)a_3^2}{(a_1 b_3 - b_1 a_3)^2}\mathrm{sgn}(a_1)$

Substituting (5.9) into the right-hand side of (5.6), evaluating it for (5.5), and equating the resulting expression to the right-hand side of (5.8), we obtain a complex equation which we can solve for the two real-valued functions $a(|z|^2, \mathrm{Re}\, z^3, \tau)$ and $b(|z|^2, \mathrm{Re}\, z^3, \tau)$. A tedious but straightforward computation of derivatives of these functions leads to

$$a = 24(C_1 + C_2 - T/4), \quad a_1 = 48(C_1 + C_2 - T/16),$$

$$a_2 = 16(C_1 - C_2 - T/32), \quad a_3 = -6, \quad (5.10)$$

$$b = 24(C_1 - C_2 - T/8), \quad b_1 = 24(C_1 - C_2 - T/32),$$

$$b_2 = (32/5)(C_1 + C_2 - T/64), \quad b_3 = -3. \quad (5.11)$$

By $(5.10)_1$, the defining condition $a = 0$ requires

$$T = 4(C_1 + C_2), \quad (5.12)$$

which gives the value of the load at the bifurcation point. Substituting (5.12) into $(5.11)_1$ shows that $b = 0$ if and only if

$$C_1 = 3C_2. \quad (5.13)$$

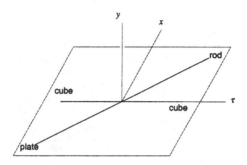

Fig. 4. Bifurcation diagram of $-\tau z + \bar{z}^2 = 0$.

We first consider the case when $C_1 > 3C_2$, which includes the neo-Hookean material as a special subcase. The normal form of the bifurcation equation in this case is

$$-\tau z + \bar{z}^2 = 0.$$

This equation has four solution branches

$$\begin{cases} x = 0 \\ y = 0, \end{cases} \qquad \begin{cases} x = \tau \\ y = 0, \end{cases} \qquad \begin{cases} x = -\frac{1}{2}\tau \\ y = \pm\sqrt{3}x. \end{cases}$$

These solution branches are sketched in Figure 4. To avoid an overcrowded figure, only the trivial solution branch $x = y = 0$ and the non-trivial solution branch $x = \tau, y = 0$ are plotted. The other two non-trivial solution branches can be obtained by rotating the plotted solution branch through angles $2\pi/3$ and $4\pi/3$ about the τ-axis. Assume that the undeformed body is a cube. It is clear from (5.5) that the trivial solution corresponds to the symmetric identity deformation with the shape of the cube remaining unchanged. It also follows from (5.5) that when $\tau < 0$, the plotted non-trivial solution corresponds to a deformation for which $\lambda_1 < 1$ and $\lambda_2 = \lambda_3 > 1$. That is, the deformed body is plate-like. On the other hand, the deformed body is rod-like with $\lambda_1 > 1$ and $\lambda_2 = \lambda_3 < 1$ when $\tau > 0$. Similar conclusions hold for the other two non-trivial solution branches.

We note that these bifurcation solutions occur for the neo-Hookean material. This shows that the present bifurcation problem features different characteristics from that of the equi-biaxial stretch discussed in the previous section, for which the neo-Hookean material does not permit bifurcation.

The bifurcation diagram when $C_1 < 3C_2$ is similar to the case where $C_1 > 3C_2$. We now turn our attention to the case where (5.13) holds, i.e. where

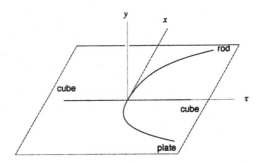

Fig. 5. Bifurcation diagram of (5.14).

$b = 0$. By (5.10) and (5.11), the coefficients of the second normal form in Table 4 are found to be

$$\epsilon_1 = 1, \ \epsilon_2 = \epsilon_3 = -1, \ \Delta = 2.$$

The normal form of the bifurcation equation is then

$$(|z|^2 - \tau)z + (-|z|^2 + 2\mathrm{Re}\, z^3)\bar{z}^2 = 0. \qquad (5.14)$$

This equation again has four solution branches

$$\begin{cases} x = 0 \\ y = 0, \end{cases} \qquad \begin{cases} x^2 - x^3 + 2x^4 = \tau \\ y = 0, \end{cases} \qquad \begin{cases} 4x^2 + 8x^3 + 32x^4 = \tau \\ y = \pm\sqrt{3}x. \end{cases}$$

In Figure 5, we plot the trivial solution branch and the non-trivial solution branch that lies in the (τ, x) plane. The other two solution branches can be obtained by rotating the plotted non-trivial solution branch by $2\pi/3$ and $4\pi/3$ about the τ-axis. It is observed that the two plotted solution branches form a pitchfork bifurcation. The trivial solution again corresponds to the identity deformation of the cube. The part of the non-trivial solution branch with positive x corresponds to rod-like deformations, and the part with negative x to plate-like deformations. The other two non-trivial solution branches have similar structures.

In reality, it is unlikely that the equality (5.13) holds exactly. This may seem to diminish the significance of the solution (5.14). However, by studying the universal unfolding of (5.14), not only can one determine the bifurcation diagram when (5.13) holds approximately, it is also possible to gain information about the bifurcation solutions which is unavailable from (5.14).

A universal unfolding of (5.14) is

$$(|z|^2 - \tau)z + [-|z|^2 + (2 + \alpha_1)\mathrm{Re}\, z^3 + \alpha_2]\bar{z}^2 = 0. \qquad (5.15)$$

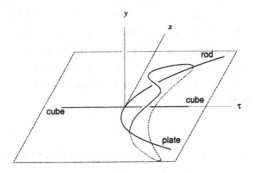

Fig. 6. Bifurcation diagram of the universal unfolding (5.15) when $\alpha_1 = 0, \alpha_2 > 0$.

This universal unfolding describes, in particular, the bifurcation diagrams for a Mooney-Rivlin material when C_1 is near $3C_2$, since we can interpret a departure of the values of material constants from those specified in (5.13) as a perturbation to the bifurcation equation (5.14). The parameter α_1 in (5.15) plays an inessential role in the bifurcation diagram of the universal unfolding. The parameter α_2, on the other hand, can change the bifurcation diagram drastically. The four solution branches in (9.5) now become

$$\begin{cases} x = 0 \\ y = 0, \end{cases} \qquad \begin{cases} \alpha_2 x + x^2 - x^3 + (2 + \alpha_1)x^4 = \tau \\ y = 0, \end{cases}$$

$$\begin{cases} -2\alpha_2 x + 4x^2 + 8x^3 + 16(2 + \alpha_1)x^4 = \tau \\ y = \pm\sqrt{3}x. \end{cases}$$

A noticeable feature of the universal unfolding (5.15) is the existence of a fifth solution branch when $\alpha_2 > 0$, given by

$$\begin{cases} -x^2 - y^2 + (2 + \alpha_1)(x^3 - 3xy^2) + \alpha_2 = 0 \\ x^2 + y^2 - \tau = 0. \end{cases} \tag{5.16}$$

This solution branch is plotted in Figure 6, along with the trivial solution branch and the non-trivial solution branch in the (τ, x) plane. It is observed that this solution branch is connected to the other three non-trivial solution branches. Therefore, it represents a secondary bifurcation. It is further observed that the solution (5.16) possesses no symmetry, i.e. the three principal stretches are distinct.

It is worth noting that the analysis of universal unfolding is still local. The perturbation from the original normal form must be sufficiently small. In the context of the present example, this is interpreted as $C_1 - 3C_2$ being sufficiently small. However, this statement is not quantified here. For individual problems, the conclusions of such local analyses may well be valid for perturbations that are not considered to be small in a practical measure. This is because the qualitative behavior of the solution could remain unchanged when the perturbation parameters vary in a certain range. From a practical point of view, the special features predicted by a local analysis may serve as indications of the global behavior of bifurcation solutions.

9.6 Bifurcation of inflation of spherical membranes

In the last section of this chapter, we present the bifurcation analysis of an infinite-dimensional problem. We consider axisymmetric deformations of a spherical membrane inflated by an enclosed controlled mass of gas. The equations of equilibrium are a pair of ordinary differential equations. By the Liapunov-Schmidt reduction, a reduced bifurcation equation with one state variable is derived. Two normal forms of the singular function are studied. One corresponds to the pitchfork in which two non-spherical solution branches bifurcate from the spherical solution branch. The other normal form corresponds to an isola. The universal unfolding of the latter normal form suggests a bifurcation diagram in which two non-spherical solution branches bifurcate from the spherical solution branch, and then, as the amount of the gas increases further, return to the spherical solution branch. This behavior agrees with the experimental observation reported by Alexander (1971), as well as the numerical analysis of Haughton (1980) for an Ogden material. The analysis of this section is developed from the work of Chen and Healey (1991).

The theory of elastic membranes is discussed in Chapter 7 of this volume. We consider an initially spherical elastic membrane of unit radius. In spherical coordinates, an axisymmetric deformation of the membrane is represented by

$$r = r(\Theta), \ \theta = \theta(\Theta), \ \phi = \Phi.$$

The functions $r(\Theta)$ and $\theta(\Theta)$ satisfy the boundary conditions

$$r(0) = r(\pi) = 0, \ \theta(0) = 0, \ \theta(\pi) = \pi, \ \theta(\pi/2) = \pi/2. \tag{6.1}$$

Equation $(6.1)_5$ is imposed to eliminate rigid-body translations. The principal stretches of the deformed membrane are given by

$$\lambda_1 = \frac{r \sin \theta}{\sin \Theta}, \quad \lambda_2 = \sqrt{r'^2 + r^2 \theta'^2}, \tag{6.2}$$

where a prime denotes the derivative with respect to Θ.

The membrane is homogeneous and isotropic, and is associated with a strain-energy function $W(\lambda_1, \lambda_2)$ which is symmetric in its two arguments. The membrane is inflated by an ideal gas, of which the mass m is taken as the control parameter. The equations of equilibrium are given by

$$p(r \sin \theta)^2 \lambda_2 + 2(r \cos \theta)' W_2 \sin \Theta = 0, \quad (r \sin \theta)' W_1 - \lambda_2 (W_2 \sin \Theta)' = 0, \quad (6.3)$$

where a subscript i on W again denotes the derivative with respect to λ_i, and p is the pressure of the gas, given by

$$p = \frac{km}{V}, \tag{6.4}$$

k being a positive gas constant, and V the volume of the gas enclosed by the deformed membrane, given by

$$V = -\pi \int_0^\pi (r \sin \theta)^2 (r \cos \theta)' d\Theta. \tag{6.5}$$

A spherical deformation is given by

$$r(\Theta) = \lambda, \quad \theta(\Theta) = \Theta, \tag{6.6}$$

where λ is the principal stretch of the deformed membrane. The equations of equilibrium for the spherical deformation become

$$m = \frac{8\pi \lambda W_1}{3k}. \tag{6.7}$$

Here and henceforth, the derivatives of W are evaluated at $(\lambda_1, \lambda_2) = (\lambda, \lambda)$, unless otherwise stated.

A bifurcation solution at the spherical deformation (6.6) exists only if the Fréchet derivative of the nonlinear differential operator associated with (6.3) is not invertible at (6.6). This leads to the linearized equations of equilibrium

$$W_1(u_1' \cot \Theta + 2u_1 + u_2 \cot \Theta) - \lambda W_{11}(u_1 + u_2') - \lambda W_{12}(u_1 + u_2 \cot \Theta)$$

$$- \frac{3}{2} W_1 \int_0^\pi u_1 \sin \Theta d\Theta = 0, \tag{6.8}$$

$$(W_1 - \lambda W_{12})(u_1' - u_2) - \lambda W_{11}(u_2'' + u_1' + u_2' \cot \Theta - u_2 \cot^2 \Theta) = 0, \quad (6.9)$$

where $u_1(\Theta)$ and $u_2(\Theta)$ correspond to the radial and transverse components of the displacement, which are related to $r(\Theta)$ and $\theta(\Theta)$ through

$$r(\Theta) = \lambda + u_1(\Theta), \quad \theta(\Theta) = \Theta + \frac{u_2(\Theta)}{\lambda}. \tag{6.10}$$

The functions u_1 and u_2 satisfy the boundary conditions

$$u_1'(0) = u_1'(\pi) = u_2(0) = u_2(\pi) = u_2(\pi/2) = 0. \tag{6.11}$$

Equations (6.8) and (6.9) can be converted, through a change of variables, to an inhomogeneous Legendre equation whose solution consists of Legendre polynomials. It can be shown that the non-spherical solution of lowest order that satisfies the boundary conditions (6.11) exists when

$$W_1 - \lambda(W_{11} + W_{12}) = 0, \qquad (6.12)$$

and is given by

$$u_1 = \cos\Theta, \ u_2 = 0.$$

This solution, corresponding to the so-called mode-one bifurcation, suggests a pear-shaped deformation for which the two principal stretches are monotone increasing from one pole to the other. In the remainder of this section, we shall consider bifurcation solutions from the spherical solution (6.6) with λ satisfying (6.12), the corresponding mass at the bifurcation point being given by (6.7).

We now use Liapunov-Schmidt reduction to derive an algebraic equation which is equivalent to the differential equations (6.3). The analysis below follows the procedure described in Section 9.2.2. We first define

$$\mathcal{X} \equiv \{u \in C^2([0,\pi];\mathbf{R}^2) : u_1'(0) = u_1'(\pi) = u_2(0) = u_2(\pi) = u_2(\pi/2) = 0\},$$

and then define $f : \mathcal{X} \times \mathbf{R} \to C^0([0,\pi];\mathbf{R}^2)$ by

$$\begin{aligned}
f(u,\mu) \equiv \ & (k(m+\mu)(r\sin\theta)^2\lambda_2 \\
& -2\pi(r\cos\theta)'W_2(\lambda_1,\lambda_2)\sin\Theta \int_0^\pi (r\sin\theta)^2(r\cos\theta)'d\Theta, \\
& (r\sin\theta)'W_1(\lambda_1,\lambda_2) - \lambda_2[W_2(\lambda_1,\lambda_2)\sin\Theta]'), \qquad (6.13)
\end{aligned}$$

where m is given by (6.7), λ_1 and λ_2 by (6.2), and r and θ by (6.10) with λ again satisfying (6.12). In the definition of $f(u,\mu)$, the bifurcation point has been moved to the origin, which corresponds to the spherical deformation specified by (6.6) and (6.12). The state variable u stands for the displacement from the spherical deformation, and the bifurcation parameter μ for the mass increment from that required by the spherical deformation.

The Fréchet derivative L of f with respect to u at $(u,\mu) = (0,0)$ is given by

$$\begin{aligned}
Lu = \ & (\frac{4}{3}\pi\lambda^3\sin^2\Theta[W_1(2u_1 + 2u_1'\cot\Theta - 3\int_0^\pi u_1\sin\Theta d\Theta) \\
& +2\lambda W_{11}(u_2\cot\Theta - u_2')], \\
& -\lambda W_{11}\sin\Theta(u_2 - u_2\cot^2\Theta + u_2'\cot\Theta + u_2''))
\end{aligned}$$

It has been shown that the linear operator L is Fredholm of index 0 with one-dimensional kernel and co-kernel (complement of range L) spanned, respectively, by

$$e(\Theta) = (\cos\Theta, 0), \ e^*(\Theta) = (0, \sin\Theta).$$

Let P be the orthogonal projection on range L, given by

$$Py(\Theta) = y(\Theta) - [\frac{2}{\pi}\int_0^\pi e^*(t) \cdot y(t)dt]e^*(\Theta).$$

Equations (6.3) are now written as the equivalent equations

$$Pf(ze(\Theta) + w(\Theta), \mu) = 0, \tag{6.14}$$

$$\int_0^\pi e^*(\Theta) \cdot f(ze(\Theta) + w(\Theta), \mu)d\Theta = 0, \tag{6.15}$$

where $z \in \mathbf{R}$ and $w : [0, \pi] \to \mathbf{R}^2$ is orthogonal to e, i.e.

$$\int_0^\pi e(\Theta) \cdot w(\Theta)d\Theta = 0.$$

In (6.14) and (6.15), the scalar variable z is the component of the displacement $u(\Theta)$ in the direction of ker L, and $w(\Theta)$ is the remaining displacement.

The Fréchet derivative of the left-hand side of (6.14) with respect to w is an invertible linear mapping from the complement of ker L onto range L. By the implicit function theorem, one can solve equation (6.14) locally for w, and write

$$w = W(z, \mu). \tag{6.16}$$

Substituting (6.16) into (6.15) yields the reduced bifurcation equation

$$g(z, \mu) \equiv \int_0^\pi e^* \cdot f(ze + W(z, \mu), \mu)d\Theta = 0.$$

The derivatives of $g(z, \mu)$ can be found by following the analysis leading to (2.19)–(2.24). Note that here ker L and $\mathcal{R}$ have dimension one and that we have used z for the state variable and μ for the bifurcation parameter in the reduced bifurcation equation. Using (2.19) and (2.21)–(2.23), along with a lengthy but routine calculation, we find that

$$g_z = 0, \ g_\mu = (e^*, f_\mu) = 0, \ g_{zz} = (e^*, f_{uu}ee) = 0, \tag{6.17}$$

$$g_{z\mu} = (e^*, -f_{uu}e(L^{-1}Pf_\mu) + f_{u\mu}e) = \frac{k\lambda(W_{111} + 3W_{112})}{4\pi W_1}. \tag{6.18}$$

We note that f defined in (6.13) has a symmetry:

$$f(\gamma u(\pi - \Theta), \mu) = \gamma f(u(\Theta), \mu), \quad \gamma \equiv \begin{pmatrix} 1 & 0 \\ 0 & -1 \end{pmatrix}.$$

This symmetry is subsequently inherited by the functions $W(z, \mu)$ and $g(z, \mu)$. In particular, g is odd in z:

$$g(-z, \mu) = -g(z, \mu).$$

Such a function is said to have Z_2-symmetry†. As a result of Z_2-symmetry, the value of g, its derivatives with respect to μ, and its even-order derivatives with respect to z all vanish at the origin, in agreement with, for example, (6.17).

The solutions of several recognition problems for Z_2-symmetric functions are given in Golubitsky and Schaeffer (1985, VI. 5). In Table 5, we list two of them which are of interest to the particular elasticity problem at hand. The first normal form is a pitchfork, while the second normal form consists of the trivial solution branch and either a simple bifurcation or an isola center.

Table 5. Solution of the Recognition Problem for Two Z_2-symmetric Singular Functions

Normal Form	Defining Conditions	Nondegeneracy Conditions
$\epsilon z^3 + \delta \mu z$	$g_z = 0$	$\epsilon = \operatorname{sgn}(g_{zzz}), \ \delta = \operatorname{sgn}(g_{z\mu})$
$\epsilon z^3 + \delta \mu^2 z$	$g_z = g_{z\mu} = 0$	$\epsilon = \operatorname{sgn}(g_{zzz}), \ \delta = \operatorname{sgn}(g_{z\mu\mu})$

Once again we turn to specific strain-energy functions. First we examine the strain-energy function of Mooney-Rivlin form given by (4.14). A simple calculation shows that

$$W_1 - \lambda(W_{11} + W_{12}) = -\frac{4[3C_1 + C_2(2\lambda^2 + \lambda^8)]}{\lambda^5} < 0.$$

That is, the Mooney-Rivlin material does not permit mode-one bifurcation. Since the deformations corresponding to such a bifurcation have been observed in experiment with spherical neoprene balloons, this result serves as an indication of the inadequacy of the Mooney-Rivlin model for rubber-like materials at large deformation.

A more accurate form of strain-energy function for a rubber-like material,

† Z_2-symmetry for functions from $\mathbf{R}^2$ to $\mathbf{R}^2$ is discussed in Section 9.4.

proposed by Ogden (1972), is given by

$$W(\lambda_1, \lambda_2) = \sum_{n=1}^{3} \frac{\mu_n}{\alpha_n} [\lambda_1^{\alpha_n} + \lambda_2^{\alpha_n} + \frac{1}{\lambda_1^{\alpha_n} \lambda_2^{\alpha_n}} - 3], \qquad (6.19)$$

where the dimensionless material parameters are given by

$$\mu_1 = 1.491, \ \mu_2 = 0.003, \ \mu_3 = -0.024, \ \alpha_1 = 1.3, \ \alpha_2 = 5.0, \ \alpha_3 = -2.0. \quad (6.20)$$

Substituting (6.19) and (6.20) into (6.12), we find that there are two mode-one solutions at

$$\lambda^{(1)} = 1.778, \quad \lambda^{(2)} = 2.514.$$

By (6.7), the corresponding masses of gas at the bifurcation points are

$$m^{(1)} = \frac{25.99}{k}, \quad m^{(2)} = \frac{50.79}{k}.$$

Furthermore, by (6.18), the coefficients of the first normal form in Table 5 are found to be

$$g_{zzz}^{(1)} > 0, \quad g_{zzz}^{(2)} > 0, \quad g_{z\mu}^{(1)} < 0, \quad g_{z\mu}^{(2)} > 0. \qquad (6.21)$$

Hence, these two solutions correspond to pitchfork bifurcation. By the signs of the coefficients in (6.21), these two pitchforks are found to be facing each other, as shown in Figure 7.

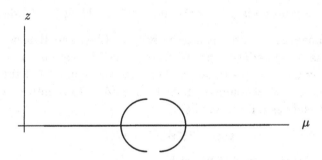

Figure 7. Bifurcation diagram with two pitchforks.

Using numerical analysis, Haughton (1980) has shown that these two pitchfork solution branches are actually connected to each other, forming a closed loop. The first normal form itself does not yield this result, since the analysis is local. However, this result can be confirmed by a study of the universal unfolding of the second normal form. Indeed, by adjusting the values of the material parameters in (6.20), it is possible to make the two bifurcation points coalesce with $g_{z\mu} = 0$ at the new bifurcation point. In the case where $g_{zzz} > 0$ and $g_{z\mu\mu} > 0$, the bifurcation equation has the second normal form in Table 5, i.e.

$$z^3 + \mu^2 z = 0. \tag{6.22}$$

A universal unfolding of (6.22) is

$$z^3 + \mu^2 z + \alpha z = 0. \tag{6.23}$$

For negative α, the bifurcation diagram of (6.23) is sketched in Figure 8, which consists of a closed loop and the trivial solution branch. This gives a qualitative description of the solutions when the values of the material parameters are near the adjusted values.

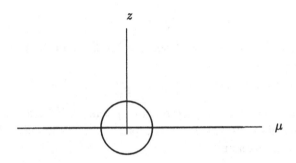

Figure 8. Bifurcation diagram of the universal unfolding (6.23) when $\alpha < 0$.

This argument is carried out alternatively in Chen and Healey (1991) for a general class of materials. The structure of the bifurcation solutions allows for a simple graphical representation. Let ξ be the reciprocal of the principal stretch of the spherical solution. Substituting (6.5)–(6.7) into (6.4), we can write the pressure as a function of ξ:

$$p(\xi) \equiv 2\xi^2 W_1(1/\xi, 1/\xi). \tag{6.24}$$

Successive differentiation of (6.24) yields

$$p'(\xi) = 4\xi W_1 - 2(W_{11} + W_{12}), \quad p''(\xi) = 4W_1 - \frac{4}{\xi}(W_{11} + W_{12}) + \frac{2}{\xi^2}(W_{111} + 3W_{112}).$$

It then follows that equation (6.12) holds at a point if and only if the tangent line of the (p, ξ) curve at that point passes through the origin. Moreover, the coefficient $g_{z\mu}$ given by (6.18) vanishes at this point if and only if the curvature of the (p, ξ) curve is zero there. This give a graphical characterization of the materials which have mode-one bifurcation solutions with one of the normal

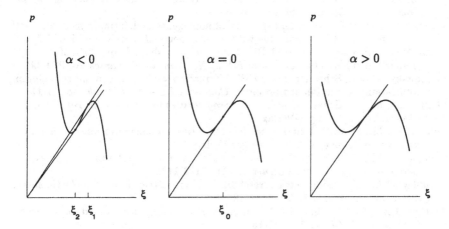

Fig. 9. Several (p, ξ) curves.

forms in Table 5. In Figure 9, we plot three (p, ξ) curves. It is observed that there are two pitchfork bifurcation points ξ_1 and ξ_2 for the material with (p, ξ) curve shown in Figure 9(a), there is one bifurcation point ξ_0 with normal form (6.22) for Figure 9(b), and no mode-one bifurcation point for Figure 9(c).

With this structure in mind, we consider a one-parameter family of strain-energy functions $W(\lambda_1, \lambda_2, \alpha)$. The above bifurcation analysis can be carried out with α being treated as a second bifurcation parameter. This leads to a reduced bifurcation equation

$$g(z, \mu, \alpha) = 0. \tag{6.25}$$

It is shown by Chen and Healey (1991) that the (p, ξ) curves of this family of materials are as shown in Figure 9 for various values of α, and that the normal form of (6.25) is precisely (6.23). We thus conclude that when α is negative and sufficiently close to zero, there are two pitchfork bifurcation solution branches that are connected to each other to form a closed loop, as shown in Figure 8.

References

Alexander, H. 1971 Tensile instability of initially spherical balloons. *Int. J. Engng. Sci.* **9**, 151-162.

Arnold, V.I. 1976 Local normal forms of functions. *Invent. Math.* **35**, 87-109.

Arnold, V.I. 1981 *Singularity Theory*. London Mathematical Society Lecture Notes Series, 53, Cambridge University Press, Cambridge.

Ball, J.M. and Schaeffer, D.G. 1983 Bifurcation and stability of homogeneous
    equilibrium configurations of an elastic body under dead-load tractions. *Math.
    Proc. Camb. Phil. Soc.* **94**, 315-339.
Chen, Y.C. 1991 Bifurcation and stability of homogeneous deformations of an elastic
    body under dead load tractions with Z_2 symmetry. *J. Elasticity* **25**, 117-136.
Chen, Y.C. and Healey, T.J. 1991 Bifurcation to pear-shaped equilibria of
    pressurized spherical membranes. *Int. J. Non-Linear Mechanics* **26**, 279-291.
Golubitsky, M. and Schaeffer, D.G. 1982 Bifurcation with O(3) symmetry including
    applications to the Bénard problem. *Commun. Pure Appl. Math.* **35**, 81-111.
Golubitsky, M. and Schaeffer, D.G. 1985 *Singularities and Groups in Bifurcation
    Theory*, Vol. I. Springer-Verlag.
Golubitsky, M., Stewart, I. and Schaeffer, D.G. 1988 *Singularities and Groups in
    Bifurcation Theory*, Vol. II. Springer-Verlag.
Haughton, D.M. 1980 Post-bifurcation of perfect and imperfect spherical elastic
    membranes. *Int. J. Solids Structures* **16**, 1123-1133.
Kearsley, E.A. 1986 Asymmetric stretching of a symmetrically loaded elastic sheet.
    Int. J. Solids Structures **22**, 111-119.
Mather, J.H. 1968 Stability of C^∞ mappings, III. Finitely determined map germs.
    Publ. Math. I.H.E.S., **35**, 127-156.
Mather, J.H. 1969a Stability of C^∞ mappings, II. Infinitesimal stability implies
    stability. *Ann. Math.* **89**, 254-291.
Mather, J.H. 1969b Right equivalence. Lecture Notes, University of Warwick.
MacSithigh, G.P. and Chen, Y.C. 1992a Bifurcation and stability of an
    incompressible elastic body under homogeneous dead loads with symmetry.
    Part I: General isotropic materials. *Quart. J. Mech. Appl. Math.* **45**, 277-291.
MacSithigh, G.P. and Chen, Y.C. 1992b Bifurcation and stability of an
    incompressible elastic body under homogeneous dead loads with symmetry.
    Part II: Mooney-Rivlin materials. *Quart. J. Mech. Appl. Math.* **45**, 293-313.
Ogden, R.W. 1972 Large deformation isotropic elasticity–on the correlation of
    theory and experiment for incompressible rubberlike solids. *Proc. R. Soc. Lond.
    A.* **326**, 565-584.
Ogden, R.W. 1985 Local and global bifurcation phenomena n plane strain finite
    elasticity. *Int. J. Solids Structures* **21**, 121-132.
Rivlin, R.S. 1948 Large elastic deformations of isotropic materials. II. Some
    uniqueness theorems for pure, homogeneous, deformations. *Phil. Trans. Roy.
    Soc. London* **A 240**, 491-508.
Rivlin, R.S. 1974 Stability of pure homogeneous deformations of an elastic cube
    under dead loading. *Quart. Appl. Math.* **32**, 265-271.
Timoshenko, S.P. and Gere, J.M. 1961 *Theory of Elastic Stability.* McGraw-Hill.
Treloar, L.R.G. 1948 Stresses and birefringence in rubber subjected to general
    homogeneous strain. *Proc. Phys. Soc.* **60**, 135-144.
Whitney, H. 1943 Differentiable even functions. *Duke Math. J.* **10**, 159-160.

10

Perturbation methods and nonlinear stability analysis

Y.B. Fu
Department of Mathematics
University of Keele
Staffordshire ST5 5BG, U.K.
Email: y.fu@maths.keele.ac.uk

In this chapter we discuss applications of the perturbation approach to stability analysis of elastic bodies subjected to large deformations. Various ideas commonly used in the perturbation approach are explained by using simple examples. Two types of bifurcations are distinguished: bifurcations at a non-zero critical mode number and bifurcations at a zero critical mode number. For each type we first explain with the aid of a model problem how stability analysis can be carried out and then explain how the analysis could be extended to problems in Finite Elasticity. Although the present analysis focuses on the perturbation approach, the dynamical systems approach is also discussed briefly and references are made to the literature where more details can be found. In the final section, we carry out a detailed analysis for the necking instability of an incompressible elastic plate under stretching.

10.1 Introduction

This chapter is concerned with nonlinear stability analysis of elastic bodies subjected to large elastic deformations. A typical problem we have in mind is the stability of a cylindrical rubber tube that is compressed either by an external pressure or by forces at the two flat ends. In general terms, we consider an elastic body which has an undeformed configuration $\mathcal{B}_r$ in a three-dimensional Euclidean point space. This elastic body is then subjected to some external forces. It is now customary to refer to such an elastic body as *pre-stressed* in the Finite Elasticity literature (the pre-stress considered in the present context is not therefore that induced in a manufacturing process). Very often one solution for the resulting deformation in the elastic body can easily be found. For instance, when an elastic plate is compressed along one direction, one solution is a uni-axial compression. We refer to this first solution as a *primary* deformation. We denote by $\mathcal{B}$ the corresponding configuration of the elastic body,

and by $x = \chi(X)$ the deformation, where X and x are the position vectors of
a representative material particle in B_r and B, respectively. In many engineering situations it is important to know if this primary deformation is stable, in
other words whether other (adjacent) deformations/configurations are possible
and are preferred by the elastic body (this approach is known as the *method
of adjacent equilibria* in the Finite Elasticity literature). For instance, when
we compress an elastic strut we can see that the straight line configuration is
the only possible configuration when the force is small enough, but when the
force exceeds a certain critical value, another (buckled) configuration is also
possible and is in fact preferred by the strut to the straight line configuration.
Thus, to determine whether a pre-stressed elastic body is stable or not, two
questions must be resolved. Firstly, are there other (static) solutions that are
mathematically possible? The answer to this question also depends on boundary conditions. Throughout this chapter, we shall be concerned with *dead-load*
boundary conditions by which we mean that if any part of the elastic body is
subjected to a traction then the resultant of this traction is held fixed in the
searching for other solutions. The stability of a spherical shell in deep water
is not a dead-load problem since the resultant acting on the shell is dependent
on the shape of the buckled configuration. Secondly, if another (static) configuration is mathematically admissible, is it preferred by the elastic body? In
this chapter, our focus will be on the first question. Therefore this chapter is
essentially concerned with local bifurcation analysis.

   In order to determine whether other configurations are possible under dead-
load conditions, we first assume that another (buckling) solution is indeed
possible and we denote it by $\tilde{x} = \tilde{\chi}(x)$, where $\tilde{x}$ is the position vector, in the
buckled configuration B_t, of a material particle whose position vector in B is x.
Since $\tilde{x}$ and x satisfy the same governing equations and boundary conditions,
the incremental displacement u, defined by $u = \tilde{x} - x$, should satisfy a nonlinear
eigenvalue problem which always admits the trivial solution $u = 0$.

   We denote the deformation gradients from B_r to B and from B_r to B_t by
F and $\tilde{F}$, respectively. The dead-load traction boundary condition is given by
$(\tilde{S}^T - S^T)N = 0$, where $\tilde{S}$ is the nominal stress in the buckled configuration,
S its value associated with the primary deformation, and N the normal to a
material surface in B_r where dead-load tractions are prescribed. With the use of
Nanson's formula (2.8) in Chapter 1, the dead-load traction boundary condition
may also be written as $(\tilde{S}^T - S^T)F^Tn = 0$, where n is the normal to the same
material surface in B. This motivates the introduction of an incremental stress
tensor Σ through

$$\Sigma = J^{-1}F(\tilde{S} - S),$$

where $J = \det(\mathbf{F})$ is inserted to preserve the useful property $\operatorname{div}(J^{-1}\mathbf{F}) = \mathbf{0}$ (cf. equation (2.11) in Chapter 1). The above definition of Σ is the same as that given by equation (5.4) in Chapter 1. In terms of this incremental stress tensor, the above-mentioned nonlinear eigenvalue problem consists of solving the incremental equilibrium equation

$$\Sigma_{ji,j} = 0, \tag{1.1}$$

and the incompressibility condition

$$u_{i,i} = \frac{1}{2}u_{m,n}u_{n,m} - \frac{1}{2}(u_{i,i})^2 - \det(u_{i,j}), \tag{1.2}$$

subjected to the dead-load boundary conditions

$$\Sigma_{ji}n_j = 0, \tag{1.3}$$

where all the dependent variables are functions of material coordinates (x_i) in $\mathcal{B}$ and a comma signifies differentiation with respect to the indicated coordinate. The incremental stress tensor (Σ_{ji}) is given, up to cubic terms, by

$$\Sigma_{ji} = \mathcal{A}^1_{jilk}u_{k,l} + \frac{1}{2}\mathcal{A}^2_{jilknm}u_{k,l}u_{m,n} + \frac{1}{6}\mathcal{A}^3_{jilknmqp}u_{k,l}u_{m,n}u_{p,q}$$

$$+ p(u_{j,i} - u_{j,k}u_{k,i} + u_{j,k}u_{k,l}u_{l,i}) - p^*(\delta_{ji} - u_{j,i} + u_{j,k}u_{k,i}) + \cdots, \tag{1.4}$$

where p and p^* are the pressure in the primary deformation and the incremental pressure, respectively, and expressions for the moduli $\mathcal{A}^1_{jilk}$, $\mathcal{A}^2_{jilknm}$ and $\mathcal{A}^3_{jilknmqp}$ in terms of principal stretches in the primary deformation, together with a derivation of (1.4), can be found in Fu and Ogden (1999). The linearized form of (1.4) appears as equation (5.6) in Chapter 1 (here we have omitted the subscript "0" in $\mathcal{A}^1_{0jilk}$ to simplify notation). If the elastic body is compressible, we drop the incompressibility condition (1.2) and all those terms involving p and p^* in (1.4). The equations have been given here in rectangular coordinates since these are what we shall use in the present chapter. The corresponding equations in cylindrical or spherical polar coordinates can be found in Appendix A at the end of this chapter.

Thus whether other buckled configurations are possible is reduced to the question of whether the above nonlinear eigenvalue problem has non-trivial solutions.

The first step in understanding solution properties of this eigenvalue problem is to conduct a linear analysis. We first linearize the governing equations and boundary conditions. By looking for a normal mode solution, the partial differential equations are reduced to ordinary differential equations, and the value of pre-stress at which the reduced problem has a non-trivial solution can then be determined. This value is usually a function of a mode number. The minimum of this function is usually referred to as the *critical pre-stress* and the corresponding mode number as the *critical mode number*.

Such linear analyses have been conducted over the past five decades or so for different geometries (plates, tubes, spherical shells, half-spaces etc.), different material constitutions and different forms of pre-stress. Without discussing these contributions individually, we have listed them in Appendix B at the end of this chapter. In most of these studies, the geometries and forms of pre-stress are simple enough so that the differential equations in the reduced linear eigenvalue problem can be solved analytically. When the differential equations have variable coefficients, as for instance in the case of a cylindrical tube compressed by an external pressure, the eigenvalue problem cannot be solved analytically. An effective numerical method that can be used in such a case is the *compound matrix method*; see Ng and Reid (1979) for a clear description of this method and Bridges (1999) for an interpretation of the method from a differential-geometric point of view. It is good practice to combine numerical calculations with a WKB analysis since the latter can be used to find an asymptotic expression for the critical pre-stress valid in the large critical mode number limit. A description of the WKB method can be found in almost any book on asymptotic analysis. Recently this method has been used in the linear stability analysis of a spherical shell of arbitrary thickness (Fu, 1998) and of the eversion of a cylindrical tube (Fu and Sanjarani Pour, 2000).

In sharp contrast with the large body of literature on linear stability analysis, studies of nonlinear solutions of the above-mentioned eigenvalue problem are scarce. This is presumably due to the algebraic complexity of the equations involved. However, with the appearance of powerful symbolic manipulation packages, such as Mathematica and Maple, the algebraic complexity is no longer a major obstacle. Progress can now be made on the understanding of post-buckling properties of elastic bodies subjected to large deformations. Since the elastic materials that we have in mind can sustain large deformations and the buckling deformation can be as large as the primary deformation, Finite Elasticity provides a fertile ground for nonlinear analysis.

In this chapter we collect together a number of methods of nonlinear stability analysis that have been well developed in Finite Elasticity and in other disciplines. Our aim is to convey to the reader the essential ideas of these methods with minimum effort. Thus, simple examples are used in the explanation of these ideas. We remark that our analysis is biased towards the perturbation approach. For alternative approaches, we refer the reader to Chapter 9 and to the book by Antman (1995). We note that the perturbation approach can be used to study a variety of problems, not only static bifurcation problems, but also for instance, problems concerning propagation of nonlinear waves. Whichever problem is studied, the techniques used are the same. Most of these techniques should become transparent in the course of our analysis of various model problems.

Stability problems in the normal mode approach can be divided into two categories: problems where the critical mode number, k_{cr} say, is non-zero and problems where k_{cr} is zero. Roughly speaking, when k_{cr} is non-zero, the linearized operator is (strongly) dispersive in the sense that if $e^{ik_{cr}x}$ is a solution of the linear operator, where x is the variable along which the solution is assumed to be sinusoidal, then $e^{ink_{cr}x}$ ($n \neq \pm 1$) are in general not solutions of the linear operator. As a result, amplitude equations are usually derived by imposing a solvability condition at third order of the successive approximations in the perturbation approach. When k_{cr} is zero, the solution depends on x through a far distance variable and a solvability condition has to be imposed at the second order of the successive approximations. Sections 10.2 and 10.3, respectively, are devoted to these two types of problems. In Section 10.4, we study the necking instability associated with an incompressible elastic plate under stretching. The chapter is concluded with some additional remarks concerning other aspects of stability analysis.

10.2 Bifurcation at a non-zero mode number

10.2.1 A model problem

We shall use the simple problem

$$u_{tt} + u_{xxxx} + Pu_{xx} + u - u^2 = 0, \quad |x| < \infty, \qquad (2.1)$$

to explain how weakly nonlinear analysis can be conducted, using the perturbation approach, for problems where bifurcation takes place at a non-zero mode number. In (2.1) u is a function of x and t, a subscript denotes differentiation and the constant P is a bifurcation parameter. The trivial solution $u = 0$ is clearly a solution for all values of P. Our aim is to find non-trivial solutions (and the conditions under which they exist).

Equation (2.1) is the scaled form of an approximate model equation governing the dynamic motion/vibration of an infinitely long beam supported by an array of nonlinear elastic springs. The parameter P is the (scaled) compressive force applied at $x = \pm \infty$ and u denotes the transverse deflection.

With the u_{tt} term neglected, (2.1) reduces to

$$u_{xxxx} + Pu_{xx} + u - u^2 = 0, \qquad (2.2)$$

which governs the deflection of a beam from its straight line configuration when the beam buckles (statically). Whether buckling can actually occur depends on whether (2.2) admits a non-trivial solution.

Equations (2.1), (2.2) and their variations (with, *e.g.*, the quadratic term replaced by a cubic term) have been studied extensively by engineers and applied mathematicians as well as by applied analysts. See, for example, Lange and Newell (1971), Potier-Ferry (1987), Wadee *et al.* (1997), Champneys and Toland (1993), Champneys (1998). Equation (2.2) also arises as the amplitude equation for some problems where the bifurcation parameter, as a function of mode number k, expands like $P_0 + P_2 k^4 + \cdots$ near $k = 0$ (instead of $P_0 + P_1 k^2 + \cdots$). See, *e.g.*, Buffoni *et al.* (1996).

Amplitude equation for a single mode

In the linear analysis, we neglect the nonlinear term and look for a normal mode solution

$$u = A e^{ikx} + c.c., \tag{2.3}$$

where A is a complex constant, k the mode number and *c.c.* denotes the complex conjugate of the preceding term. On substituting (2.3) into the linearized form of (2.2) we find that (2.3) is a possible solution for arbitrary A as long as the bifurcation parameter P and k satisfy

$$P = k^2 + k^{-2}. \tag{2.4}$$

The pressure P attains a minimum 2 when $k = 1$. We write

$$P_{cr} = 2, \quad k_{cr} = 1.$$

We note that in the linear analysis the amplitude A is left undetermined.

In a weakly nonlinear analysis, we are interested in the relation between the buckling amplitude A and the *small* pressure increment $P - P_{cr}$. Since (2.3) with (2.4) is obtained with the nonlinear term neglected, we expect that (2.3) is valid when u is small and that (2.3) represents the leading-order term of a perturbation solution. Thus, we expect

$$u = \epsilon u^{(1)} + \epsilon^2 u^{(2)} + \epsilon^3 u^{(3)} + \cdots,$$

$$u^{(1)} = AE + c.c., \quad P = P_{cr} + \epsilon^2 P_1, \tag{2.5}$$

to be a possible solution for the nonlinear equation (2.2), where

$$E \equiv e^{ik_{cr}x},$$

and ϵ is a small positive parameter characterizing the amplitude of the buckling solution. Throughout this chapter, we use "$\equiv$" to mean "defined as".

Equation (2.5) implies that the amplitude of the buckling solution should be of order $\sqrt{|P - P_{cr}|}$. Determination of such an order relation is the first task in a perturbation analysis. As will become clear shortly, this order relation is determined by the requirement that P_1 should appear in the amplitude equation for A. We expect that the amplitude equation for A will be derived by imposing a solvability condition at order ϵ^3. This then requires that $(P - P_{cr})u_{xx} = O(u^3)$, that is $P - P_{cr} = O(u^2)$.

On substituting (2.5) into (2.2) and equating the coefficients of ϵ, ϵ^2 and ϵ^3, we find that

$$\mathcal{L}[u^{(1)}] \equiv u^{(1)}_{xxxx} + 2u^{(1)}_{xx} + u^{(1)} = 0,$$

$$\mathcal{L}[u^{(2)}] = (u^{(1)})^2 = 2A\bar{A} + A^2 E^2 + \bar{A}^2 \bar{E}^2, \qquad (2.6)$$

$$\mathcal{L}[u^{(3)}] = 2u^{(1)}u^{(2)} - P_1 u^{(1)}_{xx},$$

where a bar denotes complex conjugation. We note that P_1 first appears in the $O(\epsilon^3)$ equation, as we required.

The linear operator $\mathcal{L}$ defined above is a typical dispersive operator. We see that $\mathcal{L}[E^n]$ vanishes only when $n = \pm 1$. Thus, $(2.6)_1$ is automatically satisfied and none of the terms on the right-hand side of $(2.6)_2$ satisfy the homogeneous equation $\mathcal{L}[f] = 0$. Because of this property of $\mathcal{L}$ the obvious particular integral for $(2.6)_2$ is a linear combination of E^0, E^2 and $\bar{E}^2$. We have

$$u^{(2)} = 2A\bar{A} + \frac{1}{9}(A^2 E^2 + \bar{A}^2 \bar{E}^2), \qquad (2.7)$$

where we have neglected the complementary function since it can be absorbed into $u^{(1)}$. Equation $(2.6)_3$ then becomes

$$\mathcal{L}[u^{(3)}] = (P_1 A + \frac{38}{9}A^2\bar{A})E + \frac{1}{9}A^3 E^3 + c.c. \qquad (2.8)$$

Since $\mathcal{L}[E] = 0$, the first term on the right-hand side of (2.8) would give rise to a term in $u^{(3)}$ that is proportional to xE. The latter product would make the expansion $(2.5)_1$ non-uniform in the sense that the third term $\epsilon^3 u^{(3)}$ would be of the same order as the second term $\epsilon^2 u^{(2)}$ when x becomes as large as $1/\epsilon$. Terms with this property are usually referred to as *secular terms*. To eliminate such a secular term, we must set

$$P_1 A + \frac{38}{9}A^2\bar{A} = 0, \qquad (2.9)$$

which is the desired amplitude equation for A. We may also view (2.9) as a necessary condition for the perturbation solution to be periodic. Bifurcations described by an equation of the form (2.9) are said to be of *pitch-fork* type. An amplitude equation such as (2.9) is called a *normal form*: it is algebraic and it contains the minimum number of terms sufficient to describe the near-critical behaviour.

Without loss of generality, we may assume A to be real (since the origin of x in (2.3) can always be re-defined to make the amplitude real). Equation (2.9) then has three solutions:

$$A = 0 \quad \text{and} \quad A = \pm\sqrt{-9P_1/38}. \tag{2.10}$$

We note that the non-trivial solutions are only possible if $P_1 < 0$, that is if $P < P_{cr}$. Bifurcations of this nature are said to be *sub-critical*. Bifurcations which occur when the bifurcation parameter is above the critical value are said to be *super-critical*.

Effects of imperfections

An important property associated with sub-critical bifurcations is *imperfection sensitivity*. Imperfections can take various forms. For instance the Young's modulus of the beam may not be constant, the beam may not be completely straight or the load P may not be entirely along the axis (so that deflection occurs as soon as P is non-zero). Imperfections that vary along the x-direction like the critical buckling mode (but with much smaller amplitude) are referred to as *modal imperfections*. Such imperfections can easily be incorporated into the above asymptotic analysis. We simply assume that the imperfection is of such an order that its effect is felt at order $O(\epsilon^3)$. An extra term would then appear in the amplitude equation (2.9). However, a study of each such imperfection is not necessary because there is a very elegant result in Singularity Theory (cf. Chapter 9) that tells us that, no matter what imperfections the system has, the most general amplitude equation describing the near-critical behaviour is of the form

$$P_1 A + \frac{38}{9} A^3 + c_1 + c_2 A^2 = 0, \tag{2.11}$$

where c_1 and c_2 are constants. Equation (2.11) is said to be a *universal unfolding* of the normal form (2.9). Since two independent parameters are required to do this, the bifurcation governed by (2.9) is said to be of *co-dimension* 2.

Imperfections have a drastic effect on sub-critical bifurcations. For instance, if $c_2 = 0$, equation (2.11) reduces to

$$\frac{P - P_{cr}}{\epsilon^2} A + \frac{38}{9} A^3 + c_1 = 0, \tag{2.12}$$

where we have used $(2.5)_3$. Equation (2.12) has three solution branches for A against P, but the branch which is physically most relevant satisfies $A \to 0$ as

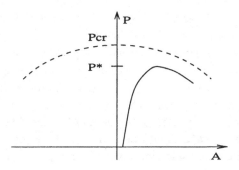

Fig. 1. Relationship between P and A with (solid line) or without (dashed line) imperfections.

$P \to 0$. This behaviour is captured by balancing the first and third terms in (2.12), and we have $A = c_1\epsilon^2/(P_{cr} - P) + \cdots$, as $P \to 0$. As P is increased gradually from 0, A is initially of order ϵ^2 (due to the imperfection). The branch has a turning point at a certain value P^* of P (cf. Figure 1). This value P^* is determined by the condition $dP/dA = 0$, that is, with the use of (2.12),

$$\frac{P - P_{cr}}{\epsilon^2} + \frac{38}{3}A^2 = 0. \tag{2.13}$$

Solving (2.12) and (2.13) simultaneously for P and A yields the coordinates at the turning point. In particular, the P coordinate is given by

$$P^* = P_{cr} - (57/2)^{1/3} \cdot (c_1\epsilon^3)^{2/3}, \tag{2.14}$$

where $c_1\epsilon^3$ characterizes the amplitude of the initial imperfection. This fact is known as Koiter's two-thirds power law (see Koiter 1945 or Hutchinson and Koiter 1970). It shows that for sub-critical bifurcations, the presence of imperfections can reduce the real critical value by an amount that is proportional to the two thirds power of the imperfection amplitude. Because of this, structures that suffer sub-critical bifurcations are said to be imperfection sensitive.

Stability of post-buckling states

The stability of the post-buckling solutions can be determined with the aid of the dynamic form of the amplitude equation (2.9). To derive this dynamic amplitude equation from (2.1) we first look for a travelling wave solution of the form $u = A\exp\{ik(x - vt)\} + c.c.$, where v is the wave speed. Substitution of this expression into the linearized form of (2.1) yields the *dispersion relation*

$$v^2 = k^2 + k^{-2} - P. \tag{2.15}$$

With P given by $(2.5)_3$ and $k = k_{cr} = 1$, this relation gives $v^2 = -\epsilon^2 P_1$. Thus, the critical mode depends on time t through the product ϵt. In other words, variation of the critical mode with respect to time is very slow and it becomes significant only when t becomes of order $1/\epsilon$ in magnitude. Thus, it is appropriate to scale t by introducing a *slow time* variable τ through $\tau = \epsilon t$. The u_{tt} term in (2.1) becomes $\epsilon^2 u_{\tau\tau}$. This extra term does not affect the leading and second-order equations $(2.6)_{1,2}$, but A in (2.5) should be allowed to depend on τ. An extra term $-u_{\tau\tau}$ should be added to the right-hand side of $(2.6)_3$, and, as a result, the amplitude equation (2.9) is replaced by

$$A_{\tau\tau} = P_1 A + \frac{38}{9} A^3. \tag{2.16}$$

To test the stability of $(2.10)_2$ we let $A = \pm\sqrt{-9P_1/38} + B(\tau)$, where $B(\tau)$ is a small perturbation. Substituting this expression into (2.16) and linearizing in terms of $B(\tau)$ we obtain $B_{\tau\tau} = -2P_1 B$, which shows that any small amplitude perturbation will grow exponentially (recall that the non-trivial solutions are only possible if $P_1 < 0$). Thus the non-trivial solutions $(2.10)_2$ are unstable.

Effects of side-band modes

From the dispersion relation (2.15) we see that, in the neighbourhood of $P = 2$ and $k = 1$, v^2 expands as

$$v^2 = 4(k - 1)^2 - (P - 2) + \cdots, \tag{2.17}$$

which indicates that those modes with mode numbers differing from k_{cr} by an $O(\epsilon)$ quantity (called *near-critical* or *side-band* modes) also evolve on an $O(1/\epsilon)$ time scale. This indicates that we may allow for an arbitrary number of such near-critical modes in our leading-order buckling solution. Thus, we may write

$$u^{(1)} = \sum_{n=0}^{\infty} A_n(\tau) e^{i(1+\epsilon k_n)x} + c.c., \tag{2.18}$$

where $k_n (n = 0, 1, 2, \ldots)$ are $O(1)$ constants. Since the above expression can be rewritten as

$$u^{(1)} = \left\{ \sum_{n=0}^{\infty} A_n(\tau) e^{ik_n \epsilon x} \right\} E + c.c.,$$

it is clear that a better formulation is to consider a solution of the form

$$u^{(1)} = A(X, \tau) E + c.c., \quad X \equiv \epsilon x. \tag{2.19}$$

The variable X defined above takes care of all possible near-critical modes and is sometimes referred to as a *far distance* variable (meaning that effects described by this variable are only significant for large distances).

The expression in (2.19) gives rise to a coordinate transformation $(x, t) \rightarrow (x, X, \tau)$. Thus,

$$\frac{\partial}{\partial t} \rightarrow \epsilon \frac{\partial}{\partial \tau}, \quad \frac{\partial}{\partial x} \rightarrow \frac{\partial}{\partial x} + \epsilon \frac{\partial}{\partial X}, \quad etc.,$$

and (2.1) becomes

$$\epsilon^2 \frac{\partial^2 u}{\partial \tau^2} + \left(\frac{\partial^4}{\partial x^4} + 4\epsilon \frac{\partial^4}{\partial x^3 \partial X} + 6\epsilon^2 \frac{\partial^4}{\partial x^2 \partial X^2} + \cdots \right) u$$

$$+ P \left(\frac{\partial^2}{\partial x^2} + 2\epsilon \frac{\partial^2}{\partial x \partial X} + \epsilon^2 \frac{\partial^2}{\partial X^2} \right) u + u - u^2 = 0. \tag{2.20}$$

On substituting (2.5) with (2.19) into (2.20), we find that the two equations obtained by equating the coefficients of ϵ and ϵ^2 are the same as in $(2.6)_{1,2}$. By equating the coefficients of ϵ^3 and then setting the coefficient of E to zero in the resulting equation (to ensure that the asymptotic expansion for u is uniform), we obtain

$$A_{\tau\tau} - 4A_{XX} = P_1 A + \frac{38}{9} A^2 \bar{A}. \tag{2.21}$$

The static form of (2.21), for which $A_{\tau\tau} = 0$, admits both periodic and localized solutions. Periodic solutions that satisfy a variety of boundary conditions (when the beam is long but finite) have been discussed in Lange and Newell (1971) and Potier-Ferry (1987). The localized solution is given by

$$A = A_{\text{loc}} \equiv \sqrt{-9P_1/19} \operatorname{sech}(\frac{1}{2}\sqrt{-P_1}X). \tag{2.22}$$

It can be shown (see, *e.g.*, Cai and Fu 1999) that this localized solution is unstable to small amplitude localized perturbations. More general discussions of stability of localized solutions admitted by (2.2) can be found in Sandstede (1997).

We note that $A = e^{i\phi} A_{\text{loc}}$ is also a solution of (2.21) for arbitrary constant ϕ. Corresponding to this solution, (2.19) gives

$$u^{(1)} = 2A_{\text{loc}} \cos(x + \phi). \tag{2.23}$$

The constant ϕ determines the relative positions of the maximum of A_{loc} and the maxima of $\cos(x + \phi)$. It is shown in Yang and Akylas (1997) that (2.23) is a uniformly valid localized solution only if $\phi = 0$ or π. For other values of ϕ in $[0, 2\pi]$, some exponentially small terms, that are beyond all orders of the standard asymptotic expansion $(2.5)_1$, becomes dominant for large values of x, thus making the expansion $(2.5)_1$ non-uniform. This serves as a cautionary example demonstrating that a solution of a model equation (obtained by a

perturbation procedure) is not necessarily a valid approximation to a solution
of the original problem. Similar analysis has been conducted by Wadee and
Bassom (1998) for the case when the quadratic term in (2.2) is replaced by a
cubic term. The same conclusion can also be reached using dynamical systems
theory (see, *e.g.*, Iooss and Pérouéme 1993).

We remark that the amplitude equation

$$A_{\tau\tau} - c_0 A_{XX} = c_1 A + c_2 A^2 \bar{A}, \tag{2.24}$$

of which (2.21) is a special case, is a universal amplitude equation for stability
problems in which (i) bifurcation is of pitch-fork type; (ii) bifurcation occurs
at a minimum of the bifurcation parameter; and (iii) near the critical mode
number the bifurcation parameter, P say, expands as $P_{cr} + P_1(k - k_{cr})^2 + \cdots$ with
$P_1 \neq 0$. The coefficients c_0 and c_1 in the linear terms are determined entirely
by the dispersion relation. For instance, for the problem under consideration,
the linearized form of (2.24) admits a solution

$$A = e^{iK(X-V\tau)}, \quad \text{with } V^2 = c_0 - c_1/K^2. \tag{2.25}$$

It then follows that

$$AE = e^{ix+iK(X-V\tau)} = e^{i(1+K\epsilon)x-iKV\epsilon t} = e^{i(1+K\epsilon)(x-KV\epsilon t)+O(\epsilon^2)}.$$

Thus, $v = \epsilon K V$ and $k = 1 + \epsilon K$ should satisfy the dispersion relation (2.17),
that is $V^2 = 4 - P_1/K^2$. Comparing this relation with $(2.25)_2$ yields $c_0 = 4$,
$c_1 = P_1$, as in (2.21). Since determination of c_0 and c_1 does not require
a nonlinear analysis, there is no need to include the variable X in a near-
critical analysis; its role can be restored according to (2.24) with the aid of the
dispersion relation if required.

Good agreement between the above asymptotic solutions and numerical solutions obtained by integrating (2.2) directly has been shown by Wadee *et al.*
(1997).

Fully nonlinear solutions

Fully nonlinear periodic solutions of (2.2) can also be obtained by using the
following relatively simple procedure.

We look for a periodic solution of the form

$$u(x) = \sum_{k=-\infty}^{\infty} A_k e^{iakx}, \quad A_k = \bar{A}_k, \tag{2.26}$$

where a is the wavenumber of the fundamental mode. This expression represents a periodic solution with period $2\pi/a$. Localized solutions can be obtained
as periodic solutions with infinite period, that is by taking the limit $a \to 0$.

On substituting (2.26) into (2.2), we obtain the following infinite system of quadratic equations for the Fourier amplitudes A_k ($k = 0, 1, 2, \cdots$):

$$(a^4 k^4 - Pa^2 k^2 + 1)A_k = A_0 A_k + \sum_{m=1}^{\infty} A_m(A_{m-p} + A_{m+p}), \quad (k = 0, 1, 2, \cdots).$$

(2.27)

Assuming that all the Fourier amplitudes except A_0, A_1 and A_2 are zero, we obtain

$$A_0 - A_0^2 - 2A_1^2 - 2A_2^2 = 0,$$

$$(a^4 - Pa^2 + 1)A_1 - 2A_0 A_1 - 2A_1 A_2 = 0, \qquad (2.28)$$

$$(16a^4 - 4Pa^2 + 1)A_2 - 2A_0 A_2 - A_1^2 = 0.$$

The amplitude equation (2.9) can be recovered from (2.28) under the assumptions $a = 1$, $|P - 2| \ll 1$, in which case, equation (2.28)$_2$ implies that $|A_0| \ll 1$ and $|A_2| \ll 1$. Equations (2.28)$_{1,3}$ then imply that $A_0 \approx 2A_1^2$, $A_2 \approx A_1^2/9$, respectively, and it then follows from (2.28)$_2$ that $(2 - P)A_1 - 38A_1^3/9 \approx 0$, which reduces to (2.9) after the substitutions $A_1 \to \epsilon A, P - 2 \to \epsilon^2 P_1$.

The infinite system (2.27) can be solved using the following procedure. We first replace the infinite system by a finite system of M equations, that is we assume that $A_k = 0$ ($k = M, M + 1, \ldots$). We start with $M = 2$, in which case the two simultaneous equations can be solved exactly. Using each of these solutions as a starting solution, we increase M in unit steps until a convergence criterion is satisfied. As each step, the finite system of quadratic equations is solved with the aid of the Nag routine C05NBF or the subroutine SNSQE, which can be downloaded from http://gams.nist.gov. This procedure provides a simple alternative to a fully numerical integration as used by Wadee *et al.* (1997). A benchmark solution that can be used to validate the above procedure is the following exact solution of (2.2) given by Abramian (2000):

$$u = a/[\cosh(bx)]^4, \quad b = 1/(2\sqrt{6}), \quad a = 840b^4, \quad P = -52b^2. \qquad (2.29)$$

This approach of expanding the solution in terms of some base functions and then solving the resulting infinite system of algebraic equations is often used in Fluid Dynamics for finding fully nonlinear solutions; see, *e.g.*, Nagata (1986, 1988).

10.2.2 Applications to problems in Finite Elasticity

In principle, stability problems in Finite Elasticity can be analysed following the same ideas as explained in Section 10.2.1. We would look for an asymptotic solution and derive a hierachy of equations, similar to those in (2.6), for the solutions at various orders. The amplitude equation is derived by imposing a solvability condition at third order. However, secular terms now appear not only in the governing equations but also in the associated boundary conditions. Derivation of the amplitude equation is not as straightforward as for the model problem discussed in the previous section.

One way to derive the amplitude equation when the problem involves boundary conditions is to use the so-called *projection method*. We now use (2.8) to illustrate the idea. To derive the amplitude equation, we multiple (2.8) by $\bar{E}$ (a linear solution with mode number $-k_{cr}$) and integrate the resulting equation from 0 to 2π (i.e. over one period). We note that only the secular term survives the integration on the right-hand side. Under the assumption that $u^{(3)}$ should be a periodic function with period 2π (this is equivalent to elimination of secular terms), the left-hand side becomes

$$\int_0^{2\pi} \bar{E}\mathcal{L}[u^{(3)}]dx = \int_0^{2\pi} u^{(3)}\mathcal{L}[\bar{E}]dx = 0, \qquad (2.30)$$

the idea being to integrate by parts repeatedly until $\bar{E}$ and $u^{(3)}$ exchange places. This procedure yields the same amplitude equation (2.9). For problems in Finite Elasticity, the above integrals would be replaced by multiple integrals and the integration by parts would also involve the associated boundary conditions. This method was used by Fu and Rogerson (1995) in their study of the stability of a pre-stressed plate. The method is straightforward, but it has the disadvantage that one has first to write down the governing equations and boundary conditions at third order and then to single out those terms that are proportional to E.

Another more efficient method for deriving amplitude equations that overcomes the above-mentioned disadvantage is the *virtual work method* (see Fu 1995, Fu and Devenish 1996). Again we use the model problem in the previous section to illustrate the idea. Instead of multiplying the third-order equation (2.8) by $\bar{E}$, we may multiply the original equation (2.2) by $\bar{E}$ and then integrate from 0 to 2π. Hence,

$$\int_0^{2\pi} \bar{E}\left\{\mathcal{L}[u] + (P - P_{cr})u_{xx} - u^2\right\}dx = 0. \qquad (2.31)$$

We now substitute (2.5) into (2.31) and equate the coefficients of ϵ, ϵ^2 and ϵ^3 to obtain

$$\int_0^{2\pi} \bar{E}\mathcal{L}[u^{(1)}]dx = 0, \qquad \int_0^{2\pi} \bar{E}\left\{\mathcal{L}[u^{(2)}] - (u^{(1)})^2\right\}dx = 0, \qquad (2.32)$$

and

$$\int_0^{2\pi} \bar{E}\left\{\mathcal{L}[u^{(3)}] + P_1 u_{xx}^{(1)} - 2u^{(1)}u^{(2)}\right\} dx = 0. \tag{2.33}$$

Equations $(2.32)_{1,2}$ are automatically satisfied. Although (2.33) contains the unknown third-order solution $u^{(3)}$, the term containing $u^{(3)}$ vanishes by the identity (2.30). Evaluating what remains then yields the same amplitude equation (2.9).

We see that if the quadratic term in (2.2) were replaced by a cubic term, the expression for $u^{(2)}$ would not be required in the evaluation of (2.33), and hence in the derivation of the amplitude equation.

For problems in Finite Elasticity that are governed by (1.1)–(1.3), we may begin the derivation by multiplying the equilibrium equation (1.1) by $\hat{u}_i$ and integrate over the domain on which the elastic body is defined, where $\hat{u}_i$ plays the role of $\bar{E}$ above and is a linear solution with mode number $-k_{cr}$. One variable in the integration should be the one along which the solution is assumed to be periodic with period $2\pi/k_{cr}$. Thus, we have

$$0 = \int_D \hat{u}_i \Sigma_{ji,j} dv = \int_{\partial D} \hat{u}_i \Sigma_{ji} n_j dS - \int_D \hat{u}_{i,j} \Sigma_{ji} dv.$$

With the use of the boundary conditions (1.3) this equation reduces to

$$\int_D \hat{u}_{i,j} \Sigma_{ji} dv = 0. \tag{2.34}$$

Clearly, the above equation could also have been obtained by multiplying the boundary condition (1.3) by $\hat{u}_i$, integrating along the boundary, and then applying the divergence theorem. This equivalent approach was originally used by Fu (1995) and Fu and Devenish (1996) when they proposed the virtual work method.

We next substitute the asymptotic expansions for u_i and p^* into (2.34) (with Σ_{ji} given by (1.4)) and equate the coefficients of ϵ, ϵ^2, ϵ^3. It can be shown that the equations obtained from the coefficients of ϵ and ϵ^2 are automatically satisfied. The equation obtained from the coefficient of ϵ^3 initially contains the unknown third-order solution, but it can be eliminated in a similar way to $u^{(3)}$ in (2.30). The amplitude equation is then obtained from the reduced equation after evaluation of certain integrals. This method has recently been applied to a variety of stability and wave propagation problems. See Fu and Ogden (1999), Cai and Fu (1999) and Fu and Zheng (1997). The major advantage of this method is that we obtain a single general expression for the nonlinear coefficient that can be evaluated effectively with the aid of a computer algebra package such as Mathematica.

Y.B. Fu

10.3 Bifurcations at a zero mode number

In this section we use a simple model problem to show how weakly nonlinear analysis can be conducted for bifurcation problems where the critical mode number is zero. We consider

$$u_{xx} + u_{yy} + Pu - u^2 = 0, \quad |y| < 1/2, \ |x| < \infty,$$

$$u(x, \pm 1/2) = 0, \tag{3.1}$$

where u is a function of x and y, P is a bifurcation parameter, and subscripts x and y signify partial differentiations. This is a special case of a more general problem that has been studied by Kirchgässner (1982) using the dynamical systems approach. The latter approach will be sketched in Section 10.3.2.

As in the previous section, the trivial solution $u = 0$ is a solution for all values of P. Our interest is in finding non-trivial solution that can bifurcate from the trivial solution at particular values of P.

We first consider the linearized problem and look for a periodic solution of the form

$$u = H(y)e^{ikx} + c.c., \tag{3.2}$$

where k is the mode number and the function $H(y)$ is to be determined. On substituting (3.2) into the linearized form of (3.1), we find that $H(y)$ must satisfy the linear eigenvalue problem

$$H''(y) + (P - k^2)H(y) = 0, \quad H(\pm 1/2) = 0. \tag{3.3}$$

It is straightforward to show that there exist two sets of non-trivial solutions, namely

$$H(y) = \cos(2n - 1)\pi y, \quad P = k^2 + (2n - 1)^2\pi^2, \tag{3.4}$$

and

$$H(y) = \sin 2n\pi y, \quad P = k^2 + 4n^2\pi^2, \tag{3.5}$$

where $n = 1, 2, \ldots$. The critical value P_{cr} (defined as the minimum of P) corresponds to the first branch ($n = 1$) of the symmetric mode (3.4), i.e.

$$P = k^2 + \pi^2, \tag{3.6}$$

and is obtained at $k = k_{cr} = 0$. We have $P_{cr} = \pi^2$. Our interest is in the behaviour of solutions near the critical point $P = P_{cr}$.

10.3.1 The perturbation approach

We let

$$P = \pi^2 + \epsilon P_1, \tag{3.7}$$

where P_1 is an $O(1)$ constant and ϵ is again a small parameter. Relation (3.6) implies that near-critical modes have $k = O(\epsilon^{1/2})$. The linear solution then indicates that dependence of the solution on x should be through the product $\sqrt{\epsilon}x$. Thus, it is appropriate to define a far-distance variable X by

$$X = \sqrt{\epsilon}x. \tag{3.8}$$

Then, the u_{xx} in (3.1) becomes ϵu_{XX} and, to leading order, $(3.1)_1$ reduces to

$$\mathcal{L}[u] \equiv u_{yy} + \pi^2 u = 0, \tag{3.9}$$

where the linear operator $\mathcal{L}$ should not be confused with the operator defined in the previous section. We note that this linear operator is non-dispersive in the sense that $u = e^{ikx}\cos\pi y$ is a solution of (3.9) for arbitrary mode number k. We then expect that a solvability condition will have to be satisfied at second order when we look for an asymptotic solution for the nonlinear problem. We would like the term ϵu_{XX} to appear in the amplitude equation. This can be achieved by requiring $\epsilon u_{XX} = O(u^2)$, from which we deduce that $u = O(\epsilon)$. Thus, we look for a perturbation solution of the form

$$u(x,y) = \epsilon u^{(1)}(X,y) + \epsilon^2 u^{(2)}(X,y) + \cdots. \tag{3.10}$$

On substituting (3.10) into (3.1) and equating the coefficients of ϵ and ϵ^2, we obtain

$$\mathcal{L}[u^{(1)}] = 0, \quad u^{(1)}(X, \pm 1/2) = 0, \tag{3.11}$$

and

$$\mathcal{L}[u^{(2)}] = -u_{XX}^{(1)} - P_1 u^{(1)} + (u^{(1)})^2, \quad u^{(2)}(X, \pm 1/2) = 0. \tag{3.12}$$

The leading-order problem (3.11) can easily be solved to yield

$$u^{(1)} = A(X)\cos\pi y, \tag{3.13}$$

where the amplitude function $A(X)$ is to be determined. On substituting (3.13) into $(3.12)_1$, we obtain

$$\mathcal{L}[u^{(2)}] = -(A_{XX} + P_1 A)\cos\pi y + A^2 \cos^2\pi y. \tag{3.14}$$

The general solution of this inhomogeneous equation is given by

$$u^{(2)} = C_1(X)\cos\pi y + C_2(X)\sin\pi y + I(X,y), \tag{3.15}$$

where C_1, C_2 are arbitrary functions and the particular integral I is given by

$$I(X, y) = -\frac{1}{2\pi}(A_{XX} + P_1 A)y \sin \pi y + \frac{1}{2\pi^2}A^2(1 - \frac{1}{3}\cos 2\pi y). \qquad (3.16)$$

On substituting (3.15) into (3.12)$_2$, we obtain $C_2(X) + I(X, 1/2) = 0$ and $-C_2(X) + I(X, -1/2) = 0$. It then follows that $I(X, 1/2) + I(X, -1/2) = 0$, from which we obtain the amplitude equation

$$A_{XX} + P_1 A - \frac{8}{3\pi}A^2 = 0. \qquad (3.17)$$

The above procedure for obtaining the amplitude equation works well for the present simple problem. However, for more complex problems such as those in Finite Elasticity it may be difficult to write down the particular integral for the second-order problem explicitly. Despite this difficulty, this approach was used in the derivation of evolution equations in early studies of nonlinear surface waves; see, *e.g.*, Lardner (1986).

A better procedure, which does not require solution of the second-order problem at all, is the projection method, which we have already discussed in the previous section. In this method we multiply (3.14) by $\cos \pi y$ (in other words we project (3.14) onto the kernel of the linear operator) and integrate the resulting equation from $-1/2$ to $1/2$ repeatedly. After making use of the identity

$$\int_{-1/2}^{1/2} \cos \pi y \mathcal{L}[u^{(2)}]dy = \int_{-1/2}^{1/2} u^{(2)} \mathcal{L}[\cos \pi y]dy = 0, \qquad (3.18)$$

we arrive at the same amplitude equation (3.17).

We now show how to apply the virtual work method to derive the same amplitude equation with minimum effort. We first multiply (3.1)$_1$ by $\cos \pi y$ to obtain

$$\int_{-1/2}^{1/2} \cos \pi y(\epsilon u_{XX} + \mathcal{L}[u] + (P - \pi^2)u - u^2)dy = 0. \qquad (3.19)$$

Equation (3.18) still holds if we replace $u^{(2)}$ by u. With the use of this relation and (3.7), (3.19) reduces to

$$\int_{-1/2}^{1/2} \cos \pi y(\epsilon u_{XX} + \epsilon P_1 u - u^2)dy = 0. \qquad (3.20)$$

Every term in the above equation is of order ϵ^2. On substituting (3.10) together with (3.13) into (3.20) and evaluating the integral, we again obtain (3.17). We note that in the above derivation we do not even need to write down the $O(\epsilon^2)$ problem satisfied by $u^{(2)}$!

10.3.2 The dynamical systems approach

Although so far we have consistently used the perturbation approach in our analysis, we sketch in this section an alternative approach in which the static problem is first converted into a dynamical system and then methods from dynamical systems theory are used to derive the amplitude equations. This approach has been given the name "spatial dynamics" in the literature. It was first conceived in Kirchgässner (1982) and later developed further by Mielke (1986, 1988a, 1991). It has the same degree of rigour as the Liapunov-Schmidt reduction scheme (cf. Chapter 9) and yet is able to describe a wider range of solutions (such as localized solutions). The dynamical systems approach has been used extensively in the study of steady nonlinear water waves (see, *e.g.*, Mielke 1991, Dias and Iooss 1993, Buffoni *et al.* 1996, Buffoni and Groves 1999), Saint-Venant's problem (Mielke 1988b, c) and the necking problem (Mielke 1991). The idea of viewing a space variable as a time-like variable and treating a given static problem as a dynamical problem has also been noted by Thompson and his co-workers (see, *e.g.*, Thompson and Virgin 1988).

In the dynamical systems approach, we view x as a time-like variable and rewrite (3.1) as an infinite dimensional dynamical system (i.e. a system of ordinary differential equations in an infinite dimensional function space):

$$w_x = \mathcal{K}(P)w + F(w), \quad w_1(x, \pm 1/2) = 0, \tag{3.21}$$

where

$$\mathcal{K}(P) \equiv \begin{pmatrix} 0 & 1 \\ -\frac{\partial^2}{\partial y^2} - P & 0 \end{pmatrix}, \quad w = \begin{pmatrix} u \\ u_x \end{pmatrix} \equiv \begin{pmatrix} w_1 \\ w_2 \end{pmatrix}, \quad F(w) = \begin{pmatrix} 0 \\ w_1^2 \end{pmatrix}.$$

For the linear system $w_x = \mathcal{K}(P)w$ we look for a solution of the form

$$w = e^{\gamma x} \begin{pmatrix} v_1(y) \\ v_2(y) \end{pmatrix}, \tag{3.22}$$

where γ is usually referred to as an eigenvalue of the linear operator $\mathcal{K}(P)$, and $(v_1, v_2)^T$ the corresponding eigenvector. On substituting (3.22) into the linearized form of (3.21) we obtain

$$v_{1yy} + (P + \gamma^2)v_1 = 0, \quad v_1(\pm 1/2) = 0. \tag{3.23}$$

Thus, there are two sets of eigenvalues and eigenvectors given by

$$\gamma^2 = -P + (2n-1)^2\pi^2, \quad v_1 = \cos(2n-1)\pi y, \quad v_2 = \gamma v_1,$$

and

$$\gamma^2 = -P + 4n^2\pi^2, \quad v_1 = \cos 2n\pi y, \quad v_2 = \gamma v_1,$$

where $n = 1, 2, \ldots$. We see that the eigenvalue pair that is closest to the origin (in the complex γ-plane) is given by $\gamma^2 = -P + \pi^2$. As P is increased across the critical value $P_{cr} = \pi^2$, the two eigenvalues meet at the origin $\gamma = 0$ along the real axis and then move to the imaginary axis. All the other eigenvalues remain on the real axis. Because of this qualitative change of properties of the eigenvalues, we expect to see bifurcation of solutions as $P = \pi^2$ is crossed. When $P = P_{cr} = \pi^2$, $\gamma = 0$ is a repeated eigenvalue. The corresponding generalized eigenvectors $v^{(1)}$ and $v^{(2)}$ are defined by

$$\mathcal{K}_0 v^{(1)} = 0 \quad \mathcal{K}_0 v^{(2)} = v^{(1)},$$

and are given by

$$v^{(1)} = \left(\begin{array}{c} \sqrt{2}\cos \pi y \\ 0 \end{array} \right), \quad v^{(2)} = \left(\begin{array}{c} 0 \\ \sqrt{2}\cos \pi y \end{array} \right), \qquad (3.24)$$

where $\mathcal{K}_0 \equiv \mathcal{K}(\pi^2)$. The eigenspace spanned by $v^{(1)}$ and $v^{(2)}$ is a $\mathcal{K}_0$-invariant space in the sense that $\mathcal{K}_0 v$ lies in this space whenever v does. We shall need to use the adjoint operator $\mathcal{K}_0^*$ of $\mathcal{K}_0$, which is defined by

$$< f, \mathcal{K}_0 g > = < g, \mathcal{K}_0^* f >, \quad \text{where} \quad < f, g > \equiv \int_{-1/2}^{1/2} f \cdot g \, dy, \qquad (3.25)$$

with $f \cdot g$ denoting the vector dot product. It can easily be shown that equation (3.25) holds if $\mathcal{K}_0^*$ is the transpose of $\mathcal{K}_0$ and f, g both vanish on $y = \pm 1/2$. The generalized eigenfunctions $\hat{v}^{(1)}$ and $\hat{v}^{(2)}$ of $\mathcal{K}_0^*$ $(= \mathcal{K}_0^T)$ are defined by

$$\mathcal{K}_0^* \hat{v}^{(2)} = 0, \quad \mathcal{K}_0^* \hat{v}^{(1)} = \hat{v}^{(2)},$$

subject to the condition that $\hat{v}^{(1)}$ and $\hat{v}^{(2)}$ both vanish on $y = \pm 1/2$. The above definition is chosen so as to make it possible to satisfy the orthogonality property $< v^{(i)}, \hat{v}^{(j)} > = \delta_{ij}$ $(i, j = 1, 2)$. We find that the eigenfunctions satisfying this property is given by

$$\hat{v}^{(1)} = v^{(1)}, \quad \hat{v}^{(2)} = v^{(2)}. \qquad (3.26)$$

We define the projections

$$S_0 w = \sum_{k=1}^{2} < w, \hat{v}^{(k)} > v^{(k)}, \quad S_1 = I - S_0, \qquad (3.27)$$

where I is the identity projection, and write

$$w = S_0 w + S_1 w \equiv A(x) v^{(1)} + B(x) v^{(2)} + \tilde{w}(x, y), \qquad (3.28)$$

It can be shown that the present problem satisfies certain conditions (see Kirchgässner 1982, or Mielke 1991) so that in a small neighbourhood of $(A, B) =$

$(0,0), P = \pi^2$, $\tilde{w}(x,y)$ can be expressed in terms of A and B, and is of order $A^2 + B^2$. We note that the $\tilde{w}(x,y)$ defined above is orthogonal to the eigenspace spanned by $\hat{v}^{(1)}$ and $\hat{v}^{(2)}$, that is

$$< \hat{v}^{(1)}, \tilde{w} > = 0, \quad < \hat{v}^{(2)}, \tilde{w} > = 0. \tag{3.29}$$

For convenience, we use a small parameter ϵ to characterize the size of $A(x)$ and $B(x)$. On substituting (3.28) into (3.21), we obtain

$$A_x v^{(1)} + B_x v^{(2)} + \tilde{w}_x = \mathcal{K}(P)(Av^{(1)} + Bv^{(2)}) + \mathcal{K}(P)\tilde{w} + F(w),$$

$$= Bv^{(1)} + [\mathcal{K}(P) - \mathcal{K}_0](Av^{(1)} + Bv^{(2)}) + \mathcal{K}_0\tilde{w}$$

$$+ F(Av^{(1)} + Bv^{(2)}) + O\left((P - \pi^2)\epsilon^2, \epsilon^3\right). \tag{3.30}$$

The right-hand side of the last equation can further be written as

$$Bv^{(1)} + \begin{pmatrix} 0 \\ (\pi^2 - P)A\cos\pi y \end{pmatrix} + \mathcal{K}_0\tilde{w} + \begin{pmatrix} 0 \\ A^2\cos^2\pi y \end{pmatrix} + O\left((P - \pi^2)\epsilon^2, \epsilon^3\right).$$
$$\tag{3.31}$$

On forming the inner product of (3.30) with $\hat{v}^{(1)}$ and $\hat{v}^{(2)}$, respectively, we obtain

$$A_x = B + O\left((P - \pi^2)\epsilon^2, \epsilon^3\right), \quad B_x = (\pi^2 - P)A + \frac{8}{3\pi}A^2 + O\left((P - \pi^2)\epsilon^2, \epsilon^3\right), \tag{3.32}$$

where we have made use of the fact that

$$< \hat{v}^{(1)}, \mathcal{K}_0\tilde{w} > = < \mathcal{K}_0^*\hat{v}^{(1)}, \tilde{w} > = < \hat{v}^{(2)}, \tilde{w} > = 0,$$

$$< \hat{v}^{(2)}, \mathcal{K}_0\tilde{w} > = < \mathcal{K}_0^*\hat{v}^{(2)}, \tilde{w} > = < 0, \tilde{w} > = 0.$$

It can be shown that the quadratic terms in (3.32) cannot be eliminated by near-identity transformations (see, *e.g.*, Wiggins 1990, p. 220), so that equations (3.32) are already in a normal form. On eliminating B from (3.32), we obtain

$$A_{xx} = (\pi^2 - P)A + \frac{8}{3\pi}A^2 + O\left((P - \pi^2)\epsilon^2, \epsilon^3\right), \tag{3.33}$$

which reduces to (3.17) after the substitutions $A \to \epsilon A$, $\pi^2 - P \to -\epsilon P_1$.

For problems where bifurcation takes place at a non-zero mode number, the counterpart of (3.32) would not be in a normal form and further reductions would be required. We now use (2.2) to illustrate the use of the dynamical systems approach for such problems. As for (3.1), we first write (2.2) as a dynamical system:

$$w_x = L(P)w + N(w), \tag{3.34}$$

where $w = (w_1, w_2, w_3, w_4)^T \equiv (u, u_x, u_{xx}, u_{xxx})^T$, $N(w) = (0,0,0,w_1^2)^T$ and $L(P)$ is a 4×4 matrix whose non-zero elements are $L_{12} = L_{23} = L_{34} = 1$, $L_{41} = -1$, $L_{43} = -P$.

It is easy to show that the four eigenvalues of the linear system $w_x = L(P)w$ coalesce to $\pm i$ when $P = 2$. The corresponding eigenspace is spanned by the generalized (complex) eigenvectors $v^{(1)}, v^{(2)}, v^{(3)}, v^{(4)}$, where

$$L_0 v^{(1)} = i\, v^{(1)}, \quad L_0 v^{(2)} = i\, v^{(2)} + v^{(1)}, \quad v^{(3)} = \bar{v}^{(1)}, \quad v^{(4)} = \bar{v}^{(2)}, \quad (3.35)$$

and $L_0 = L(2)$, a bar signifying complex conjugation. We denote the generalized eigenvectors of L_0^* ($\equiv \bar{L}_0^T$) by $\hat{v}^{(1)}, \hat{v}^{(2)}, \hat{v}^{(3)}, \hat{v}^{(4)}$ so that

$$L_0^* \hat{v}^{(2)} = -i\, \hat{v}^{(2)}, \quad L_0^* \hat{v}^{(1)} = -i\, \hat{v}^{(1)} + \hat{v}^{(2)}, \quad \hat{v}^{(3)} = \bar{\hat{v}}^{(1)}, \quad \hat{v}^{(4)} = \bar{\hat{v}}^{(2)}. \quad (3.36)$$

We may choose $v^{(i)}$ and $\hat{v}^{(i)}$ such that

$$< \hat{v}^{(i)}, v^{(j)} > \equiv \hat{v}^{(i)} \cdot \bar{v}^{(j)} = \delta_{ij},$$

where the first relation defines the scalar product $< \cdot, \cdot >$ for the present problem.

Since $v^{(1)}, v^{(2)}, v^{(3)}, v^{(4)}$ span the phase space for the present four-dimensional dynamical system, we may write the solution of (3.34) as

$$w = A(x)v^{(1)} + B(x)v^{(2)} + \bar{A}(x)v^{(3)} + \bar{B}(x)v^{(4)}, \quad (3.37)$$

where $A(x)$ and $B(x)$ are to be determined. On substituting (3.37) into (3.34), forming the scalar product of the resulting equation with $\hat{v}^{(1)}$ and $\hat{v}^{(2)}$, respectively, and making use of the property that $< L_0 w, \hat{v}^{(j)} > = < w, L_0^* \hat{v}^{(j)} >$, we obtain two first-order differential equations for A and B (counterparts of (3.32) for the present problem). However, in contrast with (3.32), these two equations are not in a normal form because their linear part shows that all the quadratic terms can be eliminated by a near-identity transformation $(A, B) \to (\hat{A}, \hat{B})$, see, e.g., Elphick et al. (1987). In terms of the new coordinates, (3.37) becomes

$$w = A(x)v^{(1)} + B(x)v^{(2)} + \bar{A}(x)v^{(3)} + \bar{B}(x)v^{(4)} + \Phi(P - 2, A, B, \bar{A}, \bar{B}), \quad (3.38)$$

where we have dropped the hats on A and B to simplify notation and Φ consists of higher-order terms in $A, B, \bar{A}, \bar{B}$. We may write

$$\Phi(0, A, B, \bar{A}, \bar{B}) = (A^2 \Phi_{2000} + c.c.) + |A|^2 \Phi_{1100} + (AB\Phi_{1010} + c.c.)$$
$$+ (\bar{A}B\Phi_{0110} + c.c.) + (B^2 \Phi_{0020} + c.c.) + |B|^2 \Phi_{0011} + \cdots, \quad (3.39)$$

where Φ_{2000} etc. are constant vectors. The normal form for this type of problems is well known (see, e.g., Elphick et al. 1987) and is given by

$$A_x = iA + B + i\, AN_1, \quad B_x = iB + iBN_1 + AN_2, \quad (3.40)$$

where N_1 and N_2 are both functions of $P-2$, $|A|^2$ and $(i/2)(A\bar{B} - \bar{A}B)$. To leading order we have

$$N_1 = \xi_1(P-2) + \xi_2|A|^2 + \xi_3(A\bar{B} - \bar{A}B),$$
$$N_2 = \xi_4(P-2) + \xi_5|A|^2 + \xi_6(A\bar{B} - \bar{A}B), \qquad (3.41)$$

where the constants ξ_1 to ξ_6 are determined as follows. By substituting (3.38) into (3.34), eliminating A_x, B_x with the aid of (3.40) and then equating the vector coefficients of monomials of $A, \bar{A}, B, \bar{B}$, we first obtain a hierachy of matrix equations, each of which is of the form $(L_0 - i\,I)\Psi = f$. Since $(L_0 - i\,I)$ is singular, the f on the right-hand side, which involves the unknown constants, must satisfy a solvability condition. Solving these solvability conditions then yields the unknown constants.

The above calculation can be simplified by noting that the constants ξ_1 and ξ_4 can in fact be determined from the linear eigenvalue problem and in determining the remaining constants we may set $P-2 = 0$ in the Taylor expansion of Φ (cf. (3.39)). We shall not present the details here. The interested reader may consult Dias and Iooss (1993) where a detailed calculation is presented for a water wave problem that has the same bifurcation properties.

If the matrix $L(P)$ in (3.34) is replaced by a differential operator, the problem becomes infinite dimensional, but the analysis remains similar. The L_0^* would be replaced by an operator adjoint to L_0. Another term would need to be added to the right-hand side of (3.37), just as in (3.28), since now the eigenspace spanned by the eigenvectors corresponding to the purely imaginary eigenvalues does not span the whole phase space. However, this extra term can be absorbed into the function Φ in (3.38) so that no extra action needs to be taken. The matrix equations referred to in the previous paragraph are replaced by differential equations. Thus, as far as the determination of the normal form (3.40) is concerned, there are only minor computational differences between finite- and infinite-dimensional dynamical systems.

10.4 Necking of an elastic plate under stretching

Necking instability of an elastic rod or strip has been studied by a number of researchers using finite elasticity theory. See, *e.g.*, Coleman (1983), Owen (1987), Mielke (1991) and the references therein. All these studies except that of Mielke (1991) have been based on the analysis of simplified models. Coleman's (1983) analysis was for a thin rod and was based on the assumption $T = \tau(\lambda) + \beta(\lambda)\lambda_z^2 + \gamma(\lambda)\lambda_{zz}$, where T is the total tensile force at a cross-section, λ is the local stretch ratio along the axial z-direction and the functions

τ, β and γ satisfy certain constitutive requirements. Owen's (1987) analysis was also for a thin rod but was based on the assumption that $z = f(Z)$, $x = g(Z)x$, $y = g(Z)y$ where (X, Y, Z) and (x, y, z) are the coordinates, relative to a common rectangular coordinate system, of material particles in the unstressed and current configurations, respectively, with z, Z measuring distances along the axial direction. Mielke (1991) considered an elastic strip and did not make any simplying assumptions. His exact analysis showed that Owen's (1987) assumption is asymptotically not consistent.

Motivated by Mielke's (1991) analysis based on a Hamiltonian dynamical systems approach, we consider in this section the necking instability of an incompressible elastic strip/plate using the perturbation approach. We note that for incompressible materials the Hamiltonian formulation is yet to be developed and Mielke's (1991) approach cannot be applied directly to incompressible materials (although the dynamical systems approach is still possible). The perturbation approach only involves elementary analysis and has the advantage of incorporating dynamical effects with little extra work.

We consider an incompressible elastic plate the strain-energy function of which takes the form

$$W = 2\mu(\lambda_1^m + \lambda_2^m + \lambda_3^m - 3)/m^2, \qquad (4.1)$$

where μ is the shear modulus and m is a constant. Equation (4.1) is a special form of the more general strain-energy function given by (2.106) in Chapter 1. Without loss of generality, we take μ and the density to be unity and the plate thickness to be 2 (since otherwise we can achieve this by non-dimensionalization). We choose our coordinate system so that in the uniformly stretched configuration the plate is defined by $|x_1| < \infty$, $|x_2| < 1$.

To simplify the analysis, we assume that the plate is in a state of plane strain, so that $\lambda_3 = 1$ and $\lambda_2 = 1/\lambda_1$. Under a uniaxial tension the principal Cauchy stresses are given by (cf. equation (2.98) in Chapter 1)

$$\sigma_1 = \lambda_1 \frac{\partial W}{\partial \lambda_1} - p = \frac{2}{m}\lambda_1^m - p, \quad \sigma_2 = \lambda_2 \frac{\partial W}{\partial \lambda_2} - p = \frac{2}{m}\lambda_2^m - p, \qquad (4.2)$$

where p is the pressure satisfying the general relation $p = \mathcal{A}_{2121} - \mathcal{A}_{2112} - \sigma_2$. The assumption of uniaxial tension implies that $\sigma_2 = 0$ which yields

$$p = \frac{2}{m}\lambda_2^m = \frac{2}{m}\lambda_1^{-m} = \mathcal{A}_{2121} - \mathcal{A}_{2112}. \qquad (4.3)$$

The principal nominal stress along the x_1-direction is given by

$$t_1 = \lambda_1^{-1}\sigma_1 = \frac{2}{m}\lambda_1^{m-1} - \frac{2}{m}\lambda_1^{-m-1}, \qquad (4.4)$$

and it describes the force per unit area in the unstressed configuration. We have

$$\frac{dt_1}{d\lambda_1} = \frac{2}{m}\left\{(m-1)\lambda_1^{m-2} + (1+m)\lambda_1^{-m-2}\right\}. \tag{4.5}$$

It is well known that necking instability can only occur if the (λ_1, t_1) curve has a local maximum, i.e. if there exists a value of λ_1 at which $dt_1/d\lambda_1 = 0$. Equation (4.5) shows that this is possible only if $m - 1 < 0$. Thus, from now on our attention will be focused on the case $m = 1/2$. Equation (4.5) then shows that the maximum of t_1 equals $8/(3\sqrt{3})$ and occurs when $\lambda_1 = 3$.

For the material (4.1) with $m = 1/2$, the non-zero first-order elastic moduli can be calculated with the aid of equations (4.27) and (4.28) in Chapter 1 (note that the $\mathcal{A}^1_{jilk}$ in the present chapter correspond to $\mathcal{A}^1_{0jilk}$ in Chapter 1). We have

$$\mathcal{A}^1_{1111} = -2\lambda_1^{1/2}, \quad \mathcal{A}^1_{2222} = -2\lambda_2^{1/2}, \quad \mathcal{A}^1_{1212} = \frac{4\lambda_1^2}{\lambda_1^2 - \lambda_2^2}(\lambda_1^{1/2} - \lambda_2^{1/2}),$$

$$\mathcal{A}^1_{2121} = \frac{4\lambda_2^2}{\lambda_2^2 - \lambda_1^2}(\lambda_2^{1/2} - \lambda_1^{1/2}), \quad \mathcal{A}^1_{1221} = \mathcal{A}^1_{2112} = \frac{4}{\lambda_1^2 - \lambda_2^2}(\lambda_2^2\lambda_1^{1/2} - \lambda_1^2\lambda_2^{1/2}).$$

$$\tag{4.6}$$

We shall also need the second-order moduli $\mathcal{A}^2_{111111}$ and $\mathcal{A}^2_{222222}$ whose expressions can be found in Fu and Ogden (1999). We have

$$\mathcal{A}^2_{111111} = \lambda_1^3 \partial^3 W/\partial\lambda_1^3 = 3\lambda_1^{1/2}, \quad \mathcal{A}^2_{222222} = \lambda_2^3 \partial^3 W/\partial\lambda_2^3 = 3\lambda_2^{1/2}. \tag{4.7}$$

We note that the pre-stress is determined by λ_1 alone. We shall take it as the bifurcation parameter and from now on write it simply as λ.

The general governing equations and boundary conditions are given by (1.1), (1.2) and (1.3) with the right-hand side of (1.1) replaced by $\rho\ddot{u}_i$ for time dependent incremental deformations, where ρ is the density. For the present plane strain problem, the governing equations reduce to

$$\mathcal{A}^1_{j1lk}u_{k,lj} + \mathcal{A}^2_{j1lknm}u_{m,n}u_{k,lj} - p^*_{,1} + p^*_{,j}u_{j,1} = \ddot{u}_1, \tag{4.8}$$

$$\mathcal{A}^1_{j2lk}u_{k,lj} + \mathcal{A}^2_{j2lknm}u_{m,n}u_{k,lj} - p^*_{,2} + p^*_{,j}u_{j,2} = \ddot{u}_2, \tag{4.9}$$

$$u_{1,1} + u_{2,2} = u_{1.2}u_{2,1} - u_{1,1}u_{2,2}, \tag{4.10}$$

and the boundary conditions reduce to

$$\mathcal{A}^1_{21lk}u_{k,l} + \frac{1}{2}\mathcal{A}^2_{21lknm}u_{k,l}u_{m,n} + p(u_{2,1} - u_{2,k}u_{k,1}) + p^*u_{2,1} = 0, \tag{4.11}$$

$$\mathcal{A}^1_{22lk}u_{k,l} + \frac{1}{2}\mathcal{A}^2_{22lknm}u_{k,l}u_{m,n} + p(u_{2,2} - u_{2,k}u_{k,2}) - p^*(1 - u_{2,2}) = 0 \tag{4.12}$$

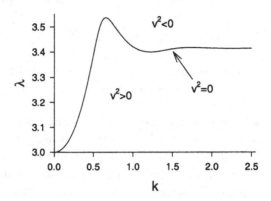

Fig. 2. Relationship between λ and k when $v^2 = 0$ for a plate under stretching .

on $x_2 = \pm 1$, where we have neglected all the cubic and higher-order terms since they do not enter our weakly nonlinear analysis.

We expect that at the critical principal stretch $\lambda = 3$, extensional travelling waves propagating in the plate will have zero wave speed. To verify this and to derive the asymptotic expression for this wave speed when λ is close to but not equal to 3, we first look for a travelling wave solution of the form

$$(u_1, u_2, p^*) = (U_1(x_2), U_2(x_2), P(x_2))e^{ik(x_1-vt)}. \qquad (4.13)$$

On substituting (4.13) into the linearized forms of (4.8)–(4.12), a linear eigenvalue problem is obtained for U_1, U_2 and P. It can be shown (see, *e.g.*, Rogerson and Fu 1995) that the dispersion relation between v and k for the extensional modes is given by

$$F(q_1)G(q_2)\tanh(kq_1) = F(q_2)G(q_1)\tanh(kq_2), \qquad (4.14)$$

where

$$F(x) = \gamma(x^2 + 1), \quad G(x) = (v^2 - 2\beta - \gamma)x + \gamma x^3,$$

$$\alpha = \mathcal{A}^1_{1212}, \quad \gamma = \mathcal{A}^1_{2121}, \quad 2\beta = \mathcal{A}_{1111} + \mathcal{A}_{2222} - 2\mathcal{A}_{1221}, \qquad (4.15)$$

and q_1^2, q_2^2 are the two roots of

$$\gamma q^4 + (v^2 - 2\beta)q^2 + \alpha - v^2 = 0. \qquad (4.16)$$

For any fixed value of λ, the dispersion relation can be solved (numerically) to find the dependence of v^2 on k. As $k \to 0$, relation (4.14) reduces to

$F(q_1)G(q_2)q_1 - F(q_2)G(q_1)q_2 = 0$, from which we find

$$v^2 = \mathcal{A}_{1111} + \mathcal{A}_{2222} + 2p = -2\lambda^{-1/2}(\lambda - 3). \qquad (4.17)$$

From this expression we find that $v^2 = 0$ when $\lambda = 3$ which confirms our earlier expectation.

Setting $v = 0$ in (4.14) and solving the resulting bifurcation condition numerically, we find that the bifurcating value of λ, as a function of k, has a global minimum at $k = 0$ (cf. Figure 2). Thus, this problem is similar to the model problem studied in Section 10.3. It is easy to show that in the double limit $\lambda \to 3$ and $k \to 0$, v^2 expands as

$$v^2 = -\frac{2}{\sqrt{3}}(\lambda - 3) + \frac{8}{3\sqrt{3}}k^2 + \cdots. \qquad (4.18)$$

This asymptotic expansion indicates that if

$$\lambda = 3 + \epsilon\lambda_0, \qquad (4.19)$$

where λ_0 is an $O(1)$ constant and ϵ is a small parameter, as before, then all modes with $k = O(\sqrt{\epsilon})$ will have $v = O(\sqrt{\epsilon})$. Equation (4.13) then shows that such near-critical modes evolve on a time scale satisfying $kvt = O(\epsilon t) = O(1)$. Thus the appropriate slow time and far-distance variables are

$$t = \epsilon t, \quad X = \sqrt{\epsilon}x_1. \qquad (4.20)$$

Corresponding to (4.19), the pressure p given by (4.3) expand as

$$p = p_0 + \epsilon p_1 + \cdots, \qquad (4.21)$$

where $p_0 = 4/\sqrt{3}, p_1 = -2\lambda_0/(3\sqrt{3})$, and the first-order elastic moduli expands as

$$\mathcal{A}_{jilk}^1 = \mathcal{A}_{jilk} + \epsilon\lambda_0\mathcal{A}_{jilk}^d + \cdots, \qquad (4.22)$$

where

$$\mathcal{A}_{jilk} = \mathcal{A}_{jilk}^1|_{\lambda=3}, \quad \mathcal{A}_{jilk}^d = \left.\frac{d\mathcal{A}_{jilk}^1}{d\lambda}\right|_{\lambda=3}.$$

From the linear eigenrelations for U_1, U_2 and P in (4.13) (which could easily be obtained but we have not written them out), we may deduce the order relations

$$u_2 = O(ku_1) = O(\sqrt{\epsilon}u_1), \quad p^* = O(ku_1) = O(\sqrt{\epsilon}u_1) \qquad (4.23)$$

in the limit $k \to 0$. The first relation for u_2 could also be deduced from the linear incompressibility relation $u_{1,1} + u_{2,2} = 0$. Guided by the analysis of the model problem in Section 10.3.1, we expect that the amplitude equation will be derived from quadratic terms. Thus, in order that the $\ddot{u}_1$ on the right-hand

372

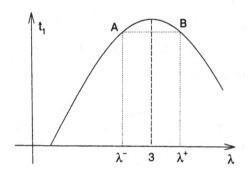

Fig. 3. Relation between λ and t_1 in uniform stretching as given by (4.4) with $m = 1/2$.

side of (4.8) should appear in the amplitude equation, it must be of the same order as the largest quadratic term on the left-hand side. It can be shown that the largest quadratic term is $u_{1,1}u_{1,11}$. Thus, we require

$$u_{tt} = \epsilon^2 u_{\tau\tau} = O(u_{1,1}u_{1,11}) = O(\epsilon^{3/2}u_1^2),$$

from which we deduce $u_1 = O(\epsilon^{1/2})$. With the use of this result and (4.23), we obtain

$$u_1 = O(\sqrt{\epsilon}), \quad u_2 = O(\epsilon), \quad p^* = O(\epsilon). \tag{4.24}$$

The size of u_1 can also be determined using the following argument. Since $\lambda = 3 + \epsilon\lambda_0$ corresponds to a uniformly stretched near-critical state (the possible bifurcation of which into another state we try to determine), this value of λ must correspond to either point A or B in Figure 3, that is $3 + \epsilon\lambda_0 = \lambda^-$ or λ^+. If $3 + \epsilon\lambda_0 = \lambda^-$, our final amplitude equation must be able to admit $\lambda = \lambda^+$ as a bifurcation solution. Likewise, if $3 + \epsilon\lambda_0 = \lambda^+$, then $\lambda = \lambda^-$ would be a possible bifurcation solution. Since $\lambda^+ - \lambda^- = O(\epsilon)$, if, for instance, $3 + \epsilon\lambda_0 = \lambda^-$, we must have $u_{1,1} = O(\lambda^+ - \lambda^-) = O(\epsilon)$. Since $u_{1,1} = \sqrt{\epsilon}\partial u_1/\partial X$, we deduce that $u_1 = O(\sqrt{\epsilon})$.

With all the scalings established, we are now in a position to assume the following form of asymptotic solution for the nonlinear system (4.8)–(4.12):

$$u_1 = \epsilon^{1/2} \left\{ u_1^{(1)} + \epsilon u_1^{(2)} + \epsilon^2 u_1^{(3)} + \cdots \right\},$$

$$u_2 = \epsilon \left\{ u_2^{(1)} + \epsilon u_2^{(2)} + \epsilon^2 u_2^{(3)} + \cdots \right\}, \tag{4.25}$$

$$p^* = \epsilon \left\{ p^{(1)} + \epsilon p^{(2)} + \epsilon^2 p^{(3)} + \cdots \right\},$$

where all the functions on the right-hand sides are functions of X, x_2 and τ. Because of the coordinate transformation $(x_1, x_2, t) \to (X, x_2, \tau)$, we have

$$\frac{\partial}{\partial x_1} \to \epsilon^{1/2} \frac{\partial}{\partial X}, \quad \frac{\partial}{\partial t} \to \epsilon \frac{\partial}{\partial \tau}. \tag{4.26}$$

On substituting (4.19), (4.20) and (4.25) into (4.8)–(4.12) and equating the coefficients of like powers of ϵ, we obtain the following hierachy of equations.

Leading order:

For $|x_2| < 1$,

$$\mathcal{A}_{2121} u_{1,22}^{(1)} = 0, \tag{4.27}$$

$$\mathcal{A}_{2222} u_{2,22}^{(1)} + \mathcal{A}_{1221} u_{1,2X}^{(1)} - p_{,2}^{(1)} = 0, \tag{4.28}$$

$$u_{1,X}^{(1)} + u_{2,2}^{(1)} = 0. \tag{4.29}$$

On $x_2 = \pm 1$,

$$\mathcal{A}_{2121} u_{1,2}^{(1)} = 0, \quad (\mathcal{A}_{2222} + p_0) u_{2,2}^{(1)} - p^{(1)} = 0. \tag{4.30}$$

Second order:

For $|x_2| < 1$,

$$\mathcal{A}_{2121} u_{1,22}^{(2)} = -\lambda_0 \mathcal{A}_{2121}^d u_{1,22}^{(1)} - \mathcal{A}_{1111} u_{1,XX}^{(1)} - \mathcal{A}_{1221} u_{2,2X}^{(1)} + p_{,X}^{(1)}, \tag{4.31}$$

$$\mathcal{A}_{2222} u_{2,22}^{(2)} + \mathcal{A}_{1221} u_{1,2X}^{(2)} - p_{,2}^{(2)} = -\lambda_0 \mathcal{A}_{2222}^d u_{2,22}^{(1)} - \lambda_0 \mathcal{A}_{1221}^d u_{1,2X}^{(1)} - \mathcal{A}_{1212} u_{2,XX}^{(1)}, \tag{4.32}$$

$$u_{1,X}^{(2)} + u_{2,2}^{(2)} = -u_{1,X}^{(1)} u_{2,2}^{(1)}. \tag{4.33}$$

On $x_2 = \pm 1$,

$$\mathcal{A}_{2121} u_{1,2}^{(2)} = -\lambda_0 \mathcal{A}_{2121}^d u_{1,2}^{(1)} - (\mathcal{A}_{1221} + p_0) u_{2,X}^{(1)}, \tag{4.34}$$

$$(\mathcal{A}_{2222} + p_0) u_{2,2}^{(2)} - p^{(2)} = -(\lambda_0 \mathcal{A}_{2222}^d + p_1) u_{2,2}^{(1)} + (p_0 - \frac{1}{2} \mathcal{A}_{222222}^2)(u_{2,2}^{(1)})^2 - p^{(1)} u_{2,2}^{(1)}. \tag{4.35}$$

Third order:

For $|x_2| < 1$,

$$\begin{aligned} \mathcal{A}_{2121} u_{1,22}^{(3)} = {}& -\lambda_0 \mathcal{A}_{2121}^d u_{1,22}^{(2)} - \mathcal{A}_{1111} u_{1,XX}^{(2)} - \lambda_0 \mathcal{A}_{1111}^d u_{1,XX}^{(1)} \\ & - \mathcal{A}_{1221} u_{2,2X}^{(2)} - \lambda_0 \mathcal{A}_{1221}^d u_{2,2X}^{(1)} + p_{,X}^{(2)} - \mathcal{A}_{111111}^2 u_{1,X}^{(1)} u_{1,XX}^{(1)} \\ & - \mathcal{A}_{211211}^2 u_{1,X}^{(1)} u_{2,2X}^{(1)} - \mathcal{A}_{211222}^2 u_{2,2}^{(1)} u_{2,2X}^{(1)} \\ & + (\mathcal{A}_{212122}^2 - \mathcal{A}_{212111}^2) u_{1,X}^{(1)} u_{1,22}^{(1)} - p_{,X}^{(1)} u_{1,X}^{(1)} + u_{1,\tau\tau}^{(1)}. \end{aligned} \tag{4.36}$$

On $x_2 = \pm 1$,

$$
\begin{aligned}
\mathcal{A}_{2121}u_{1,2}^{(3)} = &\ - \lambda_0 \mathcal{A}_{2121}^d u_{1,2}^{(2)} - (\mathcal{A}_{2112} + p_0)u_{2,X}^{(2)} - (\lambda_0 \mathcal{A}_{2112}^d + p_1)u_{2,X}^{(1)} \\
& - \frac{1}{2}(\mathcal{A}_{211121}^2 - \mathcal{A}_{212221}^2 + \mathcal{A}_{212111}^2 - \mathcal{A}_{212122}^2)u_{1,2}^{(2)}u_{1,X}^{(1)} \\
& - \frac{1}{2}(\mathcal{A}_{211112}^2 - \mathcal{A}_{212212}^2 + \mathcal{A}_{211211}^2 - \mathcal{A}_{211222}^2)u_{2,X}^{(1)}u_{1,X}^{(1)} \\
& - p^{(1)}u_{2,X}^{(1)}.
\end{aligned}
\tag{4.37}
$$

In the above expressions, all the second-order elastic moduli are evaluated at $\lambda = 3$, and $u_{1,X}^{(1)} = \partial u_1^{(1)}/\partial X$, $u_{1,\tau}^{(1)} = \partial u_1^{(1)}/\partial \tau$, etc. We note that although we have written down equations up to and including third order, only quadratic terms are involved. Also, at third order, we need only the first equation of motion and the first boundary condition for the purpose of deriving the amplitude equation.

We now proceed to solve the hierachy of equations (4.27)–(4.37). First, the leading-order problem (4.27)–(4.30) can easily be solved to yield

$$
u_1^{(1)} = A(X,\tau), \quad u_2^{(1)} = -x_2 A_X(X,\tau) + B(X,\tau), \quad p^{(1)} = -(\mathcal{A}_{2222} + p_0)A_X,
\tag{4.38}
$$

where the functions A and B are to be determined at higher orders, and $A_X = \partial A/\partial X$.

In solving the second-order problem (4.31)–(4.35) we find that it has a solution only if the following solvability conditions are satisfied:

$$
\mathcal{A}_{1111} + \mathcal{A}_{2222} + 2p_0 = 0, \quad B_{XX} = 0.
\tag{4.39}
$$

From (4.17) we see that the first relation in (4.39) simply implies that the velocity should vanish to leading order and it is therefore satisfied automatically. The second relation in (4.39) implies that B must be linear in X. By imposing the conditions that the buckling solution is either periodic or localized in X and that $u_2 = 0$ when $x_2 = 0$, we obtain $B = 0$. When (4.39) are satisfied and $B = 0$ is assumed, the second-order solution is given by

$$
\begin{aligned}
u_1^{(2)} &= \frac{1}{2}x_2^2 A_{XX} + C(X,\tau), \\
u_2^{(2)} &= -\frac{1}{6}x_2^3 A_{XXX} - x_2 C_X + F + x_2 A_X^2, \\
p^{(2)} &= \mathcal{A}_{2222}A_X^2 + \frac{1}{2}(\mathcal{A}_{1221} - \mathcal{A}_{2222} - \mathcal{A}_{1212})x_2^2 A_{XXX} \\
&\quad + (\mathcal{A}_{1221} - \mathcal{A}_{2222})C_X + E(X,\tau),
\end{aligned}
\tag{4.40}
$$

where the arbitrary functions C, E and F arise from integrations and are to be determined at higher orders.

Finally, integrating (4.36) to obtain an expression for $u_{1,2}^{(3)}$ and then substituting it into the boundary conditions (4.37), we obtain, after making use of (4.38), (4.40) and simplifying,

$$A_{\tau\tau} + c_1\lambda_0 A_{XX} + c_2 A_{XXXX} + c_3 A_X A_{XX} = 0, \qquad (4.41)$$

where

$$c_1 = 2\mathcal{A}_{1221}^d - 2\mathcal{A}_{2121}^d - \mathcal{A}_{1111}^d - \mathcal{A}_{2222}^d, \qquad c_2 = -\frac{1}{3}(\mathcal{A}_{2121} - \mathcal{A}_{1212}),$$

$$c_3 = 4p_0 + 6\mathcal{A}_{2222} + 2(\mathcal{A}_{2121} - \mathcal{A}_{1221}) + \mathcal{A}_{222222}^2 - \mathcal{A}_{111111}^2. \qquad (4.42)$$

Recalling relations (4.3) and (4.17), we see that the c_1 given above is simply $-\partial v^2/\partial\lambda$ evaluated at $k = 0, \lambda = 3$. We note that with A_X taken as the new dependent variable, the amplitude equation (4.41) can be converted, by a further scaling of variables, into the Boussinesq equation (see, for example, Ablowitz and Clarkson 1991, p. 52).

With the use of the expressions (4.6) and (4.7), we find that

$$c_1 = 2/\sqrt{3}, \quad c_2 = 8/(3\sqrt{3}), \quad c_3 = 2\sqrt{3}. \qquad (4.43)$$

The linear part of the amplitude equation (4.41) can be checked against the dispersion relation (4.18) as follows. On substituting

$$A = e^{ik(x_1 - vt)} = e^{(ik/\epsilon)(\epsilon^{1/2}X - v\tau)}$$

into the linearized form of (4.41) and making use of (4.19), we obtain

$$v^2 = -c_1\lambda_0\epsilon + c_2 k^2 + \cdots = -c_1(\lambda - 3) + c_2 k^2 + \cdots, \qquad (4.44)$$

which is, in view of (4.43), the same dispersion relation (4.18). This provides a partial check on our derivations.

We now discuss static solutions of (4.41). Upon setting $A_{\tau\tau} = 0$ in (4.41) and integrating once with respect to X, we obtain

$$A_{XXX} + \frac{3}{4}\lambda_0 A_X + \frac{9}{8}A_X^2 = D, \qquad (4.45)$$

where D is a constant.

Since, from (4.4), we have

$$t_1 = \frac{8}{3\sqrt{3}} - \frac{1}{9\sqrt{3}}(\lambda - 3)^2 + \cdots, \quad \text{as } \lambda \to 3, \qquad (4.46)$$

the λ^+ and λ^- in Figure 3 must have equal distance from $\lambda = 3$ in order to correspond to the same value of t_1. Thus, if $3 + \epsilon\lambda_0 = \lambda^+$, then $3 - \epsilon\lambda_0 = \lambda^-$, and likewise if $3 + \epsilon\lambda_0 = \lambda^-$, then $3 - \epsilon\lambda_0 = \lambda^+$. Hence, $\lambda = 3 - \epsilon\lambda_0 =$

$3 + \epsilon\lambda_0 - 2\epsilon\lambda_0$ is a solution that must be admitted by (4.45). To check this, we must first find the relation between A_X and $-2\epsilon\lambda_0$. To this end, we first note that with $\tilde{x}_1 = \lambda X_1 + u_1$, $\tilde{x}_2 = \lambda^{-1}X_2 + u_2$, where X_1 and X_2 are coordinates of material particles in the undeformed configuration, the deformation gradient in the buckled state is given by

$$\tilde{\mathbf{F}} \equiv \left(\frac{\partial \tilde{x}_i}{\partial x_j}\right) = \left(\begin{array}{cc} \lambda(1 + u_{1,1}) & \lambda^{-1}u_{1,2} \\ \lambda u_{2,1} & \lambda^{-1}(1 + u_{2,2}) \end{array} \right) =$$

$$\left(\begin{array}{cc} \lambda(1 + \epsilon A_X) + O(\epsilon^{3/2}) & O(\epsilon^{3/2}) \\ O(\epsilon^{3/2}) & \lambda^{-1}(1 - \epsilon A_X) + O(\epsilon^{3/2}) \end{array} \right). \qquad (4.47)$$

Thus, the principal stretches in the buckled configuration, denoted by $\tilde{\lambda}_1, \tilde{\lambda}_2$, are given by

$$\tilde{\lambda}_1 = \lambda(1 + \epsilon A_X) + O(\epsilon^{3/2}) = 3 + \epsilon\lambda_0 + 3\epsilon A_X + O(\epsilon^{3/2}),$$

$$\tilde{\lambda}_2 = \lambda^{-1}(1 - \epsilon A_X) + O(\epsilon^{3/2}) = \frac{1}{3} - \frac{1}{9}\epsilon\lambda_0 - \frac{1}{3}\epsilon A_X + O(\epsilon^{3/2}), \qquad (4.48)$$

where use has been made of (4.19). Thus, we conclude that $3\epsilon A_X = -2\epsilon\lambda_0$, that is $A_X = -2\lambda_0/3$, must satisfy (4.45) for arbitrary λ_0. We see that this is indeed the case if $D = 0$. Thus, this consideration not only provides a check on our derivations but it also determines the integration constant D.

It is convenient to introduce a new variable Q through

$$Q = A_X + \lambda_0/3. \qquad (4.49)$$

In terms of Q, (4.45) becomes

$$Q_{XX} = \frac{1}{8}\lambda_0^2 - \frac{9}{8}Q^2, \qquad (4.50)$$

which is of the same form as that obtained by Mielke (1991, p. 118) for a compressible strip using the dynamical systems approach. We note that in terms of Q, the $\tilde{\lambda}_1, \tilde{\lambda}_2$ given by (4.48) becomes

$$\tilde{\lambda}_1 = 3 + 3\epsilon Q + O(\epsilon^{3/2}), \quad \tilde{\lambda}_2 = 1/3 - \epsilon Q/3 + O(\epsilon^{3/2}), \qquad (4.51)$$

and $\lambda^\pm$ in Figure 3 correspond to $Q = \pm|\lambda_0|/3$, which are the equilibrium points of (4.50).

The solution behaviour of (4.50) can best be understood with the aid of its phase portrait, which is shown in Figure 4. The equilibrium point $Q = |\lambda_0|/3$ is a centre whereas the other point $Q = -|\lambda_0|/3$ is a saddle. The homoclinic

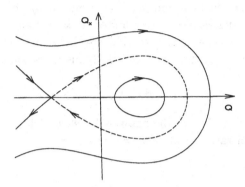

Fig. 4. Phase portrait for (4.50).

orbit, shown by dotted line in Figure 4, is given by

$$Q = \frac{2|\lambda_0|}{1 + \cosh\frac{\sqrt{3|\lambda_0|}}{2}X} - \frac{|\lambda_0|}{3}. \tag{4.52}$$

The thickness of the plate in the x_2-direction is approximately proportional to the principal stretch $\tilde{\lambda}_2$ given by (4.51). Corresponding to (4.52), $\tilde{\lambda}_2$ has a minimum $(3 - \epsilon 5|\lambda_0|)/9$ at $X = 0$ and it tends to a larger value $(3 + \epsilon|\lambda_0|)/9$ exponentially as $X \to \pm\infty$. Thus, the homoclinic solution (4.52) corresponds to a necking configuration in which necking is localized in an exponentially thin region around $x_1 = 0$. We note that the smaller value λ^- in Figure 3 is preferred as the asymptotic value of $\tilde{\lambda}_1$ at $x = \pm\infty$. The periodic orbits inside the homoclinic orbit in Figure 4 correspond to necking configurations for which necking is periodic along the x_1-direction.

Each of these necking solutions is observable only if it is stable to small amplitude perturbations. The stability of these necking solutions can be studied with the aid of the dynamic amplitude equation (4.41). Relevant results will be published elsewhere.

Finally, we remark that although we have only considered the class of materials with the strain-energy function given by (4.1) and $m = 1/2$, the amplitude equation (4.41) should also be valid for other materials that exhibit similar softening behaviour in uni-axial extension. The corresponding constants c_1, c_2 and c_3 in (4.41) can be determined simply as follows. First, comparing (4.44) with the general Taylor expansion of v^2, we obtain

$$c_1 = -\frac{\partial v^2}{\partial \lambda}, \quad c_2 = \frac{\partial v^2}{\partial(k^2)},$$

where v^2, as a function of λ and k^2, is determined by the linear dispersion relation, and both derivatives are evaluated at $k = 0$ and at the value of λ where the tension t_1 has a maximum. The coefficient c_3 of the quadratic term is determined by the fact that $A_X = -2\lambda_0/3$ must be a solution of

$$c_1\lambda_0 A_X + c_2 A_{XXX} + \frac{1}{2}c_3 A_X^2 = D,$$

the general form of (4.45); see the paragraph below (4.48). This equation can be satisfied for arbitrary λ_0 only if $D = 0$ and $c_3 = 3c_1$.

10.5 Conclusion

In this chapter we have explained a number of methods that can be used to conduct nonlinear stability analysis in Finite Elasticity. These methods have been developed in different disciplines and usually in more complex settings. Our aim in this chapter has been to explain these methods by using the simplest examples possible so that the reader can glean the essence of each method without being troubled by heavy algebra.

We conclude this chapter with a few remarks. Firstly, although these methods have been explained in the context of bifurcation analysis, they are equally applicable to the analysis of problems of nonlinear wave propagation. In fact, if one looks for a steady nonlinear wave solution with dependence on a spatial variable, x_1 say, and time t through the variable $\theta = x_1 - ct$, where c is the nonlinear wave speed, then the dynamical problem reduces to a static problem immediately with c playing the role of a bifurcation parameter. Again there are two major types of nonlinear wave problems: problems concerned with dispersive waves and those with weakly dispersive waves. The counterpart of our equation (2.21) would be the well-known nonlinear Schrödinger equation which governs the evolution of the envelop of a wave packet (the dominant wave in the wave packet has a non-zero wave number and the nonlinear term in the amplitude equation is cubic). The counterpart of our equation (4.41) would be the well-known Korteweg-de Vries equation, which governs, among other phenomena, the evolution of long wavelength water waves in a channel (long wavelength means that to leading order the wave number is zero and the nonlinear term in the amplitude equation is quadratic). We refer the reader to Newell (1985) and Drazin and Johnson (1989) for more details.

Secondly, our focus in this chapter has been on the normal mode approach. But stability analysis does not always have to follow this approach. Notable exceptions are analysis of bifurcations from one uniform state into another uniform state (see Ball and Schaeffer, 1983, and Chen, 1996, for an analysis of the bifurcation of a cube into a plate or a rod) and analysis of cavitation of elastic spheres (see Ball, 1982, Polignone and Horgan, 1993a,b, and the references therein). Another approach is to develop simplified models for elastic bodies with specific geometries and then to analyze the reduced governing equations directly to uncover bifurcation solutions; see Antman (1995) and the references therein.

Finally, although, following tradition, we have used the terms "stability" and "instability" frequently in this chapter, our analysis is essentially a local bifurcation analysis, as we have already emphasized in the introduction. Our dynamic amplitude equations (2.21) and (4.41) can be used to assess the stability of the static solutions of the corresponding equations. But the class of perturbations allowed by these dynamic equations is restricted. Consideration of stability in Finite Elasticity also requires the understanding of some fundamental issues. A selection of papers in this area is Ball (1977), Ball and Marsden (1984), Simpson and Spector (19987, 1989), Sivaloganathan (1989), Mielke and Sprenger (1998), Wang and Aron (1996) and Aron (2000) from which more references can be traced.

Acknowledgements

Some of the results in Section 10.4 were obtained when the author was visiting Mathematics Institute A, Stuttgart University, Germany, supported by a Royal Society travel grant under its European Science Exchange Programme. The author thanks Professor Mielke, Dr Groves and Professor Kirchgässner for their hospitality and for some stimulating discussions.

References

Ablowitz, M.J. and Clarkson, P.A. 1991 *Solitons, Nonlinear Evolution Equations and Inverse Scattering*. Cambridge University Press: Cambridge.

Abramian, A.K. 2000, private communication.

Aron, M. 2000 Some implications of empirical inequalities for the plane strain deformations of unconstrained, isotropic, hyperelastic solids. *Math. Mech. Solids* **5**, 19-30.

Antman, S.S. 1995 *Problems of Nonlinear Elasticity*. New York: Springer.

Ball, J.M. 1977 Convexity conditions and existence theorems in nonlinear elasticity. *Arch. Rat. Mech. Anal.* **63**, 337-403.

Ball, J.M. and Marsden, J.E. 1984 Quasiconvexity at the boundary, positivity of the second variation and elastic stability. *Arch. Rat. Mech. Anal.* **86**, 251-277.

Ball, J.M. and Schaeffer, D.G. 1983 Bifurcation and stability of homogeneous equilibrium configurations of an elastic body under dead-load tractions. *Math. Proc. Camb. Phil. Soc.* **94**, 315-339.

Bridges, T.J. 1999 The Orr-Sommerfeld equation on a manifold. *Proc. R. Soc. Lond.* **A455**, 3019-3040.

Buffoni, B., Groves, M.D. and Toland, J.F. 1996 A plethora of solitary gravity-capillary water waves with nearly critical Bond and Froude numbers. *Phil. Trans. R. Soc. Lond.* **A354**, 575-607.

Buffoni, B. and Groves, M.D. 1999 A multiplicity result for solitary gravity-capillary waves in deep water via critical-point theory. *Arch. Rat. Mech. Anal.* **146**, 183-220.

Cai, Z.X. and Fu, Y.B. 1999 On the imperfection sensitivity of a coated elastic half-space. *Proc. R. Soc. Lond.* **455**, 3285-3309.

Chadwick, P. 1999 *Continuum Mechanics: Concise Theory and Problems*. Dover: New York.

Champneys, A.R. 1998 Homoclinic orbits in reversible systems and their applications in mechanics, fluids and optics. *Physica* **D112**, 158-186.

Champneys, A.R. and Toland, J.F. 1993 Bifurcation of a plethora of multi-modal homoclinic orbits for autonomous Hamiltonian systems. *Nonlinearity* **6**, 665-721.

Coleman, B.D. 1983 Necking and drawing in polymeric fibres under tension. *Arch. Rat. Mech. Anal.* **83**, 115-137.

Dias, F. and Iooss, G. 1993 Capillary-gravity solitary waves with damped oscillations. *Physica* **D65**, 399-423.

Drazin, P.G. and Johnson, R.S. 1989 *Solitons: An Introduction*. Cambridge University Press: Cambridge.

Elphick, C., Tirapsegui, E., Brachet, M.E., Coullet, P. and Iooss, G. 1987 A simple global characterization for normal forms of singular vector fields. *Physica* **D29**, 95-127.

Fu, Y.B. 1995 Resonant-triad instability of a pre-stressed elastic plate. *J. Elasticity* **41**, 13-37

Fu, Y.B. 1998 Some asymptotic results concerning the buckling of a spherical shell of arbitrary thickness. *Int. J. Non-Linear Mech.* **33**, 1111-1122.

Fu, Y.B. and Devenish B. 1996 Effects of pre-stresses on the propagation of nonlinear surface waves in an elastic half-space. *Q. J. Mech. Appl. Math.* **49**, 65-80.

Fu, Y.B. and Ogden, R.W. 1999 Nonlinear stability analysis of pre-stressed elastic bodies. *Continuum Mech. Thermodyn.* **11**, 141-172.

Fu, Y.B. and Rogerson, G.A. 1994 A nonlinear analysis of instability of a pre-stressed incompressible elastic plate. *Proc. R. Soc. Lond.* **A446**, 233-254.

Fu, Y.B. and Sanjarani Pour, M. 2000 WKB method with repeated roots and its application to the stability analysis of an everted cylindrical tube. To be submitted.

Fu, Y.B. and Zheng, Q.-S. 1997 Nonlinear travelling waves in a neo-Hookean plate subjected to a simple shear. *Math. Mech. Solids* **2**, 27-48.

Hunt, G.W., Bolt, H.M. and Thompson, J.M.T. 1989 Structural localization phenomena and the dynamical phase-space analogy. *Proc. R. Soc. Lond.* **A425**, 245-267.

Hunt, G.W. and Wadee, M.K. 1991 Comparative Lagrangian formulations for localized buckling. *Proc. R. Soc. Lond.* **A434**, 485-502.

Hutchinson, J.W. and Koiter, W.T. 1970 Postbuckling theory. *Applied Mechanics Reviews* **23**, 1353-1365.

Iooss, G. and Pérouéme, M.C. 1993 Perturbed homoclinic solutions in reversible 1:1 resonance vector fields. *J. Diff. Eqns* **102**, 62-88.

Kirchgässner, K. 1982 Wave solutions of reversible systems and applications. *J. Diff. Eqns.* **45**, 113-127.

Koiter, W.T. 1945 *On the stability of elastic equilibrium.* Thesis Delft (in Dutch), English translation: NASA TT F-10, 1967, **833**.

Lange, C.G. and Newell, A.C. 1971 The post-buckling problem for thin elastic shells. *SIAM J. Appl. Math.* **21**, 605-629.

Lardner, R.W. 1986 Nonlinear surface acoustic waves on an elastic solid of general anisotropy. *J. Elasticity* **16**, 63-73.

Mielke, A. 1986 A reduction principle for nonautonomous systems in infinite dimensional spaces. *J. Diff. Eqns.* **65**, 68-88.

Mielke, A. 1988a Reduction of quasilinear elliptic equations in cylindrical domains with applications. *Math. Meth. Appl. Sci.* **10**, 51-66.

Mielke, A. 1988b On Saint-Venant's problem for an elastic strip. *Proc. R. Soc. Edin.* **110A**, 161-181.

Mielke, A. 1988c Saint-Venant's problem and semi-inverse solutions in nonlinear elasticity. *Arch. Rat. Mech. Anal.* **102**, 205-229.

Mielke, A. 1991 *Hamiltonian and Lagrangian Flows on Center Manifolds.* Berlin: Springer (Lecture Notes in Mathematics vol. 1489).

Mielke, A. and Sprenger, P. 1998 Quasiconvexity at the boundary and a simple variational formulation of Agmon's condition. *J. Elasticity* **51**, 23-41.

Nagata, M. 1986 Bifurcations in Couette flow between almost corotating cylinders. *J. Fluid Mech.* **169**, 229-250.

Nagata, M. 1988 On wavy instabilities of the Taylor-vortex flow between corotating cylinders. *J. Fluid Mech.* **188**, 585-598.

Newell, A.C. 1985 *Solitons in Mathematics and Physics.* SIAM: Philadelphia.

Ng, B.S. and Reid, W.H. 1979 An initial-value method for eigenvalue problems using compound matrices. *J. Comp. Phys* **30**, 125-136.

Owen, N. 1987 Existence and stability of necking deformations for nonlinearly elastic rods. *Arch. Rat. Mech. Anal.* **98**, 357-384.

Polignone, D.A. and Horgan, C.O. 1993 Cavitation for incompressible anisotropic nonlinearly elastic spheres. *J. Elasticity* **33**, 27-65.

Polignone, D.A. and Horgan, C.O. 1993 Effects of material anisotropy and inhomogeneity on cavitation for composite incompressible anisotropic nonlinear elastic spheres. *Int. J. Solids Structures* **30**, 3381-3416.

Potier-Ferry, M. 1987 Foundations of elastic post-buckling theory, in *Buckling and Post-buckling*, ed. J. Arbocz *et al.*, Lecture Notes in Physics Vol. 288, 1-82, Berlin:Springer

Rogerson, G.A. and Fu, Y.B. 1995 An asymptotic analysis of the dispersion relation of a pre-stressed incompressible elastic plate. *Acta Mechanica* **111**, 59-74.

Sandstede, B. 1997 Instability of localized buckling modes in a one-dimensional strut model. *Phil. Trans. R. Soc. Lond.* **A355**, 2083-2097.

Simpson, H.C. and Spector, S.J. 1987 On the positivity of the second variation in finite elasticity. *Arch. Rat. Mech. Anal.* **98**, 1-30.

Simpson, H.C. and Spector, S.J. 1989 Necessary conditions at the boundary for minimizers in finite elasticity. *Arch. Rat. Mech. Anal.* **107**, 105-125.

Sivaloganathan, J. 1989 The generalized Hamilton-Jacobi inequality and the stability of equilibria in nonlinear elasticity. *Arch. Rat. Mech. Anal.* **107**, 347-369.

Thompson, J.M.T. and Virgin, L.N. 1988 Spatial chaos and localization phenomena in nonlinear elasticity. *Physics Letters* **A126**, 491-496.

Yang, T.-S. and Akylas, T.R. 1997 On asymmetric gravity-capillary solitary waves.

J. Fluid Mech. 330, 215-232.

Wadee, M.K. and Bassom, A.P. 1998 Effects of exponentially small terms in the perturbation approach to localized buckling. Proc. R. Soc. Lond. A455, 2351-2370.

Wadee, W.K., Hunt, G.W. and Whiting, A.I.M. 1997 Asymptotic and Rayleigh-Ritz routes to localized buckling solutions in an elastic instability problem. Proc. R. Soc. Lond. A453, 2085-2107.

Wang, Y. and Aron, M. 1996 A reformulation of the strong ellipticity conditions for unconstrained hyperelastic media. J. Elasticity 44, 89-96.

Wiggins, S. 1990 Introduction to Applied Nonlinear Dynamical Systems and Chaos. Springer-Verlag: New York.

Appendix A: Incremental equations in cylindrical and spherical polar coordinates

The incremental equations in rectangular coordinates are given by (1.1)–(1.4). To derive the counterparts of these equations in cylindrical and spherical polar coordinates, we make use of a well-known result in tensor analysis, namely that a tensor relation holds in any coordinate system. We shall first write down the tensorial forms of (1.1)–(1.4) and then express them in terms of cylindrical and spherical polar coordinates.

The tensorial forms of (1.1)–(1.4) are obtained by replacing each comma by a semi-colon. A semi-colon signifies covariant differentiation and the notation is defined as follows.

First, let e_1, e_2, e_3 be the three orthonormal basis vectors (i.e. $e_m \cdot e_n = \delta_{mn}$) for a curvilinear coordinate system (cylindrical or spherical polar), and x_1, x_2, x_3 the three corresponding coordinates measuring distance along the curvilinear directions. By referring the displacement vector u and the incremental stress tensor Σ to such a coordinate system, we do not need to distinguish between covariant and contravariant derivatives/components (we believe that choosing a triad of non-orthonormal basis vectors and then having to make the distinctions unnecessarily complicates the derivations). We may write

$$u = u_m e_m, \quad \Sigma = \Sigma_{mn} e_m \otimes e_n. \tag{A1}$$

We define the displacement gradient grad u as

$$\frac{\partial}{\partial x_m}(u_n e_n) \otimes e_m.$$

Thus,

$$
\begin{aligned}
\text{grad } u &= u_{n,m} e_n \otimes e_m + u_n e_{n,m} \otimes e_m \\
&= (u_{n,m} + u_p e_n \cdot e_{p,m}) e_n \otimes e_m \\
&\equiv u_{n;m} e_n \otimes e_m.
\end{aligned}
$$

The last relation defines the covariant derivative $u_{n;m}$. To simplify notation, we denote $u_{n;m}$ by η_{nm}. We therefore have

$$\eta_{nm} \equiv u_{n;m} = u_{n,m} + u_p e_n \cdot e_{p,m}. \tag{A2}$$

The incremental equilibrium equation (1.1) can also be written as div $\Sigma = 0$. For a general curvilinear coordinate system, we define div Σ as

$$e_m \cdot \frac{\partial}{\partial x_m} (\Sigma_{pq} e_p \otimes e_q),$$

and $c \cdot (a \otimes b) = (c \cdot a)b$ for vectors a, b, c. Thus,

$$
\begin{aligned}
\text{div } \Sigma &= \Sigma_{pq,p} e_q + \Sigma_{pq} \{e_{q,p} + e_q(e_{p,m} \cdot e_m)\} \\
&= (\Sigma_{pq,p} + \Sigma_{mp} e_q \cdot e_{p,m} + \Sigma_{pq} e_m \cdot e_{p,m}) e_q \\
&\equiv \Sigma_{pq;p} e_q.
\end{aligned}
$$

The last relation defines $\Sigma_{pq;p}$. Thus, the tensorial form of (1.1) yields the following form of equilibrium equation that is valid for curvilinear coordinate systems:

$$\Sigma_{ji,j} + \Sigma_{jm} e_i \cdot e_{m,j} + \Sigma_{ji} e_m \cdot e_{j,m} = 0. \tag{A3}$$

The tensorial forms of (1.2)–(1.4) are

$$\eta_{ii} = \frac{1}{2} \eta_{mn} \eta_{nm} - \frac{1}{2} (\eta_{ii})^2 - \det(\eta_{ij}), \tag{A4}$$

$$\Sigma_{ji} n_j = 0, \tag{A5}$$

$$\Sigma_{ji} = \mathcal{A}^1_{jilk} \eta_{kl} + \frac{1}{2} \mathcal{A}^2_{jilknm} \eta_{kl} \eta_{mn} + \frac{1}{6} \mathcal{A}^3_{jilknmqp} \eta_{kl} \eta_{mn} \eta_{pq}$$
$$+ p(\eta_{ji} - \eta_{jk}\eta_{ki} + \eta_{jk}\eta_{kl}\eta_{li}) - p^*(\delta_{ji} - \eta_{ji} + \eta_{jk}\eta_{ki}) + \cdots. \tag{A6}$$

What remains to be done is to evaluate $e_i \cdot e_{m,j}$ for each coordinate system.

Cylindrical polar coordinates

In this case, we take

$$e_1 = \hat{r}, \quad e_2 = \hat{\theta}, \quad e_3 = \hat{z},$$

where $\hat{r}$, $\hat{\theta}$, $\hat{z}$ are unit vectors along the r-, θ-, and z-directions, respectively, (r, θ, z) being the usual cylindrical polar coordinates. We have

$$dx_1 = dr, \quad dx_2 = r\,d\theta, \quad dx_3 = dz,$$

$$\frac{\partial}{\partial x_1} = \frac{\partial}{\partial r}, \quad \frac{\partial}{\partial x_2} = \frac{1}{r}\frac{\partial}{\partial \theta}, \quad \frac{\partial}{\partial x_3} = \frac{\partial}{\partial z}.$$

With the aid of the standard relations

$$\hat{r} = \cos\theta\, i + \sin\theta\, j, \quad \hat{\theta} = -\sin\theta\, i + \cos\theta\, j, \quad \hat{z} = k,$$

where i, j, k are the unit basis vectors of a rectangular coordinate system, we have

$$e_{1,1} = \frac{\partial\hat{r}}{\partial r} = 0, \quad e_{1,2} = \frac{1}{r}\frac{\partial\hat{r}}{\partial\theta} = \frac{1}{r}\hat{\theta}, \quad e_{1,3} = \frac{\partial\hat{r}}{\partial z} = 0,$$

$$e_{2,1} = \frac{\partial\hat{\theta}}{\partial r} = 0, \quad e_{2,2} = \frac{1}{r}\frac{\partial\hat{\theta}}{\partial\theta} = -\frac{1}{r}\hat{r}, \quad e_{2,3} = 0,$$

$$e_{3,1} = \frac{\partial\hat{z}}{\partial r} = 0, \quad e_{3,2} = \frac{1}{r}\frac{\partial\hat{z}}{\partial\theta} = 0, \quad e_{3,3} = \frac{\partial\hat{z}}{\partial z} = 0.$$

The non-zero components of $e_i \cdot e_{m,j}$ are then given by

$$e_2 \cdot e_{1,2} = \frac{1}{r}, \quad e_1 \cdot e_{2,2} = -\frac{1}{r}.$$

Thus,

$$
\begin{aligned}
\eta_{11} &= u_{1,1} + u_p e_1 \cdot e_{p,1} = \frac{\partial u_1}{\partial r}, \\[4pt]
\eta_{12} &= u_{1,2} + u_p e_1 \cdot e_{p,2} = \frac{1}{r}\left(\frac{\partial u_1}{\partial\theta} - u_2\right), \\[4pt]
\eta_{13} &= u_{1,3} + u_p e_1 \cdot e_{p,3} = \frac{\partial u_1}{\partial z}, \\[4pt]
\eta_{21} &= u_{2,1} + u_p e_2 \cdot e_{p,1} = \frac{\partial u_2}{\partial r}, \\[4pt]
\eta_{22} &= u_{2,2} + u_p e_2 \cdot e_{p,2} = \frac{1}{r}\left(\frac{\partial u_2}{\partial\theta} + u_1\right), \\[4pt]
\eta_{23} &= u_{2,3} + u_p e_2 \cdot e_{p,3} = \frac{\partial u_2}{\partial z}, \\[4pt]
\eta_{31} &= u_{3,1} + u_p e_3 \cdot e_{p,1} = \frac{\partial u_3}{\partial r}, \\[4pt]
\eta_{32} &= u_{3,2} + u_p e_3 \cdot e_{p,2} = \frac{1}{r}\frac{\partial u_3}{\partial\theta}, \\[4pt]
\eta_{33} &= u_{3,3} + u_p e_3 \cdot e_{p,3} = \frac{\partial u_3}{\partial z}.
\end{aligned}
\tag{A7}
$$

The equilibrium equation (A3) yields

$$\Sigma_{j1,j} + \frac{1}{r}(\Sigma_{11} - \Sigma_{22}) = 0,$$

$$\Sigma_{j2,j} + \frac{1}{r}(\Sigma_{21} + \Sigma_{12}) = 0, \tag{A8}$$

$$\Sigma_{j3,j} + \frac{1}{r}\Sigma_{13} = 0.$$

The equilibrium equations in terms of the displacement components can be obtained by substituting (A6) together with (A7) into (A8). The linearized forms of these equations can be found in Haughton and Ogden (1979) or Haughton and Orr (1995). These references are listed in Appendix B.

Spherical polar coordinates

In this case, we take

$$e_1 = \hat{r}, \quad e_2 = \hat{\theta}, \quad e_3 = \hat{\phi},$$

where $\hat{r}$, $\hat{\theta}$ and $\hat{\phi}$ are the unit basis vectors along the r-, θ- and ϕ-directions, respectively, (r, θ, ϕ) being the usual spherical polar coordinates with $0 \leq \theta \leq 2\pi$ and $-\pi/2 \leq \phi \leq \pi/2$. We have

$$\hat{r} = \sin\theta\cos\phi\,\mathbf{i} + \sin\theta\sin\phi\,\mathbf{j} + \cos\theta\,\mathbf{k},$$

$$\hat{\theta} = \cos\theta\cos\phi\,\mathbf{i} + \cos\theta\sin\phi\,\mathbf{j} - \sin\theta\,\mathbf{k},$$

$$\hat{\phi} = -\sin\phi\,\mathbf{i} + \cos\phi\,\mathbf{j},$$

where $\mathbf{i}, \mathbf{j}, \mathbf{k}$ are the unit basis vectors of a rectangular coordinate system. Let $\mathbf{r} = r\hat{r}$ be the position vector of a representative point. We have

$$
\begin{aligned}
d\mathbf{r} &= dr\hat{r} + rd\hat{r} = dr\hat{r} + r\left(\frac{\partial\hat{r}}{\partial\theta}d\theta + \frac{\partial\hat{r}}{\partial\phi}d\phi\right) \\
&= dr\,\hat{r} + rd\theta\,\hat{\theta} + r\sin\theta d\phi\,\hat{\phi}.
\end{aligned}
$$

Thus,

$$dx_1 = dr, \quad dx_2 = rd\theta, \quad dx_3 = r\sin\theta d\phi,$$

$$\frac{\partial}{\partial x_1} = \frac{\partial}{\partial r}, \quad \frac{\partial}{\partial x_2} = \frac{1}{r}\frac{\partial}{\partial\theta}, \quad \frac{\partial}{\partial x_3} = \frac{1}{r\sin\theta}\frac{\partial}{\partial\phi}.$$

It then follows that

$$\mathbf{e}_{1,1} = \frac{\partial\hat{r}}{\partial r} = 0, \quad \mathbf{e}_{1,2} = \frac{1}{r}\frac{\partial\hat{r}}{\partial\theta} = \frac{1}{r}\hat{\theta}, \quad \mathbf{e}_{1,3} = \frac{1}{r\sin\theta}\frac{\partial\hat{r}}{\partial\phi} = \frac{1}{r}\hat{\phi},$$

$$\mathbf{e}_{2,1} = \frac{\partial\hat{\theta}}{\partial r} = 0, \quad \mathbf{e}_{2,2} = \frac{1}{r}\frac{\partial\hat{\theta}}{\partial\theta} = -\frac{1}{r}\hat{r}, \quad \mathbf{e}_{2,3} = \frac{1}{r\sin\theta}\frac{\partial\hat{\theta}}{\partial\phi} = \frac{\cot\theta}{r}\hat{\phi},$$

$$e_{3,1} = \frac{\partial \hat{\phi}}{\partial r} = 0, \quad e_{3,2} = \frac{1}{r}\frac{\partial \hat{\phi}}{\partial \theta} = 0, \quad e_{3,3} = \frac{1}{r\sin\theta}\frac{\partial \hat{\phi}}{\partial \phi} = -\frac{1}{r}\hat{r} - \frac{\cot\theta}{r}\hat{\theta}.$$

The non-zero components of $e_i \cdot e_{m,j}$ are then given by

$$e_1 \cdot e_{2,2} = -\frac{1}{r}, \quad e_1 \cdot e_{3,3} = -\frac{1}{r}, \quad e_2 \cdot e_{1,2} = \frac{1}{r},$$

$$e_2 \cdot e_{3,3} = -\frac{\cot}{r}, \quad e_3 \cdot e_{1,3} = \frac{1}{r}, \quad e_3 \cdot e_{2,3} = \frac{\cot}{r}.$$

Therefore, the components of the displacement gradient are given by

$$
\begin{aligned}
\eta_{11} &= u_{1,1} + u_p e_1 \cdot e_{p,1} = \frac{\partial u_1}{\partial r}, \\
\eta_{12} &= u_{1,2} + u_p e_1 \cdot e_{p,2} = \frac{1}{r}\left(\frac{\partial u_1}{\partial \theta} - u_2\right), \\
\eta_{13} &= u_{1,3} + u_p e_1 \cdot e_{p,3} = \frac{1}{\sin\theta}\frac{\partial u_1}{\partial \phi} - \frac{u_3}{r}, \\
\eta_{21} &= u_{2,1} + u_p e_2 \cdot e_{p,1} = \frac{\partial u_2}{\partial r}, \\
\eta_{22} &= u_{2,2} + u_p e_2 \cdot e_{p,2} = \frac{1}{r}\left(\frac{\partial u_2}{\partial \theta} + u_1\right), \\
\eta_{23} &= u_{2,3} + u_p e_2 \cdot e_{p,3} = \frac{1}{r\sin\theta}\frac{\partial u_2}{\partial \phi} - \frac{\cot\theta}{r}u_3, \\
\eta_{31} &= u_{3,1} + u_p e_3 \cdot e_{p,1} = \frac{\partial u_3}{\partial r}, \\
\eta_{32} &= u_{3,2} + u_p e_3 \cdot e_{p,2} = \frac{1}{r}\frac{\partial u_3}{\partial \theta}, \\
\eta_{33} &= u_{3,3} + u_p e_3 \cdot e_{p,3} = \frac{1}{r\sin\theta}\frac{\partial u_3}{\partial \phi} + \frac{u_1}{r} + \frac{\cot\theta}{r}u_2.
\end{aligned}
\tag{A9}
$$

The equilibrium equation (A3) yields

$$\Sigma_{j1,j} + \frac{1}{r}(2\Sigma_{11} - \Sigma_{22} - \Sigma_{33}) + \frac{\cot\theta}{r}\Sigma_{21} = 0,$$

$$\Sigma_{j2,j} + \frac{1}{r}(\Sigma_{21} + 2\Sigma_{12}) + \frac{\cot\theta}{r}(\Sigma_{22} - \Sigma_{33}) = 0, \tag{A10}$$

$$\Sigma_{j3,j} + \frac{1}{r}(\Sigma_{31} + 2\Sigma_{13}) + \frac{\cot\theta}{r}(\Sigma_{32} + \Sigma_{23}) = 0.$$

The equilbrium equations in terms of the displacement components can be obtained by substituting (A6) together with (A9) into (A10). The linearized forms of these equations can be found in Haughton and Ogden (1978), or Haughton and Chen (1999). These references are listed in Appendix B.

Appendix B: Contributions to linear stability analysis in Finite Elasticity

1 Adeleke, S.A. 1980 Stability of some states of plane deformation. *Arch. Rat. Mech. Anal.* **72**, 243-263.

2 Beatty, M.F. A theory of elastic stability for incompressible hyperelastic bodies. *Int. J. Solids. Structures* **3**, 23-37.

3 Beatty, M.F. 1987 Topics in finite elasticity: hyperelasticity of rubber, elastomers, and biological tissues – with examples. *Appl. Mech. Rev.* **40**, 1699-1734.

4 Beatty, M.F. 1990 Instability of a fibre-reinforced thick slab under axial loading. *Int. J. Non-Linear Mech.* **25**, 343-362.

5 Beatty, M.F. and Pan, F.X. 1998 Stability of an internally constrained, hyperelastic slab. *Int. J. Non-Linear Mech.* **33**, 867-906.

6 Biot, M.A. 1963 Internal buckling under initial stress in finite elasticity. *Proc. R. Soc.* **A27**, 306-328.

7 Biot, M.A. 1963 Exact theory of buckling of a thick slab. *Appl. Sci. Res.* **A12**, 182-198.

8 Biot, M.A. 1965 *Mechanics of Incremental Deformations.* Wiley: New York.

9 Biot, M.A. 1968 Edge buckling of a laminated medium. *Int. J. Solids Structures* **4**, 125-137.

10 Burgess, I.W. and Levinson, M. 1972 The instability of slightly compressible rectangular rubber-like solids under biaxial loading. *Int. J. Solids and Structures* **8**, 133-148.

11 Chen, P.J. and Gurtin, M.E. 1974 On wave propagation in inextensible elastic bodies. *Int. J. Solids and Structures* **10**, 275-281.

12 Connor, P. and Ogden, R.W. 1995 The effect of shear on the propagation of elastic surface waves. *Int. J. Eng. Sci.* **33**, 973-982.

13 Connor, P. and Ogden, R.W. 1996 The influence of shear strain and hydrostatic stress on stability and elastic waves in a layer. *Int. J. Eng. Sci.* **34**, 375-397.

14 Davies, P.J. 1989 Buckling and barrelling instabilities in finite elasticity. *J. Elasticity* **21**, 147-192.

15 Dowaikh, M.A. and Ogden, R.W. 1990 On surface waves and deformations in a pre-stressed incompressible elastic solid. *IMA J. Appl. Math.* **44**, 261-284.

16 Dowaikh, M.A. and Ogden, R.W. 1991 On surface waves and deformations in a compressible elastic half-space. *Stability and Appl. Anal. Cont. Media* **1**, 27-45.

17 Dowaikh, M.A. and Ogden, R.W. 1991 Interfacial waves and deformations in pre-stressed elastic media. *Proc. R. Soc. Lond.* **A433**, 313-328.

18 Dryburgh, G. and Ogden, R.W. 1999 Bifurcation of an elastic surface-coated incompressible isotropic elastic block subject to bending. *ZAMP* **50**, 822-838.

19 Fu, Y.B. 1998 Some asymptotic results concerning the buckling of a spherical shell of arbitrary thickness. *Int. J. Non-Linear Mech.* **33**, 1111-1122.

20 Green, A.E., Rivlin, R.S. and Shield, R.T. 1952 General theory of small elastic deformations superimposed on large elastic deformations. *Proc. R. Soc. Lond.* **A21**, 128-154.

21 Green, A.E. and Spencer, A.J.M. 1959 The stability of a circular cylinder under finite extension and torsion. *J. Math. Phys* **37**, 316-338.

22 Guo, Z.H. 1962 The problem of stability and vibration of a circular plate subjected to finite initial deformation. *Arch. Mech. Stos.* **14**, 239-252.

23 Guo, Z.H. 1962 Vibration and stability of a cylinder subject to finite deformation. *Arch. Mech. Stos.* **14**, 757-768.

24 Haughton, D.M. 1982 Wave speeds in rotating elastic cylinders at finite deformation. *Q. J. Mech. Appl. Math.* **35**, 125-139.

25 Haughton, D.M. 1987 Inflation and bifurcation of thick-walled compressible elastic spherical shells. *IMA J. Appl. Math.* **39**, 259-272.

26 Haughton, D.M. 1996 Further results for the eversion of highly compressible elastic cylinders. *Math. Mech. Solids* **1**, 355-367.

27 Haughton, D.M. 1999 Flexure and compression of incompressible elastic plates. *Int. J. Eng. Sci.* **37**, 1693-1708.

28 Haughton, D.M. and Y.-C. Chen 1999 On the eversion of incompressible elastic spherical shells. *ZAMP* **50**, 312-326.

29 Haughton, D.M. and McKay, B.A. 1996 Wrinkling of inflated elastic cylindrical membranes under flexure. *Int. J. Eng. Sci.* **34**, 1531-1550.

30 Haughton, D.M. and Ogden, R.W. 1978 On the incremental equations in nonlinear elasticity. I. Membrane theory. *J. Mech. Phys Solids* **26**, 93-110.

31 Haughton, D.M. and Ogden, R.W. 1978 On the incremental equations in nonlinear elasticity. II. bifurcation of pressurized spherical shells. *J. Mech. Phys Solids* **26**, 111-138.

32 Haughton, D.M. and Ogden, R.W. 1979 Bifurcation of inflated circular cylinders of elastic material under axial loading. I. Membrane theory for thin-walled tubes. *J. Mech. Phys Solids* **27**, 179-212.

33 Haughton, D.M. and Ogden, R.W. 1979 Bifurcation of inflated circular cylinders of elastic material under axial loading. II. Exact theory for thick-walled tubes. *J. Mech. Phys Solids* **27**, 489-512.

34 Haughton, D.M. and Ogden, R.W. 1980 Bifurcation of finitely deformed rotating elastic cylinders. *Q. J. Mech. Appl. Math.* **33**, 251-265.

35 Haughton, D.M. and Ogden, R.W. 1980 Bifurcation of rotating thick-walled elastic tubes. *J. Mech. Phys Solids* **28**, 59-74.

36 Haughton, D.M. and Ogden, R.W. 1980 Bifurcation of rotating circular cylindrical elastic membranes. *Math. Proc. Camb. Phil. Soc.* **87**, 357-376.

37 Haughton, D.M and Orr, A. 1995 On the eversion of incompressible elastic cylinders. *Int. J. Non-Linear Mech.* **30**, 81-95.

38 Haughton, D.M and Orr, A. 1997 On the eversion of compressible elastic cylinders. *Int. J. Solids Structures* **34**, 1893-1914.

39 Hayes, M. 1963 Wave propagation and uniqueness in pre-stressed elastic solids. *Proc. R. Soc. Lond.* **A274**, 500-506.

40 Hill, J.M. 1975 Buckling of long thick-walled circular cylindrical shells of isotropic incompressible hyperelastic materials under uniform external pressure. *J. Mech. Phy. Solids* **23**, 99-112.

41 Hill, J.M. 1976 Critical pressure for the buckling of thick-walled shells under uniform external pressure. *Q. J. Mech. Appl. Math.* **29**, 179-196.

42 Hill, J.M. 1976 Closed form solutions for small deformations superimposed upon the simultaneous inflation and extension of a cylindrical tube. *J. Elasticity* **6**, 113-123.

43 Hill, J.M. 1976 Closed form solutions for small deformations superimposed upon the symmetrical expansion of a spherical shell. *J. Elasticity* **6**, 125-136.

44 Hill, R. 1957 On uniqueness and stability in the theory of finite elastic strain. *J. Mech. Phys Solids* **5**, 229-241.

45 Hill, R. and Hutchinson, J.W. 1975 Bifurcation phenomena in the plane tension test. *J. Mech. Phys Solids* **23**, 239-264.

46 John, F. 1960 Plain strain problems for a perfectly elastic material of harmonic type. *Comm. Pure. Appl. Math.* **13**, 239-296.

47 Kao, B.C. and Pipkin, A.C. 1972 Finite buckling of fibre-reinforced columns. *Acta Mech.* **13**, 265-285.

48 Kearsley, E.A. 1986 Asymmetric stretching of a symmetrically loaded elastic sheet. *Int. J. Solids Structures* **22**, 111-119.

49 Kerr, A.D. and Tang, S. 1967 The instability of a rectangular elastic solid. *Acta Mech.* **4**, 43-63.

50 Kurashige, M. 1979 Instability of a fibre-reinforced elastic slab subjected to axial loads. *J. Appl. Mech.* **46**, 838-843.

51 Kurashige, M. 1981 Instability of a transversely isotropic elastic slab subjected to axial loads. *J. Appl. Mech.* **48**, 351-356.

52 Kurashige, M. 1982 Instability of an axially fibre-reinforced elastic slab under axial loads. *Bull. JSME* **25**, 1873-1875.

53 Kurashige, M. 1983 Instability of an obliquely fibre-reinforced elastic slab under axial loads. *Bull. JSME* **26**, 347-350.

54 Leroy, Y.M. 1993 Spatial patterns and size effects in shear zones: a hyperelastic model with higher-order gradients. *J. Mech. Phys Solids* **41**, 631-663.

55 Levinson, M. 1968 Stability of a compressed neo-Hookean rectangular parallelepiped. *J. Mech. Phy. Solids* **16**, 403-415.

56 Maddocks, J.H. 1984 Stability of nonlinearly elastic rods. *Arch. Rat. Mech. Anal.* **85**, 311-354.

57 Nowinski, J.L. 1969 On the elastic stability of thick columns. *Acta Mech.* **7**, 279-286.

58 Nowinski, J.L. 1969 On the surface instability of an isotropic highly elastic half-space. *Indian J. Math. Mech.* **18**, 1-10.

59 Nowinski, J.L. 1969 Surface instability of a half-space under high two-dimensional compression. *J. Franklin Inst.* **288**, 367-376.

60 Nowinski, J.L. 1969 Instability of a thick nonhomogeneous layer under high initial stress. *J. Appl. Mech.* **36**, 639-659.

61 Nowinski, J.L. 1992 On the internal instability of an infinite three-dimensionally stressed highly elastic medium. *Int. J. Eng. Sci.* **30**, 1315-1321.

62 Nowinski, J.L. and Shahinpoor, M. 1969 Stability of an elastic circular tube of arbitrary wall thickness subjected to an external pressure. *Int. J. Non-Linear Mech.* **4**, 143-158.

63 Ogden, R.W. 1984 *Non-Linear Elastic Deformations.* Ellis Horwood: Chichester.

64 Ogden, R.W. 1984 On non-uniqueness in the traction boundary-value problem for a compressible elastic solid. *Q. Appl. Math.* **42**, 337-344.

65 Ogden, R.W. 1985 Local and global bifurcation phenomena in plane-strain finite elasticity. *Int. J. Solids Structures* **21**, 121-132.

66 Ogden, R.W. and Roxburgh, D.G. 1993 The effect of pre-stress on the vibration and stability of elastic plates. *Int. J. Eng. Sci.* **31**, 1611-1639.

67 Ogden, R.W. and Sotiropoulos, D.A. 1995 On interfacial waves in pre-stressed layered incompressible elastic solids. *Proc. R. Soc. Lond.* **A450**, 319-341.

68 Ogden, R.W. and Sotiropoulos, D.A. 1996 The effect of pre-stress on guided ultrasonic waves between a surface layer and a half-space. *Ultrasonics* **34**, 491-494.

69 Ogden, R.W., Steigmann, D.J. and Haughton, D.M. 1997 The effect of elastic surface coating on the finite deformation and bifurcation of a pressurized circular annulus. *J. Elasticity* **47**, 121-145.

70 Owen, N.C. 1990 Some remarks on the stability of the homogeneous deformations for an elastic bar. *J. Elasticity* **23**, 113-125.

71 Pan, F.X. and Beatty, M.F. 1997 Instability of a bell constrained cylindrical tube under end thrust - Part 1: Theoretical development. *Math. Mech. Solids* **2**, 243-273.

72 Pan, F.X. and Beatty, M.F. 1997 Remarks on the instability of an incompressible and isotropic hyperelastic, thick-walled cylindrical tube. *J. Elasticity* **48**, 217-239.

73 Pan, F.X. and Beatty, M.F. 1999 Instability of a bell constrained cylindrical tube under end thrust - Part 2: Examples, thin tube analysis. *Math. Mech. Solids* **4**, 227-250.

74 Pan, F.X. and Beatty, M.F. 1999 Instability of an internally constrained hyperelastic material. *Int. J. Non-Linear Mech.* **34**, 169-177.

75 Patterson, J.C. 1976 Stability of an elastic thick walled tube under end thrust and external pressure. *Int. J. Non-Linear Mech.* **11**, 385-390.

76 Patterson, J.C. and Hill, J.M. 1977 The stability of a solid rotating neo-Hookean cylinder. *Mech. Res. Comm.* **4**, 69-74.

77 Pence, T.J. and Song, J. 1991 Buckling instabilities in a thick elastic three-ply composite plate under thrust. *Int. J. Solids Structures* **27**, 1809-1828.

78 Qiu, Y., Kim, S. and Pence, T.J. 1994 Plane strain buckling and wrinkling of neo-Hookean laminates. *Int. J. Solids Structures* **31**, 1149-1178.

79 Reddy, B.D. 1982 Surface instabilities on an equibiaxially stretched half-space. *Math. Proc. Camb. Phil. Soc.* **91**, 491-501.

80 Reddy, B.D. 1983 The occurrence of surface instabilities and shear bands in plane-strain deformation of an elastic half-space. *Q. J. Mech. Appl. Math.* **36**, 337-350.

81 Rivlin, R.S. 1948 Large elastic deformations of isotropic materials. II. Some uniqueness theorems for pure, homogeneous, deformations. *Phil. Trans. R. Soc. Lond.* **A240**, 491-508.

82 Rivlin, R.S. 1974 Stability of pure homogeneous deformations of an elastic cube under dead loading. *Q. J. Appl. Math.* **32**, 265-271.

83 Rivlin, R.S. 1977 Some research directions in finite elasticity. *Rheol. Acta* **16**, 101-112.

84 Roxburgh, D.G. and Ogden, R.W. 1994 Stability and vibration of prestressed compressible elastic plates. *Int. J. Eng. Sci.* **32**, 427-454.

85 Sawyers, K.N. 1976 Stability of an elastic cube under dead loading: two equal forces. *Int. J. Non-Linear Mech.* **11**, 11-23.

86 Sawyers, K.N. 1977 Material stability and bifurcation in finite elasticity. In *Finite Elasticity*, Applied Mechanics Symposia Series, vol. 27 (ed. R.S. Rivlin), 103-123, ASME: New York.

87 Sawyers, K.N. and Rivlin, R.S. 1974 Bifurcation conditions for a thick elastic plate under thrust. *Int. J. Solids Structures* **10**, 483-501.

88 Sawyers, K.N. and Rivlin, R.S. 1976 The flexural bifurcation condition for a thin plate under thrust. *Mech. Res. Comm.* **3**, 203-207.

89 Sawyers, K.N. and Rivlin, R.S. 1982 Stability of a thick elastic plate under thrust. *J. Elasticity* **12**, 101-125.

90 Sensenig, C.B. 1964 Instability of thick elastic solids. *Comm. Pure Appl. Math.* **17**, 451-491.

91 Simpson, H.C. and Spector, S.J. 1984 On the barrelling instabilities in finite elasticity. *J. Elasticity* **14**, 103-125.

92 Simpson, H.C. and Spector, S.J. 1984 On the barrelling for a special material in finite elasticity. *Q. Appl. Math.* **42**, 99-111.

93 Song, J. and Pence, T.J. 1992 On the design of three-ply nonlinearly elastic composite plates with optimal resistance to buckling. *Structural Optimization* **5**, 45-54.

94 Steigmann, D.J. and Ogden, R.W. 1997 Plane deformations of elastic solids with intrinsic boundary elasticity. *Proc. R. Soc. Lond.* **A453**, 853-877.

95 Triantafyllidis, N. and Abeyaratne, R. 1983 Instability of a finitely deformed fibre-reinforced elastic material. *J. Appl. Mech.* **50**, 149-156.

96 Usmani, S.A. and Beatty, M.F. 1974 On the surface instability of a highly elastic half-space. *J. Elasticity* **4**, 249-263.

97 Vaughan, H. 1979 Effects of stretch on wave speed in rubberlike materials. *Q. J. Mech. Appl. Math.* **32**, 215-231.

98 Wang, A.S.D. and Ertepinar, A. 1972 Stability and vibrations of elastic thick-walled cylindrical and spherical shells subjected to pressure. *Int. J. Non-Linear Mech.* **7**, 539-555.

99 Wesolowski, Z. 1962 Stability in some cases of tension in the light of finite strain. *Arch. Mech. Stos.* **14**, 875-900.

100 Wesolowski, Z. 1962 Some stability problems in the light of the theory of finite strain. *Bull. Acad. Pol. Sci.* **10**, 123-128.

101 Wesolowski, Z. 1963 The axially symmetric problem of stability loss of an elastic bar subjected to tension. *Arch. Mech. Stos.* **15**, 383-395.

102 Wesolowski, Z. 1964 Stability of a fully elastic sphere uniformly loaded on the surface. *Arch. Mech. Stos.* **16**, 1131-1150.

103 Wesolowski, Z. 1967 Stability of an elastic, thick-walled sphere uniformly loaded by an external pressure. *Arch. Mech. Stos.* **19**, 3-44.

104 Wilkes, E.W. 1955 On the stability of a circular tube under thrust. *Q. J. Mech. Appl. Math.* **8**, 88-100.

105 Wilson, A.J. 1973 Surface and plate waves in biaxially-stressed elastic media. *Pure Appl. Geophys.* **102**, 182-192.

106 Wilson, A.J. 1973 Surface waves in restricted Hadamard materials. *Pure Appl. Geophys.* **110**, 1967-1976.

107 Wilson, A.J. 1977 Surface wave propagation in thin pre-stressed elastic plates. *Int. J. Eng. Sci.* **15**, 245-251.

108 Wu, C.H. 1979 Plane-strain buckling of a crack in a harmonic solid subjected to crack-parallel compression. *J. Appl. Mech.* **46**, 597-604.

109 Wu, C.H. 1980 Plane-strain buckling of cracks in incompressible elastic solids. *J. Elasticity* **10**, 163-177.

110 Wu, C.H. and Cao, G.Z. 1983 Buckling of an axially compressed incompressible half-space. *J. Struct. Mech.* **11**, 37-48.

111 Wu, C.H. and Cao, G.Z. 1984 Buckling problems in finite plane elasticity-harmonic materials. *Q. Appl. Math.* **41**, 461-474.

112 Wu, C.H. and Widera, O.E. 1969 Stability of a thick rubber solid subjected to pressure loads. *Int. J. Solids Strut.* **5**, 1107-1117.

113 Young, N.J.B. 1976 Bifurcation phenomena in the plane compression test. *J. Mech. Phys Solids* **23**, 77-91.

11
Nonlinear dispersive waves in a circular rod composed of a Mooney-Rivlin material

Hui-Hui Dai

Department of Mathematics
City University of Hong Kong, Kowloon, Hong Kong
Email: mahhdai@math.cityu.edu.hk

This chapter studies nonlinear dispersive waves in a Mooney-Rivlin elastic rod. We first derive an approximate one-dimensional rod equation, and then show that traveling wave solutions are determined by a dynamical system of ordinary differential equations. A distinct feature of this dynamical system is that the vector field is discontinuous at a point. The technique of phase planes is used to study this singularity (there is a vertical singular line in the phase plane). By considering the relative positions of equilibrium points, we establish the existence conditions under which a phase plane contains physically acceptable solutions. In total, we find ten types of traveling waves. Some of the waves have certain distinguished features. For instance, we may have solitary cusp waves which are localized with a discontinuity in the shear strain at the wave peak. Analytical expressions for most of these types of traveling waves are obtained and graphical results are presented. The physical existence conditions for these waves are discussed in detail.

11.1 Introduction

Traveling waves in rods have been the subject of many studies. The study of plane flexure waves has formed one focus. See, e.g., Coleman and Dill (1992), and Coleman *et al.* (1995). Another focus is the study of nonlinear axisymmetric waves that propagate axial-radial deformation in circular cylindrical rods composed of a homogeneous isotropic material. This chapter is concerned with the latter aspect for incompressible Mooney-Rivlin materials. We mention in particular three related works by Wright (1982, 1985) and Coleman and Newman (1990). Wright (1982) seems to have been the first to study such a problem and he pointed out the existence of solitary waves and periodic waves. In a subsequent paper treating incompressible rods, Wright (1985) pointed out the existence of a large variety of traveling wave solutions, including traveling shock waves. Coleman and Newman (1990) found that there is only one type of solitary wave of radial expansion and one type of periodic wave for the case of neo-Hookean materials. In addition they gave explicit solution expressions in each case. In Wright's paper (1985), it was mentioned that the method used in Aifantis and Serrin (1980) and Coleman (1983) could be used to establish all the possibilities of bounded smooth solutions, and the construction of traveling shock wave solutions was demonstrated. However, no detailed and conclusive mathematical proofs were given by Wright and the existence conditions were not discussed. Here we explore these aspects for Mooney-Rivlin materials.

The method used in Aifantis and Serrin (1980) and Coleman (1983) applies to smooth solutions. For Mooney-Rivlin materials, traveling shock waves can arise. Thus, that method is not directly applicable for our purpose. Here we employ a different approach based on phase plane techniques. We are able to establish conditions for the existence of physically acceptable solutions as represented by individual paths in the phase plane. We identify three critical values of λ, where $1/\sqrt{\lambda}$ can generally be regarded as the value of the radius at infinity. Depending on these critical values, the solutions have significantly different behaviors. For example, the amplitudes of solitary waves of radial expansion may or may not have a lower bound depending on whether λ is larger or smaller than a critical value λ_L. In contrast with neo-Hookean materials, for which there are only two types of traveling waves (Coleman and Newman 1990), Mooney-Rivlin materials admit ten types of traveling wave solutions.

We also mention three papers by Antman (1973, 1974) and Antman and Liu (1979). In the first two, phase plane techniques are used to study smooth solutions in static problems. In the third paper, traveling waves are studied in the context of a general director rod theory, and a global qualitative analysis is conducted. Here we concentrate on existence conditions for solutions and on both qualitative and quantitative descriptions of traveling waves in a rod composed of a Mooney-Rivlin material.

In Sections 11.2 and 11.3 we derive the associated rod equations for a Mooney-Rivlin material and examine their specialization to traveling wave phenomena. The problem for traveling waves reduces to a single second-order ordinary differential equation. In Section 11.4 we use phase plane methods to initially develop a way to classify the different solutions that are possible. A major complication is the fact that the differential equation contains a singularity, and one must understand how this singularity manifests itself in solutions for waves in a rod. In particular it quickly becomes clear that some of the solutions of the differential equation do not correspond to physically acceptable waves for a rod, and so additional restrictions must be imposed from the physical problem. We explore such restrictions in Section 11.5. Section 11.6 is devoted to an associated linearized problem, the analysis of which is of value in exploring the fully nonlinear problem. Finally, in Sections 11.7–11.14 we examine in detail the different classes of wave motions that are "physically acceptable". In particular we show that, in addition to a qualitative description through phase portraits, it is possible to give an analytical construction of the solution using elliptic integrals. A summary of the results is presented in the concluding section.

11.2 Basic equations

We first consider the motion of a general elastic body. By a motion we mean a time-dependent placement of the body into Euclidean space, made explicit by a point transformation

$$x = x(X, t), \tag{2.1}$$

in which X is the position vector of material points in the reference placement, x is the position vector of the same points in the current placement, and t is the time.

The field equations which govern the dynamics of elastic bodies composed of a homogeneous, isotropic, incompressible Mooney-Rivlin material may be written (cf. Ogden 1984, Truesdell and Noll 1965)

$$F = \operatorname{Grad} x, \quad B = F F^T, \quad \det B = 1; \tag{2.2}$$

$$\operatorname{Div} \pi^T = \rho x_{tt}, \quad \pi = \sigma F^{-T}, \quad t = \sigma n; \tag{2.3}$$

$$\sigma = -pI + 2\mu C_1 B - 2\mu C_2 B^{-1}, \quad C_1 + C_2 = \frac{1}{2}, \quad C_1 > 0, \quad C_2 > 0. \tag{2.4}$$

Equations (2.2)–(2.4) relate to the kinematics, the kinetics, and the constitutive relations of Mooney-Rivlin elastic bodies, respectively. F is the deformation gradient, B is the Cauchy-Green strain tensor; π and σ are the Piola-Kirchhoff and Cauchy stress tensors, respectively. Equation $(2.3)_1$ is the equation of motion in the absence of body forces, while $(2.3)_3$ defines the traction vector t on a body surface element with unit normal n in the current placement. In (2.4) p is the indeterminate pressure, the reaction to the constraint of incompressibility $(2.2)_3$; C_1, C_2, and μ are constitutive constants, the latter being the shear modulus. Finally, ρ is the material density, which, as a consequence of incompressibility, homogeneity, and the principle of mass conservation, is a constant during the motion, that motion being isochoric.

We consider here axisymmetric motions of a rod whose reference placement is a circular cylinder of radius a and infinite length. In this configuration the rod is assumed to be both unloaded and undistorted; hence it is in a homogeneous and isotropic state. The axis of the rod is along the Z-axis of a rectangular Cartesian coordinate system (X, Y, Z). We introduce cylindrical polar coordinates (R, Θ, Z) in the reference placement and (r, θ, z) in the current configuration. If we preclude an overall rigid body motion of the rod, then a general axisymmetric motion is given by the map

$$r = r(R, Z, t), \quad \theta = \Theta, \quad z = z(R, Z, t), \tag{2.5}$$

which we assume to be at least C^1, and where, without loss of generality, the axes of Z and z are taken to be coincident. Clearly, the deformed rod is an axisymmetric body whose axis is always the z-axis. We seek solutions of the form (2.5) for the elastodynamic problem (2.2)–(2.4) when the lateral surface of the rod is traction-free. Thus, the motions must be maintained by forces on the circular cross sectional ends of the rod.

We introduce orthonormal bases associated with the cylindrical polar coordinates and denote these by E_R, E_Θ, E_Z and e_r, e_θ, e_z in the reference and current placements, respectively. We shall restrict consideration to physical components throughout the sequel. From (2.5) and (2.2), we find that

$$F = r_R e_r \otimes E_R + r_Z e_r \otimes E_Z + (r/R) e_\theta \otimes E_\Theta + z_R e_z \otimes E_R + z_Z e_z \otimes E_Z, \tag{2.6}$$

$$B = (r_R^2 + r_Z^2) e_r \otimes e_r + (r^2/R^2) e_\theta \otimes e_\theta + (z_R^2 + z_Z^2) e_z \otimes e_z$$

$$+ (r_R z_R + r_Z z_Z)(e_r \otimes e_z + e_z \otimes e_r), \tag{2.7}$$

while the condition of incompressibility $(2.2)_3$ reduces to

$$r_R z_Z - r_Z z_R = \varepsilon^{-1}, \quad \text{where} \quad \varepsilon = r/R, \tag{2.8}$$

where the subscripts on r and z signify differentiation with respect to the indicated coordinates. After the non-zero physical components of the symmetric Cauchy stress σ are calculated from $(2.4)_1$ and (2.7), the Piola-Kirchhoff stress π can be computed with the aid of $(2.3)_2$ and (2.6). We find that the non-zero stress components are given by

$$\pi_{rR} = 2\mu C_1 r_R - 2\mu C_2 \varepsilon^3 (z_Z^3 + z_R^2 z_Z + r_R r_Z z_R + r_Z^2 z_Z) - \varepsilon z_Z p,$$

$$\pi_{rZ} = 2\mu C_1 r_Z + 2\mu C_2 \varepsilon^3 (z_Z^2 z_R + z_R^3 + r_R^2 z_R + r_Z z_Z r_R) + \varepsilon z_R p,$$

$$\pi_{\theta\Theta} = 2\mu C_1 \varepsilon - 2\mu C_2 \varepsilon^{-3} - \varepsilon^{-1} p, \tag{2.9}$$

$$\pi_{zR} = 2\mu C_1 z_R + 2\mu C_2 \varepsilon^3 (r_Z z_Z^2 + r_R z_R z_Z + r_R^2 r_Z + r_Z^3) + \varepsilon r_Z p,$$

$$\pi_{zZ} = 2\mu C_1 z_Z - 2\mu C_2 \varepsilon^3 (r_Z z_Z z_R + r_Z z_R^2 + r_R^3 + r_Z r_R) - \varepsilon r_R p,$$

where use has been made of the incompressibility condition (2.8).

The stress components in (2.9) must satisfy the equations of motion $(2.3)_1$. For the present problem, they reduce to

$$(\pi_{rR})_R + (\pi_{rZ})_Z + (\pi_{rR} - \pi_{\theta\Theta})/R = \rho r_{tt},$$

$$(\pi_{zR})_R + (\pi_{zZ})_Z + \pi_{zR}/R = \rho z_{tt}. \qquad (2.10)$$

Equations $(2.10)_1$ and $(2.10)_2$ are the component equations of motion along the $\mathbf{e}_r$- and $\mathbf{e}_z$-directions, respectively. The $\mathbf{e}_\theta$-component of the equations reduces to the requirement that p be independent of Θ.

The traction-free boundary conditions on the lateral surface are simply given by

$$\pi_{rR}|_{R=a} = \pi_{ZR}|_{R=a} = 0, \qquad (2.11)$$

where a is the undeformed radius of the rod.

The theory as developed to this point is exact. To proceed further we require some approximations. The intention is to obtain results for thin rods, i.e. when the radius is small compared with the length. As a consequence of this and the axisymmetry, we assume a motion in which plane cross sections remain plane with normal along the rod axis. Such an assumption is standard in the development of rod theories and for the problem considered herein has been used by a number of authors (see, e.g., Wright 1982, 1985, Coleman and Newman 1990). In mathematical terms we write

$$z = \tilde{z}(Z,t). \qquad (2.12)$$

We substitute (2.12) into $(2.8)_1$, integrate with respect to R, and use the condition $r = 0$ at $R = 0$ to obtain

$$\varepsilon = \tilde{\varepsilon}(Z,t), \qquad (2.13)$$

with

$$\tilde{\varepsilon}^2 \lambda = 1 \quad \text{where} \quad \lambda = \tilde{z}_Z(Z,t). \qquad (2.14)$$

We observe that, now, one parameter, either λ or ε, defines the deformation of the rod. The quantities λ, ε, and ε_Z, all functions of (Z,t), are the axial stretch, the radial stretch, and radial stretch gradient, respectively. These quantities define the shear strain which is given explicitly by $B_{rz} = (\varepsilon_Z/\varepsilon^2)R = -(\lambda_Z/2\lambda)R$.

The standard practice in developing a rod theory is to work with stress moments rather than the stress itself. The outcome is a theory with fewer independent variables than the exact theory. In the simplest instance we may work with only averages, i.e. the zeroth moment. Here, we shall replace the balance of momentum in the z-direction, $(2.10)_2$, by the average balance of momentum in the z-direction. We simply integrate $(2.10)_2$ with respect to area over the rod cross section. Using integration by parts and the boundary condition $(2.11)_2$, we obtain

$$\pi a^2 (\bar{\pi}_{zZ})_Z = \rho \int_0^a z_{tt} R dR, \quad \text{where} \quad \bar{\pi}_{zZ} = (\pi a^2)^{-1} \int_0^a \pi_{zZ} R dR. \quad (2.15)$$

$\bar{\pi}_{zZ}$ is the average of the Piola-Kirchhoff stress components π_{zZ}.

Similarly, on averaging $(2.10)_1$ with the use of (2.9), (2.12), (2.13) and (2.14), we obtain

$$p = \mu \bar{\varepsilon} (C_1 \bar{\varepsilon}_{ZZ} - C_2 \bar{\varepsilon} (\bar{\varepsilon}_Z^2 - \bar{\varepsilon} \bar{\varepsilon}_{ZZ}) - \frac{1}{2} \nu_0^{-2} \bar{\varepsilon}_{tt}) R^2 + p_1(Z, t), \quad (2.16)$$

where $p_1(Z, t)$ is an arbitrary function, and

$$\nu_0 = \sqrt{\mu/\rho} \quad (2.17)$$

is the speed of infinitesimal shear waves. Upon using $(2.11)_1$ to find an expression for $p|_R = a$, and hence to determine $p_1(Z, t)$, we obtain

$$p = \mu \bar{\varepsilon} \big\{ (C_1 \bar{\varepsilon}_{ZZ} + C_2 \bar{\varepsilon} (\bar{\varepsilon} \bar{\varepsilon}_{ZZ} - \bar{\varepsilon}_Z^2) \frac{1}{2} \nu_0^{-2} \bar{\varepsilon}_{tt}) (R^2 - a^2)$$

$$+ 2 \bar{\varepsilon} (C_1 - C_2 (\lambda^2 + \bar{\varepsilon}_Z^2 a^2)) \big\}. \quad (2.18)$$

Finally we substitute $(2.9)_5$ and (2.18) into $(2.15)_1$ to obtain

$$2 C_1 \bar{z}_{ZZ} - (2 C_1 \bar{\varepsilon}^4 + 2 C_2 \bar{\varepsilon}^6 - \frac{1}{2} a^2 C_1 \bar{\varepsilon}^3 \bar{\varepsilon}_{ZZ}$$

$$- \frac{1}{2} C_2 (a^2 \bar{\varepsilon}^5 \bar{\varepsilon}_{ZZ} + \bar{\varepsilon}^4 \bar{\varepsilon}_Z^2) + \frac{1}{4} \nu_0^{-2} \bar{\varepsilon}^3 \bar{\varepsilon}_{tt})_Z = \nu_0^{-2} \bar{z}_{tt}. \quad (2.19)$$

Equation (2.19), along with $(2.14)_1$, are the governing one-dimensional equations that we utilize to study the nonlinear dynamics of rods. We note that our governing equations are in agreement with the equations derived by Wright (1985) and by Coleman and Newman (1990) using different approaches. To be more precise, our equations are derivable from the general equations in these papers by specializing to a Mooney-Rivlin material.

11.3 Traveling waves

In this chapter we concentrate on traveling wave solutions, i.e. $\tilde{z}(Z,t)$ and $\tilde{\varepsilon}(Z,t)$ are functions of a single parameter, the *phase* ξ defined by

$$\xi = Z - Vt, \tag{3.1}$$

i.e.

$$\tilde{z}(Z,t) = \tilde{z}(\xi), \qquad \tilde{\varepsilon}(Z,t) = \tilde{\varepsilon}(\xi), \tag{3.2}$$

where V is the speed of propagation of the traveling wave. Substituting (3.1) into (2.19) and integrating once with respect to ξ, we obtain

$$-(\nu - 2C_1)\tilde{\varepsilon}^{-2} - 2C_1\tilde{\varepsilon}^4 - 2C_2\tilde{\varepsilon}^6$$

$$+\frac{1}{2}a^2(C_2\tilde{\varepsilon}^5\tilde{\varepsilon}_{\xi\xi} + C_2\tilde{\varepsilon}^4\tilde{\varepsilon}_\xi^2 - \frac{1}{2}(\nu - 2C_1)\tilde{\varepsilon}^3\tilde{\varepsilon}_{\xi\xi}) = d_1, \tag{3.3}$$

where

$$\nu = V^2/\nu_0^2, \tag{3.4}$$

and d_1 is a constant of integration. Dividing (3.3) by $\tilde{\varepsilon}^3$ and multiplying by $\tilde{\varepsilon}_\xi$, we integrate once more to obtain

$$\frac{1}{8}a^2(2C_2\tilde{\varepsilon}^2 - (\nu - 2C_1))\tilde{\varepsilon}_\xi^2 = \frac{1}{4}(\nu - 2C_1)\tilde{\varepsilon}^{-4} - \frac{1}{2}d_1\tilde{\varepsilon}^{-2} + d_2 + C_1\tilde{\varepsilon}^2 + \frac{1}{2}C_2\tilde{\varepsilon}^4, \tag{3.5}$$

where d_2 is another constant of integration. Using (2.14) we have

$$\frac{1}{8}a^2((\nu - 2C_1)\lambda - 2C_2)\lambda_\xi^2/\lambda^2 = (\nu - 2C_1)\lambda^4 + 2d_1\lambda^3 - 4d_2\lambda^2 - 4C_1\lambda - 2C_2, \tag{3.6}$$

where we have used the notation

$$\lambda = \tilde{z}_Z(Z,t) = \tilde{z}_\xi(\xi). \tag{3.7}$$

Now we introduce the following transformation:

$$\zeta = \frac{\sqrt{8}}{a} \int \lambda d\xi. \tag{3.8}$$

By substituting (3.7) into (3.8), we have

$$\zeta = \frac{\sqrt{8}}{a}\tilde{z}(\xi). \tag{3.9}$$

Apart from a scaling factor, ζ is just the current configuration coordinate z in terms of the phase ξ. Introducing (3.8) into (3.6), we have

$$((\nu - 2C_1)\lambda - 2C_2)\lambda_\zeta^2 = (\nu - 2C_1)\lambda^4 + 2d_1\lambda^3 - 4d_2\lambda^2 - 4C_1\lambda - 2C_2. \tag{3.10}$$

Integration of (3.10) gives an implicit expression for traveling wave solutions.

We note that once $\nu = \rho V^2/\mu, d_1$, and d_2 are given, solutions are uniquely determined from (3.10). But for which values of these parameters the solutions are "physically acceptable" has yet to be determined.

11.4 Phase plane analysis

We analyze (3.10) by constructing phase portraits of the solutions in the (λ, λ_ζ)-plane. The technique of phase portraits allows us to easily interpret the behavior of solutions as well as assist in the portrait constructions. As a preliminary step we introduce the change of variables

$$b_1 = \frac{2C_1}{\nu - 2C_1}, \quad b_2 = \frac{2C_2}{\nu - 2C_1}, \quad D_1 = \frac{2d_1}{\nu - 2C_1}, \quad D_2 = \frac{2d_2}{\nu - 2C_1}. \qquad (4.1)$$

Then (3.10) takes the form

$$\lambda_\zeta^2 = F(\lambda, D_2), \qquad (4.2)$$

where

$$F(\lambda, D_2) = \frac{\lambda^4 + D_1\lambda^3 - 2D_2\lambda^2 - 2b_1\lambda - b_2}{\lambda - b_2}. \qquad (4.3)$$

We have written D_2 explicitly as an argument of F because different curves in the phase plane correspond to different values of D_2. More precisely, the parameters b_1, b_2, and D_1 uniquely determine a portrait, and then D_2 determines the curves in that portrait.

The first-order system that gives rise to (4.2) can be obtained by differentiating with respect to ζ and then eliminating D_2. Let y denote the derivative λ_ζ so that (4.2) becomes

$$y^2 = F(\lambda, D_2). \qquad (4.4)$$

Denote the solution of this equation for D_2 by $D_2 = G(\lambda, y)$. Then differentiating (4.4) with respect to ζ gives the first-order system

$$\lambda_\zeta = y, \quad y_\zeta = \frac{1}{2}F'(\lambda, G(\lambda, y)), \qquad (4.5)$$

where derivatives of F with respect to λ are denoted by a prime. This system shows immediately that equilibria in the phase plane satisfy $y = 0, F'(\lambda, G(\lambda, 0)) = 0$. Equivalently, in view of the definition of G, equilibria are solutions of the simultaneous system

$$F(\lambda, D_2) = 0, \quad F'(\lambda, D_2) = 0. \qquad (4.6)$$

Eliminating D_2 and simplifying, we obtain the polynomial equation

$$2\lambda^4 + D_1\lambda^3 + 2b_1\lambda + 2b_2 = 0. \qquad (4.7)$$

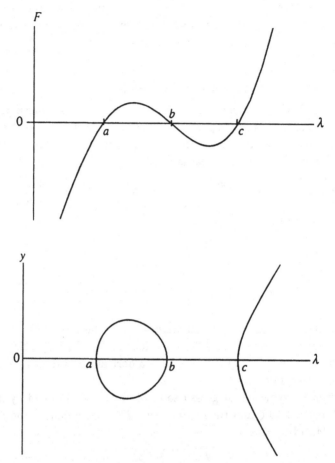

Fig. 1. A graph of F and the corresponding trajectory in the phase space.

The character of each equilibrium can be found by linearization of (4.5). If λ_e is a solution of (4.7), we set $y = Y, \lambda = \lambda_e + \Lambda$, where Y and Λ are small perturbations. Placing these in (4.5) and eliminating all but linear terms we obtain

$$\Lambda_\zeta = Y, \quad Y_\zeta = \frac{1}{2}F''(\lambda_e, D_2)\Lambda, \tag{4.8}$$

where D_2 is the parameter value representing the equilibrium point. It now follows that an equilibrium is a *center* if $F''(\lambda_e, D_2) < 0$ and a *saddle point* if $F''(\lambda_e, D_2) > 0$. In the degenerate case in which $F''(\lambda_e, D_2) = 0$ we obtain a *cusp point*.

The actual curves in the phase plane are given explicitly by (4.4). If we can easily characterize the graph of F, then by taking square roots we obtain solution curves directly. The foregoing discussion draws a connection between the location and nature of equilibria and the form of the graph of F. In particular, equilibrium points which are centers occur when the graph of F exhibits a local maximum (i.e. $F'(\lambda, D_2) = 0, F''(\lambda, D_2) < 0$) at a point on the λ-axis (i.e. $F(\lambda, D_2) = 0$). Similarly, a saddle point occurs when the graph of F exhibits a local minimum on the λ-axis. Apart from the degenerate case of a cusp point, only saddles and centers are possible.

Consider next the general character of curves in the phase plane as determined by (4.4). Each curve consists of two parts, one above the λ-axis given by $y = \sqrt{F(\lambda, D_2)}$ and the other below the λ-axis given by $y = -\sqrt{F(\lambda, D_2)}$. These two parts may be separate curves or they may be one curve if they make contact with the λ-axis. At a point of contact their slope is $dy/d\lambda = y_\zeta/\lambda_\zeta = F'(\lambda, G(\lambda, y))/(2y)$ which, except at an equilibrium point, approaches $\pm\infty$ as $y \to 0$. Thus, curves cross the λ-axis vertically except when the point of contact is an equilibrium. Clearly these curves are symmetric about the λ-axis. Above the λ-axis $y = \lambda_\zeta > 0$ and so $\lambda(\zeta)$ is increasing; below the axis $\lambda(\zeta)$ is decreasing. Finally we note that along a given curve in the phase plane only values of λ such that $F(\lambda, D_2) \geq 0$ are allowed. Consider, for example, the graph of F and the corresponding phase plane curve shown in Figure 1. Since $F < 0$ for $\lambda < a$, this part of the graph of F leads to imaginary values of y. The same is true for $b < \lambda < c$. At the end points of these intervals, i.e. at a, b, and c, the graph of F comes into contact with the λ-axis and hence the associated trajectory in the phase plane crosses the axis vertically (assuming none of these points are equilibria). We then reflect the graphs of $\sqrt{F}$ for $a \leq \lambda \leq b$ and $c \leq \lambda$ below the λ-axis to complete the curve in the phase plane. Note that this single graph of F has produced two solution curves in the phase plane. Indeed, in general a separate solution curve will arise from each portion of F that lies above the λ-axis.

We turn now to the application of the preceding techniques to the function F given by (4.3). We note first that F can cross the λ-axis at most four times, namely, at the roots of

$$\lambda^4 + D_1\lambda^3 - 2D_2\lambda^2 - 2b_1\lambda - b_2 = 0. \tag{4.9}$$

Write F in the alternate form

$$F(\lambda, D_2) = \lambda^3 + (b_2 + D_1)\lambda^2 + (b_2^2 + b_2D_1 - 2D_2)\lambda$$
$$+ (b_2^3 + b_2^2D_1 - 2b_2D_2 - 2b_1) + \frac{A(D_2)}{\lambda - b_2}, \tag{4.10}$$

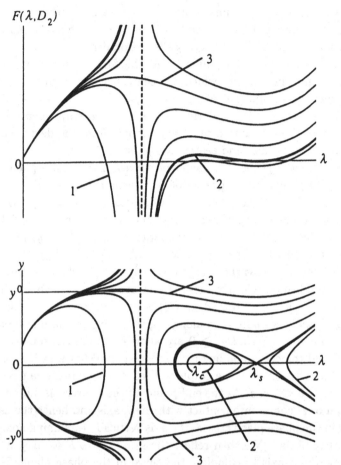

Fig. 2. The graph of $F(\lambda, D_2)$ for different values of D_2 and the associated phase portrait with $b_1 = 19.691, D_1 = -10.524, b_2 = 2, \lambda_c = 3, \lambda_s = 4$.

where

$$A(D_2) = b_2^4 + b_2^3 D_1 - 2b_2^2 D_2 - b_2 - 2b_2 b_1. \qquad (4.11)$$

This shows that $F \to \infty$ as $\lambda \to \infty$ and $F \to -\infty$ as $\lambda \to -\infty$. Also we note that the line $\lambda = b_2$ represents a singularity. If $A(D_2) > 0$, then $F \to \infty$ as $\lambda \to b_2^+$ and $F \to -\infty$ as $\lambda \to b_2^-$. The reverse occurs if $A(D_2) < 0$. A degenerate case occurs when $A(D_2) = 0$. In this case F will approach a finite limit as $\lambda \to b_2$. This gives a *singular* solution and we will see shortly that such solutions require a special interpretation in order to represent actual traveling waves for a rod. In Fig 2 we present the graphs of $F(\lambda, D_2)$ for several values of D_2 and the corresponding trajectories for each in the phase plane. These illustrate the conclusions about F and solution curves outlined previously. For example, curve 1 for F arises when $A(D_2) > 0$ since then $F \to -\infty$ as $\lambda \to b_2^-$. Since we use only that part of the graph of F where $F(\lambda, D_2) \geq 0$, we obtain the solution curve 1 in the phase plane displayed in the lower figure. The top half comes from $y = \sqrt{F}$, the bottom half from $y = -\sqrt{F}$. At the point where the two halves touch we have a vertical crossing.

Curve 2 for F arises when $A(D_2) < 0$. This is a concrete example of the curve shown in Figure 1 and it gives rise, as indicated there, to the two curves labeled 2 in the phase portrait of Figure 2. The branch to the right exhibits unbounded growth in $|y|$ as $\lambda \to \infty$. The branch to the left, in contrast, is a closed loop. It represents a periodic solution and will give rise to traveling waves of periodic type.

Consider finally the singular solution defined by the graph of F represented by curve 3. Since $A(D_2) = 0$ for this curve we see from (4.10) that F no longer exhibits a singularity. It might be assumed, because of this, that the graph of F simply proceeds across the singular line $\lambda = b_2$. However, the singularity is still present in the differential system (4.5) and in general there are no established principles that guide one toward relating trajectories on one side of a singular line to solutions on the other side. Here we will take as our guiding rule the assumption that *a singular solution must represent the limiting case of non-singular neighboring solutions*. This ensures that traveling waves arising from each singular solution will fit continuously and naturally into a class of nonsingular wave patterns. Consider the implications of this for curve 3 of F. Just below this curve we have a 1-parameter family of curves which approach 3 as D_2 decreases towards a value D_2^0 for which $A(D_2^0) = 0$. The corresponding trajectories in the phase plane have the following character near the singular line $\lambda = b_2$. Let y^0 denote the limiting value of y where the solution corresponding to curve 3 intersects the singular line in the phase plane. Then each curve in the family approaches $-y^0$ as ζ increases, then quickly (because $A(D_2)/(\lambda - b_2)$ is very large near $\lambda = b_2$) rises to the axis. Thereafter we have the reverse behavior, namely, with ζ still increasing the trajectory rises quickly from the λ-axis toward y^0 and then moves away from the singular line to the right. As D_2 moves closer to the critical value D_2^0, the rise from $-y^0$ to y^0 is ever more abrupt. This leads us to our interpretation of singular solutions, namely, the singular solution arising from curve 3 of F *never crosses the singular line* $\lambda = b_2$. Rather, as ζ increases the solution $-\sqrt{F(\lambda, D_2^0)}$ approaches the value $y = -y^0$ on the singular line, instantaneously jumps to the value $y = y^0$ on that line, and thereafter moves to the right as $\sqrt{F(\lambda, D_2^0)}$. Singular solutions to the left of the singular line experience a similar jump, this time from y^0

to $-y^0$. In essence, then, singular solutions represent a situation in which the gradient $y = \lambda_\zeta$ experiences a jump and such solutions therefore provide a type of *dynamic phase transition*.

11.5 Physically acceptable solutions

The phase portrait methods outlined previously tell us a great deal about solutions of the differential system (4.2). However, not all curves in the phase plane are interesting for the physical problem at hand, and our main task is one of characterizing those portraits, i.e. those values of b_1, b_2, and D_1 that represent physically meaningful behavior. Let us first note what is not reasonable. In practical problems the radius of the rod must not be arbitrarily large or arbitrarily small. From (2.14) we conclude that λ must not be arbitrarily large or arbitrarily small. The principal stretches must also be finite in order that the rod not break. As a result, $|y| = |\lambda_\zeta|$ must not be arbitrarily large. We are only interested, then, in solutions in the positive λ-plane for which (i) λ does not approach 0 or ∞, and (ii) $|y|$ is finite for all ζ. We take these two conditions as our definition of the class of *physically acceptable solutions*. Our main objective now is to characterize those values of b_1, b_2 and D_1 that give us physically acceptable solutions and then to describe the variety of traveling waves that arise in each case.

The first main result we establish is the following:

Theorem 1 *In order that there be a physically acceptable solution for which $A(D_2) \neq 0$, the phase plane must contain a center in the region $\lambda > 0$.*

Proof. We consider separately the cases in which an acceptable solution lies in one of the two regions $\lambda > b_2$ and $\lambda < b_2$. Suppose first that the region $\lambda > b_2$ contains a physically acceptable solution. Recall that $F \to \infty$ as $\lambda \to \infty$. Consider then what happens to F as λ decreases through the interval $\lambda > b_2$. If F is strictly positive here, it gives rise (by square roots) to a single solution $\lambda(\zeta)$ and on this solution $\lambda \to \infty$. This is unacceptable. Thus, F must cross the λ-axis at least once. Call this point $\lambda = c$. Clearly the solution produced by F for $\lambda \geq c$ is such that $\lambda \to \infty$ as ζ increases and $\lambda \to \infty$ as ζ decreases (see Figure 1). This solution is unacceptable. We conclude that F must cross the axis a second time. Let this point be $\lambda = b > 0$. If F is strictly positive in $b_2 < \lambda < b$ then the solution $\lambda(\zeta)$ arising from this portion of the graph of F is such that either $\lambda \to 0$ as ζ decreases (if $b_2 < 0$) or $y \to \infty$ as ζ decreases (if $b_2 \geq 0$, i.e. $A(D_2)$ has to be positive in this case). In either case we obtain an unacceptable solution. We conclude that F must cross the axis a third time.

Call this point $\lambda = a > 0$. As Figure 1 shows, the graph of F now leads to an acceptable solution in the interval $a \le \lambda \le b$.

Finally we note that F can be written in the form

$$F(\lambda, D_2) = \frac{\lambda^4 + D_1\lambda^3 - 2b_1\lambda - b_2}{\lambda - b_2} - D_2\frac{2\lambda^2}{\lambda - b_2}. \tag{5.1}$$

Thus, F is uniformly decreasing in $\lambda > b_2$ as a function of D_2. In particular, if we increase D_2 sufficiently from the value represented by the situation in Figure 1, the local maximum there will slowly drop and eventually go below the λ-axis. Therefore there is a value of D_2 for which this local maximum lies on the λ-axis. This means the phase plane contains a center.

Consider next the case in which $\lambda < b_2$. In order that a solution in this region be acceptable we certainly need $\lambda > 0$. Thus, it must be the case that $b_2 > 0$. Assume first that $A(D_2) < 0$. This means that $F \to \infty$ as $\lambda \to b_2^-$. Consider then what happens as λ decreases from b_2. By reasoning as in the last case we see that F must fall to a negative local minimum, then rise to a positive local maximum, and then go negative again, all in the region $\lambda > 0$. Thus, F once again resembles Figure 1, and, as before, we conclude that some point λ in the interval $a < \lambda < b$ must be a center.

If instead we have $A(D_2) > 0$, then $F \to \infty$ as $\lambda \to b_2^-$. If F is always negative for $\lambda < b_2$, no solution results. Thus, F must cross the λ-axis at least once as λ decreases. If it remains positive thereafter we obtain a curve in the phase plane which approaches $\lambda = 0$ as ζ decreases. This is unacceptable. Therefore F must again become negative, thus creating a positive local maximum between two positive crossings as in Figure 1. We again conclude that there is a center in this interval. $\qquad\square$

In order that there be physically acceptable solutions, there must then be a center on the positive λ-axis. Let this point occur at $\lambda = \lambda_c$. If we set $F(\lambda_c, D_2) = 0, F'(\lambda_c, D_2) = 0$, we can solve for values of D_1 and D_2 that will produce this center. This give us

$$D_1 = -2\frac{\lambda_c^4 + b_1\lambda_c + b_2}{\lambda_c^3}, \quad D_2 = -\frac{\lambda_c^4 + 4b_1\lambda_c + 3b_2}{2\lambda_c^2}. \tag{5.2}$$

With these as substitutions, (4.7) can now be written in the form

$$(\lambda - \lambda_c)\left[\lambda^3 - \left(\frac{b_2}{\lambda_c^3} + \frac{b_1}{\lambda_c^2}\right)\lambda^2 - \left(\frac{b_2}{\lambda_c^2} + \frac{b_1}{\lambda_c}\right)\lambda - \frac{b_2}{\lambda_c}\right] = 0. \tag{5.3}$$

Other equilibrium points are then given by the roots of

$$\lambda^3 - \left(\frac{b_2}{\lambda_c^3} + \frac{b_1}{\lambda_c^2}\right)\lambda^2 - \left(\frac{b_2}{\lambda_c^2} + \frac{b_1}{\lambda_c}\right)\lambda - \frac{b_2}{\lambda_c} = 0. \tag{5.4}$$

Finally, by computing $F''(\lambda_c, D_2)$ and substituting (5.2), we find that λ_c will be a center if

$$\frac{1}{2}F''(\lambda_c, D_2) = \frac{\lambda_c^4 - 2b_1\lambda_c - 3b_2}{\lambda_c^2(\lambda_c - b_2)} < 0. \qquad (5.5)$$

This leads to a corollary of the main theorem:

Corollary *In order to obtain physically acceptable solutions, we must have* $\nu > 2C_1$.

Proof. If the center lies in the region $\lambda < b_2$, then we must have $b_2 > 0$ since $\lambda_c > 0$. By (4.1) and the constitutive restriction (2.4) we see that $\nu > 2C_1$.

If the center lies in the region $\lambda > b_2$ and $\nu < 2C_1$, we see from (4.1) and (2.4) that $b_1 < 0, b_2 < 0$, and consequently (5.5) is violated. Thus, $\nu > 2C_1$ in this case also. □

We can now resolve the question of other equilibria. Since $\nu > 2C_1$ we see from (4.1) that $b_1 > 0$ and $b_2 > 0$. Consequently the coefficients of λ^2, λ, and the constant term in (5.4) are all negative. We easily conclude that there is a second positive root. Denoting it by λ_s, we can now rewrite (5.3) in the form

$$(\lambda - \lambda_c)(\lambda - \lambda_s)\left[\lambda^2 + \frac{b_2 + \lambda_c^2\lambda_s^2}{\lambda_c\lambda_s(\lambda_c + \lambda_s)}\lambda + \frac{b_2}{\lambda_c\lambda_s}\right] = 0, \qquad (5.6)$$

where we have eliminated b_1 using the fact that λ_s is a root:

$$b_1 = \frac{-b_2(\lambda_c^2 + \lambda_c\lambda_s + \lambda_s^2) + \lambda_c^3\lambda_s^3}{\lambda_c\lambda_s(\lambda_c + \lambda_s)}. \qquad (5.7)$$

Since b_2, λ_c, and λ_s are all positive, the quadratic expression in (5.6) can never vanish for a positive λ. Thus, there are only two equilibria in the positive λ-plane. Moreover, if we make the substitution (5.7) in $F''(\lambda, D_2)/2$ and then evaluate the result at both λ_c and λ_s, we obtain the expressions

$$\frac{1}{2}F''(\lambda_c, D_2) = \frac{(\lambda_c - \lambda_s)(2b_2\lambda_c + b_2\lambda_s + \lambda_c^4\lambda_s + 2\lambda_c^3\lambda_s^2)}{(\lambda_c - b_2)\lambda_c^2\lambda_s(\lambda_c + \lambda_s)}, \qquad (5.8)$$

$$\frac{1}{2}F''(\lambda_s, D_2) = \frac{(\lambda_s - \lambda_c)(b_2\lambda_c + 2b_2\lambda_s + 2\lambda_c^2\lambda_s^3 + \lambda_c\lambda_s^4)}{(\lambda_s - b_2)\lambda_c\lambda_s^2(\lambda_c + \lambda_s)}. \qquad (5.9)$$

From these we obtain the next result:

Theorem 2 *Either* $b_2 < \lambda_c < \lambda_s$ *or* $\lambda_s < \lambda_c < b_2$. *In either case the equilibrium at* λ_s *is a saddle point.*

Proof. We know that λ_c is a center and therefore (5.8) is negative. Since

λ_c, λ_s, and b_2 are positive we see that either $\lambda_c > b_2$ and $\lambda_s > \lambda_c$ or $\lambda_c < b_2$ and $\lambda_s < \lambda_c$. This establishes the two inequalities above. To show that λ_s is a saddle point we must show that (5.9) is positive. This is clearly the case if either $b_2 < \lambda_c < \lambda_s$ or $\lambda_s < \lambda_c < b_2$. □

Two phase portraits are regarded as equivalent if the solution curves they display are qualitatively the same. Therefore, in order to categorize different portraits we must determine those features that lead to qualitatively different solutions. The number, arrangement, and character of equilibria is an important ingredient in determining qualitative change and for this reason Theorem 2 is quite important. Another feature that often affects qualitative change is the character of those solution curves that pass through saddle points. Generally such curves divide the phase plane into regions of qualitatively similar behavior, and so if a solution through a saddle point changes qualitatively, a new portrait must be constructed.

In the system we are considering here there is a single saddle point $(\lambda_s, 0)$. The values D_1 and D_2 that correspond to the solution through this saddle are given by (5.2) provided we replace λ_c there by λ_s. Placing these back in (4.3) and simplifying, we obtain

$$\lambda_\zeta^2 = \frac{(\lambda - \lambda_s)^2 (\lambda - \lambda_n)(\lambda - \lambda_m)}{(\lambda - b_2)}, \qquad (5.10)$$

where

$$\lambda_{n,m} = \left\{ b_2 + b_1 \lambda_s \pm \sqrt{(b_2 + b_1 \lambda_s)^2 + b_2 \lambda_s^4} \right\} / \lambda_s^3 \qquad (5.11)$$

with λ_n corresponding to the "+" sign. It is clear that $\lambda_m < 0$ and therefore is not in the region of interest. For the location of λ_n there are only two options:

Theorem 3 *For physically acceptable solutions either* $0 < \lambda_n < \lambda_c < \lambda_s$ *or* $0 < \lambda_s < \lambda_c < \lambda_n$.

Proof. Consider the case $b_2 < \lambda_c < \lambda_s$. If $\lambda_n \leq b_2$ we are done. Suppose then that $\lambda_n > b_2$. Then for the values of D_1 and D_2 defining the solution through the saddle point, F vanishes at λ_n and λ_s and is positive between them. If we increase D_2, (5.1) shows that F becomes negative at both λ_n and λ_s, and for D_2 sufficiently large F is negative everywhere between λ_n and λ_s. There arises, then, a value of D_2 at which a local maximum of F occurs on the λ-axis between λ_n and λ_s. This point is λ_c, and therefore $\lambda_n < \lambda_c < \lambda_s$. The case in which $\lambda_s < \lambda_c < b_2$ is handled in the same manner. □

The precise location of λ_n relative to λ_c and b_2 plays an important role in distinguishing one phase portrait from another. When $\lambda_n = b_2$ we see from (5.10) that the singularity is absent. By comparison with (4.10) this means $A(D_2) = 0$. This case, then, describes that situation in which the solution curve through the saddle point is also a singular solution. If $b_2 < \lambda_n < \lambda_c < \lambda_s$ then the solution through the saddle point is a closed curve to the right of the singular line (as in Figure 2). If $\lambda_n < b_2 < \lambda_c < \lambda_s$, then by (5.10), the solution through the saddle point satisfies $\lambda_\zeta^2 \to \infty$ as $\lambda \to b_2$. These three cases clearly show distinct qualitative differences and so a separate phase portrait will be required for each. By adding similar considerations when $\lambda_s < \lambda_c < \lambda_n$, we find that six portraits will be required, corresponding to the following cases:

$$b_2 < \lambda_n < \lambda_c < \lambda_s, \quad \lambda_n = b_2 < \lambda_c < \lambda_s,$$

$$\lambda_n < b_2 < \lambda_c < \lambda_s, \quad \lambda_s < \lambda_c < \lambda_n < b_2, \tag{5.12}$$

$$\lambda_s < \lambda_c < \lambda_n = b_2, \quad \lambda_s < \lambda_c < b_2 < \lambda_n.$$

Phase portraits corresponding to these are plotted in the pages to follow. For a solution to be physically acceptable, it must be bounded. The curves which correspond to bounded solutions are those curves in the portraits that are either labeled or lie inside the regions bounded by the labeled curves. For simplicity we omit curves in a phase portrait in any region where there are no physically acceptable solutions.

In the following sections we discuss the individual paths in these portraits which give bounded solutions and examine each in physical terms. We also characterize, in terms of the features of the wave, the restrictions in each case. Specifically we express the restrictions (5.12) in each case in terms of V and λ_s (rather than the parameters D_1 and D_2 that do not have physical meanings), and consequently in terms of the speed of the traveling waves and the special rod radius $a/\sqrt{\lambda_s}$.

Before considering the solutions for the nonlinear problem, we first address some aspects of linear waves that will be useful in the subsequent discussion.

11.6 Linearization

Although we are interested in nonlinear waves, linearization often provides useful information. In this section we derive the linearized equation of motion and extract from it some speculations about nonlinear waves. For this purpose we introduce the notation

$$\tilde{z} = \lambda Z + W(Z, t). \tag{6.1}$$

Then

$$\tilde{z}_Z = \lambda + W_Z(Z,t), \tag{6.2}$$

and from (2.14) we obtain the following expression for the radial stretch:

$$\tilde{\varepsilon} = \frac{1}{\sqrt{\lambda + W_Z}}. \tag{6.3}$$

In the following we assume that $|W_Z|$ is small. Then

$$\tilde{\varepsilon} = \frac{1}{\sqrt{\lambda}}\left(1 - \frac{W_Z}{2\lambda}\right). \tag{6.4}$$

Thus, we are considering a small disturbance in which the stretched rod now has the radius $a/\sqrt{\lambda}$. Substituting (6.1), (6.2) and (6.4) into (2.19) and neglecting nonlinear terms, we have

$$\nu_0^{-2}W_{tt} - \frac{\nu(3b_2 + 2b_1\lambda + b_1\lambda^4)}{(1+b_1)\lambda^4}W_{ZZ} + \frac{1}{8}a^2\frac{\nu(b_2 + b_1\lambda)}{(1+b_1)\lambda^4}W_{ZZZZ}$$

$$-\frac{1}{8}a^2\nu_0^{-2}\frac{1}{\lambda^3}W_{ZZtt} = 0, \tag{6.5}$$

where we have eliminated C_1 and C_2 in favor of b_1 and b_2 using $(4.1)_{1,2}$. Equation (6.5) is the linearized equation of motion. To obtain the dispersion relation, we seek a solution of the form

$$W(Z,t) = Ae^{i(kZ-\omega t)}, \tag{6.6}$$

where A is a constant, k is the wave number, and ω is the wave frequency. Substituting (6.6) into (6.5), we obtain the following dispersion relation

$$\left(\frac{V_p}{\nu_0}\right)^2 = \frac{\nu(b_2 + b_1\lambda)}{(1+b_1)\lambda} - \frac{\nu b_2}{(1+b_1)\lambda^4}\frac{\phi(\lambda)}{1 + a^2k^2/(8\lambda^3)}, \tag{6.7}$$

where V_p is the phase velocity, i.e.

$$V_p = \frac{\omega}{k}, \tag{6.8}$$

and

$$\phi(\lambda) = \lambda^3 - 2(b_1/b_2)\lambda - 3. \tag{6.9}$$

From (6.7) it is easy to see that $(V_p/\nu_0)^2$ is an increasing function with respect to positive k if $\phi(\lambda) > 0$, and therefore

$$\min V_p = V_1 \quad \text{and} \quad \max V_p = V_0, \tag{6.10}$$

where

$$\left(\frac{V_0}{\nu_0}\right)^2 = \frac{\nu(b_2 + b_1\lambda)}{(1+b_1)\lambda} = 2C_1 + \frac{2C_2}{\lambda},$$

$$\left(\frac{V_1}{\nu_0}\right)^2 = \frac{\nu(3b_2 + 2b_1\lambda + b_1\lambda^4)}{(1+b_1)\lambda^4} = 2C_1 + \frac{4C_1}{\lambda^3} + \frac{6C_2}{\lambda^4}. \tag{6.11}$$

In this case we say equation (6.5) is of positive dispersion.

If $\phi(\lambda) = 0$ then

$$V_p = V_0 \tag{6.12}$$

which is independent of k. In this case we say equation (6.5) has no dispersion.

If $\phi(\lambda) < 0$ then $(V_p/\nu_0)^2$ is a decreasing function of k, and therefore

$$\min V_p = V_0 \quad \text{and} \quad \max V_p = V_1. \tag{6.13}$$

In this case we say equation (6.5) has negative dispersion.

$\phi(\lambda)$ has one and only one positive zero λ_L. From the above analysis we see that this value plays an important role for linear waves. Naturally we may expect it to also play some role for nonlinear waves. In later sections we show that $\lambda = \lambda_L$ is actually a critical value determining nonlinear traveling waves of different types.

Let us recall that we must satisfy the restriction $\nu > 2C_1$ for nonlinear waves. If we define the limiting wave speed V_L by

$$\left(\frac{V_L}{\nu_0}\right)^2 = 2C_1, \tag{6.14}$$

then this restriction has the simple form $V > V_L$. As suggested in Section 11.5, in the following sections we phrase some of the restrictions (5.12) in terms of the wave speed V. The inequality $V > V_L$ is the most basic of these.

11.7 Solitary waves of plane expansion

Consider first the portrait in Figure 3 that is defined by $b_2 < \lambda_n < \lambda_c < \lambda_s$ and $\nu > 2C_1$. The solution represented by curve 1 is a homoclinic orbit. On this path λ approaches λ_s as $\zeta \to \pm\infty$ and the minimum value of λ occurs at λ_n. Recall that ζ is actually a scaled current configuration coordinate of the reference configuration phase ξ. Thus, $\lambda(\zeta)$ corresponding to this homoclinic orbit represents a traveling solitary-wave solution for which λ_s is just the value of the axial stretch λ at infinity (or $1/\sqrt{\lambda_s}$ is just the value of the radial stretch at infinity). Moreover, λ_n is simply the minimum value of the axial stretch

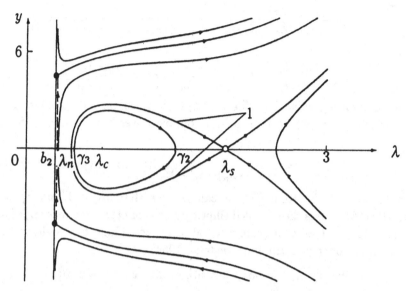

Fig. 3. Phase portrait for the case $b_2 < \lambda_n < \lambda_c < \lambda_s$ with $C_1 = 0.3, C_2 = 0.2, \lambda_s = 2.0, \nu = 2.0, \gamma_3 = 0.476, b_2 = 0.286 \ \lambda_c = 0.763, \lambda_n = 0.446, \gamma_2 = 1.5$

attained at the solitary peak. Since the lateral surface of the rod is described by $r = a/\sqrt{\lambda}$, the wave amplitude h has value

$$h = \frac{a}{\sqrt{\lambda_n}} - \frac{a}{\sqrt{\lambda_s}} > 0, \qquad (7.1)$$

which is positive since $\lambda_s > \lambda_n$. Consequently this is a *solitary wave of radial expansion*.

An explicit expression can be derived for these solitary wave solutions. Recall that the path through λ_s, i.e. the homoclinic orbit given by curve 1, is defined by the equation (5.10). Taking square roots, separating variables, and integrating, we obtain

$$\zeta - \zeta_n = \pm \int_{\lambda_n}^{\lambda} \frac{\sqrt{L - b_2}\,dL}{(\lambda_s - L)\sqrt{(L - \lambda_n)(L - \lambda_m)}}, \qquad (7.2)$$

where ξ_n is the value of ξ at which $\lambda = \lambda_n$. If we use the transformation

$$\tau^2 = \frac{L - \lambda_n}{L - b_2}, \qquad (7.3)$$

(7.2) can be written in the form

$$\zeta - \zeta_n = \pm \frac{2(\lambda_n - b_2)}{(\lambda_s - \lambda_n)\sqrt{\lambda_n - \lambda_m}} \int_0^{\tau_1} \frac{d\tau}{(1 - \beta_1^2 \tau^2)\sqrt{(1 - \tau^2)(1 - m_1^2 \tau^2)}}, \qquad (7.4)$$

where

$$\tau_1^2 = \frac{\lambda - \lambda_n}{\lambda - b_2}, \qquad \beta_1^2 = \frac{\lambda_s - b_2}{\lambda_s - \lambda_n}, \qquad m_1^2 = \frac{b_2 - \lambda_m}{\lambda_n - \lambda_m}. \qquad (7.5)$$

The integral on the right-hand side of (7.4) is the incomplete elliptic integral of the third kind $\Pi(\tau_1; \beta_1^2, m_1)$. Thus, for solitary waves of radial expansion we have

$$\zeta - \zeta_n = \pm \frac{2(\lambda_n - b_2)}{(\lambda_s - \lambda_n)\sqrt{\lambda_n - \lambda_m}} \Pi(\tau_1; \beta_1^2, m_1). \qquad (7.6)$$

Since $\lambda_s > \lambda_n > b_2$, from (7.5) it can be seen that $m_1 < 1$ and $\beta_1 > 1$. Thus, $\Pi(\tau_1; \beta_1^2, m_1)$ is the so-called elliptic integral of *hyperbolic case* (Byrd and Friedman 1954). The incomplete integral of the third kind for the hyperbolic case has the property (Byrd and Friedman 1954)

$$\Pi(\tau_1; \beta_1^2, m_1) \to +\infty \quad \text{logarithmically as} \quad \tau_1^2 \to 1/\beta_1^2. \qquad (7.7)$$

From (7.3), (7.5), and (7.7), it can be seen that ζ tends to infinity logarithmically as $\lambda \to \lambda_s$. Equivalently, as ζ tends to infinity, λ tends to λ_s exponentially. Thus, the profile of solitary waves of radial expansion is exponentially decreasing away from the wave peak.

It is useful now to relate the basic restrictions for this case, $b_2 < \lambda_n < \lambda_c < \lambda_s$, to corresponding restrictions on the wave speed V. Consider first the restriction $\lambda_c < \lambda_s$. Since λ_s and λ_c are both equilibrium values of λ, we can replace λ by λ_s in (5.4). If, in addition, we reverse the roles of λ_s and λ_c, we obtain

$$\lambda_c^3 - \left(\frac{b_2}{\lambda_s^3} + \frac{b_1}{\lambda_s^2}\right)\lambda_c^2 - \left(\frac{b_2}{\lambda_s^2} + \frac{b_1}{\lambda_s}\right)\lambda_c - \frac{b_2}{\lambda_s} = 0. \qquad (7.8)$$

This can be written in the equivalent form

$$(\lambda_c - \lambda_s)\left[\lambda_c^2 + \left(\frac{2b_2}{\lambda_s^3} + \frac{b_1}{\lambda_s^2}\right)\lambda_c + \frac{b_2}{\lambda_s^2}\right] + \left(\lambda_s - \frac{3b_2}{\lambda_s^3} - \frac{2b_1}{\lambda_s^2}\right)\lambda_c^2 = 0. \qquad (7.9)$$

We conclude that $\lambda_c < \lambda_s$ if and only if

$$\lambda_s - \frac{3b_2}{\lambda_s^3} - \frac{2b_1}{\lambda_s^2} > 0. \qquad (7.10)$$

With a little rearrangement this takes the form

$$1 > \frac{3b_2 + 2b_1\lambda_s + b_1\lambda_s^4}{(1 + b_1)\lambda_s^4}. \qquad (7.11)$$

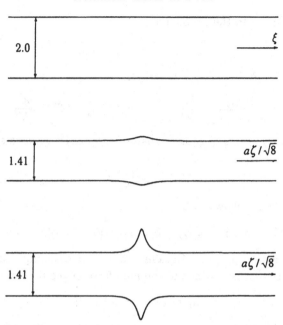

Fig. 4. Profiles of solitary waves of radial expansion with $C_1 = 0.3, C_2 = 0.2, a = 1.0$ $\lambda_s = 2.0(< \lambda_L = 2.104)$. From top to bottom: reference configuration (in terms of the phase ξ); wave profile for $\nu = 0.9, h = 0.0965$; wave profile for $\nu = 2.0, h = 0.790$.

Multiplying both sides by ν, we conclude from (3.4) and (6.11)$_2$ (with λ replaced by λ_s) that this last inequality is equivalent to

$$V > V_1. \tag{7.12}$$

This result might have been anticipated from a more careful analysis of the linear wave case. Since $V_1 > V_L$ (cf. (6.11)$_2$ and (6.14)), it follows that the basic restriction $V > V_L$ is automatically satisfied.

It is not difficult to show from general phase plane considerations that a homoclinic orbit must always contain an equilibrium point. In our case this must be λ_c, so we must have $\lambda_n < \lambda_c < \lambda_s$ if (7.12) is satisfied. It only remains, then, to consider $b_2 < \lambda_n$. By (5.11)$_1$ this is equivalent to

$$b_2\lambda_s^3 - b_2 - b_1\lambda_s < \sqrt{(b_2 + b_1\lambda_s)^2 + b_2\lambda_s^4}. \tag{7.13}$$

Squaring and simplifying we obtain the alternate expression

$$1 > \frac{b_2\lambda_s^3 - 2b_2 - b_1\lambda_s}{\lambda_s(1 + b_1)}. \tag{7.14}$$

This can be written in the simple form

$$V > V_2 \qquad (7.15)$$

if we define the new wave speed V_2 by

$$\left(\frac{V_2}{\nu_0}\right)^2 = \frac{\nu(b_2\lambda_s^3 - 2b_2 - b_1\lambda_s)}{\lambda_s(1 + b_1)} = 2C_2\lambda_s^2 - 2C_1 - \frac{4C_2}{\lambda_s}. \qquad (7.16)$$

Thus, for solitary waves of radial expansion to exist the wave speed V must satisfy

$$V > \max(V_1, V_2). \qquad (7.17)$$

A simple calculation shows that

$$V_1 \geq V_2 \quad \text{if} \quad \phi(\lambda_s) \leq 0, \quad V_1 < V_2 \quad \text{if} \quad \phi(\lambda_s) > 0, \qquad (7.18)$$

where $\phi(\lambda)$ is defined by (6.9). Recalling that λ_L is the unique positive zero of ϕ, our previous conclusion leads to the final form of the restriction:

$$V > V_1 \quad \text{if} \quad \lambda_s \leq \lambda_L, \quad V > V_2 \quad \text{if} \quad \lambda_s > \lambda_L. \qquad (7.19)$$

Condition (7.15) clearly arises due to the presence in this problem of the singular line at $\lambda = b_2$. This means that when $\lambda_s \leq \lambda_L$ the singular line does not play a role. A more refined conclusion arises by looking more closely at condition (7.19). When $\lambda_s \leq \lambda_L$ there are solitary waves whose speed of propagation V can be arbitrarily close to V_1. This means that λ_c and λ_n can be arbitrarily close to λ_s. Since the wave amplitude is given by (7.1), this means that there are solitary waves of arbitrarily small amplitude. When $\lambda_s > \lambda_L$ we need $V > V_2$. Since V_2 is automatically greater than V_1 in this case, λ_n cannot be arbitrarily close to λ_s. Indeed, from $(5.11)_1$ we see that

$$\lambda_n < \lambda_n^* = \frac{b_2^* + b_1^*\lambda_s + \sqrt{(b_2^* + b_1^*\lambda_s)^2 + b_2^*\lambda_s^4}}{\lambda_s^3} = b_2^*, \qquad (7.20)$$

where

$$b_1^* = \frac{2C_1}{(V_2/\nu_0)^2 - 2C_1}, \quad b_2^* = \frac{2C_2}{(V_2/\nu_0)^2 - 2C_1}. \qquad (7.21)$$

This means that the wave amplitude h satisfies

$$h > h^* = \frac{a}{\sqrt{b_2^*}} - \frac{a}{\sqrt{\lambda_s}}. \qquad (7.22)$$

Thus, for a given $\lambda_s > \lambda_L$ the amplitude of solitary waves must be larger than h^*, which has a finite value. This is in contrast to the case $\lambda_s \leq \lambda_L$ where the amplitude could be arbitrarily small. Since $a/\sqrt{\lambda_s}$ represents the radius of the rod at infinity, the preceding analysis shows that if at infinity the rod is thinner than $a/\sqrt{\lambda_L}$, then the amplitude of the solitary wave must be larger than h^*. This phenomenon appears to be new.

In Figure 4 we have plotted some profiles of solitary waves of radial expansion. The parameters are chosen such that $\lambda_s < \lambda_L$. For this case the amplitude of the solitary wave could be arbitrarily small. The amplitude of the wave profile in the middle figure has a very small value of 0.096 compared to the radius which has value 1. The wave profile in the bottom of Figure 4 corresponds to curve 1 in Figure 3. The profiles of solitary waves when $\lambda_s > \lambda_L$ are similar to those in Figure 4 except that the amplitude of the solitary wave cannot be arbitrarily small.

11.8 Solitary cusp waves of radial expansion

Figure 5 is the phase portrait for the case $b_2 = \lambda_n < \lambda_c < \lambda_s$. We consider here the solution represented by curves 2_μ and 2_l. Curve 2_l begins at $\zeta = -\infty$ with value $\lambda = \lambda_s$. As ζ increases along 2_l, $\lambda(\zeta)$ is a purely decreasing function approaching the singular line $\lambda = b_2$ at some value G_1. Similarly, curve 2_u ends at $\zeta = \infty$ with value $\lambda = \lambda_s$, and as ζ increases along 2_u, $\lambda(\zeta)$ is a purely increasing function which as ζ decreases approaches the singular line $\lambda = b_2$ at some value F_1. From our discussion in Section 11.4 this means that for the solution we are considering here, λ begins at $-\infty$ at λ_s, decreases steadily to the value $y = \lambda_\zeta = G_1 < 0$, "instantaneously" jumps to the value $y = \lambda_\zeta = F_1 > 0$, and thereafter λ steadily increases and approaches λ_s at ∞. The jump in λ_ζ has the effect of causing the profile of the solitary wave to exhibit a corner or cusp. Since the amplitude of the wave in this case is

$$h = \frac{a}{\sqrt{b_2}} - \frac{a}{\sqrt{\lambda_s}} > 0, \tag{8.1}$$

we refer to it as a *solitary cusp wave of radial expansion*.

From an analytic perspective, the two curves 2_u and 2_l both pass through the saddle point and therefore are solutions of (5.10). Since $\lambda_n = b_2$ in this case, (5.10) has the simpler form

$$\lambda_\zeta^2 = (\lambda - \lambda_s)^2(\lambda - \lambda_m), \quad b_2 < \lambda \leq \lambda_s. \tag{8.2}$$

To obtain 2_l we take the negative square root here, separate variables, and integrate. If we let ζ_1 denote the coordinate such that $\lambda(\zeta_1) = b_2$, then we have

$$\zeta - \zeta_1 = -\int_{b_2}^{\lambda} \frac{dL}{(\lambda_s - L)\sqrt{L - \lambda_m}}, \zeta \leq \zeta_1. \tag{8.3}$$

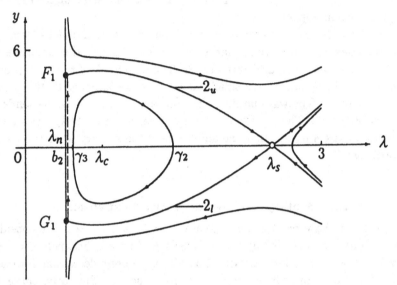

Fig. 5. Phase portrait for the case $b_2 = \lambda_n < \lambda_c < \lambda_s$ with $C_1 = 0.3, C_2 = 0.2, \lambda_s = 2.5, \nu = 1.58, \gamma_3 = 0.479, b_2 = 0.408, \lambda_c = 0.785, \gamma_2 = 1.5$.

Similarly, curve 2_u is given by

$$\zeta - \zeta_1 = \int_{b_2}^{\lambda} \frac{dL}{(\lambda_s - L)\sqrt{L - \lambda_m}}, \zeta \geq \zeta_1. \qquad (8.4)$$

These two integrals can be evaluated explicitly in the form

$$\lambda = \lambda_m + \tanh^2\left(\frac{1}{4}\sqrt{\lambda_s - \lambda_m}|\zeta - \zeta_1| + \frac{1}{2}k_0\right)(\lambda_s - \lambda_m), \qquad (8.5)$$

where

$$k_0 = \ln \frac{\sqrt{\lambda_s - \lambda_m} + \sqrt{\lambda_n - \lambda_m}}{\sqrt{\lambda_s - \lambda_m} + \sqrt{\lambda_n - \lambda_m}}. \qquad (8.6)$$

In Figure 6 we have plotted the profiles of two solitary cusp waves of radial expansion. The lower one corresponds exactly to curves 2_u and 2_l in Figure 5.

Consider finally the restrictions $b_2 = \lambda_n < \lambda_c < \lambda_s$. We have seen previously that $\lambda_n < \lambda_c < \lambda_s$ is equivalent to $V > V_1$. Hence automatically $V > V_L$. The discussion in Section 11.7 has also shown that $b_2 = \lambda_n$ is equivalent to $V = V_2$, so we also need $V_2 > V_1$. This is equivalent to $\lambda_L < \lambda_s$, where λ_L is the critical value of λ at which V_1 and V_2 agree. Note that the solitary waves in this case all propagate with the fixed velocity V_2.

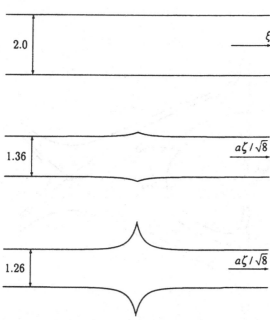

Fig. 6. Profiles of solitary cusp waves of radial expansion with $C_1 = 0.3, C_2 = 0.2, a = 1.0$. From top to bottom: reference configuration (in terms of the phase ξ); wave profile for $\lambda_s = 2.15, \nu = 0.877, h = 0.150$; wave profile for $\lambda_s = 2.5, \nu = 1.58, h = 0.933$.

11.9 Periodic cusp waves of radial expansion

Figure 7 is the phase portrait for the case $\lambda_n < b_2 < \lambda_c < \lambda_s$. Since $\lambda_n < b_2$ the trajectory through the saddle point must go to $\pm\infty$ as λ approaches the singular line. In addition, since $b_2 < \lambda_c$ there will be periodic orbits to the right of the singular line and centered on λ_c. In order to have a smooth transition from these periodic orbits to the trajectory through λ_s, there must be a trajectory, denoted in Figure 7 as curve 3, that makes "contact" with the singular line. Moreover, the place $\lambda = \gamma_2^*$ where this curve crosses the λ-axis lies between λ_c and λ_s. We consider in this case the solution represented by curve 3. Geometrically the behavior of this solution can be deduced from the portrait and our discussion in Section 11.4. Beginning at $\lambda = \gamma_2^*$ for a given ζ, $\lambda(\zeta)$ decreases until the value G_2 at the singular line is reached. Then there is an instantaneous jump in $y = \lambda_\zeta$ from G_2 to F_2. This causes a corner or cusp to appear in the wave profile. Thereafter $\lambda(\zeta)$ increases until the point $\lambda = \gamma_2^*$ is reached again. Since this point is not an equilibrium, the behavior just described repeats itself periodically. We refer to these waves, then, as *periodic cusp waves of radial expansion*.

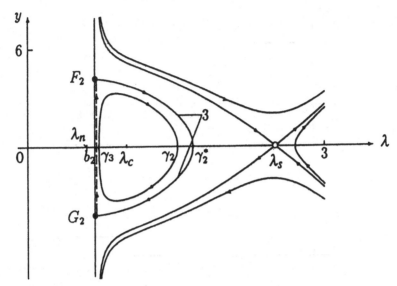

Fig. 7. Phase portrait for the case $\lambda_n < b_2 < \lambda_c < \lambda_s$ with $C_1 = 0.3, C_2 = 0.2, \lambda_s = 2.5$, $\nu = 1.2, \gamma_3 = 0.706, b_2 = 0.667, \lambda_c = 0.988, \lambda_n = 0.587, \gamma_2 = 1.5, \gamma_2^* = 1.655$.

Let us specialize the differential equation (4.2) to this trajectory. First we note that the phase plane must contain a saddle point at λ_s. This value, then, must satisfy the equation for equilibria (4.7). We use this fact to solve for D_1:

$$D_1 = -2\lambda_s - \frac{2b_1}{\lambda_s^2} - \frac{2b_2}{\lambda_s^3}. \tag{9.1}$$

Using this to eliminate D_1 from (4.2), we obtain the differential equation

$$(\lambda - b_2)\lambda_\zeta^2 = \lambda^4 - 2\lambda^3(\lambda_s + \frac{b_1}{\lambda_s^2} + \frac{b_2}{\lambda_s^3}) - 2D_2\lambda^2 - 2b_1\lambda - b_2. \tag{9.2}$$

We also know that $\lambda = b_2$ lies on curve 3 and therefore the right-hand side of (9.2) must vanish at b_2. Imposing this condition and solving for D_2 we obtain

$$D_2 = -\frac{2b_2^2}{\lambda_s^3} - \frac{b_1 b_2}{\lambda_s^2} - \frac{1}{2b_2} - \frac{b_1}{b_2} + \frac{b_2^2}{2} - b_2\lambda_s. \tag{9.3}$$

Substituting this in (9.2) gives us the reduced problem

$$\lambda_\zeta^2 = \lambda^3 + \lambda^2\left(b_2 - \frac{2b_2}{\lambda_s^3} - \frac{2b_1}{\lambda_s^2} - 2\lambda_s\right) + \lambda\frac{1 + 2b_1}{b_2} + 1. \tag{9.4}$$

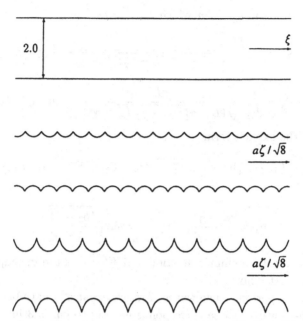

Fig. 8. Profiles of periodic cusp waves of radial expansion with $C_1 = 0.3, C_2 = 0.2, a = 1.0$. From top to bottom: reference configuration (in terms of the phase ξ); wave profile for $\lambda_s = 2.5, \nu = 1.0, T_1 = 0.578, \gamma_2^* = 1.502, \gamma_1^* = 3.307$; wave profile for $\lambda_s = 2.5, \nu = 1.2, T_1 = 0.835, \gamma_2^* = 1.655, \gamma_1^* = 3.268$.

By our previous reasoning we know that the cubic on the right-hand side here vanishes at a point γ_2^* satisfying $\lambda_c < \gamma_2^* < \lambda_s$. In addition, the cubic approaches $-\infty$ as $\lambda \to -\infty$ and has the value 1 at $\lambda = 0$. Therefore it has a negative zero which we call γ_4^*. Finally we note that the value of the cubic when $\lambda = \lambda_s$, after some manipulation, is

$$\frac{(1+b_1)(\lambda_s - b_2)}{b_2}\left(1 - \frac{b_2\lambda_s^3 - 2b_2 - b_1\lambda_s}{\lambda_s(1+b_1)}\right) \tag{9.5}$$

and since $\lambda_s > b_2$ and $b_2 > \lambda_n$ (or its equivalent, namely, the reverse of (7.14)), this value is negative. Since the cubic approaches ∞ as $\lambda \to \infty$, there is a third zero of the cubic at a point γ_1^* such that $\lambda_s < \gamma_1^*$. The product of the three roots must equal the negative of the constant term of the cubic, so we could write $\gamma_4^* = -1/(\gamma_1^*\gamma_2^*)$. In any event, the differential equation (9.4) can now be written in the form

$$\lambda_\zeta^2 = (\lambda - \gamma_1^*)(\lambda - \gamma_2^*)(\lambda - \gamma_4^*), \quad b_2 < \lambda < \gamma_2^*. \tag{9.6}$$

If we denote by ζ_3 the value for which $\lambda(\zeta_3) = b_2$, then

$$\zeta - \zeta_3 = -\int_{b_2}^{\lambda} \frac{dL}{\sqrt{(L - \gamma_1^*)(L - \gamma_2^*)(L - \gamma_4^*)}} \quad \text{if} \quad \zeta \le \zeta_3,$$

$$\zeta - \zeta_3 = \int_{b_2}^{\lambda} \frac{dL}{\sqrt{(L - \gamma_1^*)(L - \gamma_2^*)(L - \gamma_4^*)}} \quad \text{if} \quad \zeta \ge \zeta_3. \qquad (9.7)$$

This leads to the explicit form

$$\lambda = \gamma_2^* - (\gamma_2^* - \gamma_4^*)\text{cn}^2 \left(\frac{1}{4}\sqrt{\gamma_1^* - \gamma_4^*}|\zeta - \zeta_3| + F(\psi_0, n_1); n_1)\right), \qquad (9.8)$$

where

$$n_1^2 = \frac{\gamma_2^* - \gamma_4^*}{\gamma_1^* - \gamma_2^*}, \quad \psi_0 = \arcsin\sqrt{\frac{b_2 - \gamma_4^*}{\gamma_2^* - \gamma_4^*}}, \qquad (9.9)$$

$\text{cn}(\cdot\,;\,\cdot)$ is the Jacobian elliptic function and $F(\cdot\,;\,\cdot)$ is the incomplete elliptic integral of the first kind.

Two profiles of periodic cusp waves of radial expansion have been plotted in Figure 8. The lower of these corresponds exactly to curve 3 in Figure 7. In general the amplitude of these waves is given by

$$h = \frac{a}{\sqrt{b_2}} - \frac{a}{\sqrt{\gamma_2^*}}. \qquad (9.10)$$

In addition, the period of the waves is given by

$$T_1 = 2\int_{b_2}^{\gamma_2^*} \frac{dL}{\sqrt{(L - \gamma_1^*)(L - \gamma_2^*)(L - \gamma_4^*)}} = \frac{4}{\sqrt{\gamma_4^* - \gamma_4^*}}F(\phi_0, n_1), \qquad (9.11)$$

where

$$\phi_0 = \arcsin\sqrt{\frac{(\gamma_1^* - \gamma_4^*)(\gamma_2^* - b_2)}{(\gamma_2^* - \gamma_4^*)(\gamma_1^* - b_2)}}. \qquad (9.12)$$

Finally let us summarize, in terms of V and λ_s, the restrictions that define this case. In Section 11.7 we saw that $\lambda_c < \lambda_s$ was equivalent to $V > V_1$. Therefore automatically $V > V_L$. From the calculations in that section we also know that $\lambda_n < b_2$ was equivalent to $V < V_2$. We are left then to consider the inequality $b_2 < \lambda_c$. Recall that λ_c satisfies (7.8), which we can rewrite in the form

$$(\lambda_c - b_2)\left[\lambda_c^2 + \left(\frac{1}{\lambda_s b_2} + \frac{1}{\lambda_s^2} + \frac{b_1}{b_2\lambda_s}\right)\lambda_c + \frac{1}{\lambda_s}\right]$$

$$-\left[-b_2 + \frac{1}{\lambda_s b_2} + \frac{1}{\lambda_s^2} + \frac{b_1}{b_2\lambda_s} + \frac{b_2}{\lambda_s^3} + \frac{b_1}{\lambda_s^2}\lambda_s^2\right]\lambda_c^2 = 0. \qquad (9.13)$$

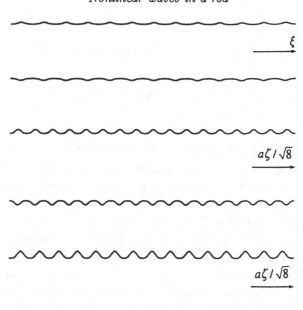

Fig. 9. Profiles of periodic waves of type I with $C_1 = 0.3, C_2 = 0.2, a = 1.0$. From top to bottom: wave profile for $\lambda_s = 2.0, \nu = 1.1, \gamma_3 = 1.2, T_3 = 0.774, \gamma_1 = 2.441, \gamma_2 = 1.359$; wave profile for $\lambda_s = 2.5, \nu = 1.58, \gamma_3 = 0.7, T_3 = 0.590, \gamma_1 = 3.835, \gamma_2 = 0.885$; wave profile for $\lambda_s = 2.5, \nu = 1.2, \gamma_3 = 0.8, T_3 = 0.657, \gamma_1 = 3.538, \gamma_2 = 1.255$.

Thus, $b_2 < \lambda_c$ if and only if

$$-b_2 + \frac{1}{\lambda_s b_2} + \frac{1}{\lambda_s^2} + \frac{b_1}{b_2 \lambda_s} + \frac{b_2}{\lambda_s^3} + \frac{b_1}{\lambda_s^2} > 0. \qquad (9.14)$$

We can write this equivalently in the form $V > V_3$ if we define V_3 to be the velocity

$$\left(\frac{V_3}{\nu_0}\right)^2 = \frac{b_1 \lambda_s - b_2 + \sqrt{(b_1 \lambda_s - b_2)^2 - 4 b_2^2 (1 - \lambda_s^3)}}{2(1 + b_1)\lambda_s} \nu,$$

$$= C_1 - \frac{C_2}{\lambda_s} + \sqrt{\left(C_1 - \frac{C_2}{\lambda_s}\right)^2 - 4 \frac{C_2^2}{\lambda_s^2}(1 - \lambda_s^3)}. \qquad (9.15)$$

Further calculations show that

$$V_2 > V_3 > V_1 \quad \text{if} \quad \lambda_s > \lambda_L,$$

$$V_2 < V_3 < V_1 \quad \text{if} \quad \lambda_s < \lambda_L, \qquad (9.16)$$

$$V_2 = V_3 = V_1 \quad \text{if} \quad \lambda_s = \lambda_L.$$

We conclude that the basic restrictions in this case, $\lambda_n < b_2 < \lambda_c < \lambda_s$, are equivalent to

$$V_2 > V > V_3 \quad \text{and} \quad \lambda_s > \lambda_L. \tag{9.17}$$

11.10 Periodic waves of type I

Within curve 1 of Figure 3, curve 2 of Figure 5 and curve 3 of Figure 7, all trajectories are closed and hence represent periodic wave solutions. We call them *periodic waves of type I* to distinguish them from another class of periodic waves we will encounter later. In each figure there is a one-parameter family of periodic solutions, and we use the minimum value of the axial stretch, denoted $\lambda = \gamma_3$, to parametrize these. In general trajectories satisfy the differential equation (9.2). In the current case γ_3 must be a zero of the right-hand side. Imposing this condition and using it to eliminate D_2, we find that periodic solutions satisfy the equation

$$(\lambda - b_2)\lambda_\zeta^2 = (\lambda - \gamma_3)\left(\lambda^3 - (2\lambda_s + \frac{2b_1}{\lambda_s^2} + \frac{2b_2}{\lambda_s^3} - \gamma_3)\lambda^2 + (\frac{2b_1}{\gamma_3} + \frac{b_2}{\gamma_3^2})\lambda + \frac{b_2}{\gamma_3}\right). \tag{10.1}$$

The closed orbit of interest also crosses the λ-axis at a second point, which we denote as $\lambda = \gamma_2$. Note that γ_2 is the maximum value of the axial stretch. Necessarily $\gamma_3 < \lambda_c < \gamma_2 < \lambda_s$. This implies that γ_2 is a zero of the cubic equation

$$\lambda^3 - (2\lambda_s + \frac{2b_1}{\lambda_s^2} + \frac{2b_2}{\lambda_s^3} - \gamma_3)\lambda^2 + (\frac{2b_1}{\gamma_3} + \frac{b_2}{\gamma_3^2})\lambda + \frac{b_2}{\gamma_3} = 0. \tag{10.2}$$

We also note that this cubic approaches $-\infty$ as $\lambda \to -\infty$ and at $\lambda = 0$ has positive value b_2/γ_3. Therefore the cubic equation has a negative root at, say, γ_4. Finally, we can show that the cubic in (10.2) has a negative value at λ_s and we easily see that it approaches ∞ as $\lambda \to \infty$. Therefore there is another zero γ_1 that is larger than λ_s. We conclude that (10.1) can be written in the simpler form

$$\lambda_\zeta^2 = \frac{(\lambda - \gamma_1)(\lambda - \gamma_2)(\lambda - \gamma_3)(\lambda - \gamma_4)}{\lambda - b_2}. \tag{10.3}$$

If ζ_3 is the coordinate where $\lambda(\zeta_3) = \gamma_3$, then the solution is

$$\zeta - \zeta_3 = \pm \int_{\gamma_3}^{\lambda} \frac{\sqrt{L - b_2}\,dL}{\sqrt{(L - \gamma_1)(L - \gamma_2)(L - \gamma_3)(L - \gamma_4)}}, \gamma_3 \leq \lambda \leq \gamma_2. \tag{10.4}$$

The period of the waves is given by

$$T_3 = 2 \int_{\gamma_3}^{\gamma_2} \frac{\sqrt{L - b_2} dL}{\sqrt{(L - \gamma_1)(L - \gamma_2)(L - \gamma_3)(L - \gamma_4)}}, \tag{10.5}$$

and their amplitude is

$$h = \frac{a}{\sqrt{\gamma_3}} - \frac{a}{\sqrt{\gamma_2}}. \tag{10.6}$$

We can read off the restrictions on V and λ_s for periodic waves of type I from the calculations in Sections 11.7–11.9. Specifically we need

$$V > V_1 \quad \text{if} \quad \lambda_s \leq \lambda_L, \quad V > V_3 \quad \text{if} \quad \lambda_s > \lambda_L. \tag{10.7}$$

(And these ensure that $V > V_L$.) There are also restrictions on γ_3, depending on which figure we are considering. These restrictions can be deduced directly from the figures:

$$\lambda_n < \gamma_3 < \lambda_c \qquad \text{in Figure 3,}$$

$$b_2 = \lambda_n < \gamma_3 < \lambda_c \qquad \text{in Figure 5,} \tag{10.8}$$

$$b_2 < \gamma_3 < \lambda_c \qquad \text{in Figure 7.}$$

In Figure 9 three periodic wave profiles are plotted. The parameters have been chosen such that from top to bottom the profiles respectively satisfy the restrictions in (10.8).

It is important to note the variety of periodic waves that can arise from these three figures. For example, if γ_3 is very close to the singular value b_2, then the waves will have very narrow wave peaks due to the sudden change that occurs in λ_ζ. In Figure 5 there will be waves with narrow wave peaks, but in addition these waves must have very long periods since they are converging as $\gamma_3 \to b_2$ to a solitary wave.

11.11 Solitary waves of radial contraction

Consider now curve 4 in Figure 10. Much of the analysis of this solution mimics the discussion in Section 11.7, so we will present only the main results. Curve 4 represents a solitary wave with axial stretch λ_s, at infinity. λ_n is now the maximum value attained at the wave peak. Since $\lambda_n > \lambda_s$, the wave represents a *solitary wave of radial contraction*.

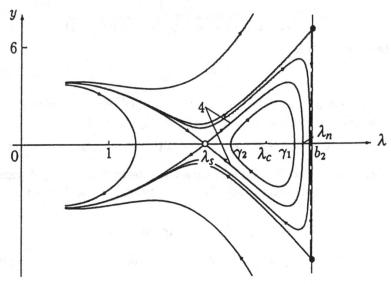

Fig. 10. Phase portrait for the case $b_2 > \lambda_n > \lambda_c > \lambda_s$ with $C_1 = 0.3, C_2 = 0.2, \lambda_s = 2.1, \nu = 0.72, \gamma_1 = 3.129, b_2 = 3.333, \lambda_c = 2.811, \lambda_n = 3.222, \gamma_2 = 2.4.$

Since curve 4 is the trajectory through the saddle point, (5.10) still applies. This gives the solution

$$\zeta - \zeta_n = \pm \int_\lambda^{\lambda_n} \frac{\sqrt{b_2 - L}\,dL}{(\lambda_s - L)\sqrt{(\lambda_n - L)(L - \lambda_m)}}, \qquad (11.1)$$

where ζ_n is once again the value of ζ at which $\lambda = \lambda_n$. With the transformation

$$\tau^2 = \frac{(b_2 - \lambda_m)(\lambda_n - L)}{(\lambda_n - \lambda_m)(b_2 - L)}, \qquad (11.2)$$

(11.1) can be written in the form

$$\zeta - \zeta_n = \pm \frac{2(b_2 - \lambda_n)}{(\lambda_n - \lambda)(\lambda_n - \lambda_m)} \int_0^{\tau_2} \frac{d\tau}{(1 - \beta_2^2 \tau^2)\sqrt{(1 - \tau^2)(1 - m_2^2 \tau^2)}}, \qquad (11.3)$$

where

$$\tau_2^2 = \frac{(b_2 - \lambda_m)(\lambda_n - \lambda)}{(\lambda_n - \lambda_m)(b_2 - \lambda)}, \quad \beta_2^2 = \frac{(\lambda_n - \lambda_m)(b_2 - \lambda_s)}{(b_2 - \lambda_m)(\lambda_n - \lambda_s)}, \quad m_2^2 = \frac{\lambda_n - \lambda_m}{b_2 - \lambda_m}. \qquad (11.4)$$

The integral on the right-hand side of (11.3) is the incomplete elliptic integral of the third kind $\Pi(\tau_2; \beta_2^2, m_2)$, and therefore the solitary waves of radial contraction are given by

$$\zeta - \zeta_n = \pm \frac{2(b_2 - \lambda_n)}{(\lambda_n - \lambda)(\lambda_n - \lambda_m)} \Pi(\tau_2; \beta_2^2, m_2). \qquad (11.5)$$

Since $b_2 > \lambda_n > \lambda_s > \lambda_m$, it is clear that $m_2 < 1$. Simple calculations show that $\beta_2^2 > 1$. Thus, $\Pi(\tau_2; \beta_2^2, m_2)$ is of hyperbolic case. As discussed in Section 11.7, λ tends to λ_s exponentially as ζ tends to infinity. Therefore the profiles of solitary waves of radial contraction are exponentially decreasing away from the wave peak.

Finally let us phrase in terms of V and λ_s the restrictions that define this case, namely, $\lambda_s < \lambda_c < \lambda_n < b_2$. And we must also account for the basic restriction $V > V_L$. Using the analyses of Section 11.7 we find that $\lambda_s < \lambda_c < \lambda_n$ is equivalent to $V < V_1$, and $\lambda_n < b_2$ is equivalent to $V < V_2$. The first of these is not surprising in view of our analysis of linear waves (cf. (6.13)). The second is clearly a product of the appearance of the singular line. Recalling that V_1 and V_2 are equal when λ has the critical value λ_L, we can write the previous two restrictions as

$$V < V_2 \quad \text{if} \quad \lambda_s < \lambda_L, \quad V < V_1 \quad \text{if} \quad \lambda_s \geq \lambda_L. \qquad (11.6)$$

Since $V_L < V_1$, we must restrict V in the second case to the interval $V_L < V < V_1$. In the former case it is possible that V_2 could be smaller than V_L, and so we will restrict λ_s so that this doesn't happen. The condition $V_2 > V_L$ is equivalent to $\psi(\lambda_s) > 0$, where

$$\psi(\lambda) = \lambda^3 - 2(b_1/b_2)\lambda - 2. \qquad (11.7)$$

This cubic has a single positive zero which we call λ_l. In terms of it, our restrictions on V take the final form

$$V_L < V < V_2 \quad \text{if} \quad \lambda_l < \lambda_s < \lambda_L, \quad V_L < V < V_1 \quad \text{if} \quad \lambda_s \geq \lambda_L. \qquad (11.8)$$

We see, then, that λ_l is a critical value, and no solitary waves of radial contraction can exist if the radius of the rod at infinity is thicker than $a/\sqrt{\lambda_l}$.

We can conclude somewhat more. If $\lambda_s \geq \lambda_L$ we see from (11.8) that there are solitary waves that travel at speeds V arbitrarily close to V_1. This would mean that λ_s is very close to λ_n and therefore the amplitude of these waves is very small. In contrast, when $\lambda_l < \lambda_s < \lambda_L$ we will have $V < V_2$. Since V_2 is smaller than V_1 in this case, we conclude that λ_s cannot get arbitrarily close to λ_n. Indeed, λ_n must always exceed b_2^* (cf. (7.20)). Therefore the wave amplitude h must satisfy the restriction

$$h = \frac{a}{\sqrt{\lambda_s}} - \frac{a}{\sqrt{\lambda_n}} > h^* = \frac{a}{\sqrt{\lambda_s}} - \frac{a}{\sqrt{b_2^*}}. \qquad (11.9)$$

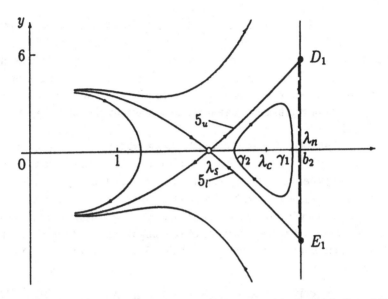

Fig. 11. Phase portrait for the case $b_2 = \lambda_n > \lambda_c > \lambda_s$ with $C_1 = 0.3, C_2 = 0.2, \lambda_s = 2.07, \nu = 0.727, \gamma_1 = 3.042, b_2 = 3.138, \lambda_c = 2.747, \gamma_2 = 2.37$.

Thus, we conclude that when the radius at infinity is thicker than the critical value $a/\sqrt{\lambda_L}$, there cannot be solitary waves with amplitudes smaller than or equal to h^*.

11.12 Solitary cusp waves of radial contraction

In this section we consider the solution represented by curves 5_u and 5_l (and $D_1 E_1$) in Figure 11. This portrait is defined by the restrictions $\lambda_s < \lambda_c < \lambda_n = b_2$ and $V > V_L$. The solution again represents a solitary wave. $\lambda(\zeta)$ begins at λ_s at $\zeta = -\infty$, then slowly increases until it approaches the positive contact value D_1 on the singular line. The solution then instantaneously jumps to the negative contact value E_1, and thereafter decreases, approaching λ_s at $\zeta = \infty$. Because of the jump in y the wave exhibits a cusp, and therefore we call it a *solitary cusp wave of radial contraction*.

The differential equation (8.2) applies to trajectories passing through the saddle point and yet contacting the singular line. Therefore it applies here, though in this case we have $\lambda_s \leq \lambda < b_2$. If ζ_1 is the coordinate for which

$\lambda(\zeta_1) = b_2$, then

$$\zeta - \zeta_1 = \int_\lambda^{b_2} \frac{dL}{(L - \lambda_s)\sqrt{L - \lambda_m}}, \quad \zeta \geq \zeta_1,$$

$$\zeta - \zeta_1 = \int_\lambda^{b_2} \frac{dL}{(L - \lambda_s)\sqrt{L - \lambda_m}}, \quad \zeta \leq \zeta_1. \tag{12.1}$$

These two integrals can be evaluated explicitly in the form

$$\lambda = \lambda_m + \coth^2\left(\frac{1}{2}k_1 - \frac{1}{4}\sqrt{\lambda_s - \lambda_m}|\zeta - \zeta_1|\right)(\lambda_s - \lambda_m), \tag{12.2}$$

where

$$k_1 = \ln \frac{\sqrt{\lambda_n - \lambda_m} + \sqrt{\lambda_s - \lambda_m}}{\sqrt{\lambda_n - \lambda_m} - \sqrt{\lambda_s - \lambda_m}}. \tag{12.3}$$

We showed in Section 11.7 that the restriction $\lambda_n = b_2$ is equivalent to $V = V_2$. Combining this with (11.8), we find that the only restrictions in this case are

$$V = V_2 \quad \text{and} \quad \lambda_l < \lambda_s < \lambda_L. \tag{12.4}$$

In particular, all solitary waves in this case travel with the fixed velocity V_2 and they will be present only if the radius of the rod at infinity is thicker than $a/\sqrt{\lambda_L}$ and thinner than $a/\sqrt{\lambda_l}$.

11.13 Periodic cusp waves of radial contraction

The portrait in Figure 12 is defined by the restrictions $\lambda_s < \lambda_c < b_2 < \lambda_n$ and $V > V_L$. We consider in this case the solution given by curve 6. This curve has the special property that it "contacts" the singular line $\lambda = b_2$ at a positive value D_2 and at a negative value E_2. From the discussion in Section 11.5, this means that the solution $\lambda(\zeta)$ starts at the singular line at $y = E_2$ at a value, say ζ_1, decreases steadily until y approaches 0 at some value $\lambda = \gamma_2^*$. Thereafter $\lambda(\zeta)$ increases steadily and eventually approaches the singular line at $y = D_2$, and then there is an instantaneous jump in y from D_2 to E_2. Thereafter the solution repeats itself in a periodic fashion. The sudden jump in y indicates a cusp, and so we call these solutions *periodic cusp waves of radial contraction*.

We note that (9.4) is the differential equation for solutions contacting the singular line, and hence applies in this case. By an analysis similar to that in Section 11.9 we can show that this equation has the alternate form

$$\lambda_\zeta^2 = (\lambda - \gamma_2^*)(\lambda - \gamma_3^*)(\lambda - \gamma_4^*), \tag{13.1}$$

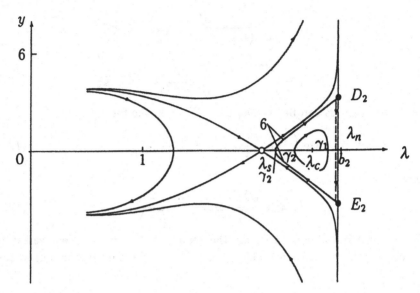

Fig. 12. Phase portrait for the case $\lambda_n > b_2 > \lambda_c > \lambda_s$ with $C_1 = 0.3, C_2 = 0.2, \lambda_s = 2.07, \nu = 0.745, \gamma_1 = 2.665, b_2 = 2.759, \lambda_c = 2.530, \lambda_n = 2.785, \gamma_2 = 2.37, \gamma_2^* = 2.198.$

where the cubic on the right-hand side of (9.4) now has a zero γ_2^* satisfying $\lambda_s < \gamma_2^* < \lambda_c$, another positive zero γ_3^* satisfying $0 < \gamma_3^* < \lambda_s$, and a negative zero γ_4^*. As before we can assume that $\gamma_4^* = -1/(\gamma_2^*\gamma_3^*)$. The natural interval in which (13.1) applies is $\gamma_2^* < \lambda < b_2$, and therefore the solution is

$$\zeta - \zeta_1 = -\int_\lambda^{b_2} \frac{dL}{\sqrt{(L - \gamma_2^*)(L - \gamma_3^*)(L - \gamma_4^*)}} \quad \text{if} \quad \zeta \leq \zeta_1,$$

$$\zeta - \zeta_1 = \int_\lambda^{b_2} \frac{dL}{\sqrt{(L - \gamma_2^*)(L - \gamma_3^*)(L - \gamma_4^*)}} \quad \text{if} \quad \zeta \geq \zeta_1. \tag{13.2}$$

Evaluating these integrals we obtain

$$\lambda = \gamma_3^* + (\gamma_2^* - \gamma_3^*)\text{cn}^{-2}\left(\tfrac{1}{4}\sqrt{\gamma_2^* - \gamma_4^*}|\zeta - \zeta_1| + F(\psi_2, n_2); n_2\right), \tag{13.3}$$

where

$$n_2^2 = \frac{\gamma_3^* - \gamma_4^*}{\gamma_2^* - \gamma_4^*}, \quad \psi_2 = \arcsin\sqrt{\frac{b_2 - \gamma_2^*}{b_2 - \gamma_3^*}}. \tag{13.4}$$

In general the amplitude of the waves is given by

$$h = \frac{a}{\sqrt{\gamma_2^*}} - \frac{a}{\sqrt{b_2}} \tag{13.5}$$

and the period of their oscillation is

$$T_2 = 2 \int_{\gamma_2^*}^{b_2} \frac{dL}{\sqrt{(L - \gamma_2^*)(L - \gamma_3^*)(L - \gamma_4^*)}} = \frac{4}{\sqrt{\gamma_2^* - \gamma_4^*}} F(\psi_2, n_2). \tag{13.6}$$

The basic restrictions in this case are $\lambda_s < \lambda_c < b_2 < \gamma_n$ and $V > V_L$. Previous analyses have shown that $\lambda_s < \lambda_c$ is equivalent to $V < V_1$ (cf. (7.12)), $b_2 < \lambda_n$ is equivalent to $V > V_2$ (cf. (7.15)), and $\lambda_c < b_2$ is equivalent to $V < V_3$ (cf. (9.15)). In particular we need $V_2 < V_1$, and this is equivalent to $\lambda_s < \lambda_L$ (cf. (7.19)). From (9.16) we conclude that V and λ_s must satisfy

$$V_2 < V < V_3 \quad \text{and} \quad \lambda_s < \lambda_L. \tag{13.7}$$

The additional restriction $V > V_L$ requires that we have $V_3 > V_L$. Using (9.15) and (6.14) we find that this inequality is equivalent to $\rho(\lambda_s) > 0$, where ρ is the cubic

$$\rho(\lambda) = \lambda^3 - (b_1/b_2)\lambda - 1. \tag{13.8}$$

$\rho(\lambda)$ has a single positive zero which we label λ_p, and in terms of it we find that $V_3 > V_L$ is equivalent to $\lambda_s > \lambda_p$. Therefore we need

$$\max(V_2, V_L) < V < V_3 \quad \text{and} \quad \lambda_p < \lambda_s < \lambda_L. \tag{13.9}$$

A comparison of V_2 and V_L shows (cf. (11.8)) that $V_2 > V_L$ if $\lambda_s > \lambda_l$ and $V_2 \leq V_L$ if $\lambda_s \leq \lambda_l$. This gives us our final form of the restrictions on V and λ_s:

$$V_2 < V < V_3 \quad \text{and} \quad \lambda_l \leq \lambda_s < \lambda_L, \quad V_L < V < V_3 \quad \text{and} \quad \lambda_p \leq \lambda_s < \lambda_l. \tag{13.10}$$

In the first case we note that it is possible for V to be arbitrarily close to V_2, and this means that λ_n can be arbitrarily close to b_2. In turn this implies that γ_2^* can be arbitrarily close to λ_s. As a consequence there can be periodic cusp waves having very long periods.

11.14 Periodic waves of type II

We consider finally the closed curves that lie inside the labeled curves in Figure 10, Figure 11, and Figure 12. These curves represent periodic solutions with smooth profiles. We call them *periodic waves of type II*. Their analysis is similar to that given in Section 11.10.

In each figure there is a one-parameter family of periodic solutions. We use the maximum value of the axial stretch, denoted γ_1, to parametrize these. This implies that the differential equation governing these periodic solutions is (10.1) provided we replace γ_3 with γ_1. In particular the points where the right-hand side vanishes, other than γ_1, are roots of the cubic equation

$$\lambda^3 - \left(2\lambda_s + \frac{2b_1}{\lambda_s^2} + \frac{2b_2}{\lambda_s^3} - \gamma_1\right)\lambda^2 + \left(\frac{2b_1}{\gamma_1} + \frac{b_2}{\gamma_1^2}\right)\lambda + \frac{b_2}{\gamma_1} = 0. \qquad (14.1)$$

One of these crossings is the minimum axial stretch. We call it γ_2. There is also a root γ_3 between 0 and λ_s, and a negative root γ_4. In terms of these (10.1) can be written in the form

$$\lambda_\zeta^2 = \frac{(\lambda - \gamma_1)(\lambda - \gamma_2)(\lambda - \gamma_3)(\lambda - \gamma_4)}{\lambda - b_2}. \qquad (14.2)$$

The solution is then given by

$$\zeta - \zeta_1 = \pm \int_\lambda^{\gamma_1} \frac{\sqrt{b_2 - L}dL}{\sqrt{(\gamma_1 - L)(L - \gamma_2)(L - \gamma_3)(L - \gamma_4)}}, \quad \gamma_2 \le \lambda \le \gamma_1, \quad (14.3)$$

where $\lambda(\zeta_1) = \gamma_1$. The period of the waves is given by

$$T_4 = 2 \int_{\gamma_2}^{\gamma_1} \frac{\sqrt{b_2 - L}dL}{\sqrt{(\gamma_1 - L)(L - \gamma_2)(L - \gamma_3)(L - \gamma_4)}}, \qquad (14.4)$$

and the wave amplitude is

$$h = \frac{a}{\sqrt{\gamma_2}} - \frac{a}{\sqrt{\gamma_1}}. \qquad (14.5)$$

The restrictions on V and λ_s for periodic waves of type II follow from the calculations in Sections 11.11–11.13. Specifically we need

$$V_L < V < V_1 \quad \text{if} \quad \lambda_s \ge \lambda_L, \quad V_L < V < V_3 \quad \text{if} \quad \lambda_p < \lambda_s < \lambda_L. \qquad (14.6)$$

In particular we note that periodic waves of type II can appear only if $\lambda_s > \lambda_p$.

11.15 Summary

To summarize the results we have obtained it is simplest to draw in the (λ_s, V)-plane the function curves of V_1, V_2, and V_3. These curves are shown in Figure 13. We draw only the region $V > V_L$ since it is only here that we find traveling wave solutions†. This part of the plane can be divided into nine regions depending on the type of waves that are observed. These regions are:

† The region R_1 can be extended to the part $0 < \lambda_s < 1$, but we omit this addition, for convenience, from Figure 13.

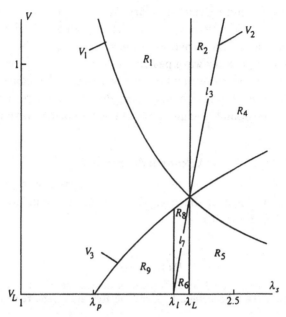

Fig. 13. The (λ_s, V) plane with $C_1 = 0.3, C_2 = 0.2, V_L = 0.775, \nu_0 = 1, \lambda_L = 2.104, \lambda_l = 2.0, \lambda_p = 1.476$.

(i) region R_1 (including the boundary $\lambda_s = \lambda_L$ but not the boundary $V = V_1$): solitary waves of radial expansion whose amplitudes have no finite lower bound; periodic waves.

(ii) region R_2 (open region): solitary waves of radial expansion whose amplitudes have a finite lower bound; periodic waves.

(iii) curve $l_3(\lambda_s > \lambda_L)$: solitary cusp waves of radial expansion; periodic waves (including those having narrow wave peaks and long periods).

(iv) region R_4 (open region): periodic cusp waves of radial expansion; periodic waves whose periods have a finite upper bound (including those having narrow wave peaks).

(v) region R_5 (including the boundary $\lambda_s = \lambda_L$ but not the boundary $V = V_1$); solitary waves of radial contraction whose amplitudes have no finite lower bound; periodic waves.

(vi) region R_6 (open region): solitary waves of radial contraction whose amplitudes have a finite lower bound; periodic waves.

(vii) curve $l_7(\lambda_l < \lambda_s < \lambda_L)$: solitary cusp waves of radial contraction; periodic waves (including those having narrow wave peaks and long periods).

432   H.-H. Dai

(viii) region R_8 (including the boundary $\lambda_s = \lambda_L$ but not the boundaries $V = V_2, V = V_3$, or $V = V_L$): periodic cusp waves of radial contraction; periodic waves whose periods have a finite upper bound (including those having narrow wave peaks).

(ix) region R_9 (open region): periodic cusp waves of radial contraction whose periods have an upper bound; periodic waves whose periods have a finite upper bound (including those having narrow wave peaks).

Acknowledgement

The work described in this chapter was fully supported by a grant from the Research Grants Council of the Hong Kong Special Administrative Region, China (Project No. 9040470).

References

Aifantis, E. D. and Serrin, J. B. 1980 Towards a mechanical theory of phase transformation. Technical Report, Corrosion Research Center, University of Minnesota.

Antman, S. S. 1973 Nonuniqueness of equilibrium states for bars in tension. *J. Math. Appl.* **44**, 333–349.

Antman, S. S. 1974 Qualitative theory of the ordinary differential equations of nonlinear elasticity. In Nemat-Nasser (Ed.), *Mechanics Today.* pp. 58–161, New York, Pergamon.

Antman, S. S. and Liu, T.-P. 1979 Traveling waves in hyperelastic rods. *Q. Appl. Math.* **36**, 377–399.

Coleman, B. D. 1983 Necking and drawing in polymeric fibers under tension. *Arch. Rational Mech. Anal.* **83**, 115–137.

Coleman, B. D. and Newman, D. C. 1990 On waves in slender elastic rods. *Arch. Rational. Mech. Anal.* **109**, 39–61.

Coleman, B. D. and Dill, E. H. 1992 Flexure waves in elastic rods. *J. Acoust. Soc. Am.* **91**, 2663–2673.

Coleman, B. D., Dill, E. H. and Swigon, D. 1995 On the dynamics of flexure and stretch in the theory of theory of elastic rods." *Arch. Rational Mech. Anal.* **129**, 147–174.

Ogden, R. W., 1984 *Nonlinear Elastic Deformations.* New York: Halsted Press.

Truesdell, C. and Noll, W. 1965 Nonlinear field theories of mechanics. In: *Handbuch der Physik*, Bd. III/3. Berlin-Heidelberg-New York, Springer-Verlag.

Wright, T. 1982 Nonlinear waves in rods. In *Proceedings of the IUTAM Symposium on Finite Elasticity*, D. E. Carlson and R. T. Shields (eds.), The Hague: Martinus Nijhoff.

Wright, T. 1985 Nonlinear waves in rods: results for incompressible elastic materials. *Stud. Appl. Math.* **72**, 149–160.

12

Strain-energy functions with multiple local minima: modeling phase transformations using finite thermoelasticity

R. Abeyaratne

Department of Mechanical Engineering
Massachusetts Institute of Technology, Cambridge, USA
Email: rohan@mit.edu

K. Bhattacharya and J. K. Knowles

Division of Engineering and Applied Science
California Institute of Technology, Pasadena, USA
Email: bhatta@its.caltech.edu and knowles@its.caltech.edu

This chapter provides a brief introduction to the following basic ideas pertaining to thermoelastic phase transitions: the lattice theory of martensite, phase boundaries, energy minimization, Weierstrass-Erdmann corner conditions, phase equilibrium, nonequilibrium processes, hysteresis, the notion of driving force, dynamic phase transitions, nonuniqueness, kinetic law, nucleation condition, and microstructure.

12.1 Introduction

This chapter provides an introduction to some basic ideas associated with the modeling of solid-solid phase transitions within the continuum theory of finite thermoelasticity. No attempt is made to be complete, either in terms of our selection of topics or in the depth of coverage. Our goal is simply to give the reader a flavor for some selected ideas.

This subject requires an intimate mix of continuum and lattice theories, and in order to describe it satisfactorily one has to draw on tools from crystallography, lattice dynamics, thermodynamics, continuum mechanics and functional analysis. This provides for a remarkably rich subject which in turn has prompted analyses from various distinct points of view. The free-energy function has multiple local minima, each minimum being identified with a distinct phase, and each phase being characterized by its own lattice. Crystallography

R. Abeyaratne et al.

plays a key role in characterizing the lattice structure and material symmetry, and restricts deformations through geometric compatibility. The thermodynamics of irreversible processes provides the framework for describing evolutionary processes. Lattice dynamics describes the mechanism by which the material transforms from one phase to the other. And eventually all of this needs to be described at the continuum scale. Since the present volume is concerned with finite elasticity, we shall strive to describe solid-solid phase transitions from this point of view and using this terminology.

The article by Schetky (1979) provides a general introduction to the subject while the book by Duerig *et al.* (1990) describes many engineering applications. For a more extensive treatment of the crystallographic and microstructural aspects, the reader is referred to Bhattacharya (2000) and James and Hane (2000); for details on dynamics, see the forthcoming monograph by Abeyaratne and Knowles (2001). An introduction from the materials science point of view can be found in Otsuka and Wayman (1999), while Christian (1975) provides a broader treatment of phase transformations. A general discussion of configurational forces can be found in Gurtin (2000); a particular example of this, the driving force on an interface, will play a major role in our discussion.

This chapter is organized as follows: Section 12.2 is devoted to explaining why we are interested in strain-energy functions with multiple local minima; the motivation we give is based on the lattice theory of martensitic transformations. Other reasons for studying such energy functions include, for example, the van der Waals type theory of two-phase fluids. In finite elasticity, the strain-energy function characterizes the material, e.g., the neo-Hookean and Mooney-Rivlin models of rubber. Accordingly in Section 12.3 we construct an explicit strain-energy function which describes a class of crystalline solids with cubic and tetragonal phases.

In preparation for solving initial-boundary-value problems, in Section 12.4 we specialize the three-dimensional theory of finite elasticity to the special case of uniaxial motions of a slab. In this way we derive a convenient one-dimensional mathematical model. Sections 12.5 and 12.6 are devoted to studying, respectively, the statics and dynamics of this slab. The static problem illustrates a number of phenomena, in particular the presence of surfaces in the body across which the deformation gradient is discontinuous. The problem also illustrates the distinction between mechanical equilibrium in the sense of balance of forces and moments, and phase equilibrium in the sense of energy extremization. The dynamic problem demonstrates the severe lack of uniqueness of solution to classically formulated initial-boundary-value problems. Attention is drawn to two distinct types of nonuniqueness. The notions of a nucleation condition and a kinetic law are then imported from materials science to show how they select the physically relevant solution from among the totality of available solutions. The quasi-static loading of a slab is also discussed in Section 12.6 and this too illustrates the role of nucleation and kinetics. Section 12.7 is devoted to trying to understand the notion of kinetics from a more general point of view. Three-dimensional thermodynamic processes are considered and the various thermodynamic driving forces and conjugate fluxes contributing to entropy production are identified. In the present setting, the thermodynamic notion of kinetics of irreversible processes is identified with the kinetics of phase transitions.

Finally in Section 12.8 we return to static problems, this time in three-dimensions. We examine various piecewise homogeneous deformations and illustrate the role played by geometric compatibility in characterizing various microstructures.

12.2 Strain-energy functions with multiple local minima: motivation from the lattice theory of martensitic transformations

(Born and Huang 1954, Ericksen 1978, 1980, 1984, Ball and James 1992, James 1992, Bhattacharya 2000)

Typically, the strain-energy function $W(F)$ in finite elasticity has an energy-well at the reference configuration, i.e. W has a local minimum at $F = I$. In this section we shall use a specific example to illustrate how, for certain materials, W has *multiple* local minima. The example pertains to a crystalline solid that can exist in a number of distinct crystalline forms, each crystal structure (lattice) being identified with a different "phase" of the material. Examples of such materials include the alloys Indium-Thallium (cubic and tetragonal phases), Gold-Cadmium (cubic and orthorhombic phases), Nickel-Titanium (cubic and monoclinic phases) and Copper-Aluminum-Nickel (cubic, orthorhombic and monoclinic phases).

Lattice model: A lattice refers to a periodic arrangement of points in space. The simplest lattice, called a *Bravais lattice*, is an infinite set of points in $I\!\!R^3$ that can be generated by the translation of a single point o through three linearly independent vectors $\{e_1, e_2, e_3\}$. Thus, if $\mathcal{L}(e, o)$ denotes a Bravais lattice, then

$$\mathcal{L}(e, o) = \left\{ X \; : \; X = \nu^i e_i + o, \text{ where } \nu^1, \; \nu^2, \; \nu^3 \text{ are integers} \right\}, \qquad (2.1)$$

where summation on the repeated index i is implied. The vectors $\{e_1, e_2, e_3\}$ are called the lattice vectors.

R. Abeyaratne et al.

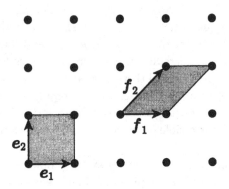

Fig. 1. A two-dimensional lattice with lattice vectors.

It is important to keep in mind that, because of the inherent symmetry of the lattice, more than one set of lattice vectors may generate the same Bravais lattice. For example, the two-dimensional lattice shown in Figure 1 is generated by both $\{e_1, e_2\}$ and $\{f_1, f_2\}$. In general, two sets of lattice vectors $\{e_1, e_2, e_3\}$ and $\{f_1, f_2, f_3\}$ generate the same lattice if and only if

$$f_i = \mu_i^j e_j, \qquad [\mu] \in I\!I; \tag{2.2}$$

here $I\!I$ denotes the set of 3×3 matrices whose elements are integers and whose determinant is ± 1:

$$I\!I = \{ [\mu] : [\mu] = 3 \times 3 \text{ matrix}, \mu_i^j = \text{integer}, \det [\mu] = \pm 1\}. \tag{2.3}$$

Next consider the *energy* stored in a Bravais lattice. Under the assumption that there are identical atoms located at each lattice point, it is natural to assume that ψ, the free-energy of the lattice, depends on the lattice vectors $\{e_1, e_2, e_3\}$ and the temperature $\theta > 0$:

$$\psi = \psi(e_1, e_2, e_3; \theta); \tag{2.4}$$

the function ψ is defined for all triplets of linearly independent vectors and temperatures. Note that due to thermal expansion of the lattice, the lattice vectors themselves depend on the temperature: $\{e_1(\theta), e_2(\theta), e_3(\theta)\}$. We will assume that the free-energy function ψ satisfies the following two requirements pertaining to frame indifference and symmetry:

1. Frame-indifference: A rigid rotation of the lattice should not change its free-

energy and so

$$\psi(\boldsymbol{Q}\boldsymbol{e}_1, \boldsymbol{Q}\boldsymbol{e}_2, \boldsymbol{Q}\boldsymbol{e}_3; \theta) = \psi(\boldsymbol{e}_1, \boldsymbol{e}_2, \boldsymbol{e}_3; \theta) \qquad \text{for all rotations } \boldsymbol{Q}. \qquad (2.5)$$

2. Lattice symmetry: The free-energy should depend on the lattice but not on the specific choice of lattice vectors, i.e. two sets of lattice vectors that generate the same lattice must have the same free-energy, and so

$$\psi(\mu_1^j \boldsymbol{e}_j, \mu_2^j \boldsymbol{e}_j, \mu_3^j \boldsymbol{e}_j; \theta) = \psi(\boldsymbol{e}_1, \boldsymbol{e}_2, \boldsymbol{e}_3; \theta) \qquad \text{for all } [\mu] \in I\!\!I. \qquad (2.6)$$

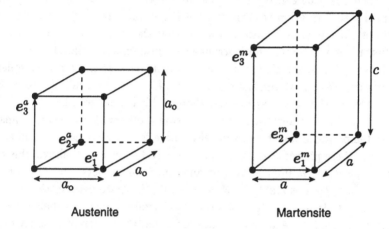

Austenite   Martensite

Fig. 2. Simple cubic lattice (austenite) with lattice parameters $a_o \times a_o \times a_o$ and lattice vectors $\{e_1^a, e_2^a, e_3^a\}$; and simple tetragonal lattice (martensite) with lattice parameters $a \times a \times c$ and lattice vectors $\{e_1^m, e_2^m, e_3^m\}$.

Two-phase martensitic materials: Now consider a *two-phase material* where the lattice has one structure at high temperatures and a different one at low temperatures. As the temperature is changed, the lattice vectors vary continuously due to thermal expansion except at the *transformation temperature* θ_T where they change *discontinuously* (a "first order" martensitic phase transformation). The high temperature lattice is characterized by the vectors $\{e_1^a(\theta), e_2^a(\theta), e_3^a(\theta)\}$ and is associated with one phase of the material (*austenite*); the low temperature lattice is described by $\{e_1^m(\theta), e_2^m(\theta), e_3^m(\theta)\}$ and corresponds to a second phase of the material (*martensite*). The discontinuous change in the lattice at the transformation temperature implies that $\{e_1^a(\theta_T), e_2^a(\theta_T), e_3^a(\theta_T)\} \neq \{e_1^m(\theta_T), e_2^m(\theta_T), e_3^m(\theta_T)\}$. Figure 2 illustrates a material with cubic and tetragonal phases (lattices).

We will be concerned with relatively modest changes in temperature from the transformation temperature. Consequently, the change in the lattice due

R. Abeyaratne et al.

to thermal expansion can be neglected in comparison with that due to phase transformation. Thus, we shall take both sets of lattice vectors $\{e_1^a, e_2^a, e_3^a\}$ and $\{e_1^m, e_2^m, e_3^m\}$ to be independent of temperature.

Consider the austenitic and martensitic lattices $\mathcal{L}(e^a, o)$ and $\mathcal{L}(e^m, o)$. Since the associated lattice vectors $\{e_1^a, e_2^a, e_3^a\}$ and $\{e_1^m, e_2^m, e_3^m\}$ are each linearly independent, there is a nonsingular tensor U_1 which relates them:

$$e_i^m = U_1 e_i^a. \tag{2.7}$$

The lattice $\mathcal{L}(e^m, o)$ can be viewed as the image of the lattice $\mathcal{L}(e^a, o)$ under the linear transformation U_1 which is called the *Bain stretch tensor*. A characteristic of a martensitic transformation is that the change of crystal structure is diffusionless, i.e. there is no rearrangement of atoms associated with the lattice change. Consequently, the kinematics of the transformation is completely characterized by U_1. Given the two lattices, U_1 can be calculated from (2.7); we will assume that U_1 is symmetric though this is not essential.

The fact that the austenite phase is usually observed above the transformation temperature and the martensite phase is observed below it, does *not* imply that these phases cease to exist, respectively, below and above, this temperature. Rather, it implies an exchange in stability between these phases: the austenite lattice is stable for $\theta > \theta_T$, while the martensite lattice is stable for $\theta < \theta_T$. Indeed, one often observes (metastable) austenite below θ_T and (metastable) martensite above θ_T. Assuming that the notion of stability here is that of energy minimization, the preceding stability properties imply that the free-energy $\psi(\cdot, \theta)$ is minimized by the lattice vectors $\{e_1^a, e_2^a, e_3^a\}$ for $\theta > \theta_T$ and by $\{e_1^m, e_2^m, e_3^m\}$ for $\theta < \theta_T$. At the transformation temperature θ_T, both lattices have equal energy. Thus, we shall assume that

$$\psi(e_1^a, e_2^a, e_3^a; \theta) \leq \psi(\ell_1, \ell_2, \ell_3; \theta) \qquad \text{for } \theta > \theta_T,$$

$$\psi(e_1^a, e_2^a, e_3^a; \theta) = \psi(e_1^m, e_2^m, e_3^m; \theta) \leq \psi(\ell_1, \ell_2, \ell_3; \theta) \qquad \text{for } \theta = \theta_T,$$

$$\psi(e_1^m, e_2^m, e_3^m; \theta) \leq \psi(\ell_1, \ell_2, \ell_3; \theta) \qquad \text{for } \theta < \theta_T,$$

$$\tag{2.8}$$

for all triplets of linearly independent vectors $\{\ell_1, \ell_2, \ell_3\}$; see Figure 3. Observe from $(2.8)_2$ that $\psi(\cdot, \theta_T)$ has (at least) two distinct energy-wells, and therefore by continuity, so does $\psi(\cdot, \theta)$ for temperatures close to θ_T.

In summary, the free-energy function ψ, which is a function of the lattice vectors and temperature, is to satisfy the requirements of frame-indifference (2.5), lattice symmetry (2.6) and energy minimization (2.8).

Continuum model: We now proceed from the preceding lattice model to a

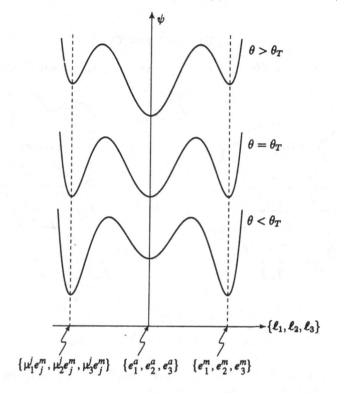

Fig. 3. Free-energy ψ as a function of lattice vectors $\{\ell_1, \ell_2, \ell_3\}$ at three different temperatures θ; the transformation temperature is denoted by θ_T. The energy has local minima at the austenite lattice vectors $\{e_1^a, e_2^a, e_3^a\}$ and the martensite lattice vectors $\{e_1^m, e_2^m, e_3^m\}$. For any $[\mu] \in \mathit{I}$, the vectors $\{\mu_1^j \ell_j, \mu_2^j \ell_j, \mu_3^j \ell_j\}$ and $\{\ell_1, \ell_2, \ell_3\}$ describe the same lattice, and so the energy has additional minima at certain symmetry related lattice vectors; this is illustrated in the figure by the third energy-well at the martensitic "variant" $\{\mu_1^j e_j^m, \mu_2^j e_j^m, \mu_3^j e_j^m\}$. Observe that because we have omitted the effect of thermal expansion, the location of each local minimum does not change with temperature.

related continuum model. Consider a crystalline solid, modeled as a continuum, which occupies some region $\mathcal{R}_o$ in a reference configuration. Suppose that underlying each point $X \in \mathcal{R}_o$, there is a Bravais lattice with lattice vectors $\{e_1^o(X), e_2^o(X), e_2^o(X)\}$. Now suppose that the continuum undergoes a deformation $x = x(X)$ and let $F(X) = \nabla x(X)$ be the associated deformation gradient. Let $\{e_1(X), e_2(X), e_3(X)\}$ be the lattice vectors of the deformed lattice at the same material point X. The Cauchy-Born hypothesis says that the lattice vectors deform according to the macroscopic deformation gradient, i.e.

$$e_i(X) = F(X) \, e_i^o(X). \tag{2.9}$$

In other words, the lattice vectors behave like "material filaments"; see Figure 4

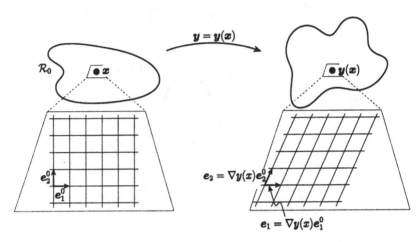

Fig. 4. The relation between the continuum and the lattice.

We can now use (2.9) to define the continuum free-energy function $W(F, \theta)$ in terms of the lattice free-energy ψ through

$$W(F, \theta) = \psi(Fe_1^o, Fe_2^o, Fe_3^o; \theta). \tag{2.10}$$

Note that the free-energy function W depends on the choice of reference configuration because the right hand side of (2.10) involves the lattice vectors associated with that configuration. It follows from (2.5) and (2.10) that W inherits the frame-indifference property that was previously built into ψ:

Frame-indifference: $W(QF, \theta) = W(F, \theta)$ for all rotations Q. (2.11)

Turning next to material symmetry, recall that the lattice free-energy ψ has the property (2.6) which ensures that it depends on the lattice but not the choice of lattice vectors. It is convenient to rewrite (2.6) in the equivalent form

$$\psi(He_1, He_2, He_3; \theta) = \psi(e_1, e_2, e_3; \theta) \qquad \text{for all } H \in \mathcal{G}(\mathcal{L}), \tag{2.12}$$

where

$$\mathcal{G}(\mathcal{L}) = \left\{ H \; : \; He_i = \mu_i^j e_j, \; [\mu] \in I\!I \right\}. \tag{2.13}$$

Observe that $\mathcal{G}(\mathcal{L})$ denotes the set of all linear transformations that map the lattice back into itself, in the sense that $\mathcal{L}(e, o) = \mathcal{L}(f, o)$ if and only if $f_i = He_i$ for some $H \in \mathcal{G}(\mathcal{L})$. The group $\mathcal{G}(\mathcal{L})$ characterizes the symmetry of the lattice and one can verify that it depends on the lattice $\mathcal{L}$ and not on the particular choice of lattice vectors on the right-hand side of (2.13). In view of (2.10) and (2.12), the continuum free-energy inherits the symmetry property

$$W(FH, \theta) = W(F, \theta) \qquad \text{for all } H \in \mathcal{G}(\mathcal{L}_o), \qquad (2.14)$$

where $\mathcal{L}_o$ is the lattice associated with the reference configuration. However, note that in addition to rotations and reflections, the symmetry group $\mathcal{G}$ contains finite shears as well; see, for example, Figure 1. Such shears cause large distortions of the lattice and are usually associated with lattice slip and plasticity. In situations where thermoelasticity provides an adequate mathematical model, it is natural therefore to exclude these large shears and restrict attention to a suitable subgroup of $\mathcal{G}$; see Pitteri (1984) and Ball and James (1992) for a precise analysis of this issue. The appropriate sub-group is the *Laue group* $\mathcal{P}$ which is the set of rotations that map a lattice back into itself:

$$\mathcal{P}(\mathcal{L}) = \{R : R \text{ is proper orthogonal, } R \in \mathcal{G}(\mathcal{L})\}. \qquad (2.15)$$

For example, the Laue group of a simple cubic lattice consists of the 24 rotations that map the unit cube back into itself. Thus, instead of (2.14), we shall require the continuum free-energy to conform to the less stringent requirement

Material Symmetry: $\quad W(FR, \theta) = W(F, \theta) \qquad \text{for all } R \in \mathcal{P}(\mathcal{L}_o). \quad (2.16)$

It is worth emphasizing that it is the symmetry of the reference lattice $\mathcal{L}_o$ that enters here.

Now consider a two-phase martensitic material. Suppose that the reference configuration coincides with a stress-free crystal of austenite at the transformation temperature. Then the reference lattice is the austenite lattice ($e^o = e^a$); moreover, $F = I$ describes stress-free austenite, and $F = U_1$ characterizes stress-free martensite where U_1 is the Bain stretch tensor introduced in (2.7). Thus, it follows from (2.7), (2.8) and (2.10) that W inherits the following multi-well structure:

$$W(I, \theta) \quad \leq \quad W(F, \theta) \qquad \text{for } \theta > \theta_T,$$

$$W(I, \theta) = W(U_1, \theta) \quad \leq \quad W(F, \theta) \qquad \text{for } \theta = \theta_T, \qquad (2.17)$$

$$W(U_1, \theta) \quad \leq \quad W(F, \theta) \qquad \text{for } \theta < \theta_T,$$

for all nonsingular F.

In summary, a continuum-scale free-energy function $W(F, \theta)$ that characterizes a two-phase material must satisfy the requirements of frame-indifference (2.11), material symmetry (2.16) and energy minimization (2.17). Observe that for such a material, the function $W(\cdot, \theta_T)$, and therefore by continuity $W(\cdot, \theta)$ for θ close to θ_T, has (at least) two distinct energy-wells.

Variants: The three properties – frame-indifference, material symmetry and energy minimization – imply certain important characteristics of W. For example for $\theta > \theta_T$, it follows from frame-indifference and the fact that $F = I$ minimizes W, that necessarily $F = Q$ also minimizes W for any rotation Q. Similarly for $\theta < \theta_T$, the free-energy is minimized by $QU_1 R$ for an arbitrary rotation Q and any rotation $R \in \mathcal{P}(\mathcal{L}_o)$.

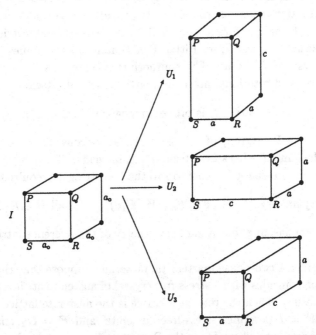

Fig. 5. Cubic austenite and three variants of tetragonal martensite. Observe that one *cannot* rigidly rotate a martensite variant in such a way as to make the atoms P,Q,R,S coincide with the locations of these same atoms in any other martensite variant.

In order to explore this systematically, we note first that by combining frame-indifference (2.11) and material symmetry (2.16)

$$W(R^T F R, \theta) = W(F, \theta) \qquad \text{for all } R \in \mathcal{P}_a, \tag{2.18}$$

where $\mathcal{P}_a = \mathcal{P}(\mathcal{L}(e^a))$ is the Laue group of the reference configuration – unstressed austenite. It follows from this that, since U_1 minimizes the free-energy for $\theta < \theta_T$, so does the tensor $R^T U_1 R$ for all $R \in \mathcal{P}_a$. As R varies over the set of all rotations in $\mathcal{P}_a$, the tensor $R^T U_1 R$ takes on various symmetric tensor values. Suppose, as is the case for most martensitic transformations, that the austenite lattice has greater symmetry than the martensite lattice in the sense that $\mathcal{P}_m \subset \mathcal{P}_a$. Then, one can show that $R^T U_1 R = U_1$ for all $R \in \mathcal{P}_m$. However, as R varies over the rotations in $\mathcal{P}_a$ which are not in $\mathcal{P}_m$, then $R^T U_1 R$ takes on a certain finite number of distinct, symmetric tensor values, which we denote by $U_2, U_3, \ldots, U_N$; these stretch tensors are said to describe the *variants* of martensite; the number of variants, N, is given by the formula

$$N = \frac{\text{the order of } \mathcal{P}_a}{\text{the order of } \mathcal{P}_m}. \tag{2.19}$$

For example, in the case of a material which is cubic in the austenite phase and tetragonal in the martensite phase, the orders of $\mathcal{P}_a$ and $\mathcal{P}_m$ are 24 and 8 respectively; thus there are three variants of martensite characterized by three stretch tensors U_1, U_2, U_3; see Figure 5. Note because of (2.18) that all variants have the same energy:

$$W(U_1, \theta) = W(U_2, \theta) = \cdots = W(U_N, \theta). \tag{2.20}$$

Thus, W has N distinct martensitic local minima at $U_1, U_2, U_3, \ldots, U_N$ all of which have the same energy. A schematic graph of W versus U would therefore have the same generic form as the curve shown in Figure 3, with the minimum at $\{e_1^m, e_2^m, e_3^m\}$ corresponding to one at U_1 and the minimum at $\{[\mu]e_1^m, [\mu]e_2^m, [\mu]e_3^m\}$ corresponding to one at some U_k.

Finally, returning to the notion of frame-indifference, it follows from (2.11) that austenite corresponds not only to the identify tensor I, but also to all rotations of the identity. Similarly, the first variant of martensite corresponds to all tensors of the form QU_1 where Q is an arbitrary rotation, and so on. Therefore, the following *sets* of tensors comprise the bottoms of the *austenite energy-well* $\mathcal{A}$ and the N *martensitic energy-wells* $\mathcal{M}_k$:

$$\left. \begin{aligned} \mathcal{A} &= \{F \colon F = Q \text{ for all rotations } Q\}, \\ \mathcal{M}_k &= \{F \colon F = QU_k \text{ for all rotations } Q\}, \quad k = 1, 2, \ldots N. \end{aligned} \right\} \tag{2.21}$$

Let $\mathcal{M}$ denote the collection of all martensite wells:

$$\mathcal{M} = \mathcal{M}_1 \bigcup \mathcal{M}_2 \bigcup \cdots \bigcup \mathcal{M}_N. \tag{2.22}$$

In *summary*, the continuum-scale free-energy $W(F, \theta)$ associated with a two-phase martensitic material has the property that W is minimized on the austenite energy-well $\mathcal{A}$ at high temperatures, it is minimized on the martensite energy-well $\mathcal{M}$ at low temperatures, and on both the austenite and martensite wells at the transformation temperature θ_T, i.e.

$$W(\boldsymbol{F},\theta) \quad > \quad W(\boldsymbol{F}',\theta) \quad \text{for all} \quad \boldsymbol{F} \notin \mathcal{A}, \quad \boldsymbol{F}' \in \mathcal{A}, \quad \theta > \theta_T,$$

$$W(\boldsymbol{F},\theta) \quad > \quad W(\boldsymbol{F}',\theta) \quad \text{for all} \quad \boldsymbol{F} \notin \mathcal{A}\bigcup\mathcal{M}, \quad \boldsymbol{F}' \in \mathcal{A}\bigcup\mathcal{M}, \quad \theta = \theta_T,$$

$$W(\boldsymbol{F},\theta) \quad > \quad W(\boldsymbol{F}',\theta) \quad \text{for all} \quad \boldsymbol{F} \notin \mathcal{M}, \quad \boldsymbol{F}' \in \mathcal{M}, \quad \theta < \theta_T.$$

$$(2.23)$$

12.3 A strain-energy function with multiple local minima: a material with cubic and tetragonal phases.

(Ericksen, 1978, 1980, 1986)

Perhaps the simplest example of a two-phase martensitic material is one that occurs as a face-centered cubic austenite phase and a face-centered tetragonal martensite phase; examples of such materials include In-Tl, Mn-Ni and Mn-Cu. Let $\{c_1, c_2, c_3\}$ denote fixed *unit* vectors in the cubic directions and let the lattice vectors associated with the austenite lattice be denoted by $\{e_1^a, e_2^a, e_3^a\}$. Observe from Figure 6 that (in contrast to a simple cubic material) for a face-centered cubic material one *cannot* choose the austenite lattice vectors to be a scalar multiple of $\{c_1, c_2, c_3\}$. Instead, if $a_o \times a_o \times a_o$ denotes the lattice parameters of the cubic phase (see figure) one acceptable choice for the lattice vectors is

$$e_1^a = \frac{a_o}{2}\,(c_2 + c_3), \quad e_2^a = \frac{a_o}{2}\,(-c_2 + c_3), \quad e_3^a = \frac{a_o}{2}\,(c_1 + c_3). \quad (3.1)$$

Similarly for the tetragonal phase, if the lattice parameters are denoted by $a \times a \times c$, the lattice vectors $\{e_1^m, e_2^m, e_3^m\}$ can be taken to be

$$e_1^m = \frac{a}{2}\,(c_2 + c_3), \quad e_2^m = \frac{a}{2}\,(-c_2 + c_3), \quad e_3^m = \frac{c}{2}\,c_1 + \frac{a}{2}\,c_3. \quad (3.2)$$

One can readily verify that the mapping $e_i^a \to e_i^m = U_1 e_i^a$ between the two sets of lattice vectors is described by the Bain stretch tensor

$$\boldsymbol{U}_1 \quad = \quad \beta c_1 \otimes c_1 + \alpha c_2 \otimes c_2 + \alpha c_3 \otimes c_3, \quad (3.3)$$

where the stretches α and β are given by

$$\alpha = \frac{a}{a_o}, \quad \beta = \frac{c}{a_o}. \quad (3.4)$$

Thus, the transformation from the cubic lattice to the tetragonal lattice is achieved by stretching the parent lattice equally in two of the cubic directions

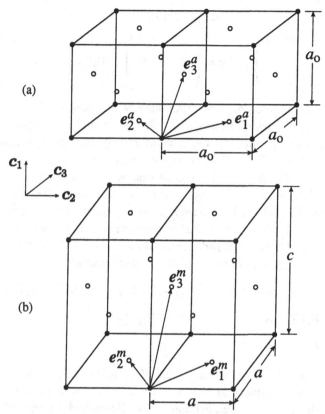

Fig. 6. Face-centered cubic lattice with lattice parameters $a_o \times a_o \times a_o$ and lattice vectors $\{e_1^a, e_2^a, e_3^a\}$, and face-centered tetragonal lattice with lattice parameters $a \times a \times c$ and lattice vectors $\{e_1^m, e_2^m, e_3^m\}$. The unit vectors $\{c_1, c_2, c_3\}$ are associated with the cubic directions. Solely for purposes of clarity, the atoms at the vertices are depicted by filled dots while those at the centers of the faces are shown by open dots.

and unequally in the third direction, the stretches associated with this being α, α and β.

As noted in the previous section, there are three variants of martensite associated with a cubic/tetragonal material. This was illustrated in Figure 5 which showed that there are three possible ways in which to stretch the cubic lattice in order to obtain the tetragonal lattice, the aforementioned one being the one where the unequal stretching occurs in the c_1 direction. More generally, one can stretch the cubic lattice by the ratio β in the c_k direction and by α in the remaining two cubic directions; this leads to the three Bain stretch tensors

$$U_k = \alpha I + [\beta - \alpha] c_k \otimes c_k, \qquad k = 1, 2, 3, \tag{3.5}$$

whose components in the cubic basis are

$$
[U_1] = \begin{pmatrix} \beta & 0 & 0 \\ 0 & \alpha & 0 \\ 0 & 0 & \alpha \end{pmatrix}, \quad
[U_2] = \begin{pmatrix} \alpha & 0 & 0 \\ 0 & \beta & 0 \\ 0 & 0 & \alpha \end{pmatrix}, \quad
[U_3] = \begin{pmatrix} \alpha & 0 & 0 \\ 0 & \alpha & 0 \\ 0 & 0 & \beta \end{pmatrix}. \tag{3.6}
$$

We now construct an explicit strain-energy function which can be used to describe a material which possesses a cubic phase and a tetragonal phase. In view of frame indifference, W depends on F only through the Cauchy-Green deformation tensor $C = F^T F$, or equivalently the Lagrangian strain tensor $E = (1/2)(C - I)$. As before, suppose that we take the reference configuration to coincide with an unstressed state of austenite at the transformation temperature. Then the reference configuration, and therefore the strain-energy function $W(E, \theta)$, must possess cubic symmetry. It is known (see, for example, Smith and Rivlin, 1958 and Green and Adkins, 1970) that, to have cubic symmetry, W must be a function of the "cubic invariants"

$$
I_1 = E_{11} + E_{22} + E_{33}, \qquad I_2 = E_{11}E_{22} + E_{22}E_{33} + E_{33}E_{11}
$$

$$
I_3 = E_{11}E_{22}E_{33}, \qquad I_4 = E_{12}E_{23}E_{31}, \tag{3.7}
$$

$$
I_5 = E_{12}^2 + E_{23}^2 + E_{31}^2, \qquad \ldots \text{etc.,}
$$

where E_{ij} refers to the i,j-component of E in the cubic basis $\{c_1, c_2, c_3\}$. The number of invariants in this list depends on the particular cubic class under consideration (see, e.g., Section 1.11 of Green and Adkins, 1970) but the analysis below is valid for all of these classes.

For temperatures close to the transformation temperature, Ericksen has argued based on experimental observations, that as a first approximation, all of the shear strain components (in the cubic basis) vanish, and additionally, that the sum of the normal strains also vanishes. Accordingly he suggested a geometrically constrained theory based on the two constraints

$$
I_1 = E_{11} + E_{22} + E_{33} = 0,
$$
$$
I_5 = E_{12}^2 + E_{23}^2 + E_{31}^2 = 0. \tag{3.8}
$$

In this case, the only two nontrivial strain invariants among those in the preceding list are I_2 and I_3 and so $W = W(I_2, I_3, \theta)$. Suppose that W is a polynomial in the components of E. In order to capture the desired multi-well character, W must be at least a quartic polynomial. It follows from (3.7) and (3.8) that the most general quartic polynomial of the form $W = W(I_2, I_3, \theta)$ is

$$
W = c_0 + c_2 I_2 + c_3 I_3 + c_{22} I_2^2, \tag{3.9}
$$

where the coefficients c_i may depend on temperature: $c_i = c_i(\theta)$.

In order to describe the cubic phase, this function W must have a local minimum at $\boldsymbol{E} = \boldsymbol{O}$, whereas to capture the tetragonal variants, W must have local minima at $\boldsymbol{E} = \boldsymbol{E}_k = (1/2)(\boldsymbol{U}_k^2 - \boldsymbol{I})$, $k = 1, 2, 3$, where the $\boldsymbol{U}_k$'s are given by (3.5). We now impose these requirements. The function W given in (3.9) automatically has an extremum at $\boldsymbol{E} = \boldsymbol{O}$, and this is a local minimum if $c_2 < 0$. Next, recall from the preceding section that, in the presence of material symmetry, if W has a local minimum at any one of the tensors $\boldsymbol{E}_k$, then W will automatically have energy-wells at the remaining two $\boldsymbol{E}_k$'s. In order for W to have an extremum at $\boldsymbol{E} = \boldsymbol{E}_1$ it is necessary that $c_2 = -pc_3 + 6p^2 c_{22}$, where we have set

$$p = \frac{1}{2}(\alpha^2 - 1). \tag{3.10}$$

It should be kept in mind that the constraint $I_1 = 0$ implies that $\text{tr } \boldsymbol{E}_1 = 0$ and therefore the lattice stretches α and β must be related by $2\alpha^2 + \beta^2 = 3$. If the extremum at $\boldsymbol{E} = \boldsymbol{E}_1$ is to be a local minimum one must also have $12pc_{22} > c_3 > 0$, where we are considering the case $p > 0$. Combining these various requirements leads to

$$c_2 = -pc_3 + 6p^2 c_{22}, \qquad 12pc_{22} > c_3 > 6pc_{22} > 0. \tag{3.11}$$

Finally, the values of W at the bottoms of the two energy-wells are $W(\boldsymbol{O}, \theta) = c_0(\theta)$ and $W(\boldsymbol{E}_1, \theta) = c_0(\theta) + p^3[c_3(\theta) - 9pc_{22}(\theta)]$. Since the martensite minimum must be lower than the austenite minimum for low temperatures, vice versa for high temperatures, and the values of W at the two minima must be the same at the transformation temperature, we require that

$$c_3'(\theta) - 9pc_{22}'(\theta) > 0, \qquad c_3(\theta_T) = 9pc_{22}(\theta_T), \tag{3.12}$$

where the prime denotes differentiation with respect to the argument. Thus, in summary, by collecting all of the preceding requirements, and introducing a more convenient pair of coefficients $d(\theta)$ and $e(\theta)$, the strain-energy function W can be expressed as

$$W = c_0(\theta) + d(\theta)\left[\frac{I_3}{p^3} - \frac{I_2}{p^2}\right] + e(\theta)\left[\left(\frac{I_2}{p^2}\right)^2 - 3\frac{I_2}{p^2} + 9\frac{I_3}{p^3}\right], \tag{3.13}$$

$$e(\theta) > 0, \quad 3e(\theta) > d(\theta) > -3e(\theta), \quad d'(\theta) > 0, \quad d(\theta_T) = 0.$$

Figure 7 shows a contour plot of the strain energy (3.13). In view of the constraints (3.8) there are only two independent strain components E_{11}, E_{22}. The figure shows the contours of $W(E_{11}, E_{22}, \theta)$, as given by (3.13), on the

E_{11}, E_{22}-plane. It shows the austenitic energy-well at the origin $\boldsymbol{E} = \boldsymbol{O}$ surrounded by the three martensitic energy-wells at $\boldsymbol{E} = \boldsymbol{E}_1, \boldsymbol{E}_2$ and $\boldsymbol{E}_3$. The values of the material parameters associated with this figure were chosen solely on the basis of obtaining a fairly clear contour plot.

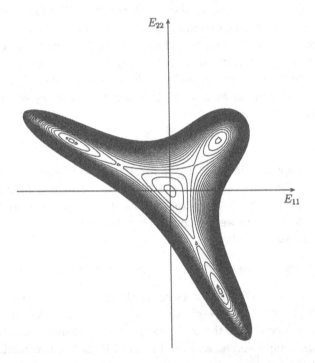

Fig. 7. Contours of constant strain energy on the E_{11}, E_{22}-plane. Observe the presence of the austenitic energy-well at $\boldsymbol{E} = \boldsymbol{O}$ surrounded by the three martensitic energy-wells at $\boldsymbol{E} = \boldsymbol{E}_1, \boldsymbol{E}_2, \boldsymbol{E}_3$.

In order to draw a one-dimensional cross-section of the energy (3.13), consider the path in strain space $\boldsymbol{E} = \boldsymbol{E}(\varepsilon)$, $-1.5 < \varepsilon < 1.5$, where

$$[E(\varepsilon)] = \begin{pmatrix} p\varepsilon^2 & 0 & 0 \\ 0 & p\varepsilon(3 - \varepsilon)/2 & 0 \\ 0 & 0 & -p\varepsilon(3 + \varepsilon)/2 \end{pmatrix}. \qquad (3.14)$$

Observe that $\boldsymbol{E}(0) = \boldsymbol{O}$, $\boldsymbol{E}(-1) = \boldsymbol{E}_2, \boldsymbol{E}(1) = \boldsymbol{E}_3$ so that this path passes through the austenite well and two of the martensite wells. Figure 8 shows the variation of energy along this path at three different temperatures: the figure displays $w(\varepsilon) = W(\boldsymbol{E}(\varepsilon), \theta)$ versus ε. The three graphs correspond to

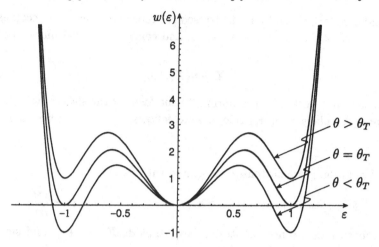

Fig. 8. A one-dimensional cross-section of the energy (3.13) along the path (3.14) in strain-space. The figure plots $w(\varepsilon) = W(\boldsymbol{E}(\varepsilon))$ versus strain ε at three different temperatures. Because we have omitted the effect of thermal expansion, the location of the local minima do not change with temperature. The material parameters underlying this plot are the same as those associated with the previous figure.

three temperatures greater than, equal to, and less than, the transformation temperature θ_T.

The strain-energy function (3.13) captures the key qualitative characteristics of a martensitic material which exists in cubic and tetragonal phases. However, due to the restrictive nature of the kinematic constraints (3.8), it fails to provide a *quantitatively* accurate model. The natural generalization of (3.13) is therefore to relax these constraints by using Lagrange multipliers c_5 and c_{11} and to replace (3.9) by

$$W = c_0 + c_2 I_2 + c_3 I_3 + c_{22} I_2^2 + c_5 I_5 + c_{11} I_1^2, \qquad c_i = c_i(\theta). \qquad (3.15)$$

James (see, for example, Kloućek and Luskin, 1994) has shown that the response predicted by this generalized form of W is in reasonable agreement with the observed behavior of In-Tl.

12.4 Uniaxial motion of a slab. Formulation

In order to explain some of the key ideas with a minimum of mathematical complexity, it is convenient to work in an essentially one-dimensional isothermal setting. In this section we formulate the basic equations pertaining to such a setting, and in the following two sections we shall make use of it.

Consider a slab, which in an unstressed reference configuration occupies the region $0 < X_1 < L, -\infty < X_2, X_3 < \infty$, and consider uniaxial motions of the form

$$x = X + u(X_1, t)e_1. \tag{4.1}$$

Here e_1 is a unit vector that is normal to the faces of the slab. Since only the coordinate X_1 plays a central role in what follows, from hereon we shall set

$$X_1 = x. \tag{4.2}$$

The deformation gradient tensor associated with (4.1) is

$$F(x, t) = I + \gamma(x, t)\, e_1 \otimes e_1, \quad \text{where} \quad \gamma(x, t) = \frac{\partial u}{\partial x}; \tag{4.3}$$

γ is a measure of the normal strain. Note that $\det F = 1 + \gamma$, and since we need $\det F > 0$, it is necessary that

$$\gamma > -1. \tag{4.4}$$

The particle velocity associated with (4.1) is

$$v = v(x, t)e_1, \quad \text{where} \quad v(x, t) = \frac{\partial u}{\partial t}. \tag{4.5}$$

Suppose that the slab is composed of an elastic material characterized by a strain-energy function $W(F)$. It is convenient for our present purposes to consider the restriction of the energy to uniaxial motions, and therefore to introduce the function w defined by

$$w(\gamma) = W(I + \gamma e_1 \otimes e_1), \quad \gamma > -1. \tag{4.6}$$

The first Piola-Kirchhoff stress tensor S is given by the constitutive relation

$$S = \left. \frac{\partial W}{\partial F} \right|_{F = I + \gamma e_1 \otimes e_1}. \tag{4.7}$$

We assume that the symmetry of the material is such that, in the uniaxial motion (4.1), all shear stress components calculated from (4.7) vanish identically (where the components are in an orthonormal basis $\{e_1, e_2, e_3\}$ with e_1 normal to the slab faces). This would be true, for example, if the material was isotropic, or even for certain anisotropic materials provided the material is suitably oriented with respect to the slab. Let σ denote the normal stress component S_{11}. It is then readily seen from (4.6), (4.7) that

$$\sigma = \hat{\sigma}(\gamma), \quad \text{where} \quad \hat{\sigma}(\gamma) = w'(\gamma), \quad \gamma > -1. \tag{4.8}$$

The only nonvanishing stresses in the slab are the normal stress components

$S_{11} (= \sigma)$, S_{22} and S_{33} and they only depend on x and t. Therefore, the equation of motion in the absence of body forces, $\mathrm{Div} S = \rho \dot{v}$, leads to the single scalar equation

$$\frac{\partial \sigma}{\partial x} = \rho \frac{\partial^2 u}{\partial t^2}, \tag{4.9}$$

where ρ denotes the mass density in the reference configuration.

In the formulation above, it has been assumed that the motion was sufficiently smooth. Suppose now that the motion is smooth on either side of a planar surface $x = s(t)$ but that the deformation gradient F and particle velocity v suffer jump discontinuities across this surface; we assume that the displacement itself remains continuous. In the context of this Chapter, $x = s(t)$ denotes the location of either a shock wave or a phase boundary, the distinction between them being that particles on either side of a shock wave belong to the same phase of the material while at a phase boundary they are in different phases. In this less smooth setting, the balance of linear momentum and the continuity of the displacement field lead, in general, to the respective jump conditions

$$[Sn] + \rho V_n [v] = o, \qquad [v] + V_n [Fn] = o, \tag{4.10}$$

where n is a unit normal vector to the surface of discontinuity and $V_n = V \cdot n$ is the normal velocity of propagation of this surface. The notation used here is that for any field quantity $g(X, t)$, $\overset{\pm}{g}$ denote its limiting values at a point on the surface of discontinuity, the limits being taken from either side of the surface, and $[g] = \overset{+}{g} - \overset{-}{g}$ is the associated jump in g. When specialized to uniaxial motions, (4.10) reduces to

$$[\sigma] = -\rho \dot{s} [v], \qquad [v] = -\dot{s} [\gamma], \tag{4.11}$$

where we have used the fact that now $V_n = \dot{s}$. By eliminating $[v]$ between the two equations in (4.11) and using (4.8) one obtains

$$\rho \dot{s}^2 = \frac{[\hat{\sigma}(\gamma)]}{[\gamma]} \tag{4.12}$$

indicating that $\rho \dot{s}^2$ equals the slope of the chord joining the points $(\overset{-}{\gamma}, \overset{-}{\sigma})$ and $(\overset{+}{\gamma}, \overset{+}{\sigma})$ on the stress-strain curve. Observe that in the special case when the stress response is linear, i.e. when $\hat{\sigma}(\gamma) = \mu \gamma$, equation (4.12) gives the propagation speed to be $\dot{s} = \pm \sqrt{\mu / \rho}$.

It is well-known in the classical theory of gas dynamics, that even though an inviscid fluid cannot dissipate energy, when the flow involves a shock wave there will in fact be dissipation localized to the shock surface. A propagating strain discontinuity in an elastic solid is similar. Thus, now consider the energetics of the slab and let $\Delta(t; \xi_1, \xi_2)$ denote the *rate of dissipation* associated with a slice of the slab between $x = \xi_1$ and $x = \xi_2$:

$$\Delta(t; \xi_1, \xi_2) = \sigma(\xi_2, t)v(\xi_2, t) - \sigma(\xi_1, t)v(\xi_1, t) - \frac{d}{dt} \int_{\xi_1}^{\xi_2} \left[w(\gamma(x, t)) + \frac{1}{2}\rho\dot{v}^2 \right] dx; \tag{4.13}$$

Δ represents the difference between the rate of external working and the rate at which the kinetic and stored energies increase (both per unit area in the X_2, X_3-plane). It is natural to require that the dissipation rate be non-negative, at all instants, and during all processes:

$$\Delta(t; \xi_1, \xi_2) \geq 0. \tag{4.14}$$

This *dissipation inequality* corresponds to a mechanical version of the second law of thermodynamics. When the fields are smooth, one can readily show that the dissipation rate Δ vanishes automatically, reflecting the conservative nature of an elastic material in a smooth process. On the other hand, when the motion involves a propagating strain discontinuity this is no longer true. Suppose that there is a strain discontinuity at some location $x = s(t)$ within the slice $\xi_1 < x < \xi_2$. Then by using the field equations and jump conditions one can rewrite (4.13) and express the dissipation rate in the slab *solely* in terms of quantities *at* the strain discontinuity:

$$\Delta = [\![\sigma v]\!] + [\![w]\!]\dot{s} + [\![\frac{1}{2}\rho v^2]\!]\dot{s} \geq 0. \tag{4.15}$$

Thus, if there is any dissipation of mechanical energy in the elastic slab it can only occur at surfaces of strain discontinuity. Finally, by substituting the momentum and kinematic jump conditions (4.11) into the identity

$$\overset{+}{\sigma}\overset{+}{v} - \bar{\sigma}\bar{v} = \frac{1}{2}(\overset{+}{\sigma} + \bar{\sigma})(\overset{+}{v} - \bar{v}) + \frac{1}{2}(\overset{+}{\sigma} - \bar{\sigma})(\overset{+}{v} + \bar{v}) \tag{4.16}$$

one finds that

$$[\![\sigma v]\!] = \frac{1}{2}(\overset{+}{\sigma} + \bar{\sigma})[\![\gamma]\!]\dot{s} + \frac{1}{2}\rho\dot{s}[\![v^2]\!]. \tag{4.17}$$

Therefore, the dissipation rate (4.15) can be expressed as

$$\Delta(t) = f\dot{s} \geq 0, \tag{4.18}$$

where

$$f = w(\overset{+}{\gamma}) - w(\bar{\gamma}) - \frac{1}{2}\left[\hat{\sigma}(\overset{+}{\gamma}) + \hat{\sigma}(\bar{\gamma}) \right](\overset{+}{\gamma} - \bar{\gamma}). \tag{4.19}$$

The quantity f is called the *driving force* (per unit area) on the surface of discontinuity and is an example of Eshelby's notion of a "configurational force on a defect" (Eshelby 1956, 1970). The dissipation inequality (4.18) must hold at all discontinuities. In summary, in a *dynamic problem* we shall seek velocity, strain and stress fields, $v(x,t), \gamma(x,t), \sigma(x,t)$, which conform to given initial and boundary conditions, obey the constitutive relation $\sigma = \hat{\sigma}(\gamma) = w'(\gamma)$ and satisfy the following field equations and jump conditions:

$$\frac{\partial \sigma}{\partial x} = \rho \frac{\partial v}{\partial t}, \qquad \frac{\partial v}{\partial x} = \frac{\partial \gamma}{\partial t} \qquad \text{where the fields are smooth,}$$

$$\rho \dot{s}^2 = \frac{\overset{+}{\sigma} - \overset{-}{\sigma}}{\overset{+}{\gamma} - \overset{-}{\gamma}}, \qquad \overset{+}{v} - \overset{-}{v} = -\dot{s}(\overset{+}{\gamma} - \overset{-}{\gamma}), \qquad f\dot{s} \geq 0 \qquad \text{at each discontinuity.}$$

In the case of a *static problem*, we set all time derivatives in the preceding formulation to vanish and so the appropriate field equations and jump conditions controlling $u(x), \gamma(x)$ and $\sigma(x)$ are

$$\frac{d\sigma}{dx} = 0, \qquad \gamma = \frac{du}{dx} \qquad \text{where the fields are smooth,}$$

$$\overset{+}{\sigma} = \overset{-}{\sigma}, \qquad \overset{+}{u} = \overset{-}{u} \qquad \text{at each discontinuity.}$$

(4.20)

This completes the statement of the various field equations and jump conditions. Before turning to specific boundary-initial value problems, we need also to prescribe the constitutive relation of the material. Since our interest is in two-phase materials, the strain-energy function w must possess a two-well structure. Accordingly we shall assume that $w(\gamma)$ has *two energy-wells* located at $\gamma = 0$ and $\gamma = \gamma_T > 0$ and an intermediate local maximum at $\gamma = \gamma_*$:

$$w'(0) = w'(\gamma_*) = w'(\gamma_T) = 0; \qquad w''(0) > 0, \quad w''(\gamma_*) < 0, \quad w''(\gamma_T) > 0.$$

(4.21)

Further, we assume that there are exactly two points $\gamma = \gamma_{max}$ and $\gamma = \gamma_{min}$, ordered such that

$$0 < \gamma_{max} < \gamma_* < \gamma_{min} < \gamma_T, \tag{4.22}$$

at which the curvature of w changes sign:

$$\left. \begin{aligned} w''(\gamma) > 0 \quad \text{for} \quad -1 < \gamma < \gamma_{max}, \\ w''(\gamma) < 0 \quad \text{for} \quad \gamma_{max} < \gamma < \gamma_{min}, \\ w''(\gamma) > 0 \quad \text{for} \quad \gamma > \gamma_{min}. \end{aligned} \right\} \tag{4.23}$$

It follows that the stress response function $\hat{\sigma}(\gamma) = w'(\gamma)$ increases for $-1 <$

$\gamma < \gamma_{max}$, decreases for $\gamma_{max} < \gamma < \gamma_{min}$, and increases again for $\gamma > \gamma_{min}$; and that the stress vanishes at $\gamma = 0, \gamma_*$ and γ_T:

$$\left.\begin{aligned} &\hat{\sigma}(0) = \hat{\sigma}(\gamma_*) = \hat{\sigma}(\gamma_T) = 0, \\[4pt] &\hat{\sigma}'(\gamma) > 0 \quad \text{for} \quad -1 < \gamma < \gamma_{max}, \\[4pt] &\hat{\sigma}'(\gamma) < 0 \quad \text{for} \quad \gamma_{max} < \gamma < \gamma_{min}, \\[4pt] &\hat{\sigma}'(\gamma) > 0 \quad \text{for} \quad \gamma > \gamma_{min}. \end{aligned}\right\} \tag{4.24}$$

The two phases of this material are a *low-strain phase* corresponding to the strain interval $-1 < \gamma < \gamma_{max}$ and a *high-strain phase* corresponding to $\gamma > \gamma_{min}$. (We shall not refer to the low-strain phase as austenite, and the high-strain phase as martensite because our discussion is also valid, for example, if the low- and high-strain phase correspond to two variants of martensite.) Figure 9 shows a particular example of such functions w and $\hat{\sigma}$; in this example w has been taken to be piecewise quadratic and consequently $\hat{\sigma}$ is piecewise linear.

12.5 A static problem and the role of energy minimization
(Ericksen, 1975)

Suppose that the slab described in the preceding section is in equilibrium. The basic equations governing its kinematics, constitutive response and equilibrium are

$$\gamma(x) = u'(x), \quad \sigma(x) = \hat{\sigma}(\gamma(x)), \quad \sigma'(x) = 0, \tag{5.1}$$

which must hold at each $x \in (0, L)$ except at locations of strain discontinuity; if there is a strain jump at $x = s$, then the jump conditions

$$\overset{+}{\sigma} = \bar{\sigma}, \quad \overset{+}{u} = \bar{u} \tag{5.2}$$

apply there. On integrating $(5.1)_3$ and using $(5.1)_2$, and if necessary $(5.2)_1$ as well, one finds

$$\hat{\sigma}(\gamma(x)) = \sigma, \quad 0 < x < L, \tag{5.3}$$

where the constant σ represents the (as yet unknown) value of normal stress in the slab. Note that (5.3) must hold independently of whether or not the strain field is continuous.

Since our aim is to illustrate certain points with a minimum of mathematical complexity, it will be convenient for us to take the energy $w(\gamma)$ to be a piecewise

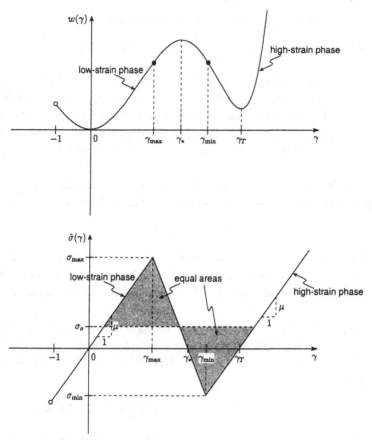

Fig. 9. Graphs of a piecewise quadratic strain-energy function $w(\gamma)$ and the corresponding piecewise linear stress-strain relation $\sigma = \hat{\sigma}(\gamma) = w'(\gamma)$.

quadratic, continuously differentiable function, so that $\hat{\sigma}(\gamma)$ is then a piecewise linear, continuous function. Specifically, we will take

$$
w(\gamma) = \begin{cases} \frac{\mu}{2}\gamma^2, \\ \frac{\mu}{2}(\gamma - \gamma_T)^2 + \sigma_o\gamma_T, \end{cases} \qquad \hat{\sigma}(\gamma) = \begin{cases} \mu\gamma, & -1 \le \gamma \le \gamma_{max}, \\ \mu(\gamma - \gamma_T), & \gamma \ge \gamma_{min}, \end{cases}
$$

(5.4)

where the modulus μ and material parameter σ_o are positive constants. Graphs of $w(\gamma)$ and $\hat{\sigma}(\gamma)$ are shown in Figure 9. The maximum and minimum values of stress, σ_{max} and σ_{min}, shown in the figure correspond to

$$\sigma_{max} = \hat{\sigma}(\gamma_{max}) = \mu\gamma_{max} > 0, \qquad \sigma_{min} = \hat{\sigma}(\gamma_{min}) = \mu(\gamma_{min} - \gamma_T) < 0.$$
$$(5.5)$$

The declining branch of the stress-strain curve corresponds to unstable states of the material and so in statics we will require the strain to lie outside the interval $(\gamma_{max}, \gamma_{min})$; for this reason we have not displayed the formulae pertaining to the interval $\gamma_{max} < \gamma < \gamma_{min}$ in (5.4). The inequality $\sigma_{min} < 0$ is not essential; it simply ensures that w has two energy-wells rather than being merely non-convex.

Suppose that the left boundary of the slab is fixed and the right boundary is subjected to a prescribed displacement δ in the x_1-direction:

$$u(0) = 0, \qquad u(L) = \delta. \qquad (5.6)$$

Given δ and the strain-energy function $w(\gamma)$, we are to determine the displacement, strain and stress fields such that the equations (5.1), jump conditions (5.2) and boundary conditions (5.6) hold. We call this the "equilibrium problem". †

Solution-1: Suppose first that the slab involves only the low-strain phase so that $\gamma(x) \in (-1, \gamma_{max})$ at every particle $x \in (0, L)$. The relevant constitutive relation is then, from (5.4)$_2$, $\sigma = \mu\gamma$. This, together with (5.3), (5.1)$_1$ and (5.6)$_1$ leads to

$$\sigma(x) = \sigma, \qquad \gamma(x) = \frac{\sigma}{\mu}, \qquad u(x) = \frac{\sigma}{\mu}x, \qquad 0 < x < L. \qquad (5.7)$$

The remaining boundary condition (5.6)$_2$ requires that the stress and elongation be related by

$$\sigma = \mu\frac{\delta}{L}. \qquad (5.8)$$

Since we have assumed that the entire slab is in the low-strain phase, i.e. that $-1 < \gamma(x) < \gamma_{max}$ for $0 < x < L$, the solution (5.7), (5.8) is valid only if

$$-1 < \frac{\delta}{L} < \frac{\sigma_{max}}{\mu}. \qquad (5.9)$$

Thus, in summary, if the prescribed elongation lies in the interval (5.9), the stress, strain and displacement fields in the slab are given by (5.7), (5.8).

Solution-2: Next, consider the case where the slab involves the high-strain

---

† It would be more natural to consider a bar, rather than a slab, since that would allow us to compare the theory with experiment. Here we have chosen to work with a slab because, in contrast to a (three-dimensional) bar, the setting of the slab leads to an exact solution within the theory of finite elasticity.

phase only, so that now $\gamma(x) > \gamma_{min}$ for $0 < x < L$. The constitutive relation is thus $\sigma = \mu(\gamma - \gamma_T)$ and we now find that

$$\sigma(x) = \sigma, \qquad \gamma(x) = \frac{\sigma}{\mu} + \gamma_T, \qquad u(x) = \left(\frac{\sigma}{\mu} + \gamma_T\right) x, \qquad 0 < x < L.$$
(5.10)

By using the boundary condition $(5.6)_2$, the stress and elongation are found to be related by

$$\sigma = \mu\left(\frac{\delta}{L} - \gamma_T\right).$$
(5.11)

Finally, the assumption that the solution involves only the high-strain phase imposes the requirement

$$\frac{\delta}{L} > \frac{\sigma_{min}}{\mu} + \gamma_T.$$
(5.12)

Summarizing, when the prescribed elongation is sufficiently small, i.e. when $-1 < \delta/L < \sigma_{max}/\mu$, there is a configuration of the slab in which the strain is uniform and every particle is in the low-strain phase. Likewise when the elongation is sufficiently large, i.e. $\delta/L > \sigma_{min}/\mu + \gamma_T$, there is a uniform configuration associated with the high-strain phase. Note that we have *not* yet found a solution for values of elongation in the intermediate range $\sigma_{max}/\mu \leq \delta/L \leq \sigma_{min}/\mu + \gamma_T$. For δ in this middle interval, there is of course a solution with uniform strain that is associated with the declining branch of the stress-strain curve; but that solution is unstable. In order to find an alternative solution when δ is in this intermediate range, it is natural to consider the possibility of a *mixture* of the low- and high-strain phases.

<u>Solution-3</u>: For some $s \in (0, L)$, suppose that the segment $0 < x < s$ of the slab is composed of the high-strain phase while the segment $s < x < L$ is composed of the low-strain phase. The interface at $x = s$ separates these two phases and is therefore a phase boundary. The stress is again constant throughout the slab (see (5.3)): $\sigma(x) = \sigma$, $0 < x < L$. The strains are determined by making use of the relevant stress-strain relations, i.e. $\sigma = \mu\gamma$ for the low-strain segment and $\sigma = \mu(\gamma - \gamma_T)$ for the high-strain segment. The displacement field then follows by integration, keeping in mind the continuity requirement $(5.2)_2$. In this way we find

$$\gamma(x) = \begin{cases} \bar{\gamma} = \dfrac{\sigma}{\mu} + \gamma_T, & 0 < x < s, \\[2mm] \overset{+}{\gamma} = \dfrac{\sigma}{\mu}, & s < x < L. \end{cases} \qquad u(x) = \begin{cases} \bar{\gamma} x, & 0 < x < s, \\[2mm] \overset{+}{\gamma} x + (\bar{\gamma} - \overset{+}{\gamma})s, & s < x < L. \end{cases}$$
(5.13)

458   R. Abeyaratne et al.

From $(5.13)_2$ and the boundary condition $(5.6)_2$ we obtain

$$\sigma = \mu \left(\frac{\delta}{L} - \frac{s}{L}\gamma_T \right), \qquad (5.14)$$

which relates the stress, elongation and phase boundary location. Finally, recall that this solution was obtained under the assumption that the strains $\bar{\gamma}$ and $\overset{+}{\gamma}$ lie in the intervals $\bar{\gamma} > \gamma_{min}$, $-1 < \overset{+}{\gamma} < \gamma_{max}$, corresponding, respectively, to the high-strain and low-strain phases. It follows by using (5.13), (5.14), that this two-phase solution exists whenever

$$\frac{\sigma_{min}}{\mu} + \frac{s}{L}\gamma_T < \frac{\delta}{L} < \frac{\sigma_{max}}{\mu} + \frac{s}{L}\gamma_T. \qquad (5.15)$$

Thus, given any $s \in (0, L)$, if the prescribed elongation δ lies in the range (5.15), then there exists a two-phase solution involving a mixture of both phases; it is given by (5.13), (5.14).

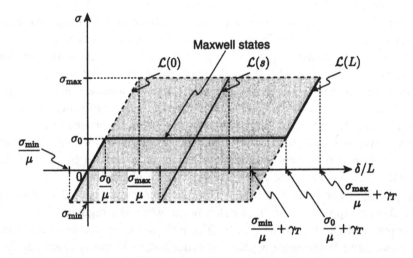

Fig. 10. The set of all two-phase equilibria described on the (δ, σ)-plane. Each point in the parallelogram can be associated with a two-phase solution (5.13)–(5.15). The left and right hand boundaries of the parallelogram correspond, respectively, to the low-strain and high-strain phase solutions. Energy extremizing two-phase equilibria, i.e. solutions which describe phase equilibrium, are associated with the horizontal line labeled Maxwell states.

The arbitrariness of the phase boundary location in Solution-3 reflects a lack of uniqueness to the equilibrium problem as posed. A useful way in which to examine the nature of this nonuniqueness is to consider the (δ, σ)-plane. In Figure 10, the straight line segment $\mathcal{L}(s)$ is defined by (5.14) and (5.15), and represents the stress-elongation response of Solution-3 at some fixed value of s. Corresponding to each $s \in (0, L)$ there is a line segment $\mathcal{L}(s)$, and as s ranges over the interval $0 < s < L$ this one-parameter family of lines fill the parallelogram shown in the figure. Each point (δ, σ) in the interior of this parallelogram can be uniquely associated with a solution of the type under consideration. The left and right boundaries of the parallelogram correspond to $s = 0$ and $s = L$ respectively. Observe by setting $s = 0$ in (5.14) and comparing it with (5.8) that the left boundary, $\mathcal{L}(0)$, in fact describes a portion of the σ, δ-curve associated with Solution-1. Similarly, the right boundary, $\mathcal{L}(L)$, describes a portion of the σ, δ-curve associated with Solution-2. It can be seen from the figure that the equilibrium problem has a two-phase solution (Solution-3) whenever the elongation δ lies in the range $\sigma_{min}/\mu < \delta/L < \sigma_{max}/\mu + \gamma_T$; moreover, for each value of δ in this interval, the problem has a one parameter family of such solutions, the (unknown) phase boundary location s being the parameter. Had we allowed for the presence of an arbitrary number of phase boundaries, the positions of each of them would have been indeterminate.

Variational approach. In an attempt to resolve this nonuniqueness, and in particular to determine the phase boundary location, we turn to an alternative notion of equilibrium, viz. equilibrium in the sense of energy extremization. The total potential energy (per unit area in the (x_2, x_3)-plane) is

$$E(u) = \int_0^L w(u'(x))dx, \qquad u \in \mathcal{A}, \qquad (5.16)$$

where the set $\mathcal{A}$ of kinematically admissible displacement fields consists of all continuous, piecewise smooth functions $u : [0, L] \to \mathbb{R}$ which satisfy the boundary conditions (5.6):

$$\mathcal{A} = \{ u \ : \ u \in C_p^2(O, L) \cap C(0, L), \ u(0) = 0, \ u(L) = \delta \ \}. \qquad (5.17)$$

Suppose that E is minimized by a particular function $u \in \mathcal{A}$, and suppose that u' is discontinuous at a location $x = s$. Consider a one-parameter family of admissible displacement fields $\bar{u}(x, \varepsilon)$, $-\varepsilon_o < \varepsilon < \varepsilon_o$, which has the property that $\bar{u}(x, 0) = u(x)$. Suppose that for each ε, the first derivative of $\bar{u}(\cdot, \varepsilon)$ is discontinuous at the location $x = \bar{s}(\varepsilon)$ and that $\bar{s}(0) = s$. Then, by setting

$$\frac{d}{d\varepsilon} E(\bar{u}(x, \varepsilon)) \bigg|_{\varepsilon=0} = \frac{d}{d\varepsilon} \int_0^{\bar{s}(\varepsilon)} w(\bar{u}'(x, \varepsilon))dx \bigg|_{\varepsilon=0}$$

$$+ \frac{d}{d\varepsilon} \int_{\bar{s}(\varepsilon)}^L w(\bar{u}'(x, \varepsilon))dx \bigg|_{\varepsilon=0} = 0 \qquad (5.18)$$

one finds in the usual way that it is necessary that u satisfy the equilibrium equation

$$\frac{d}{dx}\hat{\sigma}(u'(x)) = 0, \qquad x \neq s, \ 0 < x < L. \tag{5.19}$$

In addition, the arbitrariness of $\bar{u}_\varepsilon(s-,0)$ requires that u' satisfy the jump condition

$$[\![\hat{\sigma}(u')]\!] = 0 \qquad \text{at} \quad x = s; \tag{5.20}$$

and finally, the arbitrariness of $\bar{s}'(0)$ requires that u' satisfy a second jump condition

$$[\![w(u') - \hat{\sigma}(u')u']\!] = 0 \qquad \text{at} \quad x = s. \tag{5.21}$$

This pair of jump conditions are the so-called Weierstrass-Erdmann corner conditions of the calculus of variations (see, for example, Gelfand and Fomin, 1963). The first of these is precisely the stress continuity condition $(5.2)_1$, whereas the second condition is equivalent to the vanishing of the driving force on the phase boundary:

$$f = 0, \qquad \text{where} \quad f = w(\overset{+}{\gamma}) - w(\bar{\gamma}) - \sigma(\overset{+}{\gamma} - \bar{\gamma}); \tag{5.22}$$

cf. equation (4.19).

Thus, the energy extremization notion of equilibrium is more stringent than the force balance point of view in that it requires all of the same conditions *plus* an extra condition. This additional requirement of vanishing driving force is often referred to as a characterization of "phase equilibrium". It can be written, after using $(4.8)_2$ as

$$\int_{\bar{\gamma}}^{\overset{+}{\gamma}} \hat{\sigma}(\gamma) \, d\gamma = \sigma(\overset{+}{\gamma} - \bar{\gamma}). \tag{5.23}$$

Since $\sigma = \hat{\sigma}(\overset{+}{\gamma}) = \hat{\sigma}(\bar{\gamma})$, this can be described geometrically as saying that the stress σ cuts off equal areas of the stress-strain curve; see Figure 9. The special value of stress for which this is true is called the *Maxwell stress*.

In the case of the piecewise quadratic energy (5.4), substituting $(5.13)_1$ and (5.14) into (5.23) shows that the Maxwell stress equals σ_o, where σ_o is the material constant introduced earlier in (5.4). Thus, if a two-phase configuration is to extremize the energy, the stress must equal the Maxwell stress throughout the slab: $\sigma(x) = \sigma_o$, $0 < x < L$; the strain and displacement fields are found by replacing σ in (5.13) by σ_o:

$$u(x) = \begin{cases} \bar{\gamma} \, x, & 0 < x < s, \\ \overset{+}{\gamma} x + (\bar{\gamma} - \overset{+}{\gamma})s, & s < x < L, \end{cases} \qquad \text{where} \quad \bar{\gamma} = \frac{\sigma_o}{\mu} + \gamma_T, \quad \overset{+}{\gamma} = \frac{\sigma_o}{\mu}.$$

$$(5.24)$$

Setting $\sigma = \sigma_o$ in (5.14) allows us to solve for the previously unknown location s of the phase boundary:

$$s = \left(\frac{\delta}{L} - \frac{\sigma_o}{\mu} \right) \frac{L}{\gamma_T}. \qquad (5.25)$$

Finally, the inequality (5.15) defining the range of values of δ for which a two-phase solution exists now specializes to

$$\frac{\sigma_o}{\mu} < \frac{\delta}{L} < \frac{\sigma_o}{\mu} + \gamma_T. \qquad (5.26)$$

From among all points in the parallelogram in Figure 10, the energy-extremizing two-phase solutions are associated with points on the horizontal line $\sigma = \sigma_o$ labeled "Maxwell states".

Uniqueness. The preceding discussion concerned Solution-3. Collecting all of the solutions at hand, we now have a low-strain phase solution (Solution-1), a high-strain phase solution (Solution-2), and a two-phase (Maxwell) solution (Solution-3 with $\sigma = \sigma_o$) associated with values of δ in the respective ranges (5.9), (5.12) and (5.26). Observe from this, and perhaps more clearly from Figure 10, that the equilibrium problem now has a unique solution if the elongation lies in the interval $(-1, \sigma_o/\mu] \cup [\sigma_{max}/\mu, \sigma_{min}/\mu + \gamma_T] \cup [\sigma_o/\mu + \gamma_T, \infty)$.

However, for each $\delta \in (\sigma_o/\mu, \sigma_{max}/\mu)$ we have two solutions, a low-strain phase solution and a two-phase solution (corresponding to parts of the lines $\mathcal{L}(0)$ and $\sigma = \sigma_o$ in Figure 10); and similarly, associated with each $\delta \in (\sigma_{min}/\mu + \gamma_T, \sigma_o/\mu + \gamma_T)$, we have a high-strain phase solution and a two-phase solution (corresponding to $\mathcal{L}(L)$ and $\sigma = \sigma_o$). In order to distinguish between these solutions let us compare their energies. Evaluating (5.16) at each of the three solutions (5.7), (5.10) and (5.24) leads to the corresponding energies

$$E_1 = \frac{\mu}{2} \left(\frac{\delta}{L} \right)^2, \quad E_2 = \frac{\mu}{2} \left(\frac{\delta}{L} - \gamma_T \right)^2 + \sigma_o \gamma_T, \quad E_3 = -\frac{\sigma_o^2}{2\mu} + \sigma_o \frac{\delta}{L}, \qquad (5.27)$$

respectively. It can be readily verified that

$$E_3 - E_1 = -\frac{\mu}{2} \left(\frac{\delta}{L} - \frac{\sigma_o}{\mu} \right)^2 \leq 0, \quad \text{and} \quad E_3 - E_2 = -\frac{\mu}{2} \left(\frac{\delta}{L} - \gamma_T - \frac{\sigma_o}{\mu} \right)^2 \leq 0.$$

$$(5.28)$$

Therefore, whenever a two-phase solution and a single phase solution exist at the same value of δ, the two-phase solution has less energy.

Consequently, from among all equilibrium solutions, the one with the least energy for the various ranges of δ is:

$$-1 \quad < \quad \delta/L \quad \leq \sigma_o/\mu \qquad \text{low}-\text{strain phase solution,}$$

$$\sigma_o/\mu \quad < \quad \delta/L \quad < \sigma_o/\mu + \gamma_T \qquad \text{two}-\text{phase (Maxwell) solution,}$$

$$\sigma_o/\mu + \gamma_T \quad \leq \quad \delta/L \qquad \text{high}-\text{strain phase solution.}$$

$$(5.29)$$

From (5.8), (5.11) and $\sigma = \sigma_o$, the corresponding relationships between stress σ and elongation δ are

$$\sigma = \begin{cases} \mu\dfrac{\delta}{L} & \text{for} \quad -1 \quad < \quad \delta/L \quad \leq \sigma_o/\mu, \\ \sigma_0 & \text{for} \quad \sigma_o/\mu \quad \leq \quad \delta/L \quad \leq \sigma_o/\mu + \gamma_T, \\ \mu\left(\dfrac{\delta}{L} - \gamma_T\right) & \text{for} \qquad\qquad \delta/L \quad \geq \sigma_o/\mu + \gamma_T. \end{cases} \qquad (5.30)$$

The (σ, δ)-relation (5.30) is depicted in Figure 10 by the bold portions of $\mathcal{L}(0)$ and $\mathcal{L}(L)$ and the horizontal line $\sigma = \sigma_o$. It should be emphasized that this graph simply depicts the set of energy minimizing equilibrium solutions; as we shall discuss later, it does *not* imply that this is the curve that is traversed during a loading-unloading process.

Potential energy. Before concluding our discussion of static equilibrium it is worth making some brief observations about the *potential energy function* of the material. Recall that the local minima of the strain-energy function $w(\gamma)$ inform us about the various *stress-free* phases of the material and their relative stability. The *potential energy function* $G(\gamma, \sigma)$ plays a similar role when the stress does not vanish. In particular, given the stress σ, different phases of the material at that value of stress will be associated with different local minima of $G(\cdot, \sigma)$. It is worth pointing out that the energy E that was considered in the preceding discussion was the total (macroscopic) energy of the slab; here we are concerned with a local energy associated with the material.

The potential energy G is defined by

$$G(\gamma, \sigma) = w(\gamma) - \sigma\gamma, \qquad (5.31)$$

where σ and γ are treated as independent variables, not necessarily related

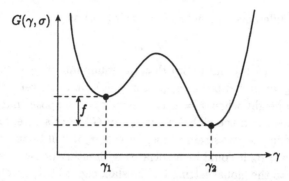

Fig. 11. The potential energy G as a function of strain at a fixed value of stress. When the stress lies in the range $\sigma_{min} < \sigma < \sigma_{max}$, $G(\cdot, \sigma)$ has two local minima. The minimum on the left is the global minimum when $\sigma_{min} < \sigma < \sigma_o$ while the one on the right is the global minimum for $\sigma_o < \sigma < \sigma_{max}$; here σ_o is the Maxwell stress.

through the stress-strain relation. Observe that

$$\frac{\partial G}{\partial \gamma} = 0 \quad \Leftrightarrow \quad \sigma = w'(\gamma) = \hat{\sigma}(\gamma), \tag{5.32}$$

and that

$$\frac{\partial^2 G}{\partial \gamma^2} = w''(\gamma) = \hat{\sigma}'(\gamma). \tag{5.33}$$

Consequently, given σ, a strain γ corresponds to a local minimum of the potential energy $G(\cdot, \sigma)$ if $\sigma = \hat{\sigma}(\gamma)$ and $\hat{\sigma}'(\gamma) > 0$; it corresponds to a local maximum if $\sigma = \hat{\sigma}(\gamma)$ and $\hat{\sigma}'(\gamma) < 0$. Thus, any point on any rising branch of the stress-strain curve can be associated with a local minimum of the potential energy function. If the stress-strain curve has the rising, falling, rising form shown in Figure 11, then G has multiple local minima for certain ranges of σ.

Consider for example the material characterized by (5.4) and described in Figure 9. For this material, when $\sigma < \sigma_{min}$ the function $G(\cdot, \sigma)$ has a single minimum and it is located in the low-strain phase at $\gamma = \gamma_1 = \sigma/\mu$; for $\sigma > \sigma_{max}$, $G(\cdot, \sigma)$ again has a single minimum but now it is located in the high-strain phase at $\gamma = \gamma_2 = \sigma/\mu + \gamma_T$. On the other hand when $\sigma_{min} < \sigma < \sigma_{max}$ the potential energy $G(\cdot, \sigma)$ has local minima at both γ_1 and γ_2. The height $G(\gamma_1, \sigma) - G(\gamma_2, \sigma)$ between these minima is

$$G(\gamma_1, \sigma) - G(\gamma_2, \sigma) = w(\gamma_1) - w(\gamma_2) - \sigma(\gamma_1 - \gamma_2). \tag{5.34}$$

Even though our discussion here has not concerned a phase boundary, (5.34) is precisely what would be the driving force if there was a a phase boundary separating the strain γ_1 from the strain γ_2, cf. $(5.22)_2$. In terms of stress, one can write this difference in height after using (5.4) as

$$f = (\sigma - \sigma_o)\gamma_T. \tag{5.35}$$

It follows from (5.34) and (5.35) that the minimum at γ_1 is lower than the minimum at γ_2 when $f < 0$ (i.e. $\sigma_{min} < \sigma < \sigma_o$), and vice versa; both minima have the same height when $f = 0$ (i.e. $\sigma = \sigma_o$). Suppose that a particle is always associated with the global minimum of G. In this case, for $\sigma < \sigma_o$ the particle will be in the low-strain phase, for $\sigma > \sigma_o$ it will be in the high-strain phase, and at $\sigma = \sigma_o$ it could be associated with either phase. Returning from this local view to the global setting for the slab, e.g., (5.30), we observe that in any configuration of the slab which minimizes the total energy E, the individual particles are associated with states that minimize the potential energy G.

12.6 A dynamic problem and the role of kinetics and nucleation
(Abeyaratne and Knowles, 1991a, 1993b, 2000b)

Thus far we have restricted attention to statics and consequently, even though we have encountered both phases of the two-phase material, we have not considered the *transformation* from one into the other. The process of phase transformation is inherently time-dependent, and in order to examine it we now turn to a dynamic problem for a slab. From Section 12.4 we recall the basic differential equations

$$\frac{\partial}{\partial x}\hat{\sigma}(\gamma) = \rho\,\frac{\partial v}{\partial t}, \qquad \frac{\partial v}{\partial x} = \frac{\partial \gamma}{\partial t}, \tag{6.1}$$

which must hold wherever the fields are smooth. If the strain and particle velocity suffer jump discontinuities at $x = s(t)$, the jump conditions

$$\rho\dot{s}^2 = \frac{\hat{\sigma}(\overset{+}{\gamma}) - \hat{\sigma}(\overline{\gamma})}{\overset{+}{\gamma} - \overline{\gamma}}, \qquad \overset{+}{v} - \overline{v} = -\dot{s}(\overset{+}{\gamma} - \overline{\gamma}), \tag{6.2}$$

and the dissipation inequality,

$$f\dot{s} \geq 0, \qquad \text{where} \qquad f = w(\overset{+}{\gamma}) - w(\overline{\gamma}) - \frac{1}{2}\left[\hat{\sigma}(\overset{+}{\gamma}) + \hat{\sigma}(\overline{\gamma})\right](\overset{+}{\gamma} - \overline{\gamma}), \tag{6.3}$$

must hold there.

Suppose that the slab is semi-infinite ($L = \infty$) and that it is initially undeformed and at rest:

$$\gamma(x,0) = 0, \quad v(x,0) = 0 \qquad \text{for } x > 0; \tag{6.4}$$

thus, the slab is composed of the unstressed low-strain phase at time $t = 0$.

Finally, suppose that the boundary $x = 0$ is subjected to a constant normal velocity $-\dot{\delta}$ for all positive time:

$$v(0,t) = -\dot{\delta} \quad \text{for } t > 0, \quad \text{where } \dot{\delta} > 0. \tag{6.5}$$

The problem at hand involves finding $\gamma(x,t), v(x,t)$ by solving the differential equations (6.1) subject to the jump conditions (6.2), the dissipation inequality (6.3), the initial conditions (6.4) and the boundary condition (6.5). Since we take $\dot{\delta} > 0$ the slab is being subjected to a sudden elongational loading. We shall refer to this as the "dynamic problem".

Several general results pertaining to the Cauchy problem associated with quasilinear systems of equations of the form (6.1) are known. In particular, if the stress response function $\hat{\sigma}(\gamma)$ is monotonically increasing, and in addition it is *either* strictly convex or strictly concave, then it is known that the Cauchy problem has a unique (weak) solution (Oleinik, 1957). In the present setting $\hat{\sigma}(\gamma)$ is characterized by (4.21)–(4.24) or more specifically by (5.4). It is neither monotonic nor does it have the convexity property required for this uniqueness result to hold. Thus, we expect that the solution to the dynamic problem may be non-unique. Nevertheless, it is worth attempting to solve this problem as posed above, with the aim of finding *all* of its solutions. The hope would then be that some characteristic of the totality of solutions might reveal the (physical) cause of the non-uniqueness and suggest a remedy.

Before attempting to solve the dynamic problem it is useful to make some preliminary observations pertaining to the character of propagating discontinuities. In contrast to the setting of statics, here a strain discontinuity can correspond to either a phase boundary or a shock wave. If the material on both sides of the discontinuity belong to the same phase then the discontinuity is a *shock wave*. If $\bar{\gamma}$ and $\overset{+}{\gamma}$ are both associated with the low-strain phase, then from (5.4) we have $\overset{\pm}{\sigma} = \mu\overset{\pm}{\gamma}$, whereas if they are both associated with the high-strain phase then $\overset{\pm}{\sigma} = \mu(\overset{\pm}{\gamma} - \gamma_T)$. In either case, the propagation speed of the shock wave is given by (6.2)$_1$, which now simplifies to

$$\dot{s} = \pm c, \quad c = \sqrt{\frac{\mu}{\rho}}. \tag{6.6}$$

Thus (for the material at hand), shock waves propagate at a speed c that is independent of the strains on either side of it; this degeneracy is special to the piecewise linear stress-strain relation. Next, if the strains on the two sides of the discontinuity belong to distinct phases, the discontinuity is a *phase boundary*. For example, suppose that $\bar{\gamma}$ is associated with the high-strain phase and $\overset{+}{\gamma}$ is associated with the low-strain phase; then $\bar{\sigma} = \mu(\bar{\gamma} - \gamma_T)$ and $\overset{+}{\sigma} = \mu\overset{+}{\gamma}$.

Substituting these into $(6.2)_1$ allows us to express the propagation speed as

$$\frac{\dot{s}^2}{c^2} = 1 - \frac{\gamma_T}{\overset{-}{\gamma} - \overset{+}{\gamma}}. \tag{6.7}$$

Note that $\overset{-}{\gamma} > \overset{+}{\gamma}$ since $\overset{-}{\gamma}$ is in the high-strain and $\overset{+}{\gamma}$ is in the low-strain phase. Thus, according to (6.7), a phase boundary propagates at a subsonic speed $|\dot{s}| < c$.

Next consider the driving force f on a discontinuity. By using the constitutive relation (5.4) and evaluating $(6.3)_2$, we find that $f = 0$ at a shock wave; again this degeneracy is peculiar to a material with a piecewise linear stress-strain relation. On the other hand, for a phase boundary where $\overset{-}{\gamma}$ is associated with the high-strain phase and $\overset{+}{\gamma}$ is associated with the low-strain phase, $(6.3)_2$ yields

$$f = \left(\frac{\overset{+}{\sigma} + \overset{-}{\sigma}}{2} - \sigma_o \right) \gamma_T = \left(\mu \frac{\overset{+}{\gamma} + \overset{-}{\gamma} - \gamma_T}{2} - \sigma_o \right) \gamma_T. \tag{6.8}$$

Note that (6.8) specializes to (5.35) when $\overset{+}{\sigma} = \overset{-}{\sigma} = \sigma$.

We now return to the dynamic problem. Since there is no length scale in the problem it is natural to seek a scale-invariant solution of the form

$$\gamma(x,t) = \gamma(x/t), \qquad v(x,t) = v(x/t). \tag{6.9}$$

Substituting this assumed solution form into the differential equations (6.1) and using the constitutive relation (5.4) shows that $\gamma(x,t)$ and $v(x,t)$ must necessarily be constant on regions of the (x,t)-plane where γ and v are smooth. Thus, on the (x,t)-plane, the solution will involve various curves corresponding to propagating shock waves and phase boundaries, and the strain and particle velocity fields will suffer jump discontinuities across these curves but will otherwise remain constant. Necessarily therefore the states on either side of a particular discontinuity are constant, and consequently by $(6.2)_1$ its propagation speed must remain constant as well. The various discontinuities are therefore described on the (x,t)-plane by straight lines.

Consequently, the solution of the dynamic problem must have the following general structure (Figure 12): there is a family of n rays, $x = c_i t$, $i = 1, 2, \ldots n$, which separate the first quadrant of the (x,t)-plane into $n + 1$ wedges. Between every pair of rays the strain and particle velocity are constant:

$$\gamma(x,t) = \gamma_j, \quad v(x,t) = v_j \quad \text{for} \quad c_j t < x < c_{j+1} t, \ j = 0, 1, 2, \ldots n. \tag{6.10}$$

Each ray corresponds to either a shock wave or a phase boundary and $0 = c_0 <$

$c_1 < c_2 \ldots < c_n < c_{n+1} = \infty$. Determining the solution thus reduces to finding all of the constants γ_k, v_k and c_k.

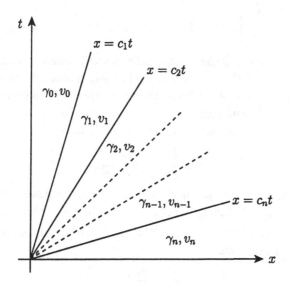

Fig. 12. General form of solution to sudden elongation problem. Piecewise homogeneous states (γ_k, v_k) separated by shock waves and phase boundaries propagating at speeds c_k.

By adapting the arguments made in Abeyaratne and Knowles (1991a) for a Riemann problem, one can establish the following useful results pertaining to the solution (6.10).

(i) No strain γ_j is associated with the declining branch of the stress-strain curve.

(ii) The solution involves at most one shock wave.

(iii) The solution involves at most one phase boundary.

Thus, in particular, a solution can exhibit at most two rays bearing discontinuities. The argument used by Abeyaratne and Knowles (1991a) in obtaining these conclusions depends heavily on the dissipation inequality. It follows from these results that the solution must necessarily have one of the two forms shown in Figure 13: Solution-A involves no phase boundaries and therefore no change of phase; in contrast Solution-B does involve a change of phase.

Solution-A (Figure 13(a)): This solution involves a single discontinuity which is a shock wave propagating at the speed $c = \sqrt{\mu/\rho}$ and has the form

$$\gamma(x,t),\ v(x,t)\ = \begin{cases} 0,\ 0 & \text{for} & x > ct, & \text{(low − strain phase)} \\ \overset{+}{\gamma},\ \overset{+}{v} & \text{for} & 0 < x < ct, & \text{(low − strain phase)} \end{cases}$$
$$(6.11)$$

where we have made use of the initial conditions to determine the state ahead of the shock. Since the slab is initially in the low-strain phase, it must necessarily remain in the low-strain phase for all time since there are no phase boundaries in this solution. By enforcing the jump condition $(6.2)_2$ and the boundary condition (6.5) we find the state behind the shock to be

$$\overset{+}{\gamma} = \frac{\dot{\delta}}{c}, \qquad \overset{+}{v} = -\dot{\delta}. \qquad (6.12)$$

Since the driving force on the shock wave vanishes, the dissipation inequality imposes no restrictions. Finally, since the slab is, by assumption, entirely in the low-strain state, we must have $-1 < \overset{+}{\gamma} < \gamma_{max}$. This requires the elongation-rate to be sufficiently small:

$$\frac{\dot{\delta}}{c} < \frac{\sigma_{max}}{\mu}. \qquad (6.13)$$

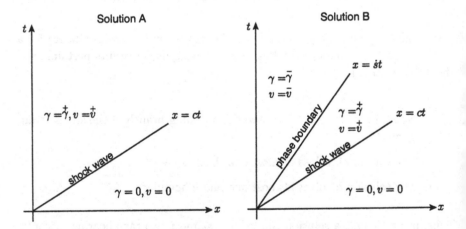

Fig. 13. Solution forms with and without phase transition.

<u>Solution-B</u> (Figure 13(b)): Here the solution involves a front running shock wave at $x = ct$ behind which there is a propagating phase boundary $x = \dot{s}t$. The material ahead of the shock wave is in the initial state (i.e. the undeformed low-strain phase at rest); the material between the shock wave and the phase

boundary is also in the low-strain phase but it involves a nonvanishing strain and particle velocity; finally, the material behind the phase boundary is in the high-strain phase. Thus, the solution has the form

$$
\gamma(x,t),\ v(x,t) = \begin{cases} 0,\ 0 & \text{for} \quad x > ct, & (\text{low}-\text{strain phase}) \\ \overset{+}{\gamma},\ \overset{+}{v} & \text{for} \quad \dot{s}t < x < ct, & (\text{low}-\text{strain phase}) \\ \overline{\gamma},\ \overline{v} & \text{for} \quad 0 < x < \dot{s}t, & (\text{high}-\text{strain phase}) \end{cases}
$$
(6.14)

where, again, we have already made use of the initial conditions. Applying the jump conditions (6.2) at both the shock wave and the phase boundary, and using the boundary condition (6.5) determines the unknown strains and particle velocities in terms of the phase boundary speed $\dot{s}$:

$$
\overset{+}{\gamma} = \frac{\dot{\delta}}{c} - \frac{c\dot{s}\gamma_T}{c^2 - \dot{s}^2}, \qquad \overline{\gamma} = \frac{\dot{\delta}}{c} + \frac{c\gamma_T}{c+\dot{s}}, \qquad \overset{+}{v} = -c\overset{+}{\gamma}, \qquad \overline{v} = -\dot{\delta}. \quad (6.15)
$$

Since the material ahead of the phase boundary is in the low-strain phase while the material behind it is in the high-strain phase, we must have $-1 < \overset{+}{\gamma} < \gamma_{max}$ and $\overline{\gamma} > \gamma_{min}$; therefore the elongation-rate must necessarily lie in the range

$$
\frac{\sigma_{min}}{\mu} + \gamma_T \frac{\dot{s}}{c+\dot{s}} < \frac{\dot{\delta}}{c} < \frac{\sigma_{max}}{\mu} + \gamma_T \frac{c\dot{s}}{c^2 - \dot{s}^2}. \quad (6.16)
$$

Finally, consider the dissipation inequality (6.3)₁. It is automatically satisfied at the shock wave since the driving force there vanishes identically. On the other hand, at the phase boundary, since $\dot{s} > 0$, we must demand $f \geq 0$. By using (6.8) and (6.15) we find that the driving force is given by

$$
f = \mu\gamma_T \left\{ \frac{\dot{\delta}}{c} - \frac{\sigma_o}{\mu} - \frac{\gamma_T}{2}\frac{\dot{s}(2c-\dot{s})}{c^2 - \dot{s}^2} \right\}. \quad (6.17)
$$

The dissipation inequality therefore requires that

$$
\frac{\dot{\delta}}{c} \geq \frac{\sigma_o}{\mu} + \frac{\gamma_T}{2}\frac{\dot{s}(2c-\dot{s})}{c^2 - \dot{s}^2}. \quad (6.18)
$$

In summary, for any arbitrarily chosen phase boundary speed $\dot{s} \in (0,c)$, if the elongation-rate $\dot{\delta} > 0$ satisfies the inequalities (6.16) and (6.18), an admissible solution to the dynamic problem is given by (6.14), (6.15). Observe that the phase boundary propagation speed $\dot{s}$ remains indeterminate. Thus, when a phase changing solution exists, there is in fact a *one-parameter family of such solutions* (parameter $\dot{s}$).

Nature of nonuniqueness. Need for nucleation and kinetics. In order to delineate the precise nature of this non-uniqueness, it is convenient to consider the $(\dot{\delta}, \dot{s})$-plane. The curves labeled $\overset{+}{\gamma} = \gamma_{max}, \overline{\gamma} = \gamma_{min}$ and $f = 0$ in Figure 14 are defined by (6.16) and (6.18) with the inequalities replaced by equalities. Each point $(\dot{\delta}, \dot{s})$ in the shaded region $\mathcal{B}$ satisfies all three inequalities in (6.16), (6.18). Each such point can therefore be associated uniquely with a solution involving a phase change (Solution-B). The bold segment $\mathcal{A}$ of the $\dot{\delta}$-axis corresponds to the interval (6.13) and delineates points which can be associated with a solution involving no phase change (Solution-A).

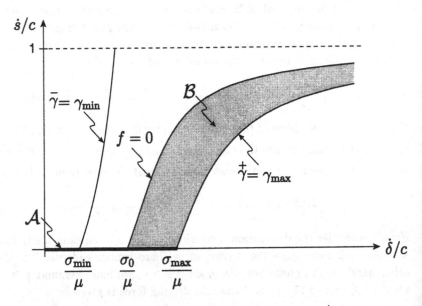

Fig. 14. The set of all solutions to the dynamic problem displayed on the $(\dot{\delta}, \sigma)$-plane. Each point in the shaded region $\mathcal{B}$ can be uniquely associated with a Solution-B which involves a phase change. Points on the line segment $\mathcal{A}$ can be associated with Solutions-A which do not involve a phase change. Observe the severe nonuniqueness of solution when the prescribed elongation-rate $\dot{\delta}/c$ exceeds σ_o/μ.

Figure 14 allows us to reach the following conclusions about the totality of (scale-invariant) solutions:

1. If the elongation-rate is sufficiently small, i.e. if $0 < \dot{\delta}/c \leq \sigma_o/\mu$, there is a unique solution (Solution-A) and it involves no phase change.

2. If the elongation-rate is sufficiently large, i.e. if $\dot{\delta}/c \geq \sigma_{max}/\mu$, then corresponding to each prescribed $\dot{\delta}$ there is a one-parameter family of solutions, each of which involves a phase change (Solution-B); the pa-

rameter is the propagation speed $\dot{s}$. There are no solutions which do not involve a phase change.

3. Finally, if the elongation-rate has a value in the intermediate range $\sigma_o/\mu < \dot{\delta}/c < \sigma_{max}/\mu$, then, corresponding to each prescribed $\dot{\delta}$, there exists a one parameter family of solutions involving a phase change (Solution-B) *as well as* a solution that does not involve a phase change (Solution-A).

The lack of uniqueness in a mathematical model of a physical problem usually reflects an incompleteness in the modeling. Observe that in the present context we have *two distinct types of nonuniqueness*: first, when a phase transforming solution exists, we need to determine the unknown propagation speed $\dot{s}$ of the phase boundary; and second, when both types of solutions exist at the same $\dot{\delta}$, we need to choose between them. The former reflects a need for information describing the propagation speed, which physically is equivalent to a description of the *rate* at which the material transforms from one phase to the other; in the materials science literature this is described by a "kinetic (or evolution) law". The second type of non-uniqueness reflects a need for information about *when* the transformation process commences, i.e. a characterization of when a new phase first "nucleates". The suggestion therefore is that the dynamic problem as posed previously should be supplemented with two additional conditions: a nucleation condition and a kinetic law.

Taking a purely continuum mechanical approach to this, it is natural to suppose that the propagation of a phase boundary is governed by the state of the material on either side of it, i.e. by a relation of the form

$$\dot{s} = V(\overset{+}{\gamma}, \overset{-}{\gamma}).\qquad(6.19)$$

The function V is determined by the material and it provides the continuum theory with an appropriate characterization of the lattice-scale dynamics underlying the transformation process. The pair of equations (6.7) and (6.8) can be solved to determine the two strains $\overset{+}{\gamma}, \overset{-}{\gamma}$ in terms of $\dot{s}$ and f. Substituting the result into (6.19) leads to an equation of the form $H(\dot{s}, f) = 0$. Assuming solvability for f, the propagation law (6.19) can therefore be written in the equivalent form

$$f = \varphi(\dot{s})\qquad(6.20)$$

relating the propagation speed and the driving force. Equation (6.20) provides the simplest example of a *kinetic law*. In order to be consistent with the dissipation inequality $f\dot{s} \geq 0$ the function φ must have the property

$$\dot{s}\varphi(\dot{s}) \geq 0\qquad(6.21)$$

but is otherwise unrestricted by the continuum theory. It is part of the characterization of the material and needs to be determined through a combination of lattice-scale modeling and laboratory experiments.

When (6.20) is enforced at the phase boundary in our dynamic problem, we find by using (6.17) that the elongation-rate and the propagation speed are related by

$$\frac{\dot{\delta}}{c} = \frac{\sigma_o}{\mu} + \frac{\gamma_T}{2}\frac{\dot{s}(2c - \dot{s})}{c^2 - \dot{s}^2} + \frac{\varphi(\dot{s})}{\mu\gamma_T}. \tag{6.22}$$

This describes a curve $\mathcal{K}$ in the $(\dot{\delta}, \dot{s})$-plane which lies in the shaded region $\mathcal{B}$. If φ increases monotonically, $\mathcal{K}$ is a monotonically rising curve, as shown in Figure 15. Thus, from among all Solutions-B the only ones which are permitted by the kinetic law are those associated with the curve $\mathcal{K}$; they are characterized by (6.14), (6.15) and (6.22).

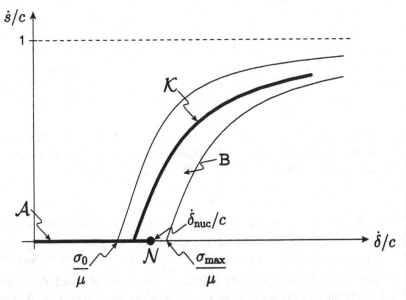

Fig. 15. Solutions to the dynamic problem which conform to the kinetic law and nucleation condition. Only points on the curve $\mathcal{K}$ can be associated two-phase solutions which conform to the kinetic law. The point $\mathcal{N}$ is determined by the nucleation condition and represents the boundary between phase change and no phase solutions: if the elongation-rate $\dot{\delta} < \dot{\delta}(\mathcal{N})$ the slab does not change phase and responds according to Solution-A; if $\dot{\delta} > \dot{\delta}(\mathcal{N})$ the slab changes phase and its response is described by Solution-B together with the kinetic law.

In the presence of the kinetic law, the totality of solutions has been reduced to those associated with the two curves $\mathcal{A}$ and $\mathcal{K}$ of the $(\dot{\delta}, \dot{s})$-plane. Thus, the remaining nonuniqueness pertains to selecting between them whenever they both exist at the same $\dot{\delta}$, i.e. when the elongation-rate lies in the range $\sigma_o/\mu +$ $\varphi(0)/\mu\gamma_T < \dot{\delta}/c < \sigma_{max}/\mu$; see Figures 14 and 15. In order to make this determination, consider the following *nucleation condition*: a phase boundary will be nucleated provided that the driving force on it soon after it nucleates exceeds some critical value $f_{nuc}(> 0)$. We can apply this to the dynamic problem by calculating $\dot{s}_{nuc}$ from $f_{nuc} = \varphi(\dot{s}_{nuc})$ and then calculating $\dot{\delta}_{nuc}$ from (6.22). This determines the threshold value of the elongation-rate for nucleation, which is depicted by the point $\mathcal{N}$ in Figure 15. In the special case where $f_{nuc} = \varphi(0)$, we have $\dot{s}_{nuc} = 0$ and

$$\frac{\dot{\delta}_{nuc}}{c} = \frac{\sigma_o}{\mu} + \frac{f_{nuc}}{\mu\gamma_T} \tag{6.23}$$

so that $\mathcal{N}$ then lies at the point of intersection of the curves $\mathcal{A}$ and $\mathcal{K}$.

In summary, the *unique solution* to the dynamic problem that is consistent with the kinetic law and nucleation condition is the following: if the elongation-rate lies in the range $0 < \dot{\delta} \leq \dot{\delta}_{nuc}$ the slab does not change phase and the solution is given by (6.11), (6.12). On the other hand, when $\dot{\delta} > \dot{\delta}_{nuc}$, the slab does change phase and the solution is given by (6.14), (6.15), (6.22).

Quasi-static processes. Before ending this section it is worth briefly considering quasi-static processes of a finite slab, $0 < x < L$. Inertial effects are now neglected and the governing problem is identical to that which described the equilibrium problem of Section 12.5 except that now all fields depend on the time as a parameter. The elongation history $\delta(t)$ is prescribed and the fields $u(x,t), \gamma(x,t), \sigma(x,t)$ are to satisfy the (static) equations, jump conditions and boundary conditions (5.1), (5.2), (5.6) at each instant t. The three families of solutions, Solutions-1, -2 and -3, carry over immediately to the present quasi-static setting, and the totality of solutions can again be described in the (δ, σ)-plane by the parallelogram with the two extended lines shown in Figure 16 (recall Figure 10). However, in contrast to the static case, there is no reason now to restrict attention to energy-minimizing solutions unless the loading-rate is infinitesimally slow. Consequently we do not have the additional phase equilibrium condition $f = 0$ at our disposal; instead we should enforce the kinetic law and nucleation condition just as in the inertial case. The driving force f associated with quasi-static processes is given by (5.35) and the dissipation inequality $f\dot{s} \geq 0$ must hold.

First, consider an instant t_1 at which the state of the slab is described by Solution-3 and is associated with a point (δ_1, σ_1) in the interior of the parallelogram (Figure 16). The slab involves a phase boundary at the location s_1 given by (5.14). Given the loading history $\delta(t)$, $t > t_1$, our task is to find the resulting loading path, $(\delta(t), \sigma(t))$, $t > t_1$, emanating from the point (δ_1, σ_1).

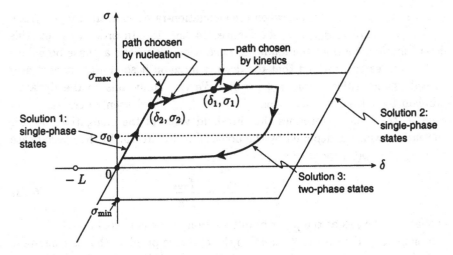

Fig. 16. The set of all solutions involved in quasi-static processes of the slab. At a two-phase state such as (δ_1, σ_1) the kinetic law determines the path to be followed during subsequent loading/unloading. At a single-phase state such as (δ_2, σ_2) the nucleation condition determines whether during subsequent loading the path is to continue up the left boundary and therefore remain a single phase solution, or whether it should enter into the parallelogram and thus nucleate a phase transformation.

This path is determined by equations (5.14), (6.20), (5.35), which involve the three unknown functions $\sigma(t), f(t)$ and $s(t)$. Eliminating f and σ between them leads to

$$\frac{\delta}{L} = \frac{\varphi(\dot{s})}{\mu \gamma_T} + \frac{s\gamma_T}{L} + \frac{\sigma_o}{\mu}. \tag{6.24}$$

Given $\delta(t)$ for $t > t_1$, this ordinary differential equation, together with the initial condition $s(t_1) = s_1$, can be solved for $s(t)$. Hence $\sigma(t)$ can be determined from (5.14). Thus, the kinetic law helps determine the loading path unambiguously.

Second, consider an instant t_2 at which the state of the slab is described by Solution-1 and is associated with a point (δ_2, σ_2) on the left boundary of the parallelogram. Note that the slab is entirely in the low-strain phase at this instant and that, therefore, it does not involve a phase boundary at which the kinetic law can be used. Again, given the elongation history, $\delta(t), t > t_2$, we wish to determine the resulting loading path. If it is known that the high-strain phase is not nucleated at the instant t_2+, then the slab remains entirely in the low-strain phase and therefore the loading path continues up the left boundary of the parallelogram. On the other hand, if we knew that the high-strain phase is nucleated at t_2+, then a phase boundary appears, say at $s = 0$, and now we can use the kinetic law (6.24) to determine the loading path as in the preceding paragraph. Thus, the decision that must be made is whether or not the new phase is nucleated at the instant t_2+. The nucleation condition answers this question and we find from (5.35) and (5.14) that the the critical values of stress and elongation at nucleation during this quasi-static process are $\sigma_{nuc} = \sigma_o + f_{nuc}/\gamma_T$ and

$$\frac{\delta_{nuc}}{L} = \frac{\sigma_o}{\mu} + \frac{f_{nuc}}{\mu\gamma_T} \tag{6.25}$$

respectively. The response during an arbitrary loading program can now be readily pieced together. The response is hysteretic and rate-dependent in general.

Conclusion. The dynamic problem describing the sudden elongation of a slab has served to illustrate the fact that when the strain-energy function has multiple minima, the usual formulation of the problem leads to a severe lack of uniqueness of solution. This lack of uniqueness was traced to the absence of information in the model describing when and how fast the transformation from one energy-well to the other takes place. This in turn pointed to the need to supplement the formulation with a nucleation condition and a kinetic law which characterize these two events. Simple models of the nucleation condition and kinetic law were presented and the role they play in determining the (unique) solution was described. Finally it has been shown how these same conditions come into play in the quasi-static loading of a slab.

Additional insight into quasi-static processes can be gained by considering the *potential energy function* of the material introduced earlier. Recall from the discussion surrounding Figure 11 that the driving force f corresponds to the height between the two potential energy-wells, and, moreover, that the low-strain energy-well lies below the high-strain energy-well when $f < 0$ and vice versa. In view of the dissipation inequality $f\dot{s} \geq 0$, this means that when the low-strain'energy-well is the lower one, then $\dot{s} \leq 0$ and so the phase boundary cannot move to the right. Thus, particles cannot transform from the phase on the right of the phase boundary to that on the left. Throughout this discussion we have taken the high-strain phase to be on the left of the phase boundary and the low-strain phase to be on the right (see, e.g., discussions surrounding (5.13) or (6.7)). The dissipation inequality thus implies that a particle cannot jump from the lower energy-well to the higher one. Observe in addition that the kinetic law is in fact a relation between the rate of transformation and the height between the two wells of the potential energy G; this is a familiar picture in the material science characterization of rate-processes.

The specific forms of the nucleation condition and kinetic law involve detailed modeling of micromechanical processes and experimentation, a topic which we shall not address here. Explicit models of kinetic laws based on various mechanisms such as thermal activation, viscosity and strain-gradient effects, and wiggly energies, can be found, for example, in Christian (1975), Truskinovsky (1985), Abeyaratne and Knowles (1991b, 1993a), Müller and Xu (1991), and Abeyaratne et al. (1996). Some experimental studies of nucleation and kinetics are described, for example, in Grujicic at al. (1985), Chu (1993), Escobar and Clifton (1993, 1995), Escobar (1995) and Kyriakides and Shaw (1995).

12.7 Nonequilibrium thermodynamic processes. Kinetics

(Heidug and Lehner, 1985 and Abeyaratne and Knowles, 1990, 2000a)

The dynamic and quasi-static problems analyzed in the preceding section both pointed to a clear need, in the continuum theory of phase transitions, for a law that characterizes the transformation rate. In the present section we take a broad approach to this issue and view it from the perspective of continuum thermodynamics. We will demonstrate, in particular, how the preceding anecdotal evidence for a kinetic law is in fact a manifestation of the more general (and well-known) need for evolution laws whenever a system undergoes an irreversible thermodynamic process.

For illustrative purposes first consider the case of a smooth thermomechanical process of a thermoelastic body. Balance of linear and angular momentum lead to the field equations

$$\text{Div } S + b = \rho \dot{v}, \qquad SF^T = FS^T, \qquad (7.1)$$

and the first law of thermodynamics requires that

$$S \cdot \dot{F} + \text{Div } q + r = \dot{\varepsilon}, \qquad (7.2)$$

where the heat flux vector q characterizes the heat flow per unit reference area, ρ is the mass density in the reference configuration, and r and ε are the heat supply and internal energy respectively, both per unit reference volume. The second law of thermodynamics requires that the total rate of entropy production associated with any subregion $\mathcal{D}$ of the body be nonnegative:

$$\Gamma(t; \mathcal{D}) = \frac{d}{dt} \int_{\mathcal{D}} \eta dV - \int_{\partial \mathcal{D}} \frac{q \cdot n}{\theta} dA - \int_{\mathcal{D}} \frac{r}{\theta} dV \geq 0, \qquad (7.3)$$

where η denotes the internal entropy per unit reference volume.

By making use of the field equations (7.1), (7.2) and the thermoelastic constitutive equations

$$S = \frac{\partial W}{\partial F}, \qquad \eta = -\frac{\partial W}{\partial \theta}, \qquad W = W(F, \theta), \qquad (7.4)$$

where $W = \varepsilon - \eta\theta$ denotes the Helmholtz free-energy per unit reference volume, one can simplify (7.3) and express the total rate of entropy production in the form

$$\Gamma(t; \mathcal{D}) = \int_{\mathcal{D}} \frac{q \cdot \nabla\theta}{\theta^2} \, dV \geq 0. \qquad (7.5)$$

This implies the pointwise inequality

$$q \cdot \nabla\theta \geq 0 \qquad (7.6)$$

which must hold at all points in the body. Observe that the entropy production rate involves the product of a flux and a quantity which "drives" that flux: one speaks of the temperature gradient $\nabla\theta$ as the *driving force* conjugate to the heat flux q. In the special case of *thermodynamic equilibrium*, the flux and the driving force both vanish:

$$q = o, \qquad \nabla\theta = o. \qquad (7.7)$$

There is nothing in the description of a thermoelastic material, or the usual principles of continuum mechanics, which provide any information about the driving force $\nabla\theta$ and the flux q when the body is *not* in thermodynamic equilibrium. According to the theory of irreversible processes (see, e.g., Truesdell, 1969, Kestin, 1979, and Callen, 1985), such information must be provided empirically, by relating the flux q to the driving force $\nabla\theta$ and other state variables. In the present setting, this would be a relation of the form

$$q = \hat{q}(\nabla\theta; F, \theta), \qquad (7.8)$$

where $\hat{q}$ is a constitutive function. We recognize the *kinetic law* (7.8) to be no more than the familiar *heat conduction law*, a common example of which is the linear relation $\hat{q} = K\nabla\theta$ with K the conductivity tensor. In general, the kinetic response function (heat conduction response function) $\hat{q}$ must be such that

$$\hat{q}(g; F, \theta) \cdot g \geq 0 \quad \text{for all vectors } g, \qquad \hat{q}(o; F, \theta) = o; \qquad (7.9)$$

the former is a consequence of the entropy inequality (7.6), while the latter reflects (7.7) and in any case follows from the former in the presence of sufficient smoothness.

The preceding discussion concerned the particular setting of smooth processes of a thermoelastic body. As discussed, for example in Callen (1985), under very general circumstances the rate of entropy production can be expressed as the sum of the products of various thermodynamic driving forces and their conjugate fluxes, and the theory of irreversible processes states that the evolution of the system is governed by kinetic laws relating the driving forces to the fluxes.

The setting of interest to us is the propagation of a phase boundary in a thermoelastic body. Let the surface S_t denote the location of the phase boundary in the reference configuration and assume that the deformation and temperature remain continuous across this surface but that the deformation gradient, particle velocity and temperature gradient are permitted to suffer jump discontinuities. Let V_n denote the normal velocity of propagation of S_t. In this case, the field equations (7.1), (7.2) must be supplemented by the jump conditions

$$[Sn] = -\rho V_n[v], \qquad [Sn \cdot v] + [q \cdot n] + V_n[\varepsilon + \rho v \cdot v/2] = 0, \qquad (7.10)$$

which are to hold at points on the phase boundary S_t. The field equations, jump conditions and constitutive relation can be used to write the total rate of entropy production as

$$\Gamma(t; \mathcal{D}) = \int_{\mathcal{D}} \frac{q \cdot \nabla \theta}{\theta^2} \, dV + \int_{S_t \cap \mathcal{D}} \frac{f V_n}{\theta} \, dA, \qquad (7.11)$$

where

$$f = [W] - \frac{\overset{+}{S} + \overset{-}{S}}{2} \cdot [F]; \qquad (7.12)$$

cf. (4.19). Thus, there are now two sources of entropy production: heat conduction and phase transformation. From (7.11) we identify $\nabla \theta$ as the driving force associated with heat conduction and q as its conjugate flux, and f as the driving force associated with phase transformation and V_n as its corresponding flux. Note that while q represents the flux of heat, ρV_n represents the flux of mass across the propagating phase boundary. If the body is in thermal equilibrium, i.e. $\theta(X, t) = $ constant, the entropy production rate due to heat conduction vanishes; similarly, if the body is in phase equilibrium, by which we mean that $f(X, t)$ vanishes on S_t, the entropy production rate due to phase transformation vanishes. Thus, $\nabla \theta$ and f are the "agents of entropy production" when a thermoelastic body is not in thermal and phase equilibrium.

As noted previously, according to the general theory, the evolution of the various fluxes in an irreversible process are governed by *kinetic laws* relating each flux to its conjugate driving force and other state variables. In the present setting, this implies the kinetic laws

$$\text{Heat conduction}: \quad q = \hat{q}(\nabla\theta; F, \theta),$$

$$\text{Phase transition}: \quad V_n = \hat{V}\left(f; \overset{+}{F}, \overset{-}{F}, \theta\right) \quad \text{on} \ \ \mathcal{S}_t. \tag{7.13}$$

The constitutive functions $\hat{q}$ and $\hat{V}$ must satisfy the requirements

$$\hat{q}(g; F, \theta) \cdot g \geq 0 \text{ for all vectors } g, \qquad \hat{V}(f; \overset{+}{F}, \overset{-}{F}, \theta)f \geq 0 \text{ for all scalars } f,$$

$$\hat{q}(o; F, \theta) = o, \qquad \hat{V}(0; \overset{+}{F}, \overset{-}{F}, \theta) = 0,$$

where the inequalities follow from the entropy inequalities $q \cdot \nabla\theta \geq 0$ and $fV_n \geq 0$, and the equalities reflect thermal and phase equilibrium.

Conclusion. The preceding discussion allows us to place our previous analyses in a more general context. In particular, we see that the driving force f, which we first encountered in Section 12.3, has a more general thermodynamic significance; that the dissipation inequality is in fact a consequence of the second law of thermodynamics; that the phase equilibrium condition $f = 0$, first encountered in Section 12.4, is consistent with the notion of thermodynamic equilibrium; and finally that the kinetic law is required because of the nonequilibrium character of the phase transformation process.

It is worth noting the following general expression for the driving force on a phase boundary,

$$f = \left(\overset{+}{\psi} - \overset{-}{\psi}\right) - \frac{\overset{+}{S} + \overset{-}{S}}{2} \cdot \left(\overset{+}{F} - \overset{-}{F}\right) + \frac{\overset{+}{\eta} + \overset{-}{\eta}}{2}\left(\overset{+}{\theta} - \overset{-}{\theta}\right), \tag{7.14}$$

which is valid in three-dimensions for any continuum (not necessarily thermoelastic) undergoing an arbitrary thermomechanical process (including an adiabatic one), Abeyaratne and Knowles (2000a). Here $S(X,t)$, $\psi(X,t)$ and $\eta(X,t)$ denote the fields of Piola-Kirchhoff stress, Helmholtz free-energy and specific entropy respectively and no constitutive relation between them is assumed. Equation (7.14) specializes for a thermoelastic material by using the constitutive characterizations $\psi = W(F, \theta)$, $S = W_F$, $\eta = -W_\theta$. In the presence of heat conduction, the temperature is continuous across the phase boundary and so the last term in (7.14) disappears; in an adiabatic process, however, the temperature can jump across the phase boundary and so this term must be retained.

12.8 Higher dimensional-static problems. The issue of geometric compatibility.

(Wechsler *et al.*, 1953, Bowles and Mackenzie, 1954, Ball and James, 1987, and Bhattacharya, 1991.)

In the equilibrium theory of finite elasticity we have the constitutive relation

$$S = \frac{\partial W(F)}{\partial F}, \tag{8.1}$$

and the equations of force and moment balance

$$\text{Div } S = o, \qquad SF^T = FS^T, \tag{8.2}$$

which must hold at points in the body where the fields are smooth. At a phase boundary, the deformation gradient tensor and stress suffer jump discontinuities but the deformation itself is continuous. The jump conditions associated with force balance and the continuity of the deformation are

$$\overset{+}{S} n = \overset{-}{S} n, \qquad \overset{+}{F} \ell = \overset{-}{F} \ell, \tag{8.3}$$

which must hold at all points X on the phase boundary; the vector n here is a unit normal to the phase boundary at X and $(8.3)_2$ must hold for all vectors ℓ which are tangent to the phase boundary at X. Finally, phase equilibrium requires that

$$f = W(\overset{+}{F}) - W(\overset{-}{F}) - \overset{\pm}{S} \cdot \left(\overset{+}{F} - \overset{-}{F} \right) = 0, \tag{8.4}$$

which is one of the Weierstrass-Erdmann corner conditions, the other being $(8.3)_1$ (see Grinfeld, 1981, James, 1981, and Abeyaratne, 1983).

Here we consider the simplest possible static problem involving a phase boundary: consider a body which occupies all of $I\!\!R^3$, and suppose that it is subjected to a piecewise homogeneous deformation

$$x = \begin{cases} \overset{+}{F}X & \text{for } X \cdot n \geq 0, \\[2mm] \overset{-}{F}X & \text{for } X \cdot n \leq 0, \end{cases} \tag{8.5}$$

where $\overset{\pm}{F}$ are constant tensors belonging to two distinct phases (or variants) of the material. The planar interface $\mathcal{S} = \{X : X \cdot n = 0\}$ corresponds to a phase boundary. Furthermore, suppose that the deformation gradient tensors $\overset{\pm}{F}$ lie at the bottoms of two energy-wells so that $\partial W(\overset{\pm}{F})/\partial F = O$. Finally, assume that the temperature is the transformation temperature so that $W(\overset{+}{F}) = W(\overset{-}{F})$.

Thus, the equilibrium requirements (8.2), (8.3)$_1$ and the phase equilibrium requirement (8.4) are all satisfied trivially. The only equation that remains to be satisfied is the kinematic requirement (8.3)$_2$, which can be expressed in the equivalent form

$$\overset{+}{F} = \overset{-}{F} + a \otimes n, \tag{8.6}$$

where $a \neq o$ is an arbitrary vector.

Keeping in mind the description of the energy-wells in (2.21)–(2.23), the questions we pose are, given two stretch tensors $\overset{+}{U}$ and $\overset{-}{U}$,

(i) under what conditions on $\overset{\pm}{U}$ does (8.6) hold for some vectors a, n and some rotation tensors $\overset{\pm}{R}$ with $\overset{\pm}{F} = \overset{\pm}{R}\overset{\pm}{U}$?, and

(ii) when (8.6) does hold, how does one determine a, n and $\overset{\pm}{R}$?

It is convenient to first rephrase the questions in a form which does not involve the rotation tensors. Observe that if $a, n, \overset{+}{R}, \overset{-}{R}$ satisfy (8.6) for some $\overset{+}{U}$ and $\overset{-}{U}$, then so do $Qa, n, Q\overset{+}{R}, Q\overset{-}{R}$ for any rotation Q. Thus, with no loss of generality we can pick one of the rotation tensors $\overset{\pm}{R}$ arbitrarily and so we take $\overset{-}{R} = I$. Next, let D, b and m be defined by

$$D = \overset{+}{F}\overset{-}{F}{}^{-1}, \qquad b = |\overset{-}{F}{}^{-T}n|a, \qquad m = \overset{-}{F}{}^{-T}n/|\overset{-}{F}{}^{-T}n|. \tag{8.7}$$

On using (8.7), equation (8.6) can be written as $D = I + b \otimes m$; since $\det D > 0$, we need $b \cdot m > -1$. Finally, set $C = D^T D$ so that C is symmetric and positive definite and note from (8.7)$_1$ that it can be written as

$$C = \overset{-}{U}{}^{-1}\overset{+}{U}{}^{2}\overset{-}{U}{}^{-1}. \tag{8.8}$$

The geometric compatibility equation now yields

$$C = (I + m \otimes b)(I + b \otimes m). \tag{8.9}$$

The questions asked above can now be posed in the following equivalent form: (i) find conditions on the symmetric positive definite tensor C under which (8.9) holds for some vectors b and m, $b \cdot m > -1$, and (ii) when these conditions hold find b and m. Note that once b and m have been found, $\overset{+}{R}$ can be determined from $D = \overset{+}{F}\overset{-}{F}{}^{-1} = \overset{+}{R}\overset{+}{U}\overset{-}{U}{}^{-1} = I + b \otimes m$ while a and n can be determined from (8.7)$_{2,3}$.

Proposition: (Ball and James, 1987) Let $0 < \lambda_1 \leq \lambda_2 \leq \lambda_3$ be the ordered eigenvalues of C and let e_1, e_2, e_3 be a set of corresponding orthonormal eigenvectors. Then a necessary and sufficient condition for there to exist vectors $b \, (\neq o)$ and m with $b \cdot m > -1$ satisfying (8.9) is that

$$\lambda_2 = 1, \qquad \lambda_1 \neq \lambda_3. \tag{8.10}$$

When (8.10) holds, there are two pairs of vectors b, m for which (8.9) holds and they are given by

$$\left.\begin{aligned}
b &= g\left\{ \sqrt{\frac{\lambda_3(1-\lambda_1)}{\lambda_3 - \lambda_1}} e_1 + \kappa\sqrt{\frac{\lambda_1(\lambda_3 - 1)}{\lambda_3 - \lambda_1}} e_3 \right\}, \\
m &= \frac{1}{g} \frac{\sqrt{\lambda_3} - \sqrt{\lambda_1}}{\sqrt{\lambda_3 - \lambda_1}} \left\{ -\sqrt{1-\lambda_1} e_1 + \kappa\sqrt{\lambda_3 - 1} e_3 \right\},
\end{aligned}\right\} \tag{8.11}$$

where g is chosen to make m a unit vector and $\kappa = \pm 1$.

The proof of this result may be found in Ball and James (1987). We now apply it to a number of examples, all of which pertain to a material which possesses cubic and tetragonal phases. As discussed in Sections 12.2 and 12.3, the cubic austenite phase is carried into the tetragonal martensite phase by subjecting it to stretches α, α, β in the cubic directions $\{c_1, c_2, c_3\}$; here $\alpha = a/a_o$, $\beta = c/a_o$, where the cubic and tetragonal lattice parameters are $a_o \times a_o \times a_o$ and $a \times a \times c$ respectively. Recall that the martensite phase comes in three variants characterized by the stretch tensors U_1, U_2, U_3 whose components in the cubic basis are given by (3.6). Since the reference configuration has been taken to coincide with unstressed austenite, it corresponds to the deformation gradient tensor I.

In the examples to follow we shall not explicitly write out the solutions $b, m, a, n, \overset{+}{R}$. We will focus instead on the implications of the condition (8.10) for piecewise homogeneous deformations. In particular, the examples will illustrate how (8.10) restricts the possibilities, in some cases disallowing such deformations, in others allowing them always, and in yet others allowing them under special circumstances.

Example 1: One variant of martensite. Consider the possibility of constructing a piecewise homogeneous deformation involving just one martensitic variant, say U_1. Thus, we take $\overset{+}{F} = \overset{+}{R} U_1$ and $\overset{-}{F} = U_1$. Then $C = I$ and so the eigenvalues of C do *not* satisfy (8.10). Hence a deformation of the assumed form does not exist. The same conclusion is reached if we had taken $\overset{+}{F} = R$ and $\overset{-}{F} = I$ corresponding to a piecewise homogeneous deformation involving austenite alone. These observations correspond to a special case of the following general result: if a continuous and piecewise smooth deformation $x(X)$ on a region $\mathcal{R}$ is such that $\nabla x(X) = R(X)U$, where U is a *constant* stretch tensor and $R(X)$ is a field of rotation tensors, then R is necessarily constant.

Example 2: Two variants of martensite. Next consider a piecewise homogeneous deformation involving two of the three martensitic variants, say U_1 and U_2. Now $\overset{+}{F} = \overset{+}{R} U_2$ and $\overset{-}{F} = U_1$ and so $C = U_1^{-1} U_2^2 U_1^{-1}$. From (3.6), C has components

$$[C] = \begin{pmatrix} \alpha^2/\beta^2 & 0 & 0 \\ 0 & \beta^2/\alpha^2 & 0 \\ 0 & 0 & 1 \end{pmatrix}, \tag{8.12}$$

whence the ordered eigenvalues of C are α^2/β^2, 1, β^2/α^2 (if $\alpha < \beta$), or β^2/α^2, 1, α^2/β^2 (if $\beta < \alpha$). Thus, the requirements (8.10) are automatically satisfied thus guaranteeing the existence of a piecewise homogeneous deformation involving two variants. A deformation involving two martensitic variants is called a *twin* or *twinning deformation*; the interface between the variants is a *twin boundary*.

Example 3: Austenite and one variant of martensite. Next consider the possibility of a piecewise homogeneous deformation involving austenite on one side of the interface and one variant of martensite, say U_1, on the other side. Thus, we take $\overset{+}{F} = \overset{+}{R} U_1$ and $\overset{-}{F} = I$ and find $C = U_1^2$. From (3.6) we get

$$[C] = \begin{pmatrix} \beta^2 & 0 & 0 \\ 0 & \alpha^2 & 0 \\ 0 & 0 & \alpha^2 \end{pmatrix}. \tag{8.13}$$

It follows that the requirement (8.10) holds if and only if

$$\alpha = 1, \qquad \beta \neq 1. \tag{8.14}$$

The condition $\alpha = 1$ is a restriction on the lattice and implies that the lattice parameters must obey $a = a_o$. The austenite $\rightarrow$ martensite transformation therefore involves a change in only *one* lattice parameter, the cubic and tetragonal lattice dimensions being $a_o \times a_o \times a_o$ and $a_o \times a_o \times c$ respectively. No known material has a lattice with this characteristic, and this explains why no sharp interfaces separating cubic austenite from one variant of tetragonal martensite have been observed.

The negative result of Example 3 suggests that we examine austenite/martensite deformations which have a more complicated structure than that in

R. Abeyaratne et al.

Example 3. A complete discussion of the next two examples in beyond the scope of this article since they must be viewed from the point of view of sequences of deformations with increasingly finer microstructure (see Ball and James, 1987, and Ball, 1989). Here we restrict attention to a discussion of the macroscopic kinematics.

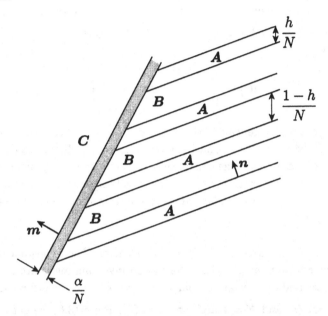

Fig. 17. An interface separating austenite from twinned martensite. The martensitic variants are characterized by the deformation gradient tensors A, B and have volume fractions h and $1 - h$. The thickness of the shaded diffuse interface tends to zero as the fineness of the twin bands increases, i.e. $N \to \infty$.

Example 4: Austenite and twinned martensite. (Ball and James, 1987) Let us now consider the deformation described schematically in Figure 17 in which there is an "interface" which separates austenite from martensite. The martensitic half-space involves fine bands of martensite which alternate between variants -1 and -2. Thus, the martensitic region is twinned, and Example 2 describes its kinematics. The deformation gradient tensors in the martensitic layers are $A = Q_2 U_2$ and $B = Q_1 U_1$ and geometric compatibility across the martensite/martensite interfaces requires

$$A - B = a' \otimes n. \tag{8.15}$$

The widths of the martensitic layers are h/N and $(1 - h)/N$, where the volume fraction h is as yet unknown and N is a positive integer. We have seen in Example 3 that neither variant of martensite (characterized by A or B) is compatible with austenite (characterized by C). Therefore, at best, we can only join austenite to martensite through an interpolation layer, corresponding to a diffuse interface of width α/N as depicted by the grey strip in Figure 17. Consider now the sequence of deformations indexed by N. As $N \to \infty$ the diffuse interface collapses to a plane. If the local deformation gradient field is to remain bounded during this limiting process, it is necessary and sufficient that the deformation gradient C in the austenite region be compatible with the *average* deformation gradient $hA + (1 - h)B$ in the martensite region. In other words, as $N \to \infty$, the austenite regions "see" not the individual martensitic variants but the average of the two. Compatibility between the austenite and the averaged martensite across the austenite/martensite interface requires

$$[hA + (1 - h)B] - C = b \otimes m. \tag{8.16}$$

Given U_1 and U_2, we wish to solve this pair of equations with $A = Q_2 U_2, B = Q_1 U_1, C = I$ and determine a', n, b, m, Q_1 and Q_2.

In order to put this into the standard form, set $R = Q_1^T Q_2$, $\bar{R} = Q_1$, $a = Q_1^T a'$ so that the two compatibility equations can be rewritten as

$$RU_2 - U_1 = a \otimes n, \qquad \bar{R}(U_1 + ha \otimes n) = I + b \otimes m. \tag{8.17}$$

The first of these is the twinning equation which we examined previously in Example 2 and found to be solvable under all conditions. The second equation leads to (8.9) provided we set

$$C = (U_1 + hn \otimes a)(U_1 + ha \otimes n). \tag{8.18}$$

If (8.9) is to be solvable, C must satisfy the conditions (8.10); in particular, it is necessary that one of the eigenvalues of C be unity, and therefore that $\det(C - I) = 0$. An explicit expression for C can be obtained by using (8.18) and substituting into it (3.6) for U_1, and the solution of Example 2 for a and n. This leads to

$$C = \frac{[h(\beta^2 - \alpha^2) - \beta^2]^2 + \alpha^2\beta^2}{\alpha^2 + \beta^2} c_1 \otimes c_1 + \frac{[h(\beta^2 - \alpha^2) + \alpha^2]^2 + \alpha^2\beta^2}{\alpha^2 + \beta^2} c_2 \otimes c_2$$

$$\pm \frac{(\beta^2 - \alpha^2)^2}{\beta^2 + \alpha^2} h(1 - h)(c_1 \otimes c_2 + c_2 \otimes c_1) + \alpha^2 c_3 \otimes c_3,$$

$$\tag{8.19}$$

which expresses C in terms of the lattice parameters α, β, the orthonormal cubic basis vectors $\{c_1, c_2, c_3\}$, and the unknown volume fraction h; the $\pm$ here arises from the fact that the twinning equation $(8.17)_1$ has two solutions.

Setting $\det(C - I) = 0$ now leads to a quadratic equation for the volume fraction h whose roots are

$$h = \frac{1}{2} \left(1 \pm \sqrt{1 + \frac{2(\alpha^2 - 1)(\beta^2 - 1)(\beta^2 + \alpha^2)}{(\beta^2 - \alpha^2)^2}} \right). \qquad (8.20)$$

Note that the sum of these two roots is unity. When (8.20) holds, the eigenvalues of C are found to be $1, \alpha^2, \alpha^2\beta^2$. The necessary and sufficient condition (8.10), together with the inequality which ensures that (8.20) yields real roots in the interval $[0, 1]$, can now be shown to be equivalent to

$$\text{either} \quad \alpha < 1 < \beta \quad \text{and} \quad \alpha^{-2} + \beta^{-2} \leq 2,$$

$$\text{or} \quad \beta < 1 < \alpha \quad \text{and} \quad \alpha^2 + \beta^2 \leq 2. \qquad (8.21)$$

Thus, in summary, an austenite-twinned martensite deformation of the assumed form (Figure 17) exists provided that the lattice parameters obey (8.21) and the variant volume fraction is given by (8.20).

Example 5: Wedge of twinned martensite in austenite. (Bhattacharya, 1991) As a final example consider the deformation illustrated in Figure 18. Here one has a wedge of martensite surrounded by austenite. The wedge involves a mid-rib, on one side of which there is a fine mixture of two martensitic variants characterized by deformation gradients A and B and volume fractions h_1 and $1 - h_1$; on the other side there is a second fine mixture of martensitic variants characterized by the deformation gradients C, D and volume fractions h_2 and $1 - h_2$.

There are five types of interfaces to be considered: two sets of twin boundaries, the two austenite-martensite boundaries, and the mid-rib. Enforcing geometric compatibility across these interfaces leads to the following five conditions:

Martensite/Martensite : $A - B = a_1 \otimes n_1,$

Martensite/Martensite : $C - D = a_2 \otimes n_2,$

Austenite/TwinnedMartensite : $[h_1 A + (1 - h_1)B] - I = b_1 \otimes m_1,$

Austenite/TwinnedMartensite : $[h_2 C + (1 - h_2)D] - I = b_2 \otimes m_2,$

Twinned Martensite/Twinned Martensite:

$$[h_1 A + (1 - h_1)B] - [h_2 C + (1 - h_2)D] = b \otimes m. \qquad (8.22)$$

The first two of these are twinning equations and we know from Example 2

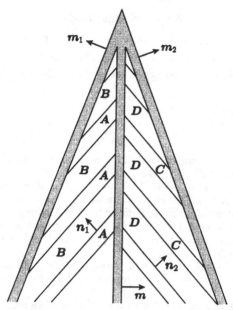

Fig. 18. Wedge of twinned martensite surrounded by austenite. The microstructure involves fives types of interfaces: two sets of twin boundaries – separating different variants of martensite, A/B and C/D –, two austenite/martensite interfaces, and the mid-rib of the wedge which separates twinned martensite from twinned martensite.

that they are solvable without the need for any restrictions. The next two equations are precisely of the type analyzed in Example 4. We know that these are solvable provided the volume fractions h_1 and h_2 of the variants are suitably chosen (and the lattice parameters satisfy some mild inequalities). This completely determines all of the free parameters in the problem. The solvability of the final equation therefore imposes a relationship between the lattice parameters which Bhattacharya (1991) has shown to be

$$\alpha^2 = \frac{(1 - \beta^2)^2 + 4\beta^2(1 + \beta^2)}{(1 - \beta^2)^2 + 8\beta^4}. \tag{8.23}$$

Thus, we conclude that a deformation of the assumed form involving a wedge of martensite can only exist in special materials where the lattice parameters satisfy the restriction (8.23). Bhattacharya (1991) has conducted an extensive survey of the literature and shown that a variety of cubic-tetragonal martensitic materials do in fact satisfy this restriction and that these materials exhibit martensitic wedges; there are also materials which do not satisfy (8.23) and they do not exhibit wedge shaped martensitic domains.

488   R. Abeyaratne et al.

Acknowledgements

Many of the results described here were obtained in the course of investigations supported by the U.S. Army Research Office, National Science Foundation and Office of Naval Research. We are grateful to Debra Blanchard for preparing all of the figures.

References

Abeyaratne, R. 1983 An admissibility condition for equilibrium shocks in finite elasticity. *Journal of Elasticity* **13**, 175–184.

Abeyaratne, R., Chu, C. and James, R.D. 1996 Kinetics of materials with wiggly energies: theory and application to the evolution of twinning microstructures in a Cu-Al-Ni shape memory alloy. *Philosophical Magazine A* **73**, 457–497.

Abeyaratne, R. and Knowles, J.K. 1990 On the driving traction acting on a surface of strain discontinuity in a continuum. *Journal of the Mechanics and Physics of Solids* **38**, 345–360.

Abeyaratne, R. and Knowles, J.K. 1991a Kinetic relations and the propagation of phase boundaries in solids. *Archive for Rational Mechanics and Analysis* **114**, 119–154.

Abeyaratne, R. and Knowles, J.K. 1991b Implications of viscosity and strain-gradient effects for the kinetics of propagating phase boundaries in solids. *SIAM Journal of Applied Mathematics* **51**, 1205–1221.

Abeyaratne, R. and Knowles, J.K. 1993a A continuum model of a thermoelastic solid capable of undergoing phase transitions. *Journal of the Mechanics and Physics of Solids* **41**, 541–571.

Abeyaratne, R. and Knowles, J.K. 1993b Nucleation, Kinetics and Admissibility Criteria for Propagating Phase Boundaries. In *Shock Induced Transitions and Phase Structures in General Media*, eds. J.E. Dunn, R.L. Fosdick and M. Slemrod. Springer-Verlag IMA Volume 52, 3–33.

Abeyaratne, R. and Knowles, J.K. 2000a A note on the driving traction acting on a propagating interface: adiabatic and non-adiabatic processes of a continuum. *ASME Journal of Applied Mechanics* accepted August 1997.

Abeyaratne, R. and Knowles, J.K. 2000b On a shock-induced martensitic phase transition. *Journal of Applied Physics* **87**, 1123–1134.

Abeyaratne, R. and Knowles, J.K. 2001 *Evolution of Phase Transitions*, Cambridge University Press. To appear.

Ball, J.M. 1989 A version of the fundamental theorem for young measures. In *Partial Differential Equations and Continuum Models of Phase Transitions Lecture Notes in Physics*, eds. M. Rascle and D. Serre and M. Slemrod. Springer-Verlag **344**, 207–215.

Ball, J.M. and James, R.D. 1987 Fine phase mixtures as minimizers of energy. *Archive for Rational Mechanics and Analysis* **100**, 13–52.

Ball, J.M. and James, R.D. 1992 Proposed experimental tests of a theory of fine microstructure and the two-well problem. *Phil. Trans. R. Soc. London A* **338**, 389–450.

Bhattacharya, K. 1991 Wedge-like microstructure in martensite. *Acta Metallurgica Materialia* **10**, 2431–2444.

Bhattacharya, K. 2000 *Theory of Martensitic Microstructure and the Shape-Memory Effect*, monograph in preparation.

Born, M. and Huang, K. 1954 *Dynamical Theory of Crystal Lattices*, Oxford University Press.

Bowles, J.S. and Mackenzie, J.K. 1954 The crystallography of martensite transformations. *Acta Metall.* **2**, 129–147.

Callen, H.B. 1985 *Thermodynamics and an Introduction to Thermostatistics*, Sections 14-2, 14-3, Wiley, New York.

Christian, J.W. 1975 *The Theory of Martensitic Transformations in Metals and Alloys, Part 1*, Pergamon.

Chu, C. 1993 Hysteresis and Microstructures: A Study of Biaxial Loading on Compound Twins of Copper-Aluminum-Nickel Single Crystals. Ph.D. dissertation, University of Minnesota, Minneapolis, MN.

Duerig, T., Melton, K., Stoeckel, D. and Wayman, C.M. 1990 *Engineering Aspects of Shape Memory Alloys*, Butterworth-Heinemann.

Eshelby, J.D. 1956 The continuum theory of lattice defects. In *Solid State Physics*, eds. F. Seitz and D. Turnbull. Academic Press, 79–144.

Eshelby, J.D. 1970 Energy relations and the energy-momentum tensor in continuum mechanics. In *Inelastic Behavior of Solids*, eds. by M.F. Kanninen *et al.* McGraw-Hill, 77–115.

Ericksen, J.L. 1975 Equilibrium of bars. *Journal of Elasticity* **5**, 191–201.

Ericksen, J.L. 1978 On the symmetry and stability of thermoelastic solids. *ASME Journal of Applied Mechanics* **45**, 740–744.

Ericksen, J.L. 1980 Some phase transitions in crystals. *Archive for Rational Mechanics and Analysis* **73**, 99–124.

Ericksen, J.L. 1984 The Cauchy and Born hypotheses for crystals. In *Phase Transformations and Material Instabilities in Solids* , ed. M.E. Gurtin. Academic Press, 61–78.

Ericksen, J.L. 1986 Constitutive theory for some constrained elastic crystals. *International Journal of Solids Structures* **22**, 951–964.

Escobar, J.C. 1995 Plate Impact Induced Phase Transformations in Cu-Al-Ni Single Crystals. Ph.D. dissertation, Brown University, Providence, RI.

Escobar, J.C. and Clifton, R.J. 1993 On pressure-shear plate impact for studying the kinetics of stress-induced phase transformations. *Journal of Materials Science and Engineering* **A170**, 125–142.

Escobar, J.C. and Clifton, R.J. 1995 Pressure-shear impact-induced phase transitions in Cu-14.4 Al-4.19 Ni single crystals. *SPIE* **2427**, 186–197.

Gelfand, I.M. and Fomin, S.V. 1963 *Calculus of Variations*. Prentice-Hall.

Green, A.E. and Adkins, J.E. 1970 *Large Elastic Deformations*. Oxford University Press.

Grinfeld, M.A. 1981 On heterogeneous equilibrium of nonlinear elastic phases ad chemical potential tensors. *Lett. Appl. Eng. Sci.* **19**, 1031–1039.

Grujicic, M., Olson G.B. and Owen, W.S. 1985 Mobility of the $\beta_1 - \gamma_1'$ martensitic interface in Cu-Al-Ni: part I. Experimental measurements. *Metallurgical Transactions A* **16A**, 1723–1734.

Gurtin, M.E. 2000 *Configuration Forces as Basic Concepts of Continuum Physics*. Springer.

Heidug, W. and Lehner, F.K. 1985 Thermodynamics of coherent phase transformations in non-hydrostatically stressed solids. *Pure Appl. Geophysics* **123**, 91–98.

James, R.D. 1981 Finite deformation by mechanical twinning. *Arch. Rational Mech. Anal.* **77**, 143–176.

James, R.D. 1992 Lecture Notes for AEM 8589: Mechanics of Crystalline Solids, (unpublished). University of Minnesota.

James, R.D. and Hane, K.F. 2000 Martensitic transformations and shape-memory materials. *Acta Materialia* **48**, 197–222.

Kestin, J. 1979 *A Course in Thermodynamics*. Section 14.3.1, Volume II. New York: McGraw-Hill.

Klouček, P. and Luskin, M. 1994 The computation of the dynamics of the martensitic transformation. *Continuum Mech. Thermodyn.* **6**, 209–240.

Kyriakides, S. and Shaw, J.A. 1995 Thermomechanical Aspects of NiTi. *Journal of the Mechanics and Physics of Solids* **43**, 1243–1281.

Müller, I. and Xu, H. 1991 On the pseudoelastic hysteresis. *Acta Metalurgica Materialia* **39**, 263–271.

Oleinik, O.A. 1957 On the uniqueness of the generalized solutions of the Cauchy problem for a nonlinear system of equations occuring in mechanics. *Uspekhi Matematicheskii Nauk (N.S.)* **12**, 169–176.

Otsuka, K., and Wayman, C.M. 1999 *Shape-memory Alloys*. Cambridge University Press.

Pitteri, M. 1984 Reconciliation of local and global symmetries of crystals. *Journal of Elasticity* **14**, 175–190.

Schetky, L.M. 1979 Shape-memory alloys. *Scientific American* **241**, 74–82.

Smith, G.F. and Rivlin, R.S. 1958 The strain energy function for anisotropic elastic materials. *Transactions of the American Mathematical Society* **88**, 175–193.

Truesdell, C. 1969 *Rational Thermodynamics*. New York: McGraw-Hill.

Truskinovsky, L. 1985 Structure of an isothermal discontinuity. *Soviet Phys. Dokl.* **30**, 945–948.

Wechsler, M.S., Lieberman, D.S. and Read, T.A. 1953 On the theory of the formation of martensite. *Transactions of the AIME Journal of Metals* **197**, 1503–1515.

13
Pseudo-elasticity and stress softening

R.W. Ogden

Department of Mathematics
University of Glasgow, Glasgow G12 8QW, UK
Email: rwo@maths.gla.ac.uk

In this chapter we describe how certain features of the nonlinear *inelastic* behaviour of solids can be described using a theory of *pseudo-elasticity*. Specifically, the quasi-static stress softening response of a material can be described by allowing the strain-energy function to change, either continuously or discontinuously, as the deformation process proceeds. In particular, the strain energy may be different on loading and unloading, residual strains may be generated and the energy dissipated in a loading/unloading cycle may be calculated explicitly. The resulting overall material response is not elastic, but at each stage of the deformation the governing equilibrium equations are those appropriate for an elastic material. The theory is described in some detail for the continuous case and then examined for an isotropic material with reference to homogeneous biaxial deformation and its simple tension, equibiaxial tension and plane strain specializations. A specific model is then examined in order to illustrate the (Mullins) stress softening effect in rubberlike materials. Two representative problems involving non-homogeneous deformation are then discussed. The chapter finishes with a brief outline of the theory for the situation in which the stress (and possibly also the strain) is discontinuous.

13.1 Introduction

For the most part the chapters in this volume are concerned with elasticity *per se*. However, there are some circumstances where elasticity theory can be used to describe certain *inelastic* behaviour. An important example is *deformation theory plasticity*, in which nonlinear elasticity theory is used to describe *loading* up to the point where a material yields and plastic deformation is initiated. Another example arises in biomechanics, where the hysteretic response of soft biological tissue in simple tension is treated on the basis of a theory of *pseudo-elasticity*. In this theory (see Fung, 1981, for example) the loading path is modelled using an elastic stress-strain law and the unloading path by a different elastic stress-strain law, but the overall response is not elastic. Elasticity theory is also used in the modelling of phase transitions, in which there is a discontinuity in the deformation gradient across the phase boundary (see, for example, Ericksen, 1991). In this situation the material response is described by a non-convex form of the (elastic) strain-energy function, and the term *pseudo-elasticity* is also used in this context. The meaning of the term *pseudo-elasticity* is different in these different situations, but there is a common underlying feature, namely that a loading-unloading cycle involves hysteresis.

In this chapter, the term *pseudo-elasticity* is used with yet another meaning. It describes situations in which there is a change in the form of strain-energy function (i.e. a change in the material properties) occasioned by, for example, deformation-induced damage. This change may occur continuously or discontinuously. The progressive stress softening arising in the so-called *Mullins effect* in rubberlike solids may be treated as an example in which the material properties change continuously. This effect is illustrated for *simple tension* in Figure 1(a), in which the nominal stress t is plotted against the stretch λ. The continuous curve is the curve that would be followed if there were no unloading, and we refer to this as the *primary loading curve*. The unloading curve shown (the dashed curve) falls below the loading curve and, for any given value of the stretch, the stress on unloading is therefore less than that on loading. This effect is referred to as *stress softening*. In Figure 1(a) no residual strain is shown, while Figure 1(b) shows a similar loading-unloading cycle but with residual strain included (this being the strain remaining when the stress has returned to zero). In each case the area between the loading and unloading curves measures the dissipation. The so-called *idealized* Mullins effect corresponds to Figure 1(a), and it is assumed in this idealization that *reloading* after initial loading and unloading follows the unloading path until the primary loading path is reached, at which point continued loading follows the primary loading path. This is the case whatever point of the primary path is the starting point for unloading. For real materials the reloading path is different from the unloading path and in general there are residual strains remaining after removal of the stress. Some typical examples of loading, unloading and reloading data in simple tension experiments for three different elastomeric materials are given in the paper by Muhr *et al.* (1999). For a detailed discussion of the Mullins effect we refer to the recent paper by Ogden and Roxburgh (1999a), in which a phenomenological theory of *pseudo-elasticity* for the description of the (idealized) Mullins effect was developed. The theory was modified by Ogden and Roxburgh (1999b) to enable residual strains to be accommodated, and an illustrative example of the application of the theory was discussed by Holzapfel *et al.* (1999).

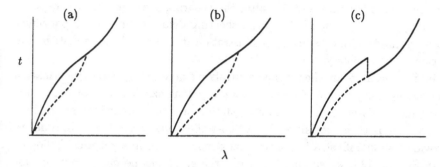

Fig. 1. Sketch of the nominal stress t against stretch λ for loading (continuous curves) and unloading (dashed curves) in simple tension: (a) continuous stress with no residual strain; (b) continuous stress with residual strain; (c) discontinuous stress.

Although we focus on continuous changes in material properties in this chapter, we note that changes in material properties may occur discontinuously, in which case abrupt partial failure of the material is accompanied by a drop in stress when a critical value of the deformation is reached. This is illustrated in Figure 1(c).

The aim of this chapter is to provide a summary of the main ingredients of the theory of pseudo-elasticity (with the term interpreted as indicated above) and some illustrative examples of its application. For the case in which the material properties change continuously, the theory is summarized in Section 13.2. This includes discussion of material symmetry and specialization to isotropy and to particular homogeneous deformations. In Section 13.3 a specific model used to describe the (idealized) Mullins effect is discussed, with particular reference to the simple tension test, and it is then shown how a simple modification of the theory allows residual strains to be included.

In Section 13.4, which is based on the recent paper by Ogden (2001a), the theory is illustrated by application to a problem involving non-homogeneous deformation. Specifically, we consider the (plane strain) azimuthal shear of a thick-walled circular cylindrical tube of incompressible initially isotropic material. The appropriate boundary-value problem for a general incompressible pseudo-elastic material for both loading and unloading is formulated. The inner circular boundary is taken to be fixed and the outer boundary rotated relative to it at fixed radius by application of a suitable shearing stress (which is assumed during loading to be monotonically increasing with the angle of rotation). Unloading is then associated with reduction in the applied shearing stress on the outer boundary and is described by an (inhomogeneous, isotropic) elastic strain-energy function different from that in loading. It is found that whether there is residual strain after the shearing stress is removed depends (locally) on the extent of the initial shearing deformation. For a particular material model, which is used for illustration, the residual strain distribution is calculated explicitly.

In Section 13.5 another problem involving non-homogeneous deformation but having a key feature different from that in the azimuthal shear problem is discussed. This is the inflation and deflation of a thick-walled spherical shell. The theory is then specialized for a thin-walled (membrane) shell in order to provide a simple illustration of residual strain induced in a spherical balloon.

In Section 13.6 we summarize briefly the equations needed to account for discontinuous stress accompanied by either continuous or discontinuous deformation.

13.2 Pseudo-elasticity

13.2.1 Basic equations

The development in this section is based severally on the papers by Ogden and Roxburgh (1999a, b) and Ogden (2000a, b). As far as possible we adopt here the notation used in Chapter 1. We consider the deformation of a continuous body which in its natural (undeformed and unstressed) configuration is taken to occupy the region B_r, and material points are identified by their position vectors $\mathbf{X}$ in B_r. After deformation the body occupies the region B, and the point $\mathbf{X}$ is deformed to the position $\mathbf{x}$. The deformation from B_r to B has gradient denoted by $\mathbf{F}$.

In *pseudo-elasticity* the strain-energy function $W(\mathbf{F})$ of elasticity theory is modified by incorporating an additional variable η into the function. Thus, we write

$$W = W(\mathbf{F}, \eta). \tag{2.1}$$

The variable η is referred to as a *damage* or *softening variable*. The inclusion of η provides a means of changing the form of the energy function during the deformation process and hence changing the character of the material properties. In general, the overall response of the material is then no longer elastic and we therefore refer to $W(\mathbf{F}, \eta)$ as a *pseudo-energy function*. The resulting theory is referred to as *pseudo-elasticity theory*.

The variable η may be active or inactive and a change from active to inactive (or conversely) effects a change in the material properties. This change may be either continuous or discontinuous. Alternatively, when η is active it may vary in such a way that the material properties change (continuously or discontinuously). Possible criteria for inducing these changes are discussed below. In this chapter we take η to be a continuous variable, except in Section 13.6, where we discuss briefly the modifications required in the theory to accommodate discontinuities in η and hence discontinuities in the material properties, as detailed in Lazopoulos and Ogden (1998, 1999) and Ogden (2000a, b).

If η is inactive we set it to the constant value unity and write

$$W_0(\mathbf{F}) = W(\mathbf{F}, 1), \tag{2.2}$$

for the resulting *strain-energy function*. For an unconstrained material the associated nominal stress is denoted $\mathbf{S}_0$ and is given by

$$\mathbf{S}_0 = \frac{\partial W_0}{\partial \mathbf{F}}(\mathbf{F}). \tag{2.3}$$

In what follows the zero subscript will be associated with the situation in which η is inactive.

When η is active it is taken to depend on the deformation gradient (in general implicitly) and we write this dependence in the form

$$C(\mathbf{F}, \eta) = 0, \tag{2.4}$$

where the function C can be regarded as a constraint between the variables $\mathbf{F}$ and η. With this constraint we associate a Lagrange multiplier, q say, so that the nominal stress $\mathbf{S}$ may be written

$$\mathbf{S} = \frac{\partial W}{\partial \mathbf{F}}(\mathbf{F}, \eta) + q\frac{\partial C}{\partial \mathbf{F}}(\mathbf{F}, \eta). \tag{2.5}$$

If (2.4) can be solved explicitly for η in terms of $\mathbf{F}$ then we write

$$\eta = \eta_e(\mathbf{F}), \tag{2.6}$$

where the subscript e indicates that (2.6) is an explicit solution of (2.4) (which, in general, may not be unique). We then define the *strain-energy function w* by

$$w(\mathbf{F}) = W(\mathbf{F}, \eta_e(\mathbf{F})), \tag{2.7}$$

and the nominal stress $\mathbf{S}$ is then given by

$$\mathbf{S} = \frac{\partial w}{\partial \mathbf{F}} = \frac{\partial W}{\partial \mathbf{F}}(\mathbf{F}, \eta) + \frac{\partial W}{\partial \eta}(\mathbf{F}, \eta)\frac{\partial \eta}{\partial \mathbf{F}}(\mathbf{F}). \tag{2.8}$$

Comparison of (2.5) and (2.8) then shows that

$$q = -W_\eta / C_\eta, \tag{2.9}$$

where use has been made of the connection

$$\frac{\partial \eta}{\partial \mathbf{F}} = -\frac{\partial C}{\partial \mathbf{F}}\frac{1}{C_\eta},$$

obtained from (2.4), and the subscript η signifies partial differentiation with respect to η.

The function C in (2.4) is essentially arbitrary and is required to satisfy only the appropriate objectivity condition (objectivity will be discussed in Section 13.2.2 below). Moreover, the dependence of W on η is also arbitrary and so no generality is lost by setting $C = W_\eta$. This amounts to setting the Lagrange multiplier q to zero. The constraint (2.4) then takes the form

$$\frac{\partial W}{\partial \eta}(\mathbf{F}, \eta) = 0. \tag{2.10}$$

The expression (2.5), or equivalently (2.8), for the nominal stress then reduces to

$$\mathbf{S} = \frac{\partial w}{\partial \mathbf{F}}(\mathbf{F}) = \frac{\partial W}{\partial \mathbf{F}}(\mathbf{F}, \eta), \tag{2.11}$$

where the right-hand side is evaluated for η given by (2.6), or implicitly by (2.10). In the theory of pseudo-elasticity developed here we adopt the implicit relationship (2.10) between $\mathbf{F}$ and η. The nominal stress is then given by (2.11) whether η is active or inactive. In the latter case (2.11) reduces to (2.3).

Equation (2.10) defines a hypersurface in the 10-dimensional $(\mathbf{F}, \eta)$-space to which values of η must be restricted, subject to the usual restrictions on $\mathbf{F}$ that

$$0 < \det \mathbf{F} < \infty. \tag{2.12}$$

The hypersurface (2.10) identifies stationary points of $W(\mathbf{F}, \eta)$ with respect to η. If η is defined uniquely in terms of $\mathbf{F}$ thereby then we may write the solution formally as (2.6), and we then use the notation w for the (unique) strain-energy function resulting, as defined by (2.7). The details of the restrictions on η depend on the particular model for the dependence of η on $\mathbf{F}$ and hence on the application considered.

We may regard equation (2.10) as a field equation, which, in the absence of body forces, is coupled with the equilibrium equation in the form

$$\mathrm{Div}\,\mathbf{S} = \mathbf{0} \quad \text{in } \mathcal{B}_r, \tag{2.13}$$

where Div denotes the divergence operator in $\mathcal{B}_r$.

Thus far we have not specified the form of the dependence of W on η, or, more particularly, the form of the function $\eta_e(\mathbf{F})$ in (2.6), i.e. we have not specified a particular model within the general framework of pseudo-elasticity.

Moreover, we have not identified a criterion for switching η on or off. Such considerations depend on the application to be considered. For example, in the application to stress softening associated with the Mullins effect (Ogden and Roxburgh 1999a, b), η is taken to be inactive during loading and to switch on during unloading (with loading and unloading being well defined relative to the energy expended during a loading path). In this example, the energy changes continuously from its value $W_0(\mathbf{F}) = W(\mathbf{F}, 1)$ to $W(\mathbf{F}, \eta)$ with η decreasing from the value 1 as unloading proceeds. The corresponding nominal stress also changes continuously. Some aspects of this theory will be examined in Section 13.3. In the applications considered by Lazopoulos and Ogden (1998, 1999), on the other hand, the criterion for a change of material properties is associated with a critical value of the deformation, at which point η and the material properties change discontinuously and there is an accompanying discontinuity in the stress. In this case (2.10) is associated with a non-unique relationship between η and $\mathbf{F}$. This will be discussed briefly in Section 13.6.

13.2.1.1 Incompressibility

For an incompressible material $\mathbf{F}$ satisfies the constraint

$$\det \mathbf{F} = 1. \qquad (2.14)$$

The nominal stress tensor associated with the strain energy (2.2) is denoted by $\mathbf{S}_0$ and is given by

$$\mathbf{S}_0 = \frac{\partial W_0}{\partial \mathbf{F}}(\mathbf{F}) - p_0 \mathbf{F}^{-1}, \qquad \det \mathbf{F} = 1, \qquad (2.15)$$

where p_0 is the Lagrange multiplier associated with the constraint (2.14).

The nominal stress tensor $\mathbf{S}$ associated with $w(\mathbf{F})$ is

$$\mathbf{S} = \frac{\partial w}{\partial \mathbf{F}}(\mathbf{F}) - p \mathbf{F}^{-1} = \frac{\partial W}{\partial \mathbf{F}}(\mathbf{F}, \eta) - p \mathbf{F}^{-1}, \qquad \det \mathbf{F} = 1, \qquad (2.16)$$

where the right-hand side is evaluated for η given by (2.6) or (2.10) and p is the counterpart of p_0 for active η. Note that for an incompressible material the Cauchy stress tensors, denoted by σ_0 and σ for inactive and active η respectively, are related to $\mathbf{S}_0$ and $\mathbf{S}$, respectively, by

$$\sigma_0 = \mathbf{F} \mathbf{S}_0, \qquad \sigma = \mathbf{F} \mathbf{S}. \qquad (2.17)$$

498   *R. W. Ogden*

13.2.2 Objectivity

For convenience, we assume *ab initio* that η is an objective scalar variable and that $W(\mathbf{F}, \eta)$ satisfies the usual objectivity condition

$$W(\mathbf{QF}, \eta) = W(\mathbf{F}, \eta) \quad \text{for all proper orthogonal } \mathbf{Q}. \qquad (2.18)$$

It follows that the dependence of η on $\mathbf{F}$ determined from (2.10), in particular in the form (2.6), is objective, and w in (2.7) is then also objective. Note, however, that objectivity of $W_0(\mathbf{F})$ does not guarantee that of $W(\mathbf{F}, \eta)$. By the same token, since we make no assumption about η in respect of material symmetry, any material symmetry ascribed to $W_0(\mathbf{F})$ is not in general inherited by $W(\mathbf{F}, \eta)$.

13.2.3 Material symmetry

Material symmetry for an elastic material with a strain-energy function $W(\mathbf{F})$ has been discussed in Section 1.2.4.1 of Chapter 1. In the present context we need to consider separately the material symmetries of a pseudo-elastic material when η is active and inactive and how these symmetries are related. In general, the symmetry of $W_0(\mathbf{F})$ is different from (and independent of) that of $w(\mathbf{F})$, as defined in (2.7). This is illustrated by the model pseudo-energy function given by

$$W(\mathbf{F}, \eta) = W_0(\mathbf{F}) + (\eta - 1)N(\mathbf{F}) + \phi(\eta). \qquad (2.19)$$

In (2.19) the function ϕ, which depends only on η, is referred to as a *damage* or *softening function* and, for consistency with (2.2), must satisfy

$$\phi(1) = 0, \qquad (2.20)$$

while $N(\mathbf{F})$ is an objective function of $\mathbf{F}$, independent of W_0, whose (material) symmetry in general differs from that of $W_0(\mathbf{F})$. The specialization of (2.10) in this case is

$$\phi'(\eta) = -N(\mathbf{F}). \qquad (2.21)$$

The nominal stress associated with (2.19) is, for an unconstrained material,

$$\mathbf{S} = \mathbf{S}_0 + (\eta - 1)\frac{\partial N}{\partial \mathbf{F}}, \qquad (2.22)$$

where $\mathbf{S}_0$ is given by (2.3), with appropriate modifications in the case of incompressibility.

If we set $N = W_0$ then equation (2.19) reduces to

$$W(\mathbf{F}, \eta) = \eta W_0(\mathbf{F}) + \phi(\eta), \qquad (2.23)$$

which, in its isotropic specialization, was used by Ogden and Roxburgh (1999a, b) in their model of the Mullins effect. Equation (2.21) then specializes to

$$\phi'(\eta) = -W_0(\mathbf{F}).\tag{2.24}$$

Equation (2.24), or, more generally, (2.21) enables η to be determined in terms of $\mathbf{F}$, at least in principle. Note, however, that in the case of (2.24) η depends on $W_0(\mathbf{F})$ and, for (2.23), the symmetry is therefore unchanged when η is switched on or off.

In respect of (2.24) the damage function ϕ was interpreted by Ogden and Roxburgh (1999a, b) as the non-recoverable part of the energy expended in a loading process. This will be discussed in Section 13.3.

In the applications of the theory of pseudo-elasticity considered so far in the literature (see, for example, Ogden and Roxburgh, 1999a, Lazopoulos and Ogden, 1999, and Ogden, 2001a) it has been assumed, for simplicity, that the material response is isotropic relative to $\mathcal{B}_r$ and that it remains isotropic when the material properties change. In general this will not be the case since, for example, the material response relative to a residually strained state is (in general) anisotropic. In the formulation based on (2.6), $\eta_e(\mathbf{F})$ can be selected so that the material symmetry of $w(\mathbf{F})$ in (2.7) either inherits the material symmetry of $W_0(\mathbf{F})$ or has material symmetry independent of that of $W_0(\mathbf{F})$. Hence switching on or off of η can change the material symmetry. Detailed discussion of the characterization of changes in material symmetry for a pseudo-elastic material will be given in the paper by Ogden (2001b). We note, in particular, that if a material is initially isotropic then the symmetry induced by the generation of residual strain will (in general) be local orthotropy. In the remainder of this chapter, however, we restrict attention to isotropic material response for ease of illustration.

13.2.4 Isotropic material response

When specialized to isotropic response (relative to $\mathcal{B}_r$) the pseudo-elastic energy function (2.1) takes the form

$$W(\lambda_1, \lambda_2, \lambda_3, \eta),\tag{2.25}$$

where $(\lambda_1, \lambda_2, \lambda_3)$ are the principal stretches associated with the deformation from $\mathcal{B}_r$. In the remainder of this paper we focus attention on incompressible materials so that the stretches satisfy the constraint

$$\lambda_1\lambda_2\lambda_3 = 1.\tag{2.26}$$

The Cauchy stress tensor σ, for an incompressible material, is related to the nominal stress $\mathbf{S}$ by $(2.17)_2$ and for an isotropic material σ is coaxial with the Eulerian principal axes, which are the principal axes of $\mathbf{FF}^T$ (see Chapter 1, Section 1.2.3.3). For the considered incompressible isotropic material it follows that the principal Cauchy stresses σ_i, $i = 1, 2, 3$, are given by

$$\sigma_i = \lambda_i \frac{\partial W}{\partial \lambda_i} - p, \quad i \in \{1, 2, 3\}, \tag{2.27}$$

whether or not η is active. Equation (2.10) specializes to

$$\frac{\partial W}{\partial \eta}(\lambda_1, \lambda_2, \lambda_3, \eta) = 0, \tag{2.28}$$

which gives η implicitly in terms of the stretches.

Since the material is incompressible it is convenient to define the modified pseudo-energy function $\hat{W}(\lambda_1, \lambda_2, \eta)$ by

$$\hat{W}(\lambda_1, \lambda_2, \eta) = W(\lambda_1, \lambda_2, \lambda_1^{-1}\lambda_2^{-1}, \eta) \tag{2.29}$$

so that, on elimination of p from (2.27),

$$\sigma_1 - \sigma_3 = \lambda_1 \hat{W}_1, \quad \sigma_2 - \sigma_3 = \lambda_2 \hat{W}_2, \tag{2.30}$$

where $\hat{W}_1$ and $\hat{W}_2$ denote the partial derivatives of $\hat{W}$ with respect to λ_1 and λ_2 respectively. Equation (2.28) is then modified to

$$\frac{\partial \hat{W}}{\partial \eta}(\lambda_1, \lambda_2, \eta) = 0 \tag{2.31}$$

so that η is now given implicitly in terms of λ_1 and λ_2 only.

We define the function $\hat{W}_0(\lambda_1, \lambda_2)$ via

$$\hat{W}_0(\lambda_1, \lambda_2) \equiv \hat{W}(\lambda_1, \lambda_2, 1), \tag{2.32}$$

which is the isotropic specialization of (2.2). This is the energy function of the perfectly elastic material for which η is inactive. From (2.30) the specialization (2.32) yields the stresses

$$\sigma_{0\alpha} - \sigma_{03} = \lambda_\alpha \hat{W}_{0\alpha}, \quad \alpha = 1, 2, \tag{2.33}$$

where the subscript zero again refers to a deformation path on which η is not active, so that (2.31) is not operative. A subscript α following the subscript 0 on $\hat{W}$ indicates partial differentiation with respect to λ_α ($\alpha = 1, 2$).

From equation (4.22) in Chapter 1 we deduce that $\hat{W}_0(\lambda_1, \lambda_2)$ satisfies

$$\hat{W}_0(1,1) = 0, \quad \hat{W}_{0\alpha}(1,1) = 0, \quad \hat{W}_{012}(1,1) = 2\mu, \quad \hat{W}_{0\alpha\alpha}(1,1) = 4\mu, \tag{2.34}$$

where μ (> 0) is the shear modulus of the material in $\mathcal{B}_r$ and the index α takes the value 1 or 2.

When η is active we suppose that equation (2.31) can be solved explicitly for η and, using the notation from (2.6), we write

$$\eta = \eta_e(\lambda_1, \lambda_2) = \eta_e(\lambda_2, \lambda_1). \tag{2.35}$$

Then, an energy function for active η, symmetrical in (λ_1, λ_2) and denoted $\hat{w}(\lambda_1, \lambda_2)$, may be defined by

$$\hat{w}(\lambda_1, \lambda_2) \equiv \hat{W}\big(\lambda_1, \lambda_2, \eta_e(\lambda_1, \lambda_2)\big). \tag{2.36}$$

From equations (2.30), (2.31) and (2.36) it follows that

$$\sigma_\alpha - \sigma_3 = \lambda_\alpha \frac{\partial \hat{w}}{\partial \lambda_\alpha} = \lambda_\alpha \frac{\partial \hat{W}}{\partial \lambda_\alpha}, \qquad \alpha = 1, 2. \tag{2.37}$$

13.2.4.1 Simple tension

For simple tension we may take $\sigma_2 = \sigma_3 = 0$ and we write $\sigma_1 = \sigma$. We also write $\lambda_1 = \lambda$, so that $\lambda_2 = \lambda_3 = \lambda^{-1/2}$, and define $\tilde{W}$ by

$$\tilde{W}(\lambda, \eta) \equiv \hat{W}(\lambda, \lambda^{-1/2}, \eta). \tag{2.38}$$

Equations (2.37) and (2.31) then simplify to

$$\sigma = \lambda \tilde{W}_\lambda(\lambda, \eta) \equiv \lambda t, \quad \tilde{W}_\eta(\lambda, \eta) = 0, \tag{2.39}$$

wherein the principal Biot stress $t \,(= t_1)$ is defined and the subscripts signify partial derivatives.

From the second (implicit) equation in (2.39), η is (in principle) determined in terms of λ, and from equations (2.39) it follows that

$$\frac{dt}{d\lambda} = \tilde{W}_{\lambda\lambda} - (\tilde{W}_{\lambda\eta})^2 / \tilde{W}_{\eta\eta}. \tag{2.40}$$

Equation (2.40) shows how the inclusion of the variable η modifies the stiffness $\tilde{W}_{\lambda\lambda}$ of the material appropriate for the case of inactive η.

By defining

$$\tilde{W}_0(\lambda) = \tilde{W}(\lambda, 1), \tag{2.41}$$

we may deduce from (2.34) the specializations

$$\tilde{W}_0(1) = \tilde{W}_0'(1) = 0, \quad \tilde{W}_0''(1) = 3\mu, \tag{2.42}$$

where the prime signifies differentiation with respect to λ.

This simple tension specialization will be examined in detail in connection with the description of stress softening in Section 13.3.

13.2.4.2 Equibiaxial deformations

In Section 13.5 we shall consider the problem of (spherically symmetric) inflation of a spherical shell. The deformation in the shell is locally an equibiaxial deformation. As a prelude to this we give here the appropriate specialization of equations (2.31) and (2.37) for equibiaxial deformations. We set $\lambda_2 = \lambda_3 = \lambda, \lambda_1 = \lambda^{-2}$ so that $\sigma_2 = \sigma_3$ and we define the function $\dot{W}$ by

$$\dot{W}(\lambda, \eta) = \hat{W}(\lambda^{-2}, \lambda, \eta). \tag{2.43}$$

It follows from (2.37) and (2.43) that

$$2(\sigma_2 - \sigma_1) = \lambda \dot{W}_\lambda(\lambda, \eta) \tag{2.44}$$

and $\dot{W}_2 = 0$, while (2.31) simplifies to

$$\dot{W}_\eta(\lambda, \eta) = 0, \tag{2.45}$$

where the subscripts λ and η again denote partial derivatives.

We now write

$$\dot{W}_0(\lambda) = \dot{W}(\lambda, 1) \tag{2.46}$$

for the case in which η is inactive. Then, the analogues of (2.42) are

$$\dot{W}_0(1) = \dot{W}_0'(1) = 0, \quad \dot{W}_0''(1) = 12\mu. \tag{2.47}$$

13.2.4.3 Plane strain and simple shear

In order to consider plane strain deformations we set $\lambda_3 = 1$ and use the notation

$$\lambda = \lambda_1, \quad \lambda_2 = \lambda^{-1}. \tag{2.48}$$

The pseudo-energy function is then given in terms of λ by the definition

$$\breve{W}(\lambda, \eta) \equiv \hat{W}(\lambda, \lambda^{-1}, \eta), \tag{2.49}$$

and equations (2.37) and (2.31) respectively specialize to

$$\sigma_1 - \sigma_2 = \lambda \breve{W}_\lambda, \quad \breve{W}_\eta = 0, \tag{2.50}$$

where, once more, the subscripts denote partial derivatives.

In preparation for the discussion, in Section 13.4, of the azimuthal shear deformation (which is locally a simple shear) we now take the plane strain to correspond to a simple shear deformation. The *amount of shear*, denoted γ, is then given in terms of λ by

$$\gamma = \lambda - \lambda^{-1}, \tag{2.51}$$

where, without loss of generality, we have taken $\lambda \geq 1$ to correspond to $\gamma \geq 0$.

To represent the pseudo-energy function in terms of γ we use the notation $\bar{W}$ defined by

$$\bar{W}(\gamma,\eta) \equiv \check{W}(\lambda,\eta). \tag{2.52}$$

Then, with the direction of shear taken as the X_1 direction in the $(1,2)$-plane the (uniform) shear stress σ_{12} has the form

$$\sigma_{12} = \bar{W}_\gamma(\gamma,\eta) \tag{2.53}$$

and we also have the *universal relation*

$$\sigma_{11} - \sigma_{22} = \gamma\sigma_{12}, \tag{2.54}$$

where σ_{11} and σ_{22} are the normal components of the Cauchy stress tensor. For a general discussion of universal relations in elasticity, see Chapter 3. The second equation in (2.50) becomes

$$\bar{W}_\eta(\gamma,\eta) = 0, \tag{2.55}$$

which, in general implicitly, gives η in terms of γ when η is active.

When η is inactive we set

$$\bar{W}_0(\gamma) = \bar{W}(\gamma,1), \tag{2.56}$$

with $\bar{W}_0$ having the properties

$$\bar{W}_0(0) = \bar{W}_0'(0) = 0, \quad \bar{W}_0''(0) = \mu, \tag{2.57}$$

these being the counterparts of (2.42) for simple shear, but with the prime here indicating differentiation with respect to γ.

13.3 A model for stress softening

13.3.1 The idealized Mullins effect

Figure 1 in Section 13.1 is associated with the idealized Mullins effect in simple tension. The essence of this effect was described in Section 13.1, but some additional comments are appropriate here. We emphasize that the continuous curve in Figure 1 models the material response on a *loading* path from the unstressed configuration of the virgin material if there is no unloading (this is the *primary loading curve*). The dashed curve is a typical unloading curve from a point on the primary loading curve. The idealized Mullins effect requires that each unloading curve passes through the origin, i.e. the stretch λ returns to 1 when the stress t is removed ($t = 0$). Moreover, if the material is reloaded after the initial unloading the unloading curve is retraced until the primary loading curve is met, after which the primary loading curve is followed on further loading. This is the second main aspect of the idealization, which, for certain materials, gives a very good approximation to the actual behaviour (Muhr *et al.*, 1999). This description applies whichever point on the primary loading curve the unloading begins from. For further discussion we refer to Ogden and Roxburgh (1999a) and references contained therein. For the original work of Mullins we refer to Mullins (1947, 1969) and Mullins and Tobin (1957).

In this section we describe a specific form of the pseudo-elastic constitutive which was used by Ogden and Roxburgh (1999a) to model the idealized Mullins effect. The material is taken to be incompressible and isotropic and we use the pseudo-energy function (2.23) in its isotropic form, subject to (2.20), together with the notation of Section 13.2.4. We focus initially on (homogeneous) biaxial deformations in (λ_1, λ_2)-space. Thus, we write

$$\hat{W}(\lambda_1, \lambda_2, \eta) = \eta \, \hat{W}_0(\lambda_1, \lambda_2) + \phi(\eta), \qquad \phi(1) = 0, \tag{3.1}$$

and, from (2.33), (2.37) and (3.1), the Cauchy stress differences are calculated as

$$\sigma_\alpha - \sigma_3 = \eta \lambda_\alpha \frac{\partial \hat{W}_0}{\partial \lambda_\alpha} = \eta(\sigma_{0\alpha} - \sigma_{03}), \qquad \alpha = 1, 2. \tag{3.2}$$

Equation (2.24) becomes

$$\phi'(\eta) = -\hat{W}_0(\lambda_1, \lambda_2), \tag{3.3}$$

which, implicitly, defines the damage parameter η in terms of the stretches.

In addition to the properties (2.34), we take $\hat{W}_0$ to have a *global minimum* of 0 at $(1,1)$ and that it has no other stationary points in (λ_1, λ_2)-space. This is appropriate since, for example, in the deformation of a thin sheet with $\sigma_{03} = 0$, any other stationary point requires that both (in-plane) stresses σ_{01} and σ_{02} in (3.2) vanish. In practice this would happen only for pathological forms of $\hat{W}_0$.

We therefore *define* a primary loading path in (λ_1, λ_2)-space as a path starting from $(1,1)$ on which $\hat{W}_0$ is *increasing*. Indeed, it can be shown that for many standard forms of strain-energy function $\hat{W}_0$ is increasing along any straight line path from $(1,1)$. A case in point is the neo-Hookean strain-energy function, which, when expressed in terms of λ_1 and λ_2 alone, has the form

$$\hat{W}_0(\lambda_1, \lambda_2) = \frac{\mu}{2}(\lambda_1^2 + \lambda_2^2 + \lambda_1^{-2}\lambda_2^{-2} - 3), \tag{3.4}$$

where μ is the shear modulus arising in (2.34). In Figure 2 contours of constant values of $\hat{W}_0$ are plotted in the (λ_1, λ_2)-plane in respect of (3.4). Note that the contours are convex.

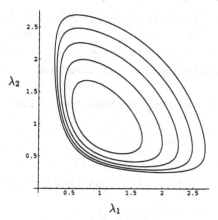

Fig. 2. Constant energy contour curves in (λ_1, λ_2)-space for the neo-Hookean strain-energy function (3.4) with values $\hat{W}_0/\mu = 0.5, 1, 1.5, 2, 2.5$.

13.3.1.1 Simple tension

On the basis of the equations in Section 13.2.4.1 the relevant specializations of the above equations are, in terms of the Biot stress t,

$$t = \eta \tilde{W}_0'(\lambda) = \eta t_0, \tag{3.5}$$

where t_0 is the Biot stress on the primary loading path at the same value of λ. For (3.5) to predict stress softening on unloading (governed by the energy function (3.1) with η active, so that η is switched on at the start of unloading) it is clear that we must have $\eta \le 1$ on the unloading path, with equality only at the point where unloading begins. Here we take $\eta > 0$, so that t remains positive on unloading until $\lambda = 1$ is reached. This means that residual strains will not arise. This condition will be relaxed in Section 13.3.2.

The simple tension specialization of (3.3) is

$$\phi'(\eta) = -\tilde{W}_0(\lambda). \tag{3.6}$$

On differentiation of (3.6) with respect to λ we obtain

$$\phi''(\eta)\frac{d\eta}{d\lambda} = -\tilde{W}_0'(\lambda). \tag{3.7}$$

In view of the stress softening requirement discussed above we associate unloading with decreasing η. Since $t_0 \equiv \tilde{W}_0'(\lambda) > 0$ for $\lambda > 1$ it follows from (3.7) that

$$\phi''(\eta) < 0, \tag{3.8}$$

and we assume henceforth that this inequality holds. We deduce that $\phi'(\eta)$ is a monotonic decreasing function of η and hence that η is uniquely determined from (3.6) as a function of $\tilde{W}_0(\lambda)$. More generally, for biaxial deformations this condition ensures that η is uniquely determined as a function of the loading energy $\hat{W}_0(\lambda_1, \lambda_2)$.

Note that on specializing (2.40) we obtain, for the model (3.1) specialized for simple tension,

$$\frac{dt}{d\lambda} = \eta \frac{dt_0}{d\lambda} - \frac{t_0^2}{\phi''(\eta)}, \tag{3.9}$$

and it follows from (3.8) and the assumption $\eta > 0$ that this is positive if

$$\frac{dt_0}{d\lambda} \equiv \tilde{W}_0''(\lambda) > 0. \tag{3.10}$$

The inequality (3.10) ensures that the material response is stable in simple tension on the primary loading path. With reference to (3.8) it then follows from (3.9) that the unloading paths are also stable.

It is important to point out that the value of η derived from (3.6) depends on the value of the principal stretch, λ_m say, attained on a primary loading path, as well as on the specific forms of $\tilde{W}_0(\lambda)$ and $\phi(\eta)$ employed. Since $\eta = 1$ at any point on the primary loading path from which unloading is initiated, it follows from equation (3.6) that

$$\phi'(1) = -\tilde{W}_0(\lambda_m) \equiv W_m, \tag{3.11}$$

wherein the notation W_m is defined. This is the current maximum value of the energy achieved on the primary loading path. In accordance with the properties of $\tilde{W}_0$, W_m increases along a primary loading path. In view of (3.11), the function ϕ depends (implicitly) on the point from which unloading begins through the energy expended on the loading path up to that point. We shall make this dependence explicit in Section 13.3.2.

When the material is fully unloaded, with $\lambda = 1$, η attains its minimum value, η_m say. This is determined by inserting these values into equation (3.6) to give, using the first equation in (2.42),

$$\phi'(\eta_m) = -\tilde{W}_0(1) = 0. \tag{3.12}$$

Since the function ϕ depends on the point where unloading begins then so does η_m, that is it depends, though W_m, on the value of λ_m.

When the material is in a fully unloaded state the pseudo-energy function (3.1) has the residual value

$$\hat{W}(1, 1, \eta_m) = \phi(\eta_m). \tag{3.13}$$

This applies for general equibiaxial deformations as well as for simple tension. Thus, the residual (non-recoverable) energy $\phi(\eta_m)$ may be interpreted as a measure of the energy required to cause the damage in the material. In simple tension $\phi(\eta_m)$ is the area between the primary loading curve and the relevant unloading curve in Figure 1(a).

Stress softening data for simple tension were examined by Ogden and Roxburgh (1999a) on the basis of the theory discussed above with particular forms of the functions $\hat{W}_0(\lambda_1, \lambda_2)$ and $\phi(\eta)$ chosen so as to fit the data. We refer to this paper for the details and for references to the background literature.

13.3.1.2 Discussion

Unloading may take place from any point on a primary loading path, and the start of unloading is taken as the signal for η to be activated, as mentioned above in respect of simple tension. Let $(\lambda_{1m}, \lambda_{2m})$ be the values of (λ_1, λ_2) at a point at which unloading begins. Then, in the notation of (2.35), $\eta_e(\lambda_{1m}, \lambda_{2m}) = 1$. This implies that the function η_e, and hence $\hat{w}$ in (2.36), depends on the point from which unloading starts.

In the simple tension considered above the point at which unloading began was identified by the value of the stretch (λ_m) at that point, or, equivalently, by the associated energy on the primary path. In the case of general biaxial deformations it is the value of the energy maximum on the primary loading path, again denoted W_m and given by $W_m = \hat{W}_0(\lambda_{1m}, \lambda_{2m})$, rather than the specific $(\lambda_{1m}, \lambda_{2m})$ pair that governs the unloading response. Thus, any other pair of (λ_1, λ_2) values corresponding to the same value of W_m could equally be taken as the starting point for unloading. The collection of all such pairs satisfies the equation

$$\hat{W}_0(\lambda_1, \lambda_2) = W_m, \qquad (3.14)$$

and, for given (constant) W_m, forms a closed contour in (λ_1, λ_2)-space (see Figure 2 for an illustration in respect of the neo-Hookean strain-energy function (3.4)).

The contour defined by the current maximum value W_m represents the current damage threshold, and $\eta = 1$ at any point on this contour. For any deformation path within this contour $\eta < 1$ and no further damage occurs. The energy required on the initial loading path to cause the damage is not required on subsequent loading up to the contour boundary. However, if the deformation path crosses the contour primary loading is again activated, $\eta = 1$ and further damage will occur. The value of $\hat{W}_0(\lambda_1, \lambda_2)$ will increase until the next maximum value W_m is reached, at which point loading terminates and a new threshold contour is established (which encloses the previous one).

We mention here that an alternative model for the Mullins effect having some features in common with the theory of pseudo-elasticity discussed here has been developed by Beatty and colleagues. For uniaxial deformations details can be found in Johnson and Beatty (1993a, b) and for general deformations in, for example, Beatty and Krishnaswamy (2000) and Krishnaswamy and Beatty (2000), which deal with incompressible and compressible materials respectively. In contrast to the model discussed here, in which the extent of the damage sustained by the material is controlled by the maximum energy state W_m attained, these models are strain based rather than energy based. This is also the case with the model of the Mullins effect developed by Govindjee and Simo (1991, 1992a, b). Related work by Miehe (1995) and Lion (1996) should also be mentioned.

13.3.2 Residual strain

In Section 13.3.1 we assumed that the damage variable was strictly positive and we noted that this condition excluded the possibility of residual strains arising when the stress vanishes. We now examine the consequences of removing this restriction in respect of the model (3.1). If we consider biaxial deformations on the basis of equations (3.2) with $\sigma_3 = 0$ then it is clear that if $\eta = 0$ then $\sigma_1 = \sigma_2 = 0$. In general the stretches λ_1 and λ_2 will not both be 1 at this point so a residual strain remains after removal of the stresses. However, it remains possible that even when $\eta = 0$ is admitted it does not reduce to zero before the stretches reach 1 on unloading. This will depend on the form of the function ϕ and the magnitude of W_m, as will be illustrated in Section 13.4 in respect of a non-homogeneous deformation. We denote the values of λ_1 and λ_2 where $\eta = 0$ by λ_{1r} and λ_{2r} respectively and set W_r to be the associated value of $\hat{W}_0$. Equation (3.3) then gives

$$\phi'(0) = -\hat{W}_0(\lambda_{1r}, \lambda_{2r}) = -W_r. \qquad (3.15)$$

Since the function ϕ depends on W_m then, for a given function ϕ, equation (3.15) defines W_r associated with the residually strained state in terms of the energy W_m associated with primary loading. Hence the residual deformation depends on the deformation at the end of the primary loading process. Note, however, that while, in view of the monotonicity of $\phi'(\eta)$, W_r is uniquely defined, the associated residual strain depends on the unloading path followed to reach this value of W_r.

In order to make explicit the dependence of ϕ on W_m we write $\Phi(\eta, W_m)$ so that (3.15) becomes

$$\Phi_\eta(0, W_m) = -W_r, \qquad (3.16)$$

where the subscript η indicates partial differentiation with respect to η. Equation (3.3) becomes

$$\Phi_\eta(\eta, W_m) = -\hat{W}_0, \qquad \hat{W}_0 \leq W_m, \qquad (3.17)$$

and we must have

$$\Phi_\eta(1, W_m) = -W_m. \qquad (3.18)$$

When $\eta = 0$ the pseudo-energy function (3.1) has the residual value

$$\phi(0) \equiv \Phi(0, W_m). \qquad (3.19)$$

This is the (non-recoverable) energy and may be interpreted as a measure of the energy required to cause damage in the material during loading. This is analogous to the interpretation of $\phi(\eta_m)$ in Section 13.3.1. In a uniaxial test such as simple tension $\phi(0)$ is the area between the primary loading curve and the relevant unloading curve and above the axis $t = 0$ (see Figure 1(b)).

With reference to the discussion in Section 13.3.1.2 of the contours in Figure 2 we remark that in the present context, for a given value of W_m, $\eta = 0$ defines a contour $\hat{W}_0(\lambda_1, \lambda_2) = W_r < W_m$ and points on this contour define possible values $(\lambda_{1r}, \lambda_{2r})$ of the residual stretches, which depend on the unloading path to the point where the stresses vanish. It should be emphasized that we have assumed here that the material response is *isotropic* relative to $\mathcal{B}_r$ in both loading and unloading. This is reflected, in particular, in the symmetry about the line $\lambda_1 = \lambda_2$ of the energy contours in Figure 2. In general, however, it cannot be expected that the material response remains isotropic when damage occurs, and, in particular, when residual strain is induced since the material symmetry is then characterized relative to an unstressed configuration that is not (in general) related to the original unstressed configuration by a pure dilatation. The unstressed (reference) configuration is being continually modified during primary loading. Characterization of the evolving anisotropic response due to damage (or other mechanisms for material property changes) is clearly important and will be addressed in the forthcoming paper by Ogden (2001b).

13.4 Azimuthal shear

In this section we discuss the (plane strain) pure azimuthal shear deformation of a circular cylindrical thick-walled tube. In large part we follow the development in Ogden (2001a) but with some differences in notation. The cross-section of the tube is defined by

$$A \leq R \leq B, \quad 0 \leq \Theta \leq 2\pi \qquad (4.1)$$

in polar coordinates (R, Θ) in the reference configuration, where A and B are constants. Let (r, θ) be the corresponding polar coordinates in the deformed configuration. Then, pure azimuthal shear is the isochoric deformation defined by

$$r = R, \qquad \theta = \Theta + g(R),$$ (4.2)

where $g(r) = g(R)$ is an unknown function to be determined by solution of the equilibrium equations. We use r as the independent variable and set $a = A, b = B$ so that $a \le r \le b$. The deformation is locally a simple shear of amount γ depending on r according to

$$\gamma = rg'(r),$$ (4.3)

where the prime indicates differentiation with respect to r. The direction of shear is locally the azimuthal (θ) direction and the radial direction is normal to the planes of shear.

With reference to the equations for simple shear discussed in Section 13.2.4.3 we may identify locally the (θ, r) axes with the $(1, 2)$ axes with θ increasing counterclockwise. Thus, the stress components are $\sigma_{r\theta} = \sigma_{12}, \sigma_{rr} = \sigma_{22}, \sigma_{\theta\theta} = \sigma_{11}$. Hence, in the notation in Section 13.2.4.3 used for simple shear, equations (2.53) and (2.54) translate to

$$\sigma_{r\theta} = \bar{W}_\gamma(\gamma, \eta), \qquad \sigma_{\theta\theta} - \sigma_{rr} = \gamma \bar{W}_\gamma(\gamma, \eta).$$ (4.4)

For the azimuthal shear problem there are two independent equilibrium equations, namely the radial equation

$$r\frac{d\sigma_{rr}}{dr} = \sigma_{\theta\theta} - \sigma_{rr} = \gamma \bar{W}_\gamma(\gamma, \eta),$$ (4.5)

and the azimuthal equation

$$\frac{d}{dr}(r^2 \sigma_{r\theta}) = 0.$$ (4.6)

On use of $(4.4)_1$ equation (4.6) integrates to give

$$\bar{W}_\gamma(\gamma, \eta) = b^2 \tau / r^2,$$ (4.7)

where τ is a constant equal to the value of $\sigma_{r\theta}$ on $r = b$.

When η is active it is determined in terms of γ from equation (2.55), which we repeat here as

$$\bar{W}_\eta(\gamma, \eta) = 0.$$ (4.8)

Suppose that equation (4.8) enables η to be determined explicitly and let the solution be written

$$\eta = \bar{\eta}(\gamma).$$ (4.9)

Then, we define

$$\bar{w}(\gamma) \equiv \bar{W}\left(\gamma, \bar{\eta}(\gamma)\right) \tag{4.10}$$

as the resulting energy as a function of γ, and the azimuthal equation (4.7) becomes

$$\bar{w}'(\gamma) \equiv \bar{W}_\gamma(\gamma, \eta) = b^2 \tau / r^2. \tag{4.11}$$

In principle equation (4.7), or its specialization for inactive η, determines γ and hence the function g, while the radial equation serves to determine σ_{rr} and hence the stress distribution. For definiteness we impose on g the boundary conditions

$$g(a) = 0, \qquad g(b) = \psi, \tag{4.12}$$

so that the inner boundary of the tube is fixed while the outer boundary is rotated through a prescribed angle ψ. We assume that τ is a monotonic increasing function of ψ and it is therefore appropriate to describe loading and unloading in terms of τ rather than ψ.

13.4.1 Loading

For loading we take η to be inactive so that equation (4.11), in the notation defined by (2.56), specializes to

$$\bar{W}_0'(\gamma) = b^2 \tau / r^2. \tag{4.13}$$

To ensure that this equation yields a unique solution for γ as a function of r for all $\tau \geq 0$ we assume that

$$\bar{W}_0''(\gamma) > 0, \qquad \bar{W}_0'(\gamma) \to \infty \quad \text{as} \quad \gamma \to \infty \tag{4.14}$$

(we are taking $\tau \geq 0$ to correspond to $\gamma \geq 0$ during loading, with $\tau = 0$ if and only if $\gamma = 0$ for $a \leq r \leq b$). For further discussion of this point and other aspects of the azimuthal shear problem we refer to the recent paper by Jiang and Ogden (1998).

Let τ_m be the maximum value of τ reached on loading and let γ_m be the corresponding value of γ, so that

$$\bar{W}_0'(\gamma_m) = b^2 \tau_m / r^2. \tag{4.15}$$

We set

$$W_m = \bar{W}_0(\gamma_m), \tag{4.16}$$

which, of course, depends, through γ_m, on r and τ_m.

13.4.1.1 Loading in the case of a neo-Hookean material

If loading is governed by the neo-Hookean strain energy then an exact solution of (4.13) is available. For the deformation considered here, the neo-Hookean strain energy has the form

$$\bar{W}_0(\gamma) = \frac{1}{2}\mu\gamma^2, \tag{4.17}$$

where μ is the shear modulus appearing in (2.57).

Equation (4.13), on use of (4.17) and (4.3), then leads to the solution

$$\gamma = \frac{b^2\tau}{\mu r^2}, \qquad g(r) = \frac{b^2\tau}{2\mu}\left(\frac{1}{a^2} - \frac{1}{r^2}\right). \tag{4.18}$$

The angle of rotation ψ of the outer surface relative to the inner one depends linearly on τ and is given by

$$\psi = g(b) = \frac{b^2\tau}{2\mu}\left(\frac{1}{a^2} - \frac{1}{b^2}\right). \tag{4.19}$$

For the neo-Hookean material we have

$$W_m = \frac{1}{2}\mu\gamma_m^2, \qquad \gamma_m = b^2\tau_m/\mu r^2. \tag{4.20}$$

13.4.2 Unloading

When $\eta = \bar{\eta}(\gamma)$ is determined from (4.8) the solution of (4.11) for γ describes the unloading deformation path as τ reduces from τ_m to 0. When $\tau = 0$ we must have

$$\bar{W}_\gamma(\gamma, \eta) = 0, \tag{4.21}$$

which has to be solved locally in conjunction with (4.8) to give a pair of values $(\gamma, \bar{\eta}(\gamma))$ at each radius r through the tube. As is illustrated below this may yield the trivial solution $\gamma = 0$ for all r, $\gamma = 0$ for some range of values of r, or $\gamma \neq 0$ for all r. These three situations correspond respectively to cases in which there is no residual strain, there is a residual value of γ for the some values of r, and there is residual strain for all r.

The equation

$$\bar{w}'(\gamma) \equiv \bar{W}_\gamma(\gamma, \bar{\eta}(\gamma)) = 0 \tag{4.22}$$

then yields the residual value of γ, denoted γ_r (which depends on r through W_m). Thus, for each value of r, either $\gamma = 0$ or $\gamma = \gamma_r$, the latter depending on W_m.

We now illustrate the results for the model (3.1), which, when specialized for azimuthal shear, has the form

$$\bar{W}(\gamma, \eta) = \eta \bar{W}_0(\gamma) + \phi(\eta), \quad \phi(1) = 0, \tag{4.23}$$

and we have

$$\phi'(\eta) = -\bar{W}_0(\gamma), \quad \phi'(1) = -W_m. \tag{4.24}$$

Equation (4.11) becomes

$$\eta \bar{W}_0'(\gamma) = b^2 \tau / r^2, \quad \tau \leq \tau_m, \tag{4.25}$$

with $\eta = \bar{\eta}(\gamma)$ determined from the first equation in (4.24). From the properties (2.57) and (4.14) we deduce that when $\tau = 0$ is reached on unloading then either $\gamma = 0$ or $\eta = 0$. When $\eta = 0$ the first equation in (4.24) becomes

$$\bar{W}_0(\gamma) = -\phi'(0), \tag{4.26}$$

which determines *uniquely* the residual value γ_r of γ locally (i.e. as a function of r).

Since ϕ, and hence $\bar{\eta}(\gamma)$, depends on r through W_m the resulting strain energy for unloading, obtained from (4.23), is *inhomogeneous* because it depends explicitly on r as well as on r through γ.

To illustrate the theory further we now select a specific form of ϕ. For simplicity, we take ϕ to be quadratic in η given by

$$\phi(\eta) = -\frac{1}{4}\mu\gamma_0^2(\eta - 1)^2 - W_m(\eta - 1), \tag{4.27}$$

so that $\phi(1) = 0$ and the second equation in (4.24) is satisfied. In (4.27) we have introduced the (positive, dimensionless) material constant γ_0. Its interpretation will be discussed below. The first equation in (4.24) now gives η explicitly in the form

$$\frac{1}{2}\mu\gamma_0^2(\eta - 1) = \bar{W}_0(\gamma) - W_m. \tag{4.28}$$

On use of (4.28) in the definition (4.10) in respect of the model (4.23) for unloading, the pseudo-energy $\bar{W}(\gamma, \eta)$ becomes the strain energy

$$\bar{w}(\gamma) = \bar{W}_0(\gamma) + \left(\bar{W}_0(\gamma) - W_m\right)^2 / \mu\gamma_0^2 \tag{4.29}$$

as a function of γ, and this shows explicitly its dependence on W_m and hence on r.

For the neo-Hookean material equation (4.28) reduces to

$$\eta = \bar{\eta}(\gamma) \equiv 1 + (\gamma^2 - \gamma_m^2)/\gamma_0^2, \tag{4.30}$$

and equation (4.25) becomes

$$[1 + (\gamma^2 - \gamma_m^2)/\gamma_0^2]\gamma = b^2\tau/\mu r^2, \qquad \tau \le \tau_m. \tag{4.31}$$

13.4.2.1 Residual strains

When $\eta = 0$ it follows from (4.28) that

$$\bar{W}_0(\gamma_r) = W_m - \frac{1}{2}\mu\gamma_0^2, \tag{4.32}$$

and from the assumed properties of $\bar{W}_0(\gamma)$ we deduce that

$$W_m > \frac{1}{2}\mu\gamma_0^2. \tag{4.33}$$

is a necessary and sufficient condition for the existence of a (non-trivial) solution of (4.32) (locally) for γ_r. If the inequality (4.33) does not hold then the trivial solution $\gamma = 0$ applies when $\tau = 0$.

For simplicity of illustration, it is henceforth assumed that the energy function for loading has the neo-Hookean form (4.17) so that η is given by (4.30) on unloading and γ by (4.31). Then, by setting $\eta = 0$ and $\gamma = \gamma_r$ in (4.30), we obtain

$$\gamma_r^2 = \gamma_m^2 - \gamma_0^2. \tag{4.34}$$

Since we have taken $\gamma \ge 0$ and $\gamma_0 > 0$, equation (4.34) has a real solution for $\gamma_r > 0$ if and only if $\gamma_m > \gamma_0$. The resulting residual strain distribution can then be calculated. The corresponding stresses are such that $\sigma_{r\theta} = 0, \sigma_{rr} = \sigma_{\theta\theta}$.

The material constant γ_0 may be regarded as a *critical value* of γ_m. If $\gamma_m \le \gamma_0$ for all r then $\gamma = 0$ is the only solution when $\tau = 0$, in which case there is no residual strain anywhere in the material. In this case there is nevertheless a stress softening effect on unloading since $\eta < 1$, as discussed by Ogden (2001a). Locally, the behaviour of the shear stress as a function of γ has the character of the stress-stretch curve shown in Figure 1(a).

In order to highlight the different situations that can arise we show in Figure 3 a plot of the function

$$f(\gamma/\gamma_0) = [1 + (\gamma^2 - \gamma_m^2)/\gamma_0^2]\gamma/\gamma_0, \tag{4.35}$$

which corresponds to the left-hand side of equation (4.31) divided by the factor γ_0, for each of the cases $\gamma_m < \gamma_0, \gamma_m = \gamma_0, \gamma_m > \gamma_0$. Also shown is the straight line $f(\gamma/\gamma_0) = \gamma/\gamma_0$ corresponding to loading for the neo-Hookean material. Note that the dashed curve (associated with the intermediate case) separates the region where there is no non-trivial residual value of γ from that where the intercept of the curve for $\gamma_m > \gamma_0$ with the horizontal axis gives the residual value γ_r of γ identified by (4.34).

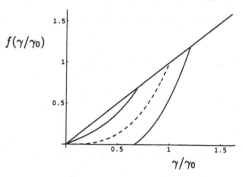

Fig. 3. Plot of the function $f(\gamma/\gamma_0)$ given by (4.35) for $\gamma_m/\gamma_0 = 0.7, 1, 1.2$ (unloading) and the line $f(\gamma/\gamma_0) = \gamma/\gamma_0$ (loading).

We now write

$$\gamma_m(r) = \gamma_m = b^2 \tau_m / \mu r^2. \tag{4.36}$$

in order to make explicit the dependence of γ_m on r. Then,

$$\gamma_m(a) = b^2 \tau_m / \mu a^2, \qquad \gamma_m(b) = \tau_m / \mu. \tag{4.37}$$

We consider separately the following three cases, which reflect the differences shown in Figure 3.

Case (i): $\gamma_m(a) \leq \gamma_0$ and $\tau_m \leq \mu \gamma_0 a^2 / b^2$. In this case there is no residual strain, $\gamma = 0$ for all r, and $g(r) = 0$.

Case (ii): $\gamma_m(b) \leq \gamma_0 \leq \gamma_m(a)$. Then $\mu \gamma_0 a^2 / b^2 \leq \tau_m \leq \mu \gamma_0$. We denote by r_0 ($a \leq r_0 \leq b$) the value of r such that $\gamma_m(r_0) = \gamma_0$. Then,

$$\gamma_m(r) = \gamma_0 r_0^2 / r^2, \tag{4.38}$$

and hence,

$$\gamma_m(r) \leq \gamma_0 \quad \text{for} \quad r_0 \leq r \leq b, \qquad \gamma_m(r) \geq \gamma_0 \quad \text{for} \quad a \leq r \leq r_0. \tag{4.39}$$

The solution for γ corresponding to $\tau = 0$ is

$$\gamma = 0 \quad \text{for} \quad r_0 \leq r \leq b,$$
$$\gamma = \gamma_r = \sqrt{\gamma_m^2 - \gamma_0^2} \quad \text{for} \quad a \leq r \leq r_0, \tag{4.40}$$

and from (4.38) we see that γ_r is given in terms of r by

$$\gamma_r = \gamma_0 \sqrt{r_0^4 - r^4} / r^2. \tag{4.41}$$

From (4.3) the solution for the residual value of $g(r)$, denoted $g_r(r)$, is then obtained as

$$g_r(r) = c \quad \text{for} \quad r_0 \leq r \leq b, \tag{4.42}$$

$$g_r(r) = c + \frac{\gamma_0}{2}\left[\cos^{-1}\left(\frac{r^2}{r_0^2}\right) - \frac{1}{r^2}\sqrt{r_0^4 - r^4}\right] \quad \text{for} \quad a \leq r \leq r_0, \tag{4.43}$$

where c is a constant.

From the boundary conditions (4.12) we then deduce that

$$c = \frac{\gamma_0}{2}\left[\frac{1}{a^2}\sqrt{r_0^4 - a^4} - \cos^{-1}\left(\frac{a^2}{r_0^2}\right)\right], \tag{4.44}$$

which is the residual value of ψ.

Case (iii): $\gamma_m(b) \geq \gamma_0$ and $\tau_m \geq \mu\gamma_0$. In this case there is a residual strain for all r and the solution is

$$g_r(r) = c + \frac{\gamma_0}{2}\left[\cos^{-1}\left(\frac{r^2}{b^2}\right) - \frac{1}{r^2}\sqrt{b^4 - r^4}\right] \quad \text{for} \quad a \leq r \leq b, \tag{4.45}$$

the constant c being

$$c = \frac{\gamma_0}{2}\left[\frac{1}{a^2}\sqrt{b^4 - a^4} - \cos^{-1}\left(\frac{a^2}{b^2}\right)\right]. \tag{4.46}$$

For further discussion of these results and associated numerical results we refer to Ogden (2001a). It is interesting to note that because of the geometry of the problem considered here vanishing of τ requires that the shear stress $\sigma_{r\theta}$ vanishes for each r and hence, for the particular model (4.23) used, that η vanishes simultaneously for all r. A different situation arises in the problem considered in Section 13.5, in which, for the same model, it is necessary to allow η to become negative in part of the material in order for residual strains to be predicted.

13.5 Inflation and deflation of a spherical shell

The problem of inflation followed by deflation of a thick-walled spherical shell of incompressible pseudo-elastic material was considered by Lazopoulos and Ogden (1999) for the case in which there is a discontinuity in the material properties and, for a different material model, by Ogden (2000a, b). Here we examine the same problem but for a pseudo-elastic model in which the material properties change continuously. This problem has also been studied by Haddow (2000) using a different form of pseudo-elasticity.

13.5.1 Geometry and equilibrium equations

We consider a spherical shell of incompressible isotropic pseudo-elastic material with initial geometry defined by

$$A \leq R \leq B, \quad 0 \leq \Theta \leq \pi, \quad 0 \leq \Phi \leq 2\pi, \tag{5.1}$$

where (R, Θ, Φ) are spherical polar coordinates.

The spherically-symmetric deformation is defined by

$$r = Rf(R), \quad \theta = \Theta, \quad \varphi = \Phi, \tag{5.2}$$

where (r, θ, φ) are spherical polar coordinates in the current configuration, the function f is defined by

$$f(R) = \left(1 + \frac{a^3 - A^3}{R^3}\right)^{1/3} \equiv \lambda, \tag{5.3}$$

where the notation $\lambda = \lambda_2 = \lambda_3$ is introduced for the (local) equibiaxial stretch associated with the principal directions corresponding to the θ and φ coordinates. Since the deformation is locally equibiaxial and the material is incompressible, we have, from Section 13.2.4.2, $\lambda_1 = \lambda^{-2}$.

We use the notations

$$\lambda_a = a/A, \quad \lambda_b = b/B \tag{5.4}$$

respectively for the values of λ at the inner and outer boundaries of the spherical shell. Then, from (5.3), we obtain

$$\lambda_a^3 - 1 = \left(\frac{R}{A}\right)^3 (\lambda^3 - 1) = \left(\frac{B}{A}\right)^3 (\lambda_b^3 - 1), \tag{5.5}$$

as in Ogden (1997). Since we are considering inflation from the initial configuration, we have

$$\lambda_a \geq \lambda \geq \lambda_b \geq 1, \tag{5.6}$$

with equality holding in (5.6) if and only if $\lambda \equiv 1$, in which case the shell is undeformed.

For the considered spherically-symmetric deformation, the only equilibrium equation not satisfied trivially is the radial equation

$$\frac{\mathrm{d}\sigma_1}{\mathrm{d}r} + \frac{2}{r}(\sigma_1 - \sigma_2) = 0, \tag{5.7}$$

where σ_1, σ_2 are the principal Cauchy stresses associated with the directions r and θ (and φ by symmetry). Using (5.2) and (5.3) we change variables from r to λ by means of the formula

$$r\frac{\mathrm{d}\lambda}{\mathrm{d}r} = -\lambda(\lambda^3 - 1) \tag{5.8}$$

and replace (5.7) by

$$\lambda\frac{d\sigma_1}{d\lambda} = 2\frac{\sigma_1 - \sigma_2}{\lambda^3 - 1}.$$ (5.9)

The boundary conditions are taken in the form

$$\sigma_1 = \begin{cases} -P & \text{on } r = a \\ 0 & \text{on } r = b, \end{cases}$$ (5.10)

corresponding to pressure loading on the inside of the sphere and zero traction on the outside.

On integration of (5.9) and use of (2.44) and the boundary conditions (5.10) we obtain, in the notation of Section 13.2.4.2,

$$P = \int_{\lambda_b}^{\lambda_a} \frac{\dot{W}_\lambda(\lambda,\eta)}{\lambda^3 - 1} d\lambda.$$ (5.11)

For an elastic material this formula was given by Haughton and Ogden (1978); see, also, Ogden (1997). Here, however, it includes the additional variable η, which depends on λ through (2.45) when η is active.

The distribution of radial stress σ_1 as a function of λ is given by

$$\sigma_1 = -\int_{\lambda_b}^{\lambda} \frac{\dot{W}_\lambda(\lambda,\eta)}{\lambda^3 - 1} d\lambda,$$ (5.12)

and hence, from (2.44), the corresponding distribution of hoop stress σ_2 may be determined. In the formulas (5.11) and (5.12) the material model remains general in that the special model (3.1), appropriately specialized, has not been used. Henceforth, we make use of the model (3.1) in these formulas, with the notation of Section 13.2.4.2 adopted for the equibiaxial specialization. Thus, (3.1) becomes

$$\dot{W}(\lambda,\eta) = \eta\dot{W}_0(\lambda) + \phi(\eta), \qquad \phi(1) = 0.$$ (5.13)

In respect of (5.13) equation (5.11) becomes

$$P = \int_{\lambda_b}^{\lambda_a} \frac{\eta\dot{W}_0'(\lambda)}{\lambda^3 - 1} d\lambda,$$ (5.14)

where η is given locally in terms of λ by

$$\phi'(\eta) = -\dot{W}_0(\lambda).$$ (5.15)

13.5.2 Loading-unloading behaviour

For loading we take $\eta = 1$ and we suppose that λ_{am} is the greatest value of λ_a reached. Let P_m be the corresponding value of P. In general this need not be the greatest value of P achieved during inflation, but for convenience we assume here that P is a monotonic increasing function of λ_a, in which case P_m *is* the largest value of P. We then assume that unloading is achieved by decreasing P monotonically from its value P_m and is accompanied by λ_a decreasing from λ_{am}.

We now consider the possibility that a residual deformation remains when P is reduced to zero. For this to happen the integrand in (5.14) must vanish for some $\lambda_a > 1$. This cannot in general be met on the same basis as for the azimuthal shear problem (i.e. with $\eta = 0$ for all r) since, unlike for the latter problem, the form of the deformation is determined by the incompressibility constraint. Vanishing of η for all points of the integrand would lead to an expression for λ incompatible with that given in (5.3). For the integral (5.14) to vanish, therefore, η, which is 1 for all r at the start of unloading, must become negative for some part of the integrand as λ_a decreases. The behaviour of the denominator $\dot{W}_\lambda(\lambda, \eta) = \eta \dot{W}_0'(\lambda)$ must therefore have the character depicted in Figure 4, in which the continuous curve corresponds to loading ($\eta = 1$) and the dashed curve to unloading. The dashed curve cuts the horizontal axis at the critical value λ_c of λ at which η vanishes. Further details of this problem will be discussed elsewhere.

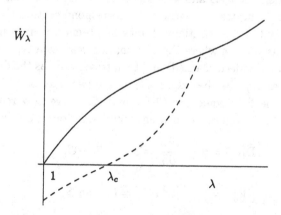

Fig. 4. Sketch of $\dot{W}_\lambda(\lambda, \eta) = \eta \dot{W}_0'(\lambda)$ as a function of λ: continuous curve (loading); dashed curve (unloading) showing the critical value λ_c of λ where the function vanishes.

In the special case in which the wall thickness of the spherical shell is very small, membrane theory is applicable and (5.14) is approximated as

$$P \approx \epsilon \eta \lambda^{-2} \dot{W}_0'(\lambda), \qquad (5.16)$$

where λ can be taken as the value of λ_a at the inner surface and $\epsilon = (B - A)/A$. In this case P vanishes when $\eta = 0$ and the critical value λ_c identified in Figure 4 is just the residual value of the stretch in this case. This model is appropriate for the description of the inflation and deflation of a virgin balloon where the stress-free radius of the balloon is increased by the inflation-deflation cycle.

13.6 Discontinuous changes in material properties

Discontinuities in material properties are associated with discontinuities in stress which in turn may be associated with discontinuous damage occurring at some critical point during the deformation process. For simple tension such a discontinuity is illustrated in Figure 1(c). Discontinuities of this type can be accommodated on the basis of the theory of pseudo-elasticity discussed in Section 13.2 if the relationship (2.10) between η and $\mathbf{F}$ is not one-to-one. There is then the possibility that a given deformation $\mathbf{F}$ may correspond to different values of η. In particular, as $\mathbf{F}$ varies a surface across which η is discontinuous may be generated in the reference configuration $\mathcal{B}_r$ as the deformation in $\mathcal{B}$ proceeds. Such a surface acts as a switch for η to change its value as the surface is crossed and hence as a switch for changing material properties. Let such a surface be denoted by $\mathcal{S}$ and we assume that $\mathbf{F}$ itself is continuous across $\mathcal{S}$. The surface $\mathcal{S}$ separates the parts of $\mathcal{B}_r$ corresponding to different forms of the strain-energy function. Initially, $\mathcal{S}$ may be absent but may appear during the deformation process where the deformation reaches a critical value (if such a critical value is indeed reached) and then traverse $\mathcal{B}_r$ as the deformation proceeds. Let η^+ and η^- be the values of η on the two sides of $\mathcal{S}$.

In addition to the field equations (2.10) and (2.13) we now require jump conditions across $\mathcal{S}$. Together, we then have four equations, namely

$$\mathrm{Div}\,\mathbf{S} = \mathbf{0}, \quad \frac{\partial W}{\partial \eta} = 0 \quad \text{in } \mathcal{B}_r - \mathcal{S}, \qquad (6.1)$$

$$[W]_-^+ = 0, \quad [\mathbf{S}^T\mathbf{N}]_-^+ = \mathbf{0} \quad \text{on } \mathcal{S}, \qquad (6.2)$$

where $\mathbf{N}$ is the unit normal to $\mathcal{S}$ (in either sense). In (6.2), $[\,\cdot\,]_-^+$ denotes the difference in the enclosed quantity on the two sides of $\mathcal{S}$. For example,

$$[W]_-^+ = W(\mathbf{F}, \eta^+) - W(\mathbf{F}, \eta^-),$$

evaluated on S.

Equation $(6.2)_2$ is simply a statement that the traction is continuous across S, while $(6.2)_1$ states that there is no jump in the strain energy across S. The simple tension example in Figure 1 (c) illustrates the continuity in energy since the energy in this case is just the area under the curve and this is unaffected as the point of discontinuity is crossed. Equation $(6.2)_1$ is important because it provides the criterion for a switch in the material properties, i.e. for a jump in η. In general, the stress will be discontinuous across S even though the traction is continuous. Appropriate boundary conditions may be given as required, but we omit details here. For further discussion we refer to Lazopoulos and Ogden (1998, 1999) and Ogden (2000a, b).

If the restriction to continuous F is removed then equation $(6.2)_1$ must be replaced by

$$[W]_-^+ - \mathrm{tr}\left[S^-(F^+ - F^-)\right] = 0, \qquad (6.3)$$

but the other equations are unchanged.

The theory in equations (6.1) and (6.2) was developed and applied to specific boundary-value problems by Lazopoulos and Ogden (1998, 1999) and Ogden (2000a, b). In particular, it was used, for example, to describe the inflation and deflation cycle for a thick-walled spherical shell in which a spherical surface (S) of (material property) discontinuity evolved. Further development of the theory and its applications is in progress.

References

Beatty, M.F. and Krishnaswamy, S. 2000 A theory of stress softening in incompressible isotropic elastic materials. *J. Mech. Phys. Solids* **48**, 1931–1965.

Ericksen, J.L. 1991 *Introduction to the Thermodynamics of Solids*. London: Chapman and Hall.

Fung, Y.C. 1981. *Biomechanics: Mechanical Properties of Living Tissue*. New York: Springer.

Govindjee, S. and Simo, J. C. 1991 A micro-mechanically based continuum damage model for carbon black-filled rubbers incorporating the Mullins' effect. *J. Mech. Phys. Solids* **39**, 87–112.

Govindjee, S. and Simo, J. C. 1992a Transition from micro-mechanics to computationally efficient phenomenology: carbon black filled rubbers incorporating Mullins' effect. *J. Mech. Phys. Solids* **40**, 213–233.

Govindjee, S. and Simo, J. C. 1992b Mullins' effect and the strain amplitude dependence of the storage modulus. *Int. J. Solids Structures* **29**, 1737–1751.

Haddow, J.B. 2000 Inflation and deflation of a pseudo-elastic spherical shell. *Int. J. Non-linear Mech.* **35**, 481–486.

Haughton, D.M. and Ogden, R.W. 1978 On the incremental equations in non-linear elasticity II. Bifurcation of pressurized spherical shells. *J. Mech. Phys. Solids* **26**, 111–138.

Holzapfel, G.A., Stadler, M. and Ogden, R.W. 1999 Aspects of stress softening in filled rubbers incorporating residual strains. In Dorfmann, A. and Muhr, A., eds., *Proceedings of the First European Conference on Constitutive Models for Rubber*, pp. 189–193. Rotterdam: Balkema.

Jiang, X. and Ogden, R.W. 1998 On azimuthal shear of a circular cylindrical tube of compressible elastic material. *Q. J. Mech. Appl. Math.* **51**, 143–158.

Johnson, M. A. and Beatty, M. F. 1993*a* The Mullins effect in uniaxial extension and its influence on the transverse vibration of a rubber string. *Continuum Mech. Thermodyn.* **5**, 83–115.

Johnson, M. A. and Beatty, M. F. 1993*b* A constitutive equation for the Mullins effect in stress controlled uniaxial extension experiments. *Continuum Mech. Thermodyn.* **5**, 301–318.

Krishnaswamy, S. and Beatty, M.F. 2000 The Mullins effect in compressible solids. *Int. J. Engng Sci.* **38**, 1397–1414.

Lazopoulos, K. A. and Ogden, R. W. 1998 Nonlinear elasticity theory with discontinuous internal variables. *Math. Mech. Solids* **3**, 29–51.

Lazopoulos, K.A. and Ogden, R.W. 1999 Spherically-symmetric solutions for a spherical shell in finite pseudo-elasticity. *European J. Mech. A/Solids* **18**, 617–632.

Lion, A. 1996 A constitutive model for carbon black filled rubber: experimental investigations and mathematical representation. *Continuum Mech. Thermodyn.* **8**, 153–169.

Miehe, C. 1995 Discontinuous and continuous damage evolution in Ogden-type large-strain elastic materials. *Eur. J. Mech. A/Solids* **14**, 697–720.

Muhr, A.H., Gough, J. and Gregory, I.H. 1999 Experimental determination of model for liquid silicone rubber: Hyperelasticity and Mullins' effect. In Dorfmann, A. and Muhr, A., eds., *Proceedings of the First European Conference on Constitutive Models for Rubber*, pp. 181–187. Rotterdam: Balkema.

Mullins, L. 1947 Effect of stretching on the properties of rubber. *J. Rubber Research* **16**, 275–289.

Mullins, L. 1969 Softening of rubber by deformation. *Rubber Chem. Technol.* **42**, 339–362.

Mullins, L. and Tobin, N. R. 1957 Theoretical model for the elastic behaviour of filler-reinforced vulcanized rubbers. *Rubber Chem. Technol.* **30**, 551–571.

Ogden, R.W. 1997 *Non-linear Elastic Deformations*. New York: Dover Publications.

Ogden, R.W. 2000*a* Elastic and pseudo-elastic instability and bifurcation. In Petryk, H., ed., *Material Instabilities in Elastic and Plastic Solids*. CISM Courses and Lectures Series no. **414**, pp. 209–259. Wien: Springer.

Ogden, R.W. 2000*b* Non-smooth changes in elastic material properties under finite deformation. In *Nonconvex and Nonsmooth Mechanics* (eds D.Y. Gao, R.W. Ogden and G.E. Stavroulakis), pp. 277–299. Dordrecht: Kluwer.

Ogden, R.W. 2001*a* Stress softening and residual strain in the azimuthal shear of a pseudo-elastic circular cylindrical tube. *Int. J. Non-linear Mech.* **36**, 477–487.

Ogden, R.W. 2001*b* On an anisotropic theory of pseudo-elasticity. Manuscript in preparation.

Ogden, R.W. and Roxburgh, D.G. 1999*a* A pseudo-elastic model for the Mullins effect in filled rubber. *Proc. R. Soc. Lond.* A **455**, 2861–2877.

Ogden, R.W. and Roxburgh, D.G. 1999*b* An energy-based model of the Mullins effect. In Dorfmann, A. and Muhr, A., eds., *Proceedings of the First European Conference on Constitutive Models for Rubber*, pp. 23–28. Rotterdam: Balkema.

Subject index

Printed in the United States
By Bookmasters